D1104960

ENCYCLOPEDIA OF MATHEMATICS AND ITS APPLICATIONS

EDITED BY G.-C. ROTA

Volume 35

Basic Hypergeometric Series

ENCYCLOPEDIA OF MATHEMATICS AND ITS APPLICATIONS

1 Luis A. Santalo Integral Geometric Probability
2 George E. Andrews The Theory of Partitions
3 Robert J. McEliece The Theory of Information and Coding: A Mathematical Framework for Communication
4 Willard Miller, Jr. Symmetry and Separation of Variables
5 David Ruelle Thermodynamic Formalism: the Mathematical Structures of Classical Equilibrium Statistical Mechanics
6 Henryk Minc Permanents
7 Fred S. Roberts Measurement Theory with Applications to Decisionmaking, Utility, and the Social Services
8 L. C. Biedenharn and J. D. Louck Angular Momentum in Quantum Physics: Theory and Application
9 L. C. Biedenharn and J. D. Louck The Racah-Wigner Algebra in Quantum Theory
10 W. Dollard and Charles N. Friedman Product Integration with Application to Differential Equations
11 William B. Jones and W. J. Thron Continued Fractions: Analytic Theory and Applications
12 Nathaniel F. G. Martin and James W. England Mathematical Theory of Entropy
13 George A. Baker, Jr. and Peter R. Graves-Morris Padé Approximants, Part I: Basic Theory
14 George A. Baker, Jr. and Peter R. Graves-Morris Padé Approximants, Part II: Extensions and Applications
15 E. C. Beltrametti and G. Cassinelli The Logic of Quantum Mechanics
16 G. D. James and A. Kerber The Representation Theory of the Symmetric Group
17 M. Lothaire Combinatorics on Words
18 H. O. Fattorini The Cauchy Problem
19 G. G. Lorentz, K. Jetter, and S. D. Riemenschneider Birkhoff Interpolation
20 Rudolf Lidl and Harald Niederreiter Finite Fields
21 William T. Tutte Graph Theory
22 Julio R. Bastida Field Extensions and Galois Theory
23 John R. Cannon The One-Dimensional Heat Equation
24 S. Wagon The Banach-Tarski Paradox
25 A. Salomaa Computation and Automata
26 N. White (ed) Theory of Matroids
27 N. Bingham, C. Goldie & J. L. Teugels Regular Variation
28 P. Petrushev, & P. Popov Rational Approximation of Real Variables
29 N. White (ed) Combinatorial Geometries
30 M. Pohst and H. Zassenhaus Algorithmic Algebraic Number Theory
31 J. Aczel & J. D. Hombres Functional Equations containing Several Variables
32 M. Kuczma, B. Chozewski & R. Ger Iterative Functional Equations
34 G. Gripenberg, S.-O. Londen and O. Staffans Volterra Integral and Functional Equations
35 G. Gasper & M. Rahman Basic Hypergeometric Series

ENCYCLOPEDIA OF MATHEMATICS AND ITS APPLICATIONS

BASIC HYPERGEOMETRIC SERIES

GEORGE GASPER
Northwestern University, Evanston, Illinois, USA

MIZAN RAHMAN
Carleton University, Ottawa, Canada

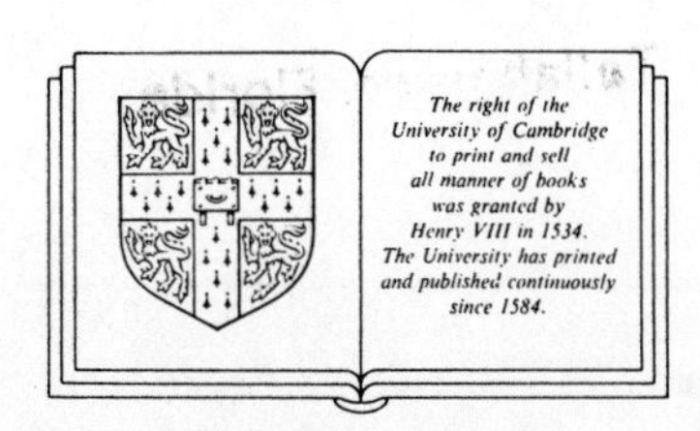
The right of the
University of Cambridge
to print and sell
all manner of books
was granted by
Henry VIII in 1534.
The University has printed
and published continuously
since 1584.

CAMBRIDGE UNIVERSITY PRESS

Cambridge
New York Port Chester
Melbourne Sydney

Published by the Press Syndicate of the University of Cambridge
The Pitt Building, Trumpington Street, Cambridge CB2 1RP
40 West 20th Street, New York, NY 10011, USA
10 Stamford Road, Oakleigh, Melbourne 3166, Australia

First published 1990

Printed in Great Britain at the University Press, Cambridge

British Library cataloguing in publication data available

Library of Congress cataloguing in publication data available

ISBN 0 521 35049 2

To

Brigitta, Karen, and Kenneth Gasper

and

Parul, Babu, and Raja Rahman

Contents

Foreword

My education was not much different from that of most mathematicians of my generation. It included courses on modern algebra, real and complex variables, both point set and algebraic topology, some number theory and projective geometry, and some specialized courses such as one on Riemann surfaces. In none of these courses was a hypergeometric function mentioned, and I am not even sure if the gamma function was mentioned after an advanced calculus course. The only time Bessel functions were mentioned was in an undergraduate course on differential equations, and the only thing done with them was to find a power series solution for the general Bessel equation. It is small wonder that with a similar education almost all mathematicians think of special functions as a dead subject which might have been interesting once. They have no idea why anyone would care about it now.

Fortunately there was one part of my education which was different. As a junior in college I read Widder's book *The Laplace Transform* and the manuscript of its very important sequel, Hirschman and Widder's *The Convolution Transform.* Then as a senior, I.I. Hirschman gave me a copy of a preprint of his on a multiplier theorem for Legendre series and suggested I extend it to ultraspherical series. This forced me to become acquainted with two other very important books, Gabor Szegő's great book *Orthogonal Polynomials*, and the second volume of *Higher Transcendental Functions*, the monument to Harry Bateman which was written by Arthur Erdélyi and his co-workers W. Magnus, F. Oberhettinger and F.G. Tricomi.

From this I began to realize that the many formulas that had been found, usually in the 18th or 19th century, but once in a while in the early 20th century, were useful, and started to learn about their structure. However, I had written my Ph.D. thesis and worked for three more years before I learned that not every fact about special functions I would need had already been found, and it was a couple of more years before I learned that it was essential to understand hypergeometric functions. Like others, I had been put off by all the parameters. If there were so many parameters that it was necessary to put subscripts on them, then there has to be a better way to solve a problem than this. That was my initial reaction to generalized hypergeometric functions, and a very common reaction to judge from the many conversations I have had on these functions in the last twenty years. After learning a little more about hypergeometric functions, I was very surprised to realize that they had occurred regularly in first year calculus. The reason for the subscripts on the parameters is that not all interesting polynomials are of degree one or two. For a generalized hypergeometric function has a series representation

$$\sum_{n=0}^{\infty} c_n \tag{1}$$

with c_{n+1}/c_n a rational function of n. These contain almost all the examples of infinite series introduced in calculus where the ratio test works easily. The ratio c_{n+1}/c_n can be factored, and it is usually written as

$$\frac{c_{n+1}}{c_n} = \frac{(n+a_1)\cdots(n+a_p)x}{(n+b_1)\cdots(n+b_q)(n+1)}. \tag{2}$$

Introduce the shifted factorial

$$\begin{aligned}(a)_0 &= 1,\\ (a)_n &= a(a+1)\cdots(a+n-1), \quad n = 1,2,\ldots .\end{aligned} \tag{3}$$

Then if $c_0 = 1$, equation (2) can be solved for c_n as

$$c_n = \frac{(a_1)_n \cdots (a_p)_n}{(b_1)_n \cdots (b_q)_n}\frac{x^n}{n!} \tag{4}$$

and

$${}_pF_q\left[\begin{matrix} a_1, \cdots, a_p \\ b_1, \cdots, b_q\end{matrix}; x\right] = \sum_{n=0}^{\infty} \frac{(a_1)_n \cdots (a_p)_n}{(b_1)_n \cdots (b_q)_n}\frac{x^n}{n!} \tag{5}$$

is the usual notation.

The first important result for a ${}_pF_q$ with $p > 2,\ q > 1$ is probably Pfaff's sum

$${}_3F_2\left[\begin{matrix} -n,\ a,\ b \\ c,\ a+b+1-c-n\end{matrix}; 1\right] = \frac{(c-a)_n(c-b)_n}{(c)_n(c-a-b)_n}, \quad n = 0,1,\ldots . \tag{6}$$

This result from 1797, see Pfaff [1797], contains as a limit when $n \to \infty$ another important result usually attributed to Gauss [1813],

$${}_2F_1\left[\begin{matrix} a,\ b \\ c\end{matrix}; 1\right] = \frac{\Gamma(c)\Gamma(c-a-b)}{\Gamma(c-a)\Gamma(c-b)}, \quad \mathrm{Re}\ (c-a-b) > 0. \tag{7}$$

The next instance is a very important result of Clausen [1828]:

$$\left\{{}_2F_1\left[\begin{matrix} a,\ b \\ a+b+\frac{1}{2}\end{matrix}; x\right]\right\}^2 = {}_3F_2\left[\begin{matrix} 2a,\ 2b,\ a+b \\ a+b+\frac{1}{2},\ 2a+2b\end{matrix}; x\right]. \tag{8}$$

Some of the interest in Clausen's formula is that it changes the square of a class of ${}_2F_1$'s to a ${}_3F_2$. In this direction it is also interesting because it was probably the first instance of anyone finding a differential equation satisfied by $[y(x)]^2$, $y(x)z(x)$ and $[z(x)]^2$ when $y(x)$ and $z(x)$ satisfy

$$a(x)y'' + b(x)y' + c(x)y = 0. \tag{9}$$

This problem was considered for (9) by Appell, see Watson [1952], but the essence of his general argument occurs in Clausen's paper. This is a common phenomenon, which is usually not mentioned when the general method is introduced to students, so they do not learn how often general methods come from specific problems or examples. See D. and G. Chudnovsky [1988] for an instance of the use of Clausen's formula, where a result for a ${}_2F_1$ is carried to a ${}_3F_2$ and from that to a very interesting set of expansions of π^{-1}. Those identities were first discovered by Ramanujan. Here is Ramanujan's most impressive example:

$$\frac{9801}{2\pi\sqrt{2}} = \sum_{n=0}^{\infty} [1103 + 26390n] \frac{(1/4)_n (1/2)_n (3/4)_n}{(1)_n (1)_n n!} \frac{1}{(99)^{4n}}. \tag{10}$$

There is another important reason why Clausen's formula is important. It leads to a large class of ${}_3F_2$'s that are nonnegative for the power series variable between -1 and 1. The most famous use of this is in the final step of de Branges' solution of the Bieberbach conjecture, see de Branges [1985]. The integral of the ${}_2F_1$ or Jacobi polynomial he had is a ${}_3F_2$, and its positivity is an easy consequence of Clausen's formula, as Gasper had observed ten years earlier. There are other important results which follow from the positivity in Clausen's identity.

Once Kummer [1836] wrote his long and important paper on ${}_2F_1$'s and ${}_1F_1$'s, this material became well-known. It has been reworked by others. Riemann redid the ${}_2F_1$ using his idea that the singularities of a function go a long way toward determining the function. He showed that if the differential equation (9) has regular singularities at three points, and every other point in the extended complex plane is an ordinary point, then the equation is equivalent to the hypergeometric equation

$$x(1-x)y'' + [c - (a+b+1)x]y' - aby = 0, \tag{11}$$

which has regular singular points at $x = 0, 1, \infty$. Riemann's work was very influential, so much so that much of the mathematical community that considered hypergeometric functions studied them almost exclusively from the point of view of differential equations. This is clear in Klein's book [1933], and in the work on multiple hypergeometric functions that starts with Appell in 1880 and is summarized in Appell and Kampé de Fériet [1926].

The integral representations associated with the differential equation point of view are similar to Euler's integral representation. This is

$$ {}_2F_1\left[\begin{matrix} a,\ b \\ c \end{matrix}; x\right] = \frac{\Gamma(c)}{\Gamma(b)\Gamma(c-b)} \int_0^1 (1-xt)^{-a} t^{b-1} (1-t)^{c-b-1} dt, \tag{12}$$

$|x| < 1$, $\operatorname{Re} c > \operatorname{Re} b > 0$, and includes related integrals with different contours. The differential equation point of view is very powerful where it works, but it does not work well for $p \geq 3$ or $q \geq 2$ as Kummer discovered. Thus there is a need to develop other methods to study hypergeometric functions.

In the late 19th and early 20th century a different type of integral representation was introduced. These two different types of integrals are best represented by Euler's beta integral

$$\int_0^1 t^{a-1}(1-t)^{b-1}dt = \frac{\Gamma(a)\Gamma(b)}{\Gamma(a+b)}, \quad \text{Re } (a,b) > 0 \tag{13}$$

and Barnes' beta integral

$$\begin{aligned} &\frac{1}{2\pi}\int_{-\infty}^{\infty} \Gamma(a+it)\Gamma(b+it)\Gamma(c-it)\Gamma(d-it)\, dt \\ &= \frac{\Gamma(a+c)\Gamma(a+d)\Gamma(b+c)\Gamma(b+d)}{\Gamma(a+b+c+d)}, \quad \text{Re } (a,b,c,d) > 0. \end{aligned} \tag{14}$$

There is no direct connection with differential equations for integrals like (14), so it stands a better chance to work for larger values of p and q.

While Euler, Gauss, and Riemann and many other great mathematicians wrote important and influential papers on hypergeometric functions, the development of basic hypergeometric functions was much slower. Euler and Gauss did important work on basic hypergeometric functions, but most of Gauss' work was unpublished until after his death and Euler's work was more influential on the development of number theory and elliptic functions.

Basic hypergeometric series are series $\sum c_n$ with c_{n+1}/c_n a rational function of q^n for a fixed parameter q, which is usually taken to satisfy $|q| < 1$, but at other times is a power of a prime. In this Foreword $|q| < 1$ will be assumed.

Euler summed three basic hypergeometric series. The one which had the largest impact was

$$\sum_{-\infty}^{\infty} (-1)^n q^{(3n^2-n)/2} = (q;q)_\infty, \tag{15}$$

where

$$(a;q)_\infty = \prod_{n=0}^{\infty}(1-aq^n). \tag{16}$$

If

$$(a;q)_n = (a;q)_\infty/(aq^n;q)_\infty \tag{17}$$

then Euler also showed that

$$\frac{1}{(x;q)_\infty} = \sum_{n=0}^{\infty} \frac{x^n}{(q;q)_n}, \quad |x| < 1, \tag{18}$$

and

$$(x;q)_\infty = \sum_{n=0}^{\infty} \frac{(-1)^n q^{\binom{n}{2}} x^n}{(q;q)_n}. \tag{19}$$

Eventually all of these were contained in the q-binomial theorem

$$(20) \qquad \frac{(ax;q)_\infty}{(x;q)_\infty} = \sum_{n=0}^{\infty} \frac{(a;q)_n}{(q;q)_n} x^n, \quad |x| < 1.$$

While (18) is clearly the special case $a = 0$, and (19) follows easily on replacing x by xa^{-1} and letting $a \to \infty$, it is not so clear how to obtain (15) from (20). The easiest way was discovered by Cauchy and many others. Take $a = q^{-2N}$, shift n by N, rescale and let $N \to \infty$. The result is called the triple product, and can be written as

$$(21) \qquad (x;q)_\infty (qx^{-1};q)_\infty (q;q)_\infty = \sum_{-\infty}^{\infty} (-1)^n q^{\binom{n}{2}} x^n.$$

Then $q \to q^3$ and $x = q$ gives Euler's formula (15).

Gauss used a basic hypergeometric series identity in his first proof of the determination of the sign of the Gauss sum, and Jacobi used some to determine the number of ways an integer can be written as the sum of two, four, six and eight squares. However, this particular aspect of Gauss' work on Gauss sums was not very influential, as his hypergeometric series work had been, and Jacobi's work appeared in his work on elliptic functions, so its hypergeometric character was lost in the great interest in the elliptic function work. Thus neither of these led to a serious treatment of basic hypergeometric series. The result that seems to have been the crucial one was a continued fraction of Eisenstein. This along with the one hundredth anniversary of Euler's first work on continued fractions seems to have been the motivating forces behind Heine's introduction of a basic hypergeometric extension of ${}_2F_1(a,b;c;x)$. He considered

$$(22) \qquad {}_2\phi_1\left[\begin{matrix} q^a, \ q^b \\ q^c \end{matrix}; q, x\right] = \sum_{n=0}^{\infty} \frac{(q^a;q)_n (q^b;q)_n}{(q^c;q)_n (q;q)_n} x^n, \quad |x| < 1.$$

Observe that

$$\lim_{q\to 1} \frac{(q^a;q)_n}{(1-q)^n} = (a)_n,$$

so

$$\lim_{q\to 1} {}_2\phi_1\left[\begin{matrix} q^a, \ q^b \\ q^c \end{matrix}; q, x\right] = {}_2F_1\left[\begin{matrix} a, \ b \\ c \end{matrix}; x\right].$$

Heine followed the pattern of Gauss' published paper on hypergeometric series, and so obtained contiguous relations and from them continued fraction expansions. He also obtained some series transformations, and the sum

$$(23) \qquad {}_2\phi_1\left[\begin{matrix} q^a, \ q^b \\ q^c \end{matrix}; q, q^{c-a-b}\right] = \frac{(q^{c-a};q)_\infty (q^{c-b};q)_\infty}{(q^c;q)_\infty (q^{c-a-b};q)_\infty}, \quad |q^{c-a-b}| < 1.$$

This sum becomes (7) when $q \to 1$.

As often happens to path breaking work, this work of Heine was to a large extent ignored. When writing the second edition of *Kugelfunctionen* (Heine [1878]) Heine decided to include some of his work on basic hypergeometric series. This material was printed in smaller type, and it is clear that Heine included it because he thought it was important, and he wanted to call attention to it, rather than because he thought it was directly related to spherical harmonics, the subject of his book. Surprisingly, his inclusion of this material led to some later work, which showed there was a very close connection between Heine's work on basic hypergeometric series and spherical harmonics. The person Heine influenced was L.J. Rogers, who is still best known as the first discoverer of the Rogers-Ramanujan identities. Rogers tried to understand this aspect of Heine's work, and one transformation in particular. Thomae [1879] had observed this transformation of Heine could be written as an extension of Euler's integral representation (12), but Rogers was unaware of this explanation, and so discovered a second reason. He was able to modify the transformation so it became the permutation symmetry in a new series. While doing this he introduced a new set of polynomials which we now call the continuous q-Hermite polynomials. In a very important set of papers which were unjustly neglected for decades, Rogers discovered a more general set of polynomials and found some remarkable identities they satisfy, see Rogers [1893, 1894, 1895]. For example, he found the linearization coefficients of these polynomials which we now call the continuous q-ultraspherical polynomials. These polynomials contain many of the spherical harmonics Heine studied. Contained within this product identity is the special case of the square of one of these polynomials as a double series. As Gasper and Rahman have observed, one of these series can be summed, and the resulting identity is an extension of Clausen's sum in the terminating case. Earlier, others had found a different extension of Clausen's identity to basic hypergeometric series, but the resulting identity was not satisfactory. The identity had the product of two functions, the same functions but one evaluated at x and the other at qx, and so was not a square. Thus the nonnegativity that is so useful in Clausen's formula was not true for the corresponding basic hypergeometric series. Rogers' result for his polynomials led directly to the better result which contains the appropriate nonnegativity. From this example and many others, one sees that orthogonal polynomials provide an alternative approach to the study of hypergeometric and basic hypergeometric functions. Both this approach and that of differential equations are most useful for small values of the degrees of the numerator and denominator polynomials in the ratio c_{n+1}/c_n, but orthogonal polynomials work for a larger class of series, and are much more useful for basic hypergeometric series. However, neither of these approaches is powerful enough to encompass all aspects of these functions. Direct series manipulations are surprisingly useful, when done by a master, or when a computer algebra system is used as an aid. Gasper and Rahman are both experts at symbolic calculations, and I regularly marvel at some of the formulas they have found. As quantum groups become better known, and as Baxter's work spreads to other parts of mathematics as it has started to do, there will be many people trying to learn how to deal with basic hypergeometric series. This book is where I would start.

For many years people have asked me what is the best book on special functions. My response was George Gasper's copy of Bailey's book, which was heavily annotated with useful results and remarks. Now others can share the information contained in these margins, and many other very useful results.

Richard Askey
University of Wisconsin

Preface

The study of basic hypergeometric series (also called q-hypergeometric series or q-series) essentially started in 1748 when Euler considered the infinite product $(q;q)_\infty^{-1} = \prod_{k=0}^{\infty}(1-q^{k+1})^{-1}$ as a generating function for $p(n)$, the number of partitions of a positive integer n into positive integers (see §8.10). But it was not until about a hundred years later that the subject acquired an independent status when Heine converted a simple observation that $\lim_{q\to 1}[(1-q^a)/(1-q)] = a$ into a systematic theory of ${}_2\phi_1$ basic hypergeometric series parallel to the theory of Gauss' ${}_2F_1$ hypergeometric series. Heine's transformation formulas for ${}_2\phi_1$ series and his q-analogue of Gauss' ${}_2F_1(1)$ summation formula are derived in Chapter 1, along with a q-analogue of the binomial theorem, Jacobi's triple product identity, and some formulas for q-analogues of the exponential, gamma and beta functions.

Apart from some important work by J. Thomae and L. J. Rogers the subject remained somewhat dormant during the latter part of the nineteenth century until F. H. Jackson embarked on a lifelong program of developing the theory of basic hypergeometric series in a systematic manner, studying q-differentiation and q-integration and deriving q-analogues of the hypergeometric summation and transformation formulas that were discovered by A. C. Dixon, J. Dougall, L. Saalschütz, F. J. W. Whipple, and others. His work is so pervasive that it is impossible to cover all of his contributions in a single volume of this size, but we have tried to include many of his important formulas in the first three chapters. In particular, a derivation of his summation formula for an ${}_8\phi_7$ series is given in §2.6. During the 1930's and 1940's many important results on hypergeometric and basic hypergeometric series were derived by W. N. Bailey. Some mathematicians consider Bailey's greatest work to be the Bailey transform (an equivalent form of which is covered in Chapter 2), but equally significant are his nonterminating extensions of Jackson's ${}_8\phi_7$ summation formula and of Watson's transformation formula connecting very-well-poised ${}_8\phi_7$ series with balanced ${}_4\phi_3$ series. Much of the material on summation, transformation and expansion formulas for basic hypergeometric series in Chapter 2 is due to Bailey.

D. B. Sears, L. Carlitz, W. Hahn, and L. J. Slater were among the prominent contributors during the 1950's. Sears derived several transformation formulas for ${}_3\phi_2$ series, balanced ${}_4\phi_3$ series, and very-well-poised ${}_{n+1}\phi_n$ series. Simple proofs of some of his ${}_3\phi_2$ transformation formulas are given in Chapter 3. Three of his very-well-poised transformation formulas are derived in Chapter 4, where we follow G. N. Watson and Slater to develop the theory of basic hypergeometric series from a contour integral point of view, an idea first introduced by Barnes in 1907.

Chapter 5 is devoted to bilateral basic hypergeometric series, where the most fundamental formula is Ramanujan's ${}_1\psi_1$ summation formula. Substantial contributions were also made by Bailey, M. Jackson, Slater and others, whose works form the basis of this chapter.

During the 1960's R. P. Agarwal and Slater each published a book partially devoted to the theory of basic hypergeometric series, and G. E. Andrews initiated his work in number theory, where he showed how useful the summation and transformation formulas for basic hypergeometric series are in the theory of partitions. Andrews gave simpler proofs of many old results, wrote review articles pointing out many important applications and, during the mid 1970's, started a period of very fruitful collaboration with R. Askey. Thanks to these two mathematicians, basic hypergeometric series is an active field of research today. Since Askey's primary area of interest is orthogonal polynomials, q-series suddenly provided him and his co-workers with a very rich environment for deriving q-extensions of beta integrals and of the classical orthogonal polynomials of Jacobi, Gegenbauer, Legendre, Laguerre and Hermite. Askey and his students and collaborators who include W. A. Al-Salam, M. E. H. Ismail, T. H. Koornwinder, W. G. Morris, D. Stanton, and J. A. Wilson have produced a substantial amount of interesting work over the past fifteen years. This flurry of activity has been so infectious that many researchers found themselves hopelessly trapped by this alluring "q-disease", as it is affectionately called.

Our primary motivation for writing this book was to present in one modest volume the significant results of the past two hundred years so that they are readily available to students and researchers, to give a brief introduction to the applications to orthogonal polynomials that were discovered during the current renaissance period of basic hypergeometric series, and to point out important applications to other fields. Most of the material is elementary enough so that persons with a good background in analysis should be able to use this book as a textbook and a reference book. In order to assist the reader in developing a deeper understanding of the formulas and proof techniques and to include additional formulas, we have given a broad range of exercises at the end of each chapter. Additional information is provided in the Notes following the Exercises, particularly in relation to the results and relevant applications contained in the papers and books listed in the References. Although the References may have a bulky appearance, it is just an introduction to the vast literature available. Appendices I, II, and III are for quick reference, so that it is not necessary to page through the book in order to find the most frequently needed identities, summation formulas, and transformation formulas. It can be rather tedious to apply the summation and transformation formulas to the derivation of other formulas. But now that several symbolic computer algebraic systems are available, persons having access to such a system can let it do some of the symbolic manipulations, such as computing the form of Bailey's ${}_{10}\phi_9$ transformation formula when its parameters are replaced by products of other parameters.

Due to space limitations, we were unable to be as comprehensive in our coverage of basic hypergeometric series and their applications as we would have liked. In particular, we could not include a systematic treatment of basic hypergeometric series in two or more variables, covering F. H. Jackson's work on basic Appell series and the works of R. A. Gustafson and S. C. Milne on $U(n)$ multiple series generalizations of basic hypergeometric series referred to in the References. But we do highlight Askey and Wilson's fundamental work on their beautiful q-analogue of the classical beta integral in Chapter 6 and develop its connection with very-well-poised ${}_8\phi_7$ series. Chapter 7 is devoted to applications to orthogonal polynomials, mostly developed by Askey and his collaborators. We conclude the book with some further applications in Chapter 8, where we present part of our work on product and linearization formulas, Poisson kernels, and nonnegativity, and we also manage to point out some elementary facts about applications to the theory of partitions and the representations of integers as sums of squares of integers. The interested reader is referred to the books and papers of Andrews and N. J. Fine for additional applications to partition theory, and recent references are pointed out for applications to affine root systems (Macdonald identities), association schemes, combinatorics, difference equations, Lie algebras and groups, physics (such as representations of quantum groups and R. J. Baxter's work on the hard hexagon model of phase transitions in statistical mechanics), statistics, etc.

We use the common numbering system of letting (k.m.n) refer to the n-th numbered display in Section m of Chapter k, and letting (I.n), (II.n), and (III.n) refer to the n-th numbered display in Appendices I, II, and III, respectively. To refer to the papers and books in the References, we place the year of publication in square brackets immediately after the author's name. Thus Bailey [1935] refers to Bailey's 1935 book. Suffixes a, b, ... are used after the years to distinguish different papers by an author that appeared in the same year. Papers that have not yet been published are referred to with the year 1989, even though they might be published later due to the backlogs of journals. Since there are three Agarwals, two Chiharas and three Jacksons listed in the References, to minimize the use of initials we drop the initials of the author whose works are referred to most often. Hence Agarwal, Chihara, and Jackson refer to R. P. Agarwal, T. S. Chihara, and F. H. Jackson, respectively.

We would like to thank the publisher for their cooperation and patience during the preparation of this book. Thanks are also due to R. Askey, W. A. Al-Salam, R. P. Boas, T. S. Chihara, B. Gasper, R. Holt, M. E. H. Ismail, T. Koornwinder, and B. Nassrallah for pointing out typos and suggesting improvements in earlier versions of the book. We also wish to express our sincere thanks and appreciation to our TEXtypist, Diane Berezowski, who suffered through many revisions of the book but never lost her patience or sense of humor.

George Gasper

Mizan Rahman

1

BASIC HYPERGEOMETRIC SERIES

1.1 Introduction

Our main objective in this chapter is to present the definitions and notations for hypergeometric and basic hypergeometric series, and to derive the elementary formulas that form the basis for most of the summation, transformation and expansion formulas, basic integrals, and applications to orthogonal polynomials and to other fields that follow in the subsequent chapters. We begin by defining Gauss' ${}_2F_1$ hypergeometric series, the ${}_rF_s$ (generalized) hypergeometric series, and pointing out some of their most important special cases. Next we define Heine's ${}_2\phi_1$ basic hypergeometric series which contains an additional parameter q, called the base, and then give the definition and notations for ${}_r\phi_s$ basic hypergeometric series. Basic hypergeometric series are called q-analogues (basic analogues or q-extensions) of hypergeometric series because an ${}_rF_s$ series can be obtained as the $q \to 1$ limit case of an ${}_r\phi_s$ series.

Since the binomial theorem is at the foundation of most of the summation formulas for hypergeometric series, we then derive a q-analogue of it, called the q-binomial theorem, and use it to derive Heine's q-analogues of Euler's transformation formulas, Jacobi's triple product identity, and summation formulas that are q-analogues of those for hypergeometric series due to Chu and Vandermonde, Gauss, Kummer, Pfaff and Saalschütz, and to Karlsson and Minton. We also introduce q-analogues of the exponential, gamma and beta functions, as well as the concept of a q-integral that allows us to give a q-analogue of Euler's integral representation of a hypergeometric function. Many additional formulas and q-analogues are given in the exercises at the end of the chapter.

1.2 Hypergeometric and basic hypergeometric series

In 1812, Gauss presented to the Royal Society of Sciences at Göttingen his famous paper (Gauss [1813]) in which he considered the infinite series

$$1+\frac{ab}{1\cdot c}z+\frac{a(a+1)b(b+1)}{1\cdot 2\cdot c(c+1)}z^2+\frac{a(a+1)(a+2)b(b+1)(b+2)}{1\cdot 2\cdot 3\cdot c(c+1)(c+2)}z^3+\cdots \quad (1.2.1)$$

as a function of a, b, c, z, where it is assumed that $c \neq 0, -1, -2, \ldots$, so that no zero factors appear in the denominators of the terms of the series. He showed that the series converges absolutely for $|z| < 1$, and for $|z| = 1$ when $\mathrm{Re}\ (c-a-b) > 0$, gave its (contiguous) recurrence relations, and derived his famous formula (eq. (1.2.11) below) for the sum of this series when $z = 1$ and $\mathrm{Re}\ (c-a-b) > 0$.

Although Gauss used the notation $F(a,b,c,z)$ for his series, it is now customary to use $F(a,b;c;z)$ or either of the notations

$$ {}_2F_1(a,\ b;\ c;\ z), \qquad {}_2F_1\left[\begin{matrix} a,b \\ c \end{matrix};z\right] $$

for this series (and for its sum when it converges), because these notations separate the numerator parameters a,b from the denominator parameter c and the variable z. In view of Gauss' paper, his series is frequently called *Gauss' series.* However, since the special case $a=1, b=c$ yields the geometric series

$$ 1+z+z^2+z^3+\cdots, $$

Gauss' series is also called the *(ordinary) hypergeometric series* or the *Gauss hypergeometric series.*

Some important functions which can be expressed by means of Gauss' series are

$$ \begin{aligned} (1+z)^a &= F(-a,b;b;-z), \\ \log(1+z) &= zF(1,1;2;-z), \\ \sin^{-1} z &= zF(1/2,1/2;3/2;z^2), \\ \tan^{-1} z &= zF(1/2,1;3/2;-z^2), \\ e^z &= \lim_{a\to\infty} F(a,b;b;z/a), \end{aligned} \tag{1.2.2} $$

where $|z|<1$ in the first four formulas. Also expressible by means of Gauss' series are the classical orthogonal polynomials, such as the *Tchebichef polynomials of the first and second kinds*

$$ T_n(x) = F(-n,n;1/2;(1-x)/2), \tag{1.2.3} $$

$$ U_n(x) = (n+1)F(-n,n+2;3/2;(1-x)/2), \tag{1.2.4} $$

the *Legendre polynomials*

$$ P_n(x) = F(-n,n+1;1;(1-x)/2), \tag{1.2.5} $$

the *Gegenbauer (ultraspherical) polynomials*

$$ C_n^\lambda(x) = \frac{(2\lambda)_n}{n!} F(-n,n+2\lambda;\lambda+1/2;(1-x)/2), \tag{1.2.6} $$

and the more general *Jacobi polynomials*

$$ P_n^{(\alpha,\beta)}(x) = \frac{(\alpha+1)_n}{n!} F(-n,n+\alpha+\beta+1;\ \alpha+1;(1-x)/2), \tag{1.2.7} $$

where $n=0,1,\ldots,$ and $(a)_n$ denotes the *shifted factorial* defined by

$$ (a)_0=1,\ (a)_n = a(a+1)\cdots(a+n-1) = \frac{\Gamma(a+n)}{\Gamma(a)}, \quad n=1,2,\ldots. \tag{1.2.8} $$

Before Gauss, Chu [1303] (see Needham [1959, p.138], Takács [1973] and Askey [1975 p.59]) and Vandermonde [1772] had proved the summation formula

$$ F(-n,b;c;1) = \frac{(c-b)_n}{(c)_n}, \quad n=0,1,\ldots, \tag{1.2.9} $$

which is now called *Vandermonde's formula* or the *Chu-Vandermonde formula*, and Euler [1748] had derived several results for hypergeometric series, including his transformation formula

$$F(a,b;c;z) = (1-z)^{c-a-b}\, F(c-a, c-b; c; z), \quad |z| < 1. \tag{1.2.10}$$

Formula (1.2.9) is the terminating case $a = -n$ of the summation formula

$$F(a,b;c;1) = \frac{\Gamma(c)\Gamma(c-a-b)}{\Gamma(c-a)\Gamma(c-b)}, \quad \mathrm{Re}(c-a-b) > 0, \tag{1.2.11}$$

which Gauss proved in his paper.

Thirty-three years after Gauss' paper, Heine [1846, 1847, 1878] introduced the series

$$1+\frac{(1-q^a)(1-q^b)}{(1-q)(1-q^c)}z+\frac{(1-q^a)(1-q^{a+1})(1-q^b)(1-q^{b+1})}{(1-q)(1-q^2)(1-q^c)(1-q^{c+1})}z^2+\cdots, \tag{1.2.12}$$

where it is assumed that $c \neq 0, -1, -2, \ldots$. This series converges absolutely for $|z| < 1$ when $|q| < 1$ and it tends (at least termwise) to Gauss' series as $q \to 1$, because

$$\lim_{q\to 1} \frac{1-q^a}{1-q} = a. \tag{1.2.13}$$

The series in (1.2.12) is usually called *Heine's series* or, in view of the base q, the *basic hypergeometric series* or *q-hypergeometric series.*

Analogous to Gauss' notation, Heine used the notation $\phi(a, b, c, q, z)$ for his series. However, since one would like to also be able to consider the case when q to the power $a, b,$ or c is replaced by zero, it is now customary to define the *basic hypergeometric series* by

$$\begin{aligned}\phi(a,b;c;q,z) &\equiv {}_2\phi_1(a,b;c;q,z) \equiv {}_2\phi_1\left[\begin{matrix}a,b\\ c\end{matrix};q,z\right]\\ &= \sum_{n=0}^{\infty} \frac{(a;q)_n(b;q)_n}{(q;q)_n(c;q)_n} z^n,\end{aligned} \tag{1.2.14}$$

where

$$(a;q)_n = \begin{cases} 1, & n = 0,\\ (1-a)(1-aq)\cdots(1-aq^{n-1}), & n = 1,2,\ldots, \end{cases} \tag{1.2.15}$$

is the *q-shifted factorial* and it is assumed that $c \neq q^{-m}$ for $m = 0, 1, \ldots$. Some other notations that have been used in the literature for the product $(a;q)_n$ are $(a)_{q,n}$, $[a]_n$, and even $(a)_n$ when (1.2.8) is not used and the base is not displayed.

Another generalization of Gauss' series is the *(generalized) hypergeometric series* with r numerator parameters $a_1, \ldots, a_r$ and s denominator parameters $b_1, \ldots, b_s$ defined by

$$\begin{aligned}{}_rF_s(a_1,a_2,\ldots,a_r;\ b_1,\ldots,b_s;z) &\equiv {}_rF_s\left[\begin{matrix}a_1,a_2,\ldots,a_r\\ b_1,\ldots,b_s\end{matrix};z\right]\\ &= \sum_{n=0}^{\infty} \frac{(a_1)_n(a_2)_n\cdots(a_r)_n}{n!(b_1)_n\cdots(b_s)_n} z^n.\end{aligned} \tag{1.2.16}$$

Some well-known special cases are the *exponential function*

$$e^z = {}_0F_0(-;-;z), \tag{1.2.17}$$

the *trigonometric functions*

$$\begin{aligned} \sin z &= z\,{}_0F_1(-;3/2;-z^2/4), \\ \cos z &= {}_0F_1(-;1/2;-z^2/4), \end{aligned} \tag{1.2.18}$$

the *Bessel function*

$$J_\alpha(z) = (z/2)^\alpha\ {}_0F_1(-;\alpha+1;-z^2/4)/\Gamma(\alpha+1), \tag{1.2.19}$$

where a dash is used to indicate the absence of either numerator (when $r = 0$) or denominator (when $s = 0$) parameters. Some other well-known special cases are the *Hermite polynomials*

$$H_n(x) = (2x)^n\ {}_2F_0(-n/2,(1-n)/2;-;-x^{-2}), \tag{1.2.20}$$

and the *Laguerre polynomials*

$$L_n^\alpha(x) = \frac{(\alpha+1)_n}{n!}\ {}_1F_1(-n;\alpha+1;x). \tag{1.2.21}$$

Generalizing Heine's series, we shall define an ${}_r\phi_s$ *basic hypergeometric series* by

$$\begin{aligned} {}_r\phi_s(a_1,a_2,\ldots,a_r;b_1,\ldots,b_s;q,z) &\equiv {}_r\phi_s\left[\begin{matrix} a_1,a_2,\ldots,a_r \\ b_1,\ldots,b_s \end{matrix};q,z\right] \\ &= \sum_{n=0}^{\infty} \frac{(a_1;q)_n(a_2;q)_n\cdots(a_r;q)_n}{(q;q)_n(b_1;q)_n\cdots(b_s;q)_n}\left[(-1)^n q^{\binom{n}{2}}\right]^{1+s-r} z^n \end{aligned} \tag{1.2.22}$$

with $\binom{n}{2} = n(n-1)/2$, where $q \neq 0$ when $r > s+1$.

In (1.2.16) and (1.2.22) it is assumed that the parameters $b_1,\ldots,b_s$ are such that the denominator factors in the terms of the series are never zero. Since

$$(-m)_n = \left(q^{-m};q\right)_n = 0, \quad n = m+1, m+2, \ldots, \tag{1.2.23}$$

an ${}_rF_s$ series terminates if one of its numerator parameters is zero or a negative integer, and an ${}_r\phi_s$ series terminates if one of its numerator parameters is of the form q^{-m} with $m = 0,1,2,\ldots$, and $q \neq 0$. Basic analogues of the classical orthogonal polynomials will be considered in Chapter 7 as well as in the exercises at the ends of the chapters.

Unless stated otherwise, when dealing with nonterminating basic hypergeometric series we shall assume that $|q| < 1$ and that the parameters and variables are such that the series converges absolutely. Note that if $|q| > 1$, then we can perform an inversion with respect to the base by setting $p = q^{-1}$ and using the identity

$$(a;q)_n = (a^{-1};p)_n(-a)^n p^{-\binom{n}{2}} \tag{1.2.24}$$

to convert the series (1.2.22) to a similar series in base p with $|p| < 1$. The inverted series will have a finite radius of convergence if the original series does.

Observe that if we denote the terms of the series (1.2.16) and (1.2.22) which contain z^n by u_n and v_n, respectively, then

$$\frac{u_{n+1}}{u_n} = \frac{(a_1+n)(a_2+n)\cdots(a_r+n)}{(1+n)(b_1+n)\cdots(b_s+n)}\, z \tag{1.2.25}$$

is a rational function of n, and

$$\frac{v_{n+1}}{v_n} = \frac{(1-a_1q^n)(1-a_2q^n)\cdots(1-a_rq^n)}{(1-q^n)(1-b_1q^n)\cdots(1-b_sq^n)}\left(-q^n\right)^{1+s-r} z \tag{1.2.26}$$

is a rational function of q^n. Conversely, if $\sum_{n=0}^{\infty} u_n$ and $\sum_{n=0}^{\infty} v_n$ are power series with $u_0 = v_0 = 1$ such that u_{n+1}/u_n is a rational function of n and v_{n+1}/v_n is a rational function of q^n, then these series are of the forms (1.2.16) and (1.2.22), respectively.

By the ratio test, the ${}_rF_s$ series converges absolutely for all z if $r \le s$, and for $|z| < 1$ if $r = s+1$. By an extension of the ratio test (Bromwich [1959, p.241]), it converges absolutely for $|z| = 1$ if $r = s+1$ and Re $[b_1 + \cdots + b_s - (a_1 + \cdots + a_r)] > 0$. If $r > s+1$ and $z \ne 0$ or $r = s+1$ and $|z| > 1$, then this series diverges, unless it terminates.

If $0 < |q| < 1$, the ${}_r\phi_s$ series converges absolutely for all z if $r \le s$ and for $|z| < 1$ if $r = s+1$. This series also converges absolutely if $|q| > 1$ and $|z| < |b_1b_2\cdots b_s|/|a_1a_2\cdots a_r|$. It diverges for $z \ne 0$ if $0 < |q| < 1$ and $r > s+1$, and if $|q| > 1$ and $|z| > |b_1b_2\cdots b_s|/|a_1a_2\cdots a_r|$, unless it terminates. As is customary, the ${}_rF_s$ and ${}_r\phi_s$ notations are also used for the sums of these series inside the circle of convergence and for their analytic continuations (called *hypergeometric functions* and *basic hypergeometric functions*, respectively) outside the circle of convergence.

Observe that the series (1.2.22) has the property that if we replace z by z/a_r and let $a_r \to \infty$, then the resulting series is again of the form (1.2.22) with r replaced by $r-1$. Because this is not the case for the ${}_r\phi_s$ series defined without the factors $\left[(-1)^n q^{\binom{n}{2}}\right]^{1+s-r}$ in the books of Bailey [1935] and Slater [1966] and we wish to be able to handle such limit cases, we have chosen to use the series defined in (1.2.22). There is no loss in generality since the Bailey and Slater series can be obtained from the $r = s+1$ case of (1.2.22) by choosing s sufficiently large and setting some of the parameters equal to zero.

An ${}_{r+1}F_r$ series is called *k-balanced* if $b_1 + b_2 + \cdots + b_r = k + a_1 + a_2 + \cdots + a_{r+1}$ and $z = 1$; a 1-balanced series is called *balanced* (or *Saalschützian*). Analogously, an ${}_{r+1}\phi_r$ series is called *k-balanced* if $b_1b_2\cdots b_r = q^k a_1a_2\cdots a_{r+1}$ and $z = q$, and a 1-balanced series is called *balanced* (or *Saalschützian*). We will first encounter balanced series in §1.7, where we derive a summation formula for such a series.

For negative subscripts, the *shifted factorial* and the *q-shifted factorials* are defined by

$$(a)_{-n} = \frac{1}{(a-1)(a-2)\cdots(a-n)} = \frac{1}{(a-n)_n} = \frac{(-1/a)^n}{(1-a)_n}, \tag{1.2.27}$$

$$(a;q)_{-n} = \frac{1}{(1-aq^{-1})(1-aq^{-2})\cdots(1-aq^{-n})} = \frac{1}{(aq^{-n};q)_n} = \frac{(-q/a)^n q^{\binom{n}{2}}}{(q/a;q)_n}, \tag{1.2.28}$$

where $n = 0, 1, \ldots$. We also define

$$(a;q)_\infty = \prod_{k=0}^{\infty}(1-aq^k) \tag{1.2.29}$$

for $|q| < 1$. Since the infinite product in (1.2.29) diverges when $a \neq 0$ and $|q| \geq 1$, whenever $(a;q)_\infty$ appears in a formula, we shall assume that $|q| < 1$. The following easily verified identities will be frequently used in this book:

$$(a;q)_n = \frac{(a;q)_\infty}{(aq^n;q)_\infty}, \tag{1.2.30}$$

$$(a^{-1}q^{1-n};q)_n = (a;q)_n(-a^{-1})^n q^{-\binom{n}{2}}, \tag{1.2.31}$$

$$(a;q)_{n-k} = \frac{(a;q)_n}{(a^{-1}q^{1-n};q)_k}\,(-qa^{-1})^k q^{\binom{k}{2}-nk}, \tag{1.2.32}$$

$$(a;q)_{n+k} = (a;q)_n(aq^n;q)_k, \tag{1.2.33}$$

$$(aq^n;q)_k = \frac{(a;q)_k(aq^k;q)_n}{(a;q)_n}, \tag{1.2.34}$$

$$(aq^k;q)_{n-k} = \frac{(a;q)_n}{(a;q)_k}, \tag{1.2.35}$$

$$(aq^{2k};q)_{n-k} = \frac{(a;q)_n(aq^n;q)_k}{(a;q)_{2k}}, \tag{1.2.36}$$

$$(q^{-n};q)_k = \frac{(q;q)_n}{(q;q)_{n-k}}(-1)^k q^{\binom{k}{2}-nk}, \tag{1.2.37}$$

$$(aq^{-n};q)_k = \frac{(a;q)_k(qa^{-1};q)_n}{(a^{-1}q^{1-k};q)_n}\,q^{-nk}, \tag{1.2.38}$$

$$(a;q)_{2n} = (a;q^2)_n(aq;q^2)_n, \tag{1.2.39}$$

$$(a^2;q^2)_n = (a;q)_n(-a;q)_n, \tag{1.2.40}$$

where n and k are integers. A more complete list of useful identities is given in Appendix I at the end of the book.

Since products of q-shifted factorials occur so often, to simplify them we shall frequently use the more compact notations

$$(a_1, a_2, \ldots, a_m;q)_n = (a_1;q)_n(a_2;q)_n\cdots(a_m;q)_n, \tag{1.2.41}$$

$$(a_1, a_2, \ldots, a_m;q)_\infty = (a_1;q)_\infty(a_2;q)_\infty\cdots(a_m;q)_\infty. \tag{1.2.42}$$

1.3 The q-binomial theorem

One of the most important summation formulas for hypergeometric series is given by the *binomial theorem*:

$$ {}_2F_1(a,c;c;z) = {}_1F_0(a;—;z) = \sum_{n=0}^{\infty} \frac{(a)_n}{n!} z^n = (1-z)^{-a}, \tag{1.3.1}$$

where $|z|<1$. We shall show that this formula has the following q-analogue

$$ {}_1\phi_0(a;—;q,z) = \sum_{n=0}^{\infty} \frac{(a;q)_n}{(q;q)_n} z^n = \frac{(az;q)_\infty}{(z;q)_\infty}, \quad |z|<1,\ |q|<1, \tag{1.3.2}$$

which was derived by Cauchy [1843], Heine [1847] and by other mathematicians.

Heine's proof of (1.3.2), which can also be found in the books Heine [1878], Bailey [1935, p.66] and Slater [1966, p.92], is better understood if one first follows Askey's [1980a] approach of evaluating the sum of the binomial series in (1.3.1), and then carries out the analogous steps for the series in (1.3.2).

Let us set

$$ f_a(z) = \sum_{n=0}^{\infty} \frac{(a)_n}{n!} z^n. \tag{1.3.3}$$

Since this series is uniformly convergent in $|z| \le \epsilon$ when $0<\epsilon<1$, we may differentiate it termwise to get

$$\begin{aligned} f_a'(z) &= \sum_{n=1}^{\infty} \frac{n(a)_n}{n!} z^{n-1} \\ &= \sum_{n=0}^{\infty} \frac{(a)_{n+1}}{n!} z^n = a f_{a+1}(z). \end{aligned} \tag{1.3.4}$$

Also

$$\begin{aligned} f_a(z) - f_{a+1}(z) &= \sum_{n=1}^{\infty} \frac{(a)_n - (a+1)_n}{n!} z^n \\ = \sum_{n=1}^{\infty} \frac{(a+1)_{n-1}}{n!}&\,[a-(a+n)]\, z^n = -\sum_{n=1}^{\infty} \frac{n(a+1)_{n-1}}{n!} z^n \\ = -\sum_{n=0}^{\infty} \frac{(a+1)_n}{n!}&\, z^{n+1} = -z f_{a+1}(z). \end{aligned} \tag{1.3.5}$$

Eliminating $f_{a+1}(z)$ from (1.3.4) and (1.3.5), we obtain the first order differential equation

$$ f_a'(z) = \frac{a}{1-z} f_a(z), \tag{1.3.6}$$

subject to the initial condition $f_a(0)=1$, which follows from the definition (1.3.3) of $f_a(z)$. Solving (1.3.6) under this condition immediately gives that $f_a(z) = (1-z)^{-a}$ for $|z|<1$.

Analogously, let us now set

$$h_a(z) = \sum_{n=0}^{\infty} \frac{(a;q)_n}{(q;q)_n} z^n, \quad |z| < 1, \ |q| < 1. \tag{1.3.7}$$

Clearly, $h_{q^a}(z) \to f_a(z)$ as $q \to 1$. Since $h_{aq}(z)$ is a q-analogue of $f_{a+1}(z)$, we first compute the difference

$$\begin{aligned} h_a(z) - h_{aq}(z) &= \sum_{n=1}^{\infty} \frac{(a;q)_n - (aq;q)_n}{(q;q)_n} z^n \\ &= \sum_{n=1}^{\infty} \frac{(aq;q)_{n-1}}{(q;q)_n} [1 - a - (1 - aq^n)] z^n \\ &= -a \sum_{n=1}^{\infty} \frac{(1-q^n)(aq;q)_{n-1}}{(q;q)_n} z^n \\ &= -a \sum_{n=1}^{\infty} \frac{(aq;q)_{n-1}}{(q;q)_{n-1}} z^n = -azh_{aq}(z), \end{aligned} \tag{1.3.8}$$

giving an analogue of (1.3.5). Observing that

$$f'(z) = \lim_{q\to 1} \frac{f(z) - f(qz)}{(1-q)z} \tag{1.3.9}$$

for a differentiable function f, we next compute the difference

$$\begin{aligned} h_a(z) - h_a(qz) &= \sum_{n=1}^{\infty} \frac{(a;q)_n}{(q;q)_n} (z^n - q^n z^n) \\ &= \sum_{n=1}^{\infty} \frac{(a;q)_n}{(q;q)_{n-1}} z^n = \sum_{n=0}^{\infty} \frac{(a;q)_{n+1}}{(q;q)_n} z^{n+1} \\ &= (1-a)zh_{aq}(z). \end{aligned} \tag{1.3.10}$$

Eliminating $h_{aq}(z)$ from (1.3.8) and (1.3.10) gives

$$h_a(z) = \frac{1-az}{1-z} h_a(qz). \tag{1.3.11}$$

Iterating this relation $n-1$ times and then letting $n \to \infty$ we obtain

$$\begin{aligned} h_a(z) &= \frac{(az;q)_n}{(z;q)_n} h_a(q^n z) \\ &= \frac{(az;q)_\infty}{(z;q)_\infty} h_a(0) = \frac{(az;q)_\infty}{(z;q)_\infty}, \end{aligned} \tag{1.3.12}$$

since $q^n \to 0$ as $n \to \infty$ and $h_a(0) = 1$ by (1.3.7), which completes the proof of (1.3.2).

One consequence of (1.3.2) is the product formula

$${}_1\phi_0(a;—;q,z)\, {}_1\phi_0(b;—;q,az) = {}_1\phi_0(ab;—;q,z), \tag{1.3.13}$$

which is a q-analogue of $(1-z)^{-a}(1-z)^{-b} = (1-z)^{-a-b}$.

In the special case $a = q^{-n}, n = 0, 1, 2, \ldots$, (1.3.2) gives

$$_1\phi_0(q^{-n}; —; q, z) = (zq^{-n}; q)_n = (-z)^n q^{-n(n+1)/2}(q/z; q)_n, \tag{1.3.14}$$

where, by analytic continuation, z can be any complex number. From now on, unless stated othewise, whenever $q^{-j}, q^{-k}, q^{-m}, q^{-n}$ appear as numerator parameters in basic series it will be assumed that j, k, m, n, respectively, are nonnegative integers.

If we set $a = 0$ in (1.3.2), we get

$$_1\phi_0(0; —; q, z) = \sum_{n=0}^{\infty} \frac{z^n}{(q; q)_n} = \frac{1}{(z; q)_\infty}, \quad |z| < 1, \tag{1.3.15}$$

which is a q-analogue of the exponential function e^z. Another q-analogue of e^z can be obtained from (1.3.2) by replacing z by $-z/a$ and then letting $a \to \infty$ to get

$$_0\phi_0(—; —; q, -z) = \sum_{n=0}^{\infty} \frac{q^{n(n-1)/2}}{(q; q)_n} z^n = (-z; q)_\infty. \tag{1.3.16}$$

Observe that if we denote the *q-exponential* in (1.3.15) and (1.3.16) by $e_q(z)$ and $E_q(z)$, respectively, then $e_q(z)E_q(-z) = 1$ and

$$\lim_{q \to 1^-} e_q(z(1-q)) = \lim_{q \to 1^-} E_q(z(1-q)) = e^z. \tag{1.3.17}$$

In deriving q-analogues of various formulas we shall sometimes use the observation that

$$\frac{(q^a z; q)_\infty}{(z; q)_\infty} = {}_1\phi_0(q^a; —; q, z) \to {}_1F_0(a; —; z) = (1-z)^{-a} \text{ as } q \to 1^-. \tag{1.3.18}$$

Thus

$$\lim_{q \to 1^-} \frac{(q^a z; q)_\infty}{(z; q)_\infty} = (1-z)^{-a}, \quad |z| < 1, \quad a \text{ real}. \tag{1.3.19}$$

By analytic continuation this holds for z in the complex plane cut along the positive real axis from 1 to ∞, with $(1-z)^{-a}$ positive when z is real and less than 1.

1.4 Heine's transformation formulas for $_2\phi_1$ series

Heine [1847], [1878] showed that

$$_2\phi_1(a, b; c; q, z) = \frac{(b, az; q)_\infty}{(c, z; q)_\infty} {}_2\phi_1(c/b, z; az; q, b), \tag{1.4.1}$$

where $|z| < 1$ and $|b| < 1$. To prove this transformation formula, first observe from the q-binomial theorem (1.3.2) that

$$\frac{(cq^n; q)_\infty}{(bq^n; q)_\infty} = \sum_{m=0}^{\infty} \frac{(c/b; q)_m}{(q; q)_m} (bq^n)^m.$$

Hence, for $|z| < 1$ and $|b| < 1$,

$$\begin{aligned}
{}_2\phi_1(a, b; c; q, z) &= \frac{(b;q)_\infty}{(c;q)_\infty} \sum_{n=0}^{\infty} \frac{(a;q)_n (cq^n;q)_\infty}{(q;q)_n (bq^n;q)_\infty} z^n \\
&= \frac{(b;q)_\infty}{(c;q)_\infty} \sum_{n=0}^{\infty} \frac{(a;q)_n}{(q;q)_n} z^n \sum_{m=0}^{\infty} \frac{(c/b;q)_m}{(q;q)_m} (bq^n)^m \\
&= \frac{(b;q)_\infty}{(c;q)_\infty} \sum_{m=0}^{\infty} \frac{(c/b;q)_m}{(q;q)_m} b^m \sum_{n=0}^{\infty} \frac{(a;q)_n}{(q;q)_n} (zq^m)^n \\
&= \frac{(b;q)_\infty}{(c;q)_\infty} \sum_{m=0}^{\infty} \frac{(c/b;q)_m}{(q;q)_m} b^m \frac{(azq^m;q)_\infty}{(zq^m;q)_\infty} \\
&= \frac{(b, az;q)_\infty}{(c, z;q)_\infty} \, {}_2\phi_1(c/b, z; az; q, b)
\end{aligned}$$

by (1.3.2), which gives (1.4.1).

Heine also showed that Euler's transformation formula

$$ {}_2F_1(a, b; c; z) = (1-z)^{c-a-b} \, {}_2F_1(c-a, c-b; c; z) \tag{1.4.2}$$

has a q-analogue of the form

$$ {}_2\phi_1(a, b; c; q, z) = \frac{(abz/c;q)_\infty}{(z;q)_\infty} \, {}_2\phi_1(c/a, c/b; c; q, abz/c). \tag{1.4.3}$$

A short way to prove this formula is just to iterate (1.4.1) as follows

$$ {}_2\phi_1(a, b; c; q, z) = \frac{(b, az;q)_\infty}{(c, z;q)_\infty} \, {}_2\phi_1(c/b, z; az; q, b) \tag{1.4.4}$$

$$ = \frac{(c/b, bz;q)_\infty}{(c, z;q)_\infty} \, {}_2\phi_1(abz/c, b; bz; q, c/b) \tag{1.4.5}$$

$$ = \frac{(abz/c;q)_\infty}{(z;q)_\infty} \, {}_2\phi_1(c/a, c/b; c; q, abz/c). \tag{1.4.6}$$

1.5 Heine's q-analogue of Gauss' summation formula

In order to derive Heine's [1847] q-analogue of Gauss' summation formula (1.2.11) it suffices to set $z = c/ab$ in (1.4.1), assume that $|b| < 1, |c/ab| < 1$, and observe that the series on the right side of

$$ {}_2\phi_1(a, b; c; q, c/ab) = \frac{(b, c/b;q)_\infty}{(c, c/ab;q)_\infty} \, {}_1\phi_0(c/ab; —; q, b)$$

can be summed by (1.3.2) to give

$$ {}_2\phi_1(a, b; c; q, c/ab) = \frac{(c/a, c/b;q)_\infty}{(c, c/ab;q)_\infty}. \tag{1.5.1}$$

By analytic continuation, we may drop the assumption that $|b| < 1$ and require only that $|c/ab| < 1$ for (1.5.1) to be valid.

For the terminating case when $a = q^{-n}$, (1.5.1) reduces to

$$ {}_2\phi_1(q^{-n}, b; c; q, cq^n/b) = \frac{(c/b;q)_n}{(c;q)_n}. \tag{1.5.2} $$

By inversion or by changing the order of summation it follows from (1.5.2) that

$$ {}_2\phi_1(q^{-n}, b; c; q, q) = \frac{(c/b;q)_n}{(c;q)_n} b^n. \tag{1.5.3} $$

Both (1.5.2) and (1.5.3) are q-analogues of Vandermonde's formula (1.2.9). These formulas can be used to derive other important formulas such as, for example, Jackson's [1910a] transformation formula

$$ \begin{aligned} {}_2\phi_1(a, b; c; q, z) &= \frac{(az;q)_\infty}{(z;q)_\infty} \sum_{k=0}^{\infty} \frac{(a, c/b;q)_k}{(q, c, az;q)_k} (-bz)^k q^{\binom{k}{2}} \\ &= \frac{(az;q)_\infty}{(z;q)_\infty} \, {}_2\phi_2(a, c/b; c, az; q, bz). \end{aligned} \tag{1.5.4} $$

This formula is a q-analogue of the Pfaff-Kummer transformation formula

$$ {}_2F_1(a, b; c; z) = (1 - z)^{-a} \, {}_2F_1(a, c - b; c; z/(z - 1)). \tag{1.5.5} $$

To prove (1.5.4), we use (1.5.2) to write

$$ \frac{(b;q)_k}{(c;q)_k} = \sum_{n=0}^{k} \frac{(q^{-k}, c/b;q)_n}{(q, c;q)_n} \left(bq^k\right)^n $$

and hence

$$ \begin{aligned} &{}_2\phi_1(a, b; c; q, z) \\ &= \sum_{k=0}^{\infty} \frac{(a;q)_k}{(q;q)_k} z^k \sum_{n=0}^{k} \frac{(q^{-k}, c/b;q)_n}{(q, c;q)_n} \left(bq^k\right)^n \\ &= \sum_{n=0}^{\infty} \sum_{k=n}^{\infty} \frac{(a;q)_k (c/b;q)_n}{(q;q)_{k-n} (q, c;q)_n} z^k (-b)^n q^{\binom{n}{2}} \\ &= \sum_{n=0}^{\infty} \sum_{k=0}^{\infty} \frac{(a;q)_{k+n} (c/b;q)_n}{(q;q)_k (q, c;q)_n} (-bz)^n z^k q^{\binom{n}{2}} \\ &= \sum_{n=0}^{\infty} \frac{(a, c/b;q)_n}{(q, c;q)_n} (-bz)^n q^{\binom{n}{2}} \sum_{k=0}^{\infty} \frac{(aq^n;q)_k}{(q;q)_k} z^k \\ &= \frac{(az;q)_\infty}{(z;q)_\infty} \sum_{n=0}^{\infty} \frac{(a, c/b;q)_n}{(q, c, az;q)_n} (-bz)^n q^{\binom{n}{2}}, \end{aligned} $$

by (1.3.2). Also see Andrews [1973]. If $a = q^{-n}$, then the series on the right side of (1.5.4) can be reversed (by replacing k and $n - k$) to yield Sears' [1951c] transformation formula

$$ \begin{aligned} &{}_2\phi_1(q^{-n}, b; c; q, z) \\ &= \frac{(c/b;q)_n}{(c;q)_n} \left(\frac{bz}{q}\right)^n {}_3\phi_2(q^{-n}, q/z, c^{-1}q^{1-n}; bc^{-1}q^{1-n}, 0; q, q). \end{aligned} \tag{1.5.6} $$

1.6 Jacobi's triple product identity and the theta functions

Jacobi's [1829] well-known *triple product identity* (see Andrews [1971])

$$(zq^{\frac{1}{2}}, q^{\frac{1}{2}}/z, q; q)_\infty = \sum_{n=-\infty}^{\infty} (-1)^n q^{n^2/2} z^n \tag{1.6.1}$$

can be easily derived by using Heine's summation formula (1.5.1).

First, set $c = bzq^{\frac{1}{2}}$ in (1.5.1) and then let $b \to 0$ and $a \to \infty$ to obtain

$$\sum_{n=0}^{\infty} \frac{(-1)^n q^{n^2/2}}{(q;q)_n} z^n = (zq^{\frac{1}{2}}; q)_\infty. \tag{1.6.2}$$

Similarily, setting $c = zq$ in (1.5.1) and letting $a \to \infty$ and $b \to \infty$ we get

$$\sum_{n=0}^{\infty} \frac{q^{n^2}}{(q, zq; q)_n} = \frac{1}{(zq;q)_\infty}. \tag{1.6.3}$$

Now use (1.6.2) to find that

$$\begin{aligned}
&(zq^{\frac{1}{2}}, q^{\frac{1}{2}}/z; q)_\infty \\
&= \sum_{m=0}^{\infty}\sum_{n=0}^{\infty} \frac{(-1)^{m+n} q^{(m^2+n^2)/2}}{(q;q)_m (q;q)_n} z^{m-n} \\
&= \sum_{n=0}^{\infty} \frac{(-1)^n q^{n^2/2}}{(q;q)_n} z^n \sum_{k=0}^{\infty} \frac{q^{k^2}}{(q, q^{n+1}; q)_k} q^{nk} \\
&\quad + \sum_{n=1}^{\infty} \frac{(-1)^n q^{n^2/2}}{(q;q)_n} z^{-n} \sum_{k=0}^{\infty} \frac{q^{k^2}}{(q, q^{n+1}; q)_k} q^{nk}.
\end{aligned} \tag{1.6.4}$$

Formula (1.6.1) then follows from (1.6.3) by observing that

$$\frac{1}{(q;q)_n} \sum_{n=0}^{\infty} \frac{q^{k^2}}{(q, q^{n+1}; q)_k} q^{nk} = \frac{1}{(q;q)_n (q^{n+1}; q)_\infty} = \frac{1}{(q;q)_\infty}.$$

An important application of (1.6.1) is that it can be used to express the *theta functions* (Whittaker and Watson [1965, Chapter 21])

$$\vartheta_1(x) = 2\sum_{n=0}^{\infty} (-1)^n q^{(n+1/2)^2} \sin(2n+1)x, \tag{1.6.5}$$

$$\vartheta_2(x) = 2\sum_{n=0}^{\infty} q^{(n+1/2)^2} \cos(2n+1)x, \tag{1.6.6}$$

$$\vartheta_3(x) = 1 + 2\sum_{n=1}^{\infty} q^{n^2} \cos 2nx, \tag{1.6.7}$$

$$\vartheta_4(x) = 1 + 2\sum_{n=1}^{\infty} (-1)^n q^{n^2} \cos 2nx \tag{1.6.8}$$

in terms of infinite products. Just replace q by q^2 in (1.6.1) and then set z equal to $qe^{2ix}, -qe^{2ix}, -e^{2ix}, e^{2ix}$, respectively, to obtain

$$\vartheta_1(x) = 2q^{1/4}\sin x \prod_{n=1}^{\infty}(1-q^{2n})(1-2q^{2n}\cos 2x + q^{4n}), \tag{1.6.9}$$

$$\vartheta_2(x) = 2q^{1/4}\cos x \prod_{n=1}^{\infty}(1-q^{2n})(1+2q^{2n}\cos 2x + q^{4n}), \tag{1.6.10}$$

$$\vartheta_3(x) = \prod_{n=1}^{\infty}(1-q^{2n})(1+2q^{2n-1}\cos 2x + q^{4n-2}), \tag{1.6.11}$$

and

$$\vartheta_4(x) = \prod_{n=1}^{\infty}(1-q^{2n})(1-2q^{2n-1}\cos 2x + q^{4n-2}). \tag{1.6.12}$$

1.7 A q-analogue of Saalschütz's summation formula

Pfaff [1797] discovered the summation formula

$${}_3F_2(a,b,-n;c,1+a+b-c-n;1) = \frac{(c-a)_n(c-b)_n}{(c)_n(c-a-b)_n}, \quad n=0,1,\ldots, \tag{1.7.1}$$

which sums a terminating balanced ${}_3F_2(1)$ series with argument 1. It was rediscovered by Saalschütz [1890] and is usually called *Saalschütz formula* or the *Pfaff-Saalschütz formula*; see Askey [1975]. To derive a q-analogue of (1.7.1), observe that since, by (1.3.2),

$$\frac{(abz/c;q)_\infty}{(z;q)_\infty} = \sum_{k=0}^{\infty}\frac{(ab/c;q)_k}{(q;q)_k}z^k$$

the right side of (1.4.3) equals

$$\sum_{k=0}^{\infty}\sum_{m=0}^{\infty}\frac{(ab/c;q)_k(c/a,c/b;q)_m}{(q;q)_k(q,c;q)_m}\left(\frac{ab}{c}\right)^m z^{k+m},$$

and hence, equating the coefficients of z^n on both sides of (1.4.3) we get

$$\sum_{j=0}^{n}\frac{(q^{-n},c/a,c/b;q)_j}{(q,c,cq^{1-n}/ab;q)_j}q^j = \frac{(a,b;q)_n}{(c,ab/c;q)_n}.$$

Replacing a,b by $c/a, c/b$, repectively, this gives the following sum of a terminating balanced ${}_3\phi_2$ series

$${}_3\phi_2(a,b,q^{-n};c,abc^{-1}q^{1-n};q,q) = \frac{(c/a,c/b;q)_n}{(c,c/ab;q)_n}, \quad n=0,1,\ldots, \tag{1.7.2}$$

which was first derived by Jackson [1910a]. It is easy to see that (1.7.1) follows from (1.7.2) by replacing a,b,c in (1.7.2) by q^a, q^b, q^c, respectively, and letting $q\to 1$. Note that letting $a\to\infty$ in (1.7.2) gives (1.5.2), while letting $a\to 0$ gives (1.5.3).

1.8 The Bailey-Daum summation formula

Bailey [1941] and Daum [1942] independently discovered the summation formula

$$
{}_2\phi_1(a,b;aq/b;q,-q/b)=\frac{(-q;q)_\infty(aq,aq^2/b^2;q^2)_\infty}{(aq/b,-q/b;q)_\infty}, \tag{1.8.1}
$$

which is a q-analogue of Kummer's formula

$$
{}_2F_1(a,b;1+a-b;-1)=\frac{\Gamma(1+a-b)\Gamma(1+\frac{1}{2}a)}{\Gamma(1+a)\Gamma(1+\frac{1}{2}a-b)}. \tag{1.8.2}
$$

Formula (1.8.1) can be easily obtained from (1.4.1) by using the identity (1.2.40) and a limiting form of (1.2.39), namely, $(a;q)_\infty=(a,aq;q^2)_\infty$, to see that

$$
\begin{aligned}
&{}_2\phi_1(a,b;aq/b;q,-q/b)\\
&=\frac{(a,-q;q)_\infty}{(aq/b,-q/b;q)_\infty}\,{}_2\phi_1(q/b,-q/b;-q;q,a)\\
&=\frac{(a,-q;q)_\infty}{(aq/b,-q/b;q)_\infty}\sum_{n=0}^{\infty}\frac{(q^2/b^2;q^2)_n}{(q^2;q^2)_n}a^n\\
&=\frac{(a,-q;q)_\infty}{(aq/b,-q/b;q)_\infty}\frac{(aq^2/b^2;q^2)_\infty}{(a;q^2)_\infty} \qquad \text{by (1.3.2)}\\
&=\frac{(-q;q)_\infty(aq,aq^2/b^2;q^2)_\infty}{(aq/b,-q/b;q)_\infty}.
\end{aligned}
$$

1.9 q-Analogues of the Karlsson-Minton summation formulas

Minton [1970] showed that if a is a negative integer and $m_1,m_2,\ldots,m_r$ are nonnegative integers such that $-a\ge m_1+\cdots+m_r$, then

$$
\begin{aligned}
&{}_{r+2}F_{r+1}\left[\begin{matrix}a,b,b_1+m_1,\ldots,b_r+m_r\\ b+1,b_1,\ldots,b_r\end{matrix};1\right]\\
&\quad=\frac{\Gamma(b+1)\Gamma(1-a)}{\Gamma(1+b-a)}\frac{(b_1-b)_{m_1}\cdots(b_r-b)_{m_r}}{(b_1)_{m_1}\cdots(b_r)_{m_r}}
\end{aligned} \tag{1.9.1}
$$

where, as usual, it is assumed that none of the factors in the denominators of the terms of the series is zero. Karlsson [1971] showed that (1.9.1) also holds when a is not a negative integer provided that the series converges, i.e., if $\text{Re}(-a)>m_1+\cdots+m_r-1$, and he deduced from (1.9.1) that

$$
{}_{r+1}F_r\left[\begin{matrix}a,b_1+m_1,\ldots,b_r+m_r\\ b_1,\ldots,b_r\end{matrix};1\right]=0,\quad \text{Re}\,(-a)>m_1+\cdots+m_r, \tag{1.9.2}
$$

$$
\begin{aligned}
&{}_{r+1}F_r\left[\begin{matrix}-(m_1+\cdots+m_r),b_1+m_1,\ldots,b_r+m_r\\ b_1,\ldots,b_r\end{matrix};1\right]\\
&\quad=(-1)^{m_1+\cdots+m_r}\frac{(m_1+\cdots+m_r)!}{(b_1)_{m_1}\cdots(b_r)_{m_r}}.
\end{aligned} \tag{1.9.3}
$$

These formulas are particularly useful for evaluating sums that appear as solutions to some problems in theoretical physics such as the Racah coefficients. They were also used by Gasper [1981b] to prove the orthogonality on $(0, 2\pi)$ of certain functions that arose in Greiner's [1980] work on spherical harmonics on the Heisenberg group. Here we shall present Gasper's [1981a] derivation of q-analogues of the above formulas. Some of the formulas derived below will be used in Chaper 7 to prove the orthogonality relation for the continuous q-ultraspherical polynomials.

Observe that if m and n are nonnegative integers with $m \geq n$, then

$$ {}_2\phi_1(q^{-n}, q^{-m}; b_r; q, q) = \frac{(b_r q^m; q)_n}{(b_r; q)_n} q^{-mn} $$

by (1.5.3), and hence

$$ \begin{aligned} &{}_{r+1}\phi_r \left[\begin{matrix} a_1, \ldots, a_r, b_r q^m \\ b_1, \ldots, b_{r-1}, b_r \end{matrix} ; q, z \right] \\ &= \sum_{n=0}^{\infty} \frac{(a_1, \ldots, a_r; q)_n}{(q, b_1, \ldots, b_{r-1}; q)_n} z^n \sum_{k=0}^{n} \frac{(q^{-n}, q^{-m}; q)_k}{(q, b_r; q)_k} q^{mn+k} \\ &= \sum_{n=0}^{\infty} \sum_{k=0}^{m} \frac{(a_1, \ldots, a_r; q)_n (q^{-m}; q)_k}{(b_1, \ldots, b_{r-1}; q)_n (q; q)_{n-k} (q, b_r; q)_k} z^n (-1)^k q^{mn+k-nk+\binom{k}{2}} \\ &= \sum_{k=0}^{m} \frac{(q^{-m}, a_1, \ldots, a_r; q)_k}{(q, b_1, \ldots, b_r; q)_k} (-zq^m)^k q^{\binom{k}{2}} \\ &\quad \cdot {}_r\phi_{r-1} \left[\begin{matrix} a_1 q^k, \ldots, a_r q^k \\ b_1 q^k, \ldots, b_{r-1} q^k \end{matrix} ; zq^{m-k} \right], \quad |z| < 1. \end{aligned} \tag{1.9.4} $$

This expansion formula is a q-analogue of a formula used by Minton [1970, (4)].

When $r = 2$, formulas (1.9.4), (1.5.1) and (1.5.3) yield

$$ \begin{aligned} {}_3\phi_2 \left[\begin{matrix} a, b, b_1 q^m \\ bq, b_1 \end{matrix} ; q, a^{-1} q^{1-m} \right] &= \frac{(q, bq/a; q)_\infty}{(bq, q/a; q)_\infty} {}_2\phi_1(q^{-m}, b; b_1; q, q) \\ &= \frac{(q, bq/a; q)_\infty (b_1/b; q)_m}{(bq, q/a; q)_\infty (b_1; q)_m} b^m, \end{aligned} \tag{1.9.5} $$

provided that $|a^{-1} q^{1-m}| < 1$. By induction it follows from (1.9.4) and (1.9.5) that if $m_1, \ldots, m_r$ are nonnegative integers and $|a^{-1} q^{1-(m_1+\cdots+m_r)}| < 1$, then

$$ \begin{aligned} &{}_{r+2}\phi_{r+1} \left[\begin{matrix} a, b, b_1 q^{m_1}, \ldots, b_r q^{m_r} \\ bq, b_1, \ldots, b_r \end{matrix} ; q, a^{-1} q^{1-(m_1+\cdots+m_r)} \right] \\ &= \frac{(q, bq/a; q)_\infty}{(bq, q/a; q)_\infty} \frac{(b_1/b; q)_{m_1} \cdots (b_r/b; q)_{m_r}}{(b_1; q)_{m_1} \cdots (b_r; q)_{m_r}} b^{m_1+\cdots+m_r} \end{aligned} \tag{1.9.6} $$

which is a q-analogue of (1.9.1). Formula (1.9.1) can be derived from (1.9.6) by replacing $a, b, b_1, \ldots, b_r$ by $q^a, q^b, q^{b_1}, \ldots, q^{b_r}$, respectively, and letting $q \to 1$.

Setting $b_r = b, m_r = 1$ and then replacing r by $r+1$ in (1.9.6) gives

$$ {}_{r+1}\phi_r \left[\begin{matrix} a, b_1 q^{m_1}, \ldots, b_r q^{m_r} \\ b_1, \ldots, b_r \end{matrix} ; q, a^{-1} q^{-(m_1+\cdots+m_r)} \right] = 0, \ |a^{-1} q^{-(m_1+\cdots+m_r)}| < 1, \tag{1.9.7} $$

while letting $b \to \infty$ in the case $a = q^{-(m_1+\cdots+m_r)}$ of (1.9.6) gives

$$
{}_{r+1}\phi_r \left[\begin{matrix} q^{-(m_1+\cdots+m_r)}, b_1 q^{m_1}, \ldots, b_r q^{m_r} \\ b_1, \ldots, b_r \end{matrix} ; q, 1 \right]
= \frac{(-1)^{m_1+\cdots+m_r} (q;q)_{m_1+\cdots+m_r}}{(b_1;q)_{m_1} \cdots (b_r;q)_{m_r}} q^{-(m_1+\cdots+m_r)(m_1+\cdots+m_r+1)/2}, \quad (1.9.8)
$$

which are q-analogues of (1.9.2) and (1.9.3). Another q-analogue of (1.9.3) can be derived by letting $b \to 0$ in (1.9.6)to obtain

$$
{}_{r+1}\phi_r \left[\begin{matrix} a, b_1 q^{m_1}, \ldots, b_r q^{m_r} \\ b_1, \ldots, b_r \end{matrix} ; q, a^{-1} q^{1-(m_1+\cdots+m_r)} \right]
= \frac{(-1)^{m_1+\cdots+m_r} (q;q)_\infty b_1^{m_1} \cdots b_r^{m_r}}{(q/a;q)_\infty (b_1;q)_{m_1} \cdots (b_r;q)_{m_r}} q^{\binom{m_1}{2}+\cdots+\binom{m_r}{2}}, \quad (1.9.9)
$$

when $|a^{-1} q^{1-(m_1+\cdots+m_r)}| < 1$.

In addition, if $a = q^{-n}$ and n is a nonnegative integer then we can reverse the order of summation of the series in (1.9.6), (1.9.7) and (1.9.9) to obtain

$$
{}_{r+2}\phi_{r+1} \left[\begin{matrix} q^{-n}, b, b_1 q^{m_1}, \ldots, b_r q^{m_r} \\ bq, b_1, \ldots, b_r \end{matrix} ; q, q \right]
= \frac{b^n (q;q)_n (b_1/b;q)_{m_1} \cdots (b_r/b;q)_{m_r}}{(bq;q)_n (b_1;q)_{m_1} \cdots (b_r;q)_{m_r}}, \quad n \geq m_1 + \cdots + m_r, \quad (1.9.10)
$$

$$
{}_{r+1}\phi_r \left[\begin{matrix} q^{-n}, b_1 q^{m_1}, \ldots, b_r q^{m_1} \\ b_1, \ldots, b_r \end{matrix} ; q, q \right] = 0, \quad n > m_1 + \cdots + m_r, \quad (1.9.11)
$$

and the following generalization of (1.9.8)

$$
{}_{r+1}\phi_r \left[\begin{matrix} q^{-n}, b_1 q^{m_1}, \ldots, b_r q^{m_r} \\ b_1, \ldots, b_r \end{matrix} ; q, 1 \right] = \frac{(-1)^n (q;q)_n q^{-n(n+1)/2}}{(b_1;q)_{m_1} \cdots (b_r;q)_{m_r}}, \quad (1.9.12)
$$

where $n \geq m_1 + \cdots + m_r$, which also follows by letting $b \to \infty$ in (1.9.10).

1.10 The q-gamma and q-beta functions

The *q-gamma* function

$$
\Gamma_q(x) = \frac{(q;q)_\infty}{(q^x;q)_\infty} (1-q)^{1-x}, \quad 0 < q < 1, \quad (1.10.1)
$$

was introduced by Thomae [1869] and later by Jackson [1904e]. Heine [1847] gave an equivalent definition, but without the factor $(1-q)^{1-x}$. When $x = n+1$ with n a nonnegative integer, this definition reduces to

$$
\Gamma_q(n+1) = 1(1+q)(1+q+q^2) \cdots (1+q+\cdots+q^{n-1}), \quad (1.10.2)
$$

which clearly approaches $n!$ as $q \to 1^-$. Hence $\Gamma_q(n+1)$ tends to $\Gamma(n+1) = n!$ as $q \to 1^-$. The definition of $\Gamma_q(x)$ can be extended to $|q| < 1$ by using the principal values of q^x and $(1-q)^{1-x}$ in (1.10.1).

To show that

$$
\lim_{q \to 1^-} \Gamma_q(x) = \Gamma(x) \quad (1.10.3)
$$

we shall give a simple proof due to Wm. Gosper; see Andrews [1986]. From (1.10.1),

$$\Gamma_q(x+1) = \frac{(q;q)_\infty}{(q^{x+1};q)_\infty}(1-q)^{-x}$$
$$= \prod_{n=1}^{\infty} \frac{(1-q^n)(1-q^{n+1})^x}{(1-q^{n+x})(1-q^n)^x}.$$

Hence

$$\lim_{q\to 1^-} \Gamma_q(x+1) = \prod_{n=1}^{\infty} \frac{n}{n+x}\left(\frac{n+1}{n}\right)^x$$
$$= x\left[x^{-1}\prod_{n=1}^{\infty}\left(1+\frac{x}{n}\right)^{-1}\left(1+\frac{1}{n}\right)^x\right]$$
$$= x\Gamma(x) = \Gamma(x+1)$$

by Euler's product formula (see Whittaker and Watson [1965, §12.11]) and the well-known functional equation for the gamma function

$$\Gamma(x+1) = x\Gamma(x), \quad \Gamma(1) = 1. \tag{1.10.4}$$

For a rigorous justification of the above steps see Koornwinder [1989a]. From (1.10.1) it is easily seen that, analogous to (1.10.4), $\Gamma_q(x)$ satisfies the functional equation

$$f(x+1) = \frac{1-q^x}{1-q} f(x), \quad f(1) = 1. \tag{1.10.5}$$

Askey [1978] derived analogues of many of the well-known properties of the gamma function, including its log-convexity (see the exercises at the end of this chapter), which show that (1.10.1) is a natural analogue of $\Gamma(x)$.

It is obvious from (1.10.1) that $\Gamma_q(x)$ has poles at $x = 0, -1, -2, \ldots$. The residue at $x = -n$ is

$$\lim_{x\to -n}(x+n)\Gamma_q(x) = \frac{(1-q)^{n+1}}{(1-q^{-n})(1-q^{1-n})\cdots(1-q^{-1})} \lim_{x\to -n} \frac{x+n}{1-q^{x+n}}$$
$$= \frac{(1-q)^{n+1}}{(q^{-n};q)_n \log q^{-1}}. \tag{1.10.6}$$

The q-gamma function has no zeros, so its reciprocal is an entire function with zeros at $x = 0, -1, -2, \ldots$. Since

$$\frac{1}{\Gamma_q(x)} = (1-q)^{x-1}\prod_{n=0}^{\infty}\frac{1-q^{n+x}}{1-q^{n+1}}, \tag{1.10.7}$$

the function $1/\Gamma_q(x)$ has zeros at $x = -n \pm 2\pi ik/\log q$, where k and n are nonnegative integers.

A q-analogue of Legendre's duplication formula

$$\Gamma(2x)\Gamma(\tfrac{1}{2}) = 2^{2x-1}\Gamma(x)\Gamma(x+\tfrac{1}{2}) \tag{1.10.8}$$

can be easily derived by observing that

$$\frac{\Gamma_{q^2}(x)\Gamma_{q^2}(x+\frac{1}{2})}{\Gamma_{q^2}(\frac{1}{2})} = \frac{(q,q^2;q^2)_\infty}{(q^{2x},q^{2x+1};q^2)_\infty}(1-q^2)^{1-2x}$$
$$= \frac{(q;q)_\infty}{(q^{2x};q)_\infty}(1-q^2)^{1-2x} = (1+q)^{1-2x}\Gamma_q(2x)$$

and hence

$$\Gamma_q(2x)\Gamma_{q^2}(\frac{1}{2}) = (1+q)^{2x-1}\Gamma_{q^2}(x)\Gamma_{q^2}(x+\frac{1}{2}). \tag{1.10.9}$$

Similarly, it can be shown that the Gauss multiplication formula

$$\Gamma(nx)(2\pi)^{(n-1)/2} = n^{nx-\frac{1}{2}}\Gamma(x)\Gamma(x+\frac{1}{n})\cdots\Gamma(x+\frac{n-1}{n}) \tag{1.10.10}$$

has a q-analogue of the form

$$\Gamma_q(nx)\Gamma_r(\frac{1}{n})\Gamma_r(\frac{2}{n})\cdots\Gamma_r(\frac{n-1}{n})$$
$$= (1+q+\cdots+q^{n-1})^{nx-1}\Gamma_r(x)\Gamma_r(x+\frac{1}{n})\cdots\Gamma_r(x+\frac{n-1}{n}) \tag{1.10.11}$$

with $r = q^n$; see Jackson [1904e, 1905d]. The q-gamma function for $q > 1$ is considered in Exercise 1.23. For other interesting properties of the q-gamma function see Askey [1978] and Moak [1980a,b] and Ismail, Lorch and Muldoon [1986].

Since the *beta function* is defined by

$$B(x,y) = \frac{\Gamma(x)\Gamma(y)}{\Gamma(x+y)}, \tag{1.10.12}$$

it is natural to define the *q-beta function* by

$$B_q(x,y) = \frac{\Gamma_q(x)\Gamma_q(y)}{\Gamma_q(x+y)}, \tag{1.10.13}$$

which tends to $B(x,y)$ as $q \to 1^-$. By (1.10.1) and (1.3.2),

$$B_q(x,y) = (1-q)\frac{(q,q^{x+y};q)_\infty}{(q^x,q^y;q)_\infty}$$
$$= (1-q)\frac{(q;q)_\infty}{(q^y;q)_\infty}\sum_{n=0}^{\infty}\frac{(q^y;q)_n}{(q;q)_n}q^{nx}$$
$$= (1-q)\sum_{n=0}^{\infty}\frac{(q^{n+1};q)_\infty}{(q^{n+y};q)_\infty}q^{nx}, \quad \text{Re } x,\ \text{Re } y > 0. \tag{1.10.14}$$

This series expansion will be used in the next section to derive a q-integral representation for $B_q(x,y)$.

1.11 The q-integral

Thomae [1869, 1870] and Jackson [1910c] introduced the *q-integral*

$$\int_0^1 f(t)\, d_qt = (1-q)\sum_{n=0}^{\infty} f(q^n)q^n \tag{1.11.1}$$

and Jackson gave the more general definition

$$\int_a^b f(t)\, d_qt = \int_0^b f(t)\, d_qt - \int_0^a f(t)\, d_qt, \tag{1.11.2}$$

where

$$\int_0^a f(t)\, d_qt = a(1-q)\sum_{n=0}^{\infty} f(aq^n)q^n. \tag{1.11.3}$$

Jackson also defined an integral on $(0,\infty)$ by

$$\int_0^\infty f(t)\, d_qt = (1-q)\sum_{n=-\infty}^{\infty} f(q^n)q^n. \tag{1.11.4}$$

The *bilateral q-integral* is defined by

$$\int_{-\infty}^\infty f(t)\, d_qt = (1-q)\sum_{n=-\infty}^{\infty} [f(q^n)+f(-q^n)]\, q^n. \tag{1.11.5}$$

If f is continuous on $[0,a]$, then it is easily seen that

$$\lim_{q\to 1}\int_0^a f(t)\, d_qt = \int_0^a f(t)\, dt \tag{1.11.6}$$

and that a similar limit holds for (1.11.4) and (1.11.5) when f is suitably restricted. By (1.11.1), it follows from (1.10.14) that

$$B_q(x,y) = \int_0^1 t^{x-1}\frac{(tq;q)_\infty}{(tq^y;q)_\infty}\, d_qt, \ \operatorname{Re} x>0, \quad y\neq 0,-1,-2,\ldots, \tag{1.11.7}$$

which clearly approaches the beta function integal

$$B(x,y) = \int_0^1 t^{x-1}(1-t)^{y-1}\, dt, \quad \operatorname{Re} x,\ \operatorname{Re} y>0, \tag{1.11.8}$$

as $q\to 1^-$. Thomae [1869] rewrote Heine's formula (1.4.1) in the q-integral form

$${}_2\phi_1(q^a,q^b;q^c;q,z) = \frac{\Gamma_q(c)}{\Gamma_q(b)\Gamma_q(c-b)}\int_0^1 t^{b-1}\frac{(tzq^a,tq;q)_\infty}{(tz,tq^{c-b};q)_\infty}\, d_qt, \tag{1.11.9}$$

which is a q-analogue of Euler's integral representation

$${}_2F_1(a,b;c;z) = \frac{\Gamma(c)}{\Gamma(b)\Gamma(c-b)}\int_0^1 t^{b-1}(1-t)^{c-b-1}(1-tz)^{-a}\, dt, \tag{1.11.10}$$

where $|\arg(1-z)|<\pi$ and $\operatorname{Re} c >\ \operatorname{Re} b>0$.

The q-integral notation is, as we shall see later, quite useful in simplifying and manipulating various formulas involving sums of series.

Exercises 1

1.1 Verify the identities (1.2.30)–(1.2.40), and show that

(i) $$(aq^{-n};q)_n = (q/a;q)_n \left(-\frac{a}{q}\right)^n q^{-\binom{n}{2}},$$

(ii) $$(aq^{-k-n};q)_n = \frac{(q/a;q)_{n+k}}{(q/a;q)_k} \left(-\frac{a}{q}\right)^n q^{-nk-\binom{n}{2}},$$

(iii) $$\frac{(qa^{\frac{1}{2}},-qa^{\frac{1}{2}};q)_n}{(a^{\frac{1}{2}},-a^{\frac{1}{2}};q)_n} = \frac{1-aq^{2n}}{1-a},$$

(iv) $$(a;q)_\infty = (a^{\frac{1}{2}},-a^{\frac{1}{2}},(aq)^{\frac{1}{2}},-(aq)^{\frac{1}{2}};q)_\infty.$$

1.2 The *q-binomial coefficient* is defined by

$$\begin{bmatrix} n \\ k \end{bmatrix}_q = \frac{(q;q)_n}{(q;q)_k(q;q)_{n-k}}$$

for $k = 0,1,\ldots,n$, and by

$$\begin{bmatrix} \alpha \\ \beta \end{bmatrix}_q = \frac{(q^{\beta+1},q^{\alpha-\beta+1};q)_\infty}{(q,q^{\alpha+1};q)_\infty} = \frac{\Gamma_q(\alpha+1)}{\Gamma_q(\beta+1)\Gamma_q(\alpha-\beta+1)}$$

for complex α and β when $|q| < 1$. Verify that

(i) $$\begin{bmatrix} n \\ k \end{bmatrix}_q = \begin{bmatrix} n \\ n-k \end{bmatrix}_q,$$

(ii) $$\begin{bmatrix} \alpha \\ k \end{bmatrix}_q = \frac{(q^{-\alpha};q)_k}{(q;q)_k}(-q^\alpha)^k q^{-\binom{k}{2}},$$

(iii) $$\begin{bmatrix} k+\alpha \\ k \end{bmatrix}_q = \frac{(q^{\alpha+1};q)_k}{(q;q)_k},$$

(iv) $$\begin{bmatrix} -\alpha \\ k \end{bmatrix}_q = \begin{bmatrix} \alpha+k-1 \\ k \end{bmatrix}_q (-q^{-\alpha})^k q^{-\binom{k}{2}},$$

(v) $$\begin{bmatrix} \alpha+1 \\ k \end{bmatrix}_q = \begin{bmatrix} \alpha \\ k \end{bmatrix}_q q^k + \begin{bmatrix} \alpha \\ k-1 \end{bmatrix}_q = \begin{bmatrix} \alpha \\ k \end{bmatrix}_q + \begin{bmatrix} \alpha \\ k-1 \end{bmatrix}_q q^{\alpha+1-k},$$

(vi) $$(z;q)_n = \sum_{k=0}^{n} \begin{bmatrix} n \\ k \end{bmatrix}_q (-z)^k q^{\binom{k}{2}},$$

when k and n are nonnegative integers.

1.3 (i) Show that the *binomial theorem*

$$(a+b)^n = \sum_{k=0}^{n} \binom{n}{k} a^k b^{n-k}$$

where $n = 0,1,\ldots,$ has a q-analogue of the form

$$(ab;q)_n = \sum_{k=0}^{n} \begin{bmatrix} n \\ k \end{bmatrix}_q b^k (a;q)_k (b;q)_{n-k}.$$

(ii) Extend the above formula to the *q-multinomial theorem*

$$(a_1 a_2 \cdots a_{m+1}; q)_n$$
$$= \sum_{\substack{0 \le k_1, \ldots, 0 \le k_m \\ k_1 + \cdots + k_m \le n}} \begin{bmatrix} n \\ k_1, \ldots, k_m \end{bmatrix}_q a_2^{k_1} a_3^{k_1+k_2} \cdots a_{m+1}^{k_1+k_2+\cdots+k_m}$$
$$\cdot (a_1; q)_{k_1} (a_2; q)_{k_2} \cdots (a_m; q)_{k_m} (a_{m+1}; q)_{n-(k_1+\cdots+k_m)},$$

where $m = 1, 2, \ldots,\ n = 0, 1, \ldots,$ and

$$\begin{bmatrix} n \\ k_1, \ldots, k_m \end{bmatrix}_q = \frac{(q;q)_n}{(q;q)_{k_1} \cdots (q;q)_{k_m} (q;q)_{n-(k_1+\cdots+k_m)}}$$

is the *q-multinomial coefficient.*

1.4 (i) Prove the inversion formula

$${}_r\phi_s \left[\begin{matrix} a_1, \ldots, a_r \\ b_1, \ldots, b_s \end{matrix} ; q, z \right]$$
$$= \sum_{n=0}^{\infty} \frac{(a_1^{-1}, \ldots, a_r^{-1}; q^{-1})_n}{(q^{-1}, b_1^{-1}, \ldots, b_s^{-1}; q^{-1})_n} \left(\frac{a_1 \cdots a_r z}{b_1 \cdots b_s q} \right)^n.$$

(ii) By reversing the order of summation, show that

$${}_{r+1}\phi_s \left[\begin{matrix} a_1, \ldots, a_r, q^{-n} \\ b_1, \ldots, b_s \end{matrix} ; q, z \right]$$
$$= \frac{(a_1, \ldots, a_r; q)_n}{(b_1 \ldots, b_s; q)_n} \left(\frac{z}{q} \right)^n \left((-1)^n q^{\binom{n}{2}} \right)^{s-r-1}$$
$$\cdot \sum_{k=0}^{n} \frac{(q^{1-n}/b_1, \ldots, q^{1-n}/b_s, q^{-n}; q)_k}{(q, q^{1-n}/a_1, \ldots, q^{1-n}/a_r; q)_k} \left(\frac{b_1 \cdots b_s}{a_1 \cdots a_r} \frac{q^{n+1}}{z} \right)^k$$

when $n = 0, 1, \ldots$.

1.5 Show that

$$\frac{(c, bq^n; q)_m}{(b;q)_m} = \frac{(b/c;q)_n}{(b;q)_n} \sum_{k=0}^{n} \frac{(q^{-n}, c; q)_k q^k}{(q, cq^{1-n}/b; q)_k} (cq^k; q)_m.$$

1.6 Prove the summation formulas

(i) ${}_2\phi_1(q^{-n}, q^{1-n}; qb^2; q^2, q^2) = \dfrac{(b^2;q^2)_n}{(b^2;q)_n} q^{-\binom{n}{2}},$

(ii) ${}_1\phi_1(a; c; q, c/a) = \dfrac{(c/a;q)_\infty}{(c;q)_\infty}.$

1.7 Show that, for $|z| < 1$,

$${}_2\phi_1(a^2, aq; a; q, z) = (1 + az) \frac{(a^2 qz; q)_\infty}{(z;q)_\infty}.$$

1.8 Show that, when $|a| < 1$ and $|bq/a^2| < 1$,

$$
\begin{aligned}
&{}_2\phi_1(a^2, a^2/b; b; q^2, bq/a^2) \\
&= \frac{(a^2, q; q^2)_\infty}{2(b, bq/a^2; q^2)_\infty}\left[\frac{(b/a;q)_\infty}{(a;q)_\infty} + \frac{(-b/a;q)_\infty}{(-a;q)_\infty}\right].
\end{aligned}
$$

(Andrews and Askey [1977])

1.9 Let $\phi(a, b, c)$ denote the series ${}_2\phi_1(a, b; c; q, z)$. Verify Heine's [1847] *q-contiguous relations*:

(i) $\phi(a, b, cq^{-1}) - \phi(a, b, c) = cz\dfrac{(1-a)(1-b)}{(q-c)(1-c)}\phi(aq, bq, cq)$,

(ii) $\phi(aq, b, c) - \phi(a, b, c) = az\dfrac{1-b}{1-c}\phi(aq, bq, cq)$,

(iii) $\phi(aq, b, cq) - \phi(a, b, c) = az\dfrac{(1-b)(1-c/a)}{(1-c)(1-cq)}\phi(aq, bq, cq^2)$,

(iv) $\phi(aq, bq^{-1}, c) - \phi(a, b, c) = az\dfrac{(1-b/aq)}{1-c}\phi(aq, b, cq)$.

1.10 Denoting ${}_2\phi_1(a, b; c; q, z)$, ${}_2\phi_1(aq^{\pm 1}, b; c, q, z)$, ${}_2\phi_1(a, bq^{\pm 1}; c; q, z)$ and ${}_2\phi_1(a, b; cq^{\pm 1}; q, z)$ by $\phi, \phi(aq^{\pm 1}), \phi(bq^{\pm 1})$ and $\phi(cq^{\pm 1})$, respectively, show that

(i) $b(1-a)\phi(aq) - a(1-b)\phi(bq) = (b-a)\phi$,

(ii) $a(1-b/c)\phi(bq^{-1}) - b(1-a/c)\phi(aq^{-1}) = (a-b)(1-abz/cq)\phi$,

(iii) $q(1-a/c)\phi(aq^{-1}) + (1-a)(1-abz/c)\phi(aq)$
$= [1+q-a-aq/c+a^2z(1-b/a)/c]\phi$,

(iv) $(1-c)(q-c)(abz-c)\phi(cq^{-1}) + (c-a)(c-b)z\phi(cq)$
$= (c-1)[c(q-c)+(ca+cb-ab-abq)z]\phi$.

(Heine [1847])

1.11 Let $g(\theta; \lambda, \mu, \nu) = (\lambda e^{i\theta}, \mu\nu; q)_\infty \, {}_2\phi_1(\mu e^{-i\theta}, \nu e^{-i\theta}; \mu\nu; q, \lambda e^{i\theta})$. Prove that $g(\theta; \lambda, \mu, \nu)$ is symmetric in λ, μ, ν and is even in θ.

1.12 Let $\mathcal{D}_q$ be the *q-derivative operator* defined for fixed q by

$$
\mathcal{D}_q f(z) = \frac{f(z) - f(qz)}{(1-q)z},
$$

and let $\mathcal{D}_q^n u = \mathcal{D}_q(D_q^{n-1}u)$ for $n = 1, 2, \ldots$. Show that

(i) $\lim_{q\to 1} \mathcal{D}_q f(z) = \dfrac{d}{dz}f(z)$ if f is differentiable at z,

(ii) $\mathcal{D}_q^n \, {}_2\phi_1(a, b; c; q, dz) = \dfrac{(a, b; q)_n d^n}{(c; q)_n(1-q)^n} \, {}_2\phi_1(aq^n, bq^n; cq^n; q, dz)$,

(iii) $\mathcal{D}_q^n \left\{\dfrac{(z;q)_\infty}{(abz/c;q)_\infty} \, {}_2\phi_1(a, b; c; q, z)\right\}$
$= \dfrac{(c/a, c/b; q)_n}{(c;q)_n(1-q)^n}\left(\dfrac{ab}{c}\right)^n \dfrac{(zq^n;q)_\infty}{(abz/c;q)_\infty} \, {}_2\phi_1(a, b; cq^n; q, zq^n)$.

1.13 Show that $u(z) = {}_2\phi_1(a, b; c; q, z)$ satisfies (for $|z| < 1$ and in the formal power series sense) the second order q-differential equation

$$z(c - abqz)\mathcal{D}_q^2 u + \left[\frac{1-c}{1-q} + \frac{(1-a)(1-b) - (1-abq)}{1-q} z\right] \mathcal{D}_q u$$
$$- \frac{(1-a)(1-b)}{(1-q)^2} u = 0,$$

where $\mathcal{D}_q$ is defined as in Ex. 1.12. By replacing a, b, c, respectively, by q^a, q^b, q^c and then letting $q \to 1^-$ show that the above equation tends to the second order differential equation

$$z(1-z)v'' + [c - (a+b+1)z]v' - abv = 0$$

for the hypergeometric function $v(z) = {}_2F_1(a, b; c; z)$, where $|z| < 1$. (Heine [1847])

1.14 Let $|x| < 1$ and let $e_q(x)$ and $E_q(x)$ be as defined in §1.3. Define

$$\sin_q(x) = \frac{e_q(ix) - e_q(-ix)}{2i} = \sum_{n=0}^{\infty} \frac{(-1)^n x^{2n+1}}{(q;q)_{2n+1}},$$

$$\cos_q(x) = \frac{e_q(ix) + e_q(-ix)}{2} = \sum_{n=0}^{\infty} \frac{(-1)^n x^{2n}}{(q;q)_{2n}}.$$

Also define

$$\mathrm{Sin}_q(x) = \frac{E_q(ix) - E_q(-ix)}{2i}, \quad \mathrm{Cos}_q(x) = \frac{E_q(ix) + E_q(-ix)}{2}.$$

Show that

(*i*) $$e_q(ix) = \cos_q(x) + i\sin_q(x),$$
(*ii*) $$E_q(ix) = \mathrm{Cos}_q(x) + i\mathrm{Sin}_q(x),$$
(*iii*) $$\sin_q(x)\mathrm{Sin}_q(x) + \cos_q(x)\mathrm{Cos}_q(x) = 1,$$
(*iv*) $$\sin_q(x)\mathrm{Cos}_q(x) - \mathrm{Sin}_q(x)\cos_q(x) = 0.$$

For these identities and other identities involving q-analogues of $\sin x$ and $\cos x$, see Jackson [1904a] and Hahn [1949c].

1.15 Prove the transformation formulas

(i) $${}_2\phi_1\left[\begin{matrix} q^{-n}, b \\ c \end{matrix}; q, z\right] = \frac{(bzq^{-n}/c;q)_\infty}{(bz/c;q)_\infty} {}_3\phi_2\left[\begin{matrix} q^{-n}, c/b, 0 \\ c, cq/bz \end{matrix}; q, q\right],$$

(ii) $${}_2\phi_1\left[\begin{matrix} q^{-n}, b \\ c \end{matrix}; q, z\right] = \frac{(c/b;q)_n}{(c;q)_n} b^n {}_3\phi_1\left[\begin{matrix} q^{-n}, b, q/z \\ bq^{1-n}/c \end{matrix}; q, z/c\right],$$

(iii) $${}_2\phi_1\left[\begin{matrix} q^{-n}, b \\ c \end{matrix}; q, z\right] = \frac{(c/b;q)_n}{(c;q)_n} {}_3\phi_2\left[\begin{matrix} q^{-n}, b, bzq^{-n}/c \\ bq^{1-n}/c, 0 \end{matrix}; q, q\right].$$

(Jackson [1905a, 1927])

1.16 Show that

$$\sum_{n=0}^{\infty} \frac{(a;q)_n}{(q;q)_n} q^{n(n+1)/2} = (-q;q)_\infty (aq;q^2)_\infty.$$

1.17 Show that

$$\sum_{k=0}^{n} \frac{(a,b;q)_k}{(q;q)_k} (-ab)^{n-k} q^{(n-k)(n+k-1)/2}$$

$$= (a;q)_{n+1} \sum_{k=0}^{n} \frac{(-b)^k q^{\binom{k}{2}}}{(q;q)_k (q;q)_{n-k} (1-aq^{n-k})}. \quad \text{(Carlitz [1974])}$$

1.18 Show that

$$\sum_{n=0}^{\infty} \frac{(a;q)_n}{(q,a^2;q)_n} q^{\binom{n}{2}} (at/z)^n \ {}_2\phi_1(q^{-n}, a; q^{1-n}/a; q, qz^2/a)$$

$$= (-zt;q)_\infty \ {}_2\phi_1(a, a/z^2; a^2; q, -zt), \quad |zt| < 1.$$

1.19 Using (1.5.4) show that

(i)
$$ {}_2\phi_2 \left[\begin{matrix} a, q/a \\ -q, b \end{matrix} ; q, -b \right] = \frac{(ab, bq/a; q^2)_\infty}{(b;q)_\infty},$$

(ii)
$$ {}_2\phi_2 \left[\begin{matrix} a^2, b^2 \\ abq^{\frac{1}{2}}, -abq^{\frac{1}{2}} \end{matrix} ; q, -q \right] = \frac{(a^2 q, b^2 q; q^2)}{(q, a^2 b^2 q; q^2)_\infty}.$$

(Andrews [1973])

1.20 Prove that if Re $x > 0$ and $0 < q < 1$, then

(i)
$$\Gamma_q(x) = (q;q)_\infty (1-q)^{1-x} \sum_{n=0}^{\infty} \frac{q^{nx}}{(q;q)_n},$$

(ii)
$$\frac{1}{\Gamma_q(x)} = \frac{(1-q)^{x-1}}{(q;q)_\infty} \sum_{n=0}^{\infty} \frac{(-1)^n q^{nx}}{(q;q)_n} q^{\binom{n}{2}}.$$

1.21 For $0 < q < 1$ and $x > 0$, show that

$$\frac{d^2}{dx^2} \log \Gamma_q(x) = (\log q)^2 \sum_{n=0}^{\infty} \frac{q^{n+x}}{1-q^{n+x}},$$

which proves that $\log \Gamma_q(x)$ is convex for $x > 0$ when $0 < q < 1$.

1.22 Conversely, prove that if $f(x)$ is a function which satisfies

$$f(x+1) = \frac{1-q^x}{1-q} f(x) \text{ for some } q, 0 < q < 1,$$

$$f(1) = 1,$$

and $\log f(x)$ is convex for $x > 0$, then $f(x) = \Gamma_q(x)$. This is Askey's [1978] q-analogue of the Bohr-Mollerup [1922] theorem for $\Gamma(x)$. For two extensions to the $q > 1$ case (with $\Gamma_q(x)$ defined as in the next exercise), see Moak [1980b].

1.23 For $q > 1$ the *q-gamma function* is defined by

$$\Gamma_q(x) = \frac{(q^{-1}; q^{-1})_\infty}{(q^{-x}; q^{-1})_\infty}(q-1)^{1-x} q^{x(x-1)/2}.$$

Show that this function also satisfies the functional equation (1.10.5) and that $\Gamma_q(x) \to \Gamma(x)$ as $q \to 1^+$. Show that for $q > 1$ the residue of $\Gamma_q(x)$ at $x = -n$ is

$$\frac{(q-1)^{n+1} q^{\binom{n+1}{2}}}{(q;q)_n \log q}.$$

1.24 Jackson [1905a,b,e] gave the following q-analogues of Bessel functions:

$$J_\nu^{(1)}(x;q) = \frac{(q^{\nu+1};q)_\infty}{(q;q)_\infty}(x/2)^\nu \; {}_2\phi_1(0,0;q^{\nu+1};q,-x^2/4),$$

$$J_\nu^{(2)}(x;q) = \frac{(q^{\nu+1};q)_\infty}{(q;q)_\infty}(x/2)^\nu \; {}_0\phi_1\left(—;q^{\nu+1};q,-\frac{x^2 q^{\nu+1}}{4}\right),$$

where $0 < q < 1$. The above notations for the *q-Bessel functions* are due to Ismail [1981, 1982].
Show that

$$J_\nu^{(2)}(x;q) = \left(-x^2/4;q\right)_\infty J_\nu^{(1)}(x;q)_\infty, \quad |x| < 2, \quad \text{(Hahn [1949c])}$$

and

$$\lim_{q\to 1} J_\nu^{(k)}(x(1-q);q) = J_\nu(x), \qquad k = 1, 2.$$

1.25 For the q-Bessel functions defined as in Exercise 1.24 prove that

(i) $$q^\nu J_{\nu+1}^{(k)}(x;q) = \frac{2(1-q^\nu)}{x} J_\nu^{(k)}(x;q) - J_{\nu-1}^{(k)}(x;q), \; k = 1, 2;$$

(ii) $$J_\nu^{(1)}(xq^{\frac{1}{2}};q) = q^{\nu/2}\left(J_\nu^{(1)}(x;q) + \frac{x}{2} J_{\nu+1}^{(1)}(x;q)\right);$$

(iii) $$J_\nu^{(1)}(xq^{\frac{1}{2}};q) = q^{-\nu/2}\left(J_\nu^{(1)}(x;q) - \frac{x}{2} J_{\nu-1}^{(1)}(x;q)\right).$$

1.26 Following Ismail [1982], let

$$f_\nu(x) = J_\nu^{(1)}(x;q) J_{-\nu}^{(1)}(xq^{\frac{1}{2}};q) - J_{-\nu}^{(1)}(x;q) J_\nu^{(1)}(xq^{\frac{1}{2}};q).$$

Show that

$$f_\nu(xq^{\frac{1}{2}}) = \left(1 + \frac{x^2}{4}\right) f_\nu(x).$$

and deduce that, for non-integral ν,

$$f_\nu(x) = q^{-\nu/2}(q^\nu, q^{1-\nu};q)_\infty / (q, q, -x^2/4; q)_\infty.$$

1.27 Show that

$$\sum_{n=-\infty}^{\infty} t^n J_n^{(2)}(x;q) = (-x^2/4;q)_\infty \, e_q(xt/2)e_q(x/2t).$$

This is a q-analogue of the generating function

$$\sum_{n=-\infty}^{\infty} t^n J_n(x) = e^{x(t-t^{-1})/2}.$$

1.28 The *continuous q-Hermite polynomials* are defined in Askey and Ismail [1983] by

$$H_n(x|q) = \sum_{k=0}^{n} \frac{(q;q)_n}{(q;q)_k(q;q)_{n-k}} e^{i(n-2k)\theta},$$

where $x = \cos\theta$; see Szegő [1926], Carlitz [1955, 1957a, 1958, 1960] and Rogers [1894, 1917]. Derive the generating function

$$\sum_{n=0}^{\infty} \frac{H_n(x|q)}{(q;q)_n} t^n = \frac{1}{(te^{i\theta}, te^{-i\theta};q)_\infty}, \quad |t| < 1. \quad \text{(Rogers [1894])}$$

1.29 The *continuous q-ultraspherical polynomials* are defined in Askey and Ismail [1983] by

$$C_n(x;\beta|q) = \sum_{k=0}^{n} \frac{(\beta;q)_k(\beta;q)_{n-k}}{(q;q)_k(q;q)_{n-k}} e^{i(n-2k)\theta},$$

where $x = \cos\theta$. Show that

$$\begin{aligned} C_n(x;\beta|q) &= \frac{(\beta;q)_n}{(q;q)_n} e^{in\theta} \, {}_2\phi_1\left[\begin{matrix} q^{-n}, \beta \\ \beta^{-1}q^{1-n} \end{matrix}; q, q\beta^{-1}e^{-2i\theta}\right] \\ &= \frac{(\beta^2;q)_n}{(q;q)_n} e^{-in\theta} \, {}_3\phi_2\left[\begin{matrix} q^{-n}, \beta, \beta e^{2i\theta} \\ \beta^2, 0 \end{matrix}; q, q\right], \end{aligned}$$

$$\lim_{q\to 1} C_n(x;q^\lambda|q) = C_n^\lambda(x),$$

and

$$\sum_{n=0}^{\infty} C_n(x;\beta|q)t^n = \frac{(\beta te^{i\theta}, \beta te^{-i\theta};q)_\infty}{(te^{i\theta}, te^{-i\theta};q)_\infty}, \quad |t| < 1. \quad \text{(Rogers [1895])}$$

1.30 Show that if $m_1, \ldots, m_r$ are nonnegative integers, then

(i)
$$\begin{aligned} &{}_{r+1}\phi_{r+1}\left[\begin{matrix} b, \; b_1q^{m_1}, \ldots, b_rq^{m_r} \\ bq, b_1, \ldots, b_r \end{matrix}; q, q^{1-(m_1+\cdots+m_r)}\right] \\ &= \frac{(q;q)_\infty (b_1/b;q)_{m_1} \cdots (b_r/b;q)_{m_r}}{(bq;q)_\infty (b_1;q)_{m_1} \cdots (b_r;q)_{m_r}} b^{m_1+\cdots+m_r}, \end{aligned}$$

(ii)
$${}_r\phi_r\left[\begin{matrix} b_1q^{m_1}, \ldots, b_rq^{m_r} \\ b_1, \ldots, b_r \end{matrix}; q, q^{m_1+\cdots+m_r}\right] = 0,$$

(iii)
$$\begin{aligned} &{}_r\phi_r\left[\begin{matrix} b_1q^{m_1}, \ldots, b_rq^{m_r} \\ b_1, \ldots, b_r \end{matrix}; q, q^{1-(m_1+\cdots+m_r)}\right] \\ &= \frac{(-1)^{m_1+\cdots+m_r}(q;q)_\infty b_1^{m_1} \cdots b_r^{m_r}}{(b_1;q)_{m_1} \cdots (b_r;q)_{m_r}} q^{\binom{m_1}{2}+\cdots+\binom{m_r}{2}}. \end{aligned}$$

(Gasper [1981a])

1.31 Let Δ_b denote the *q-difference operator* defined for a fixed q by

$$\Delta_b f(z) = bf(qz) - f(z)$$

and let $\Delta = \Delta_1$. Show that if

$$v_n(z) = \frac{(a_1, \ldots, a_r; q)_n}{(q, b_1, \ldots, b_s; q)_n} (-1)^{(1+s-r)n} q^{(1+s-r)n(n-1)/2} z^n,$$

then

$$\begin{aligned}&(\Delta\Delta_{b_1/q}\Delta_{b_2/q}\cdots\Delta_{b_s/q})v_n(z)\\&= z(\Delta_{a_1}\Delta_{a_2}\cdots\Delta_{a_r})v_{n-1}(zq^{1+s-r}), \quad n = 1, 2, \ldots .\end{aligned}$$

Use this to show that the basic hypergeometric series

$$v(z) = {}_r\phi_s(a_1, \ldots, a_r; b_1, \ldots, b_s; q, z)$$

satisfies (in the sense of formal power series) the q-difference equation

$$(\Delta\Delta_{b_1/q}\Delta_{b_2/q}\cdots\Delta_{b_s/q})v(z) = z(\Delta_{a_1}\cdots\Delta_{a_r})v(zq^{1+s-r}).$$

This is a q-analogue of the formal differential equation for generalized hypergeometric series given, e.g. in Henrici [1974, Theorem (1.5)] and Slater [1966, (2.1.2.1)].

1.32 The *little q-Jacobi polynomials* are defined by

$$p_n(x; a, b; q) = {}_2\phi_1(q^{-n}, abq^{n+1}; aq; q, qx)$$

Show that these polynomials satisfy the orthogonality relation

$$\begin{aligned}&\sum_{j=0}^{\infty} \frac{(bq; q)_j}{(q; q)_j} (aq)^j p_n(q^j; a, b; q) p_m(q^j; a, b; q)\\&= \begin{cases} 0, & \text{if } m \neq n,\\ \dfrac{(q, bq; q)_n (1 - abq)(aq)^n}{(aq, abq; q)_n (1 - abq^{2n+1})} \dfrac{(abq^2; q)_\infty}{(aq; q)_\infty}, & \text{if } m = n.\end{cases}\end{aligned}$$

(Andrews and Askey [1977])

1.33 Show for the above little q-Jacobi polynomials that the formula

$$p_n(x; c, d; q) = \sum_{k=0}^{n} a_{k,n} p_k(x; a, b; q)$$

holds with

$$a_{k,n} = \frac{(q^{-n}, aq, cdq^{n+1}; q)_k}{(q, cq, abq^{k+1}; q)_k} {}_3\phi_2 \left[\begin{matrix} q^{k-n}, cdq^{n+k+1}, aq^{k+1} \\ cq^{k+1}, abq^{2k+2} \end{matrix}; q, q \right].$$

(Andrews and Askey [1977])

1.34 (i) If $m, m_1, m_2, \ldots, m_r$ are arbitrary nonnegative integers and

$|a^{-1}q^{m+1-(m_1+\cdots+m_r)}| < 1$, show that

$$
{}_{r+2}\phi_{r+1}\left[\begin{matrix} a, b, b_1q^{m_1}, \ldots, b_rq^{m_r} \\ bq^{1+m}, b_1, \ldots, b_r \end{matrix}; q, a^{-1}q^{m+1-(m_1+\cdots+m_r)}\right]
$$

$$
= \frac{(q, bq/a; q)_\infty (bq;q)_m (b_1/b;q)_{m_1} \cdots (b_r/b;q)_{m_r}}{(bq, q/a; q)_\infty (q;q)_m (b_1;q)_{m_1} \cdots (b_r;q)_{m_r}} b^{m_1+\cdots+m_r-m}
$$

$$
\cdot {}_{r+2}\phi_{r+1}\left[\begin{matrix} q^{-m}, b, bq/b_1, \ldots, bq/b_r \\ bq/a, bq^{1-m_1}/b_1, \ldots, bq^{1-m_r}/b_r \end{matrix}; q, q\right];
$$

(ii) if $m_1, m_2, \ldots, m_r$ are nonnegative integers and $|a^{-1}q^{1-(m_1+\cdots+m_r)}| < 1$, $|cq| < 1$, show that

$$
{}_{r+2}\phi_{r+1}\left[\begin{matrix} a, b, b_1q^{m_1}, \ldots, b_rq^{m_r} \\ bcq, b_1, \ldots, b_r \end{matrix}; q, a^{-1}q^{1-(m_1+\cdots+m_r)}\right]
$$

$$
= \frac{(bq/a, cq; q)_\infty}{(bcq, q/a; q)_\infty} \frac{(b_1/b;q)_{m_1} \cdots (b_r/b;q)_{m_r}}{(b_1;q)_{m_1} \cdots (b_r;q)_{m_r}} b^{m_1+\cdots+m_r}
$$

$$
\cdot {}_{r+2}\phi_{r+1}\left[\begin{matrix} c^{-1}, b, bq/b_1, \ldots, bq/b_r \\ bq/a, bq^{1-m_1}/b_1, \ldots, bq^{1-m_r}/b_r \end{matrix}; q, cq\right].
$$

(Gasper [1981a])

1.35 Use Ex. 1.2 (v) to prove that if x and y are indeterminates such that $xy = qyx$, q commutes with x and y, and the associate law holds, then

$$
(x+y)^n = \sum_{k=0}^{n} \begin{bmatrix} n \\ k \end{bmatrix}_q y^k x^{n-k} = \sum_{k=0}^{n} \begin{bmatrix} n \\ k \end{bmatrix}_{q^{-1}} x^k y^{n-k}.
$$

(See Cigler [1979], Feinsilver [1982], Koornwinder [1989c], Schützenberger [1953], and Yang [1989])

Notes 1

§§1.1 and 1.2 For additional material on hypergeometric series and orthogonal polynomials see, e.g., the books Erdélyi [1953], Rainville [1960], Szegő [1975], Whittaker and Watson [1965], Agarwal [1963], Carlson [1977], T.S. Chihara [1978], Henrici [1974], Luke [1969], Miller [1968], Nikiforov and Uvarov [1988], Vilenkin [1968], and Watson [1952]. Some techniques for using symbolic computer algebraic systems such as Mathematica and Macsyma to derive formulas containing hypergeometric and basic hypergeometric series are discussed in Gasper [1989e].

§§1.3–1.5 The q-binomial theorem was also derived in Jacobi [1846], along with the q-Vandermonde formula. Bijective proofs of the q-binomial theorem, Heine's ${}_2\phi_1$ transformation and q-analogue of Gauss' summation formula, the q-Saalschütz formula, and of other formulas are presented in Joichi and Stanton [1987]. Bender [1971] used partitions to derived an extension of the q-Vandermonde sum in the form of a generalized q-binomial Vandermonde convolution.

§1.6 Other proofs of Jacobi's triple product identity and/or applications of it are presented in Adiga *et al.* [1985], Andrews [1965], Cheema [1964], Ewell [1981], Gustafson [1989b], Joichi and Stanton [1989], Kac [1978, 1985], Lepowsky and Milne [1978], Lewis [1984], Macdonald [1972], Menon [1965], Milne [1985a], Sudler [1966], Sylvester [1882], and Wright [1965]. Concerning theta functions, see Adiga *et al.* [1985], Askey [1989c], Bellman [1961], and Jensen's use of theta functions in Pólya [1927] to derive necessary and sufficient conditions for the Riemann hypothesis to hold.

§1.7 Some applications of the q-Saalschütz formula are contained in Carlitz [1969b] and Wright [1968].

§1.9 Formulas (1.9.3) and (1.9.8) were rediscovered by Gustafson [1987a, Theorems 3.15 and 3.18] while working on multivariable orthogonal polynomials.

§1.11 Also see Jackson [1917, 1951] and, for fractional q-integrals and q-derivatives, Al-Salam [1966] and Agarwal [1969b]. Toeplitz [1963, pp.53-55] pointed out that around 1650 Fermat used a q-integral type Riemann sum to evaluate the integral of x^k on the interval $[0, b]$.

Ex. 1.2 The q-binomial coefficient $\begin{bmatrix} n \\ k \end{bmatrix}_q$, which is also called the Gaussian binomial coefficient, counts the number of k dimensional subspaces of an n dimensional vector space over a field with q elements (Goldman and Rota [1970]), and it is the generating function, in powers of q, for partitions into at most k parts not exceeding $n - k$ (Sylvester [1882]). It arises in such diverse fields as analysis, computer programming, geometry, number theory, physics, and statistics. See, e.g., Aigner [1979], Andrews [1971a, 1976], Baker and Coon [1970], Baxter and Pearce [1983], Berman and Fryer [1972], Dowling [1973], Dunkl [1981], Garvan and Stanton [1989], Handa and Mohanty [1980], Ihrig and Ismail [1981], Jimbo [1985, 1986], van Kampen [1961], Kendall and Stuart [1979, §31.25], Knuth [1971, 1973], Pólya [1970], Pólya and Alexanderson [1970], Szegő [1975, §2.7], and Zaslavsky [1987]. Sylvester [1878] used the invariant theory that he and Cayley developed to prove that the coefficients of the Gaussian polynomial $\begin{bmatrix} n \\ k \end{bmatrix}_q = \Sigma a_j q^j$ are unimodal. A constructive proof was recently given by O'Hara [1989]. Also see Bressoud [1989], Pólya [1970], and Zeilberger [1989b,c,d]. The unimodality of the sequence $\left(\begin{bmatrix} n \\ k \end{bmatrix}_q : k = 0, 1, \ldots, n\right)$ is explicitly displayed in Aigner [1979, Proposition 3.13], and Macdonald [1979, p. 67].

Ex. 1.3 Cigler [1979] derived an operator form of the q-binomial theorem. MacMahon [1916, Arts. 105–107] showed that if a multiset is permuted, then the generating function for inversions is the q-multinomial coefficient. Also see Carlitz [1963a], Kadell [1985a], and Knuth [1973, p. 33, Ex. 16]. Gasper derived the q-multinomial theorem in part (ii) several years ago by using the q-binomial theorem and mathematical induction. G.E. Andrews observed in a 1988 letter that it can also be derived by using the expansion formula for the q-Lauricella function Φ_D stated in Andrews [1972, (4.1)] and the q-Vandermonde sum. Some sums of q-multinomial coefficients are considered in Bressoud [1978,

1981c]. See also Agarwal [1953a].

Ex. 1.8 Jain [1980c] showed that the sum in this exercise is equivalent to the sum of a certain ${}_2\psi_2$ series, and summed some other ${}_2\psi_2$ series.

Ex. 1.10 Analogous recurrence relations for ${}_1\phi_1$ series are given in Slater [1954c].

Exercises 1.12 and 1.13 The notations Δ_q, ϑ_q, and D_q are also employed in the literature for this q-derivative operator. We employed the script $\mathcal{D}_q$ operator notation to distinguish this q-derivative operator from the q-derivative operator defined in (7.7.3) and the q-difference operator defined in Ex. 1.31. Additional results involving q-derivatives are contained in Adams [1931], Agarwal [1953d], Andrews [1968, 1971a], Carmichael [1912], Hahn [1949a,c, 1950, 1952, 1953], Ismail, Merkes and Styer [1989], Jackson [1905c, 1909a, 1910b,d,e], Miller [1970], Mimachi [1989], Starcher [1931], and Trjitzinsky [1933]. For fractional q-derivatives and q-integrals see Agarwal [1969b] and Al-Salam and Verma [1975a,b]. Some "q-Taylor series" are considered in Jackson [1909b,c] and Wallisser [1985].

Ex. 1.14 For q-tangent and q-secant numbers and some of their properties, see Andrews and Foata [1980] and Foata [1981].

Ex. 1.22 A different characterization of Γ_q is presented in Kairies and Muldoon [1982].

Exercises 1.24–1.27 Other formulas involving q-Bessel functions are contained in Jackson [1904a–d, 1908], Ismail and Muldoon [1988], and Rahman [1987, 1988c, 1989b,c].

Ex. 1.28 Also see the generating functions for the continuous q-Hermite polynomials derived in Carlitz [1963b, 1972] and Bressoud [1980b], and the applications to modular forms in Bressoud [1986].

Ex. 1.32 Masuda *et al.* [1989] showed that the matrix elements that arise in the representations of certain quantum groups are expressible in terms of little q-Jacobi polynomials, and that this and a form of the Peter-Weyl theorem imply the orthogonality relation for these polynomials. Padé approximants for the moment generating function for the little q-Jacobi polynomials are employed in Andrews, Goulden and D.M. Jackson [1986] to explain and extend Shank's method for accelerating the convergence of sequences.

2

SUMMATION, TRANSFORMATION, AND EXPANSION FORMULAS

2.1 Well-poised, nearly-poised, and very-well-poised hypergeometric and basic hypergeometric series

The hypergeometric series

$$ {}_{r+1}F_r\left[\begin{matrix} a_1, a_2, ..., a_{r+1} \\ b_1, ..., b_r \end{matrix}; z\right] \tag{2.1.1}$$

is called *well-poised* if its parameters satisfy the relations

$$1 + a_1 = a_2 + b_1 = a_3 + b_2 = ... = a_{r+1} + b_r, \tag{2.1.2}$$

and it is called *nearly-poised* if all but one of the above pairs of parameters (regarding 1 as the first denominator parameter) have the same sum. The series (2.1.1) is called a *nearly-poised series of the first kind* if

$$1 + a_1 \neq a_2 + b_1 = a_3 + b_2 = \cdots = a_{r+1} + b_r, \tag{2.1.3}$$

and it is called a *nearly-poised series of the second kind* if

$$1 + a_1 = a_2 + b_1 = a_3 + b_2 = \cdots = a_r + b_{r-1} \neq a_{r+1} + b_r. \tag{2.1.4}$$

The order of summation of a terminating nearly-poised series can be changed, without altering the series, so that the resulting series is either of the first kind or of the second kind.

Kummer's summation formula (1.8.2) gives the sum of a well-poised ${}_2F_1$ series with argument -1. Another example of a summable well-poised series is provided by Dixon's [1903] formula

$$\begin{aligned} &{}_3F_2\left[a, b, c; 1 + a - b, 1 + a - c; 1\right] \\ &\quad = \frac{\Gamma(1 + \frac{1}{2}a)\Gamma(1 + a - b)\Gamma(1 + a - c)\Gamma(1 + \frac{1}{2}a - b - c)}{\Gamma(1 + a)\Gamma(1 + \frac{1}{2}a - b)\Gamma(1 + \frac{1}{2}a - c)\Gamma(1 + a - b - c)}, \end{aligned} \tag{2.1.5}$$

Re $(1 + \frac{1}{2}a - b - c) > 0$, which reduces to Kummer's formula (1.8.2) by letting $c \to -\infty$.

If the series (2.1.1) is well-poised and $a_2 = 1 + \frac{1}{2}a_1$, then it is called a *very-well-poised* series. Dougall's [1907] summation formulas

$$\begin{aligned} &{}_7F_6\left[\begin{matrix} a,\ 1 + \frac{1}{2}a,\ b,\ c,\ d,\ e,\ -n \\ \frac{1}{2}a, 1 + a - b, 1 + a - c, 1 + a - d, 1 + a - e, 1 + a + n \end{matrix}; 1\right] \\ &\quad = \frac{(1 + a)_n(1 + a - b - c)_n(1 + a - b - d)_n(1 + a - c - d)_n}{(1 + a - b)_n(1 + a - c)_n(1 + a - d)_n(1 + a - b - c - d)_n} \end{aligned} \tag{2.1.6}$$

when the series is 2-balanced (i.e, $1+2a+n=b+c+d+e$), and

$$
{}_5F_4\left[\begin{matrix} a, 1+\frac{1}{2}a,\ b,\ c,\ d \\ \frac{1}{2}a, 1+a-b, 1+a-c, 1+a-d \end{matrix}; 1\right]
= \frac{\Gamma(1+a-b)\Gamma(1+a-c)\Gamma(1+a-d)\Gamma(1+a-b-c-d)}{\Gamma(1+a)\Gamma(1+a-b-c)\Gamma(1+a-b-d)\Gamma(1+a-c-d)}, \tag{2.1.7}
$$

when Re $(1+a-b-c-d)>0$, illustrate the importance of very-well-poised hypergeometric series. Note that Dixon's formula (2.1.5) follows from (2.1.7) by setting $d=\frac{1}{2}a$.

Analogous to the hypergeometric case, we shall call the basic hypergeometric series

$$
{}_{r+1}\phi_r\left[\begin{matrix} a_1, a_2, \ldots, a_{r+1} \\ b_1, \ldots, b_r \end{matrix}; q, z\right] \tag{2.1.8}
$$

well-poised if the parameters satisfy the relations

$$
qa_1 = a_2b_1 = a_3b_2 = \cdots = a_{r+1}b_r; \tag{2.1.9}
$$

very-well-poised if, in addition, $a_2 = qa_1^{\frac{1}{2}}, a_3 = -qa_1^{\frac{1}{2}}$; a *nearly-poised series of the first kind* if

$$
qa_1 \neq a_2b_1 = a_3b_2 = \cdots = a_{r+1}b_r,
$$

and a *nearly-poised series of the second kind* if

$$
qa_1 = a_2b_1 = a_3b_2 = \cdots = a_rb_{r-1} \neq a_{r+1}b_r. \tag{2.1.10}
$$

In this chapter we shall be primarily concerned with the summation and transformation formulas for very-well-poised basic hypergeometric series. To help simplify some of the displays involving very-well-poised ${}_{r+1}\phi_r$ series which arise in the proofs in this and the subsequent chapters we shall frequently replace

$$
{}_{r+1}\phi_r\left[\begin{matrix} a_1, qa_1^{\frac{1}{2}}, -qa_1^{\frac{1}{2}}, a_4, \ldots, a_{r+1} \\ a_1^{\frac{1}{2}}, -a_1^{\frac{1}{2}}, qa_1/a_4, \ldots, qa_1/a_{r+1} \end{matrix}; q, z\right]
$$

by the more compact notation

$$
{}_{r+1}W_r\,(a_1; a_4, a_5, \ldots, a_{r+1}; q, z)\,. \tag{2.1.11}
$$

In the displays of the main formulas, however, we shall continue to use the ${}_{r+1}\phi_r$ notation, since in most applications one needs to know the denominator parameters.

2.2 A general expansion formula

Let a, b, c be arbitrary parameters and k be a nonnegative integer. Then, by the q-Saalschütz formula (1.7.2)

$$
{}_3\phi_2\left(q^{-k}, aq^k, aq/bc; aq/b, aq/c; q, q\right)
= \frac{(c, q^{1-k}/b; q)_k}{(aq/b, cq^{-k}/a; q)_k} = \frac{(b, c; q)_k}{(aq/b, aq/c; q)_k}\left(\frac{aq}{bc}\right)^k, \tag{2.2.1}
$$

so that

$$\sum_{k=0}^{n} \frac{(b,c,q^{-n};q)_k}{(q,aq/b,aq/c;q)_k} A_k$$

$$= \sum_{k=0}^{n}\sum_{j=0}^{k} \frac{(aq/bc,aq^k,q^{-k};q)_j(q^{-n};q)_k}{(q,aq/b,aq/c;q)_j(q;q)_k} q^j \left(\frac{bc}{aq}\right)^k A_k$$

$$= \sum_{j=0}^{n}\sum_{i=0}^{n-j} \frac{\left(aq/bc,aq^{i+j},q^{-i-j};q\right)_j (q^{-n};q)_{i+j}}{(q,aq/b,aq/c;q)_j(q;q)_{i+j}} q^j \left(\frac{bc}{aq}\right)^{i+j} A_{i+j}$$

$$= \sum_{j=0}^{n} \frac{\left(aq/bc,aq^j,q^{-n};q\right)_j}{(q,aq/b,aq/c;q)_j} (-1)^j q^{-\binom{j}{2}}$$

$$\cdot \sum_{i=0}^{n-j} \frac{\left(q^{j-n},aq^{2j};q\right)_i}{(q,aq^j;q)_i} q^{-ij} \left(\frac{bc}{aq}\right)^{i+j} A_{i+j}, \tag{2.2.2}$$

where $\{A_k\}$ is an arbitrary sequence. This is equivalent to Bailey's [1949] lemma. Choosing

$$A_k = \frac{(a,a_1,\ldots,a_r;q)_k}{(b_1,b_2,\ldots,b_{r+1};q)_k} z^k, \tag{2.2.3}$$

we obtain the expansion formula

$${}_{r+4}\phi_{r+3}\left[\begin{matrix} a,b,c,a_1,a_2,\ldots,a_r,q^{-n} \\ aq/b,aq/c,b_1,b_2,\ldots,b_r,b_{r+1} \end{matrix};q,z\right]$$

$$= \sum_{j=0}^{n} \frac{\left(aq/bc,a_1,a_2,\ldots,a_r,q^{-n};q\right)_j}{(q,aq/b,aq/c,b_1,\ldots,b_r,b_{r+1};q)_j} \left(-\frac{bcz}{aq}\right)^j q^{-\binom{j}{2}} (a;q)_{2j}$$

$$\cdot {}_{r+2}\phi_{r+1}\left[\begin{matrix} aq^{2j},a_1q^j,a_2q^j,\ldots,a_rq^j,q^{j-n} \\ b_1q^j,b_2q^j,\ldots,b_rq^j,b_{r+1}q^j \end{matrix};q,\frac{bcz}{aq^{j+1}}\right]. \tag{2.2.4}$$

This is a q-analogue of Bailey's formula [1935, 4.3(1)]. The most important property of (2.2.4) is that it enables one to reduce the ${}_{r+4}\phi_{r+3}$ series to a sum of ${}_{r+2}\phi_{r+1}$ series. Consequently, if the above ${}_{r+2}\phi_{r+1}$ series is summable for some values of the parameters then (2.2.4) gives a transformation formula for the corresponding ${}_{r+4}\phi_{r+3}$ series in terms of a single series.

2.3 A summation formula for a terminating very-well-poised ${}_4\phi_3$ series

Setting $b = qa^{\frac{1}{2}}, c = -qa^{\frac{1}{2}}$ and $a_k = b_k, k = 1,2,\ldots,r, b_{r+1} = aq^{n+1}$, we obtain from (2.2.4) that

$${}_4\phi_3\left[\begin{matrix} a,qa^{\frac{1}{2}},-qa^{\frac{1}{2}},q^{-n} \\ a^{\frac{1}{2}},-a^{\frac{1}{2}},aq^{n+1} \end{matrix};q,z\right]$$

$$= \sum_{j=0}^{n} \frac{\left(-q^{-1},q^{-n};q\right)_j (a;q)_{2j}}{(q,a^{\frac{1}{2}},-a^{\frac{1}{2}},aq^{n+1};q)_j} (qz)^j q^{-\binom{j}{2}}$$

$$\cdot {}_2\phi_1\left(aq^{2j},q^{j-n};aq^{j+n+1};q,-zq^{1-j}\right). \tag{2.3.1}$$

If $z = q^n$, then the above ${}_2\phi_1$ series can be summed by means of the Bailey-Daum summation formula (1.8.1), which gives

$$ {}_2\phi_1\left(aq^{2j}, q^{j-n}; aq^{j+n+1}; q, -q^{1+n-j}\right) = \frac{(-q;q)_\infty \left(aq^{2j+1}, aq^{2n+2}; q^2\right)_\infty}{(aq^{n+j+1}, -q^{1+n-j}; q)_\infty}. \tag{2.3.2} $$

Hence, using the identities (1.2.32), (1.2.39) and (1.2.40), and simplifying, we obtain the transformation formula

$$ {}_4\phi_3\left[\begin{matrix} a, qa^{\frac{1}{2}}, -qa^{\frac{1}{2}}, q^{-n} \\ a^{\frac{1}{2}}, -a^{\frac{1}{2}}, aq^{n+1} \end{matrix}; q, q^n\right] = \frac{(aq, -q; q)_n}{(qa^{\frac{1}{2}}, -qa^{\frac{1}{2}}; q)_n} {}_2\phi_1\left(q^{-n}, -q^{-1}; -q^{-n}; q, q\right). \tag{2.3.3} $$

Clearly, both sides of (2.3.3) are equal to 1 when $n = 0$. By (1.5.3) the ${}_2\phi_1$ series on the right of (2.3.3) has the sum$\left(q^{1-n}; q\right)_n \left(-q^{-1}\right)^n / \left(-q^{-n}; q\right)_n$ when $n = 0, 1, \ldots$. Since $\left(q^{1-n}; q\right)_n = 0$ unless $n = 0$, it follows that

$$ {}_4\phi_3\left[\begin{matrix} a, qa^{\frac{1}{2}}, -qa^{\frac{1}{2}}, q^{-n} \\ a^{\frac{1}{2}}, -a^{\frac{1}{2}}, aq^{n+1} \end{matrix}; q, q^n\right] = \delta_{n,0}, \tag{2.3.4} $$

where

$$ \delta_{m,n} = \begin{cases} 1, & m = n, \\ 0, & m \neq n, \end{cases} \tag{2.3.5} $$

is the Kronecker delta function. This summation formula will be used in the next section to obtain the sum of a ${}_6\phi_5$ series.

2.4 A summation formula for a terminating very-well-poised ${}_6\phi_5$ series

Let us now set $a_1 = qa^{\frac{1}{2}}$, $a_2 = -qa^{\frac{1}{2}}$, $b_1 = a^{\frac{1}{2}}$, $b_2 = -a^{\frac{1}{2}}$, $b_{r+1} = aq^{n+1}$ and $a_k = b_k$, for $k = 3, 4, \ldots, r$. Then (2.2.4) gives

$$ \begin{aligned} &{}_6\phi_5\left[\begin{matrix} a, qa^{\frac{1}{2}}, -qa^{\frac{1}{2}}, b, c, q^{-n} \\ a^{\frac{1}{2}}, -a^{\frac{1}{2}}, aq/b, aq/c, aq^{n+1} \end{matrix}; q, z\right] \\ &= \sum_{j=0}^{n} \frac{(aq/bc, qa^{\frac{1}{2}}, -qa^{\frac{1}{2}}, q^{-n}; q)_j (a;q)_{2j}}{(q, a^{\frac{1}{2}}, -a^{\frac{1}{2}}, aq/b, aq/c, aq^{n+1}; q)_j} \left(-\frac{bcz}{aq}\right)^j q^{-\binom{j}{2}} \\ &\quad \cdot {}_4\phi_3\left[\begin{matrix} aq^{2j}, q^{j+1}a^{\frac{1}{2}}, -q^{j+1}a^{\frac{1}{2}}, q^{j-n} \\ q^j a^{\frac{1}{2}}, -q^j a^{\frac{1}{2}}, aq^{j+n+1} \end{matrix}; q, \frac{bcz}{aq^{j+1}}\right]. \end{aligned} \tag{2.4.1} $$

If $z = aq^{n+1}/bc$, then we can sum the above ${}_4\phi_3$ series by means of (2.3.4) and obtain the summation formula

$$ \begin{aligned} &{}_6\phi_5\left[\begin{matrix} a, qa^{\frac{1}{2}}, -qa^{\frac{1}{2}}, b, c, q^{-n} \\ a^{\frac{1}{2}}, -a^{\frac{1}{2}}, aq/b, aq/c, aq^{n+1} \end{matrix}; q, \frac{aq^{n+1}}{bc}\right] \\ &= \frac{(aq/bc, qa^{\frac{1}{2}}, -qa^{\frac{1}{2}}, q^{-n}; q)_n \ (a;q)_{2n}}{(q, a^{\frac{1}{2}}, -a^{\frac{1}{2}}, aq/b, aq/c, aq^{n+1}; q)_n} (-1)^n q^{n(n+1)/2} \\ &= \frac{(aq, aq/bc; q)_n}{(aq/b, aq/c; q)_n}. \end{aligned} \tag{2.4.2} $$

Note the special relationship that the argument of the above ${}_6\phi_5$ series must have with the parameters of the series in order that the series be summable, namely, it is the positive square root of q times the product of the denominator parameters divided by the product of the numerator parameters, when one temporarily assumes that q and the parameters are positive. This relationship between the argument and the parameters also holds for the ${}_4\phi_3$ series in (2.3.4).

2.5 Watson's transformation formula for a terminating very-well-poised ${}_8\phi_7$ series

We shall now use (2.4.2) to prove Watson's [1929a] transformation formula for a terminating very-well-poised ${}_8\phi_7$ series as a multiple of a terminating balanced ${}_4\phi_3$ series:

$$\begin{aligned}{}_8\phi_7&\left[\begin{matrix}a, qa^{\frac{1}{2}}, -qa^{\frac{1}{2}}, b, c, d, e, q^{-n}\\ a^{\frac{1}{2}}, -a^{\frac{1}{2}}, aq/b, aq/c, aq/d, aq/e, aq^{n+1}\end{matrix}; q, \frac{a^2q^{2+n}}{bcde}\right]\\ &= \frac{(aq, aq/de; q)_n}{(aq/d, aq/e; q)_n}\, {}_4\phi_3\left[\begin{matrix}q^{-n}, d, e, aq/bc\\ aq/b, aq/c, deq^{-n}/a\end{matrix}; q, q\right]. \end{aligned} \tag{2.5.1}$$

It suffices to observe that from (2.2.4) we have

$$\begin{aligned}{}_8\phi_7&\left[\begin{matrix}a, qa^{\frac{1}{2}}, -qa^{\frac{1}{2}}, b, c, d, e, q^{-n}\\ a^{\frac{1}{2}}, -a^{\frac{1}{2}}, aq/b, aq/c, aq/d, aq/e, aq^{n+1}\end{matrix}; q, \frac{a^2q^{2+n}}{bcde}\right]\\ &= \sum_{j=0}^{n} \frac{(aq/bc, qa^{\frac{1}{2}}, -qa^{\frac{1}{2}}, d, e, q^{-n}; q)_j\,(a;q)_{2j}}{(q, a^{\frac{1}{2}}, -a^{\frac{1}{2}}, aq/b, aq/c, aq/d, aq/e, aq^{n+1}; q)_j}\left(-\frac{aq^{n+1}}{de}\right)^j q^{-\binom{j}{2}}\\ &\cdot {}_6\phi_5\left[\begin{matrix}aq^{2j}, q^{j+1}a^{\frac{1}{2}}, -q^{j+1}a^{\frac{1}{2}}, dq^j, eq^j, q^{j-n}\\ q^ja^{\frac{1}{2}}, -q^ja^{\frac{1}{2}}, aq^{j+1}/d, aq^{j+1}/e, aq^{j+n+1}\end{matrix}; q, \frac{aq^{1+n-j}}{de}\right], \end{aligned} \tag{2.5.2}$$

which gives formula (2.5.1) by using (2.4.2) to sum the above ${}_6\phi_5$ series.

Note that the argument of the ${}_8\phi_7$ series in (2.5.1) is related to the parameters in exactly the same way as stated for (2.3.4) and (2.4.2) at the end of §2.4.

2.6 Jackson's sum of a terminating very-well-poised balanced ${}_8\phi_7$ series

The ${}_8\phi_7$ series in (2.5.1) is balanced when the six parameters a, b, c, d, e and n satisfy the relation

$$a^2q^{n+1} = bcde. \tag{2.6.1}$$

For such a series Jackson [1921] showed that

$$\begin{aligned}{}_8\phi_7&\left[\begin{matrix}a, qa^{\frac{1}{2}}, -qa^{\frac{1}{2}}, b, c, d, e, q^{-n}\\ a^{\frac{1}{2}}, -a^{\frac{1}{2}}, aq/b, aq/c, aq/d, aq/e, aq^{n+1}\end{matrix}; q, q\right]\\ &= \frac{(aq, aq/bc, aq/bd, aq/cd; q)_n}{(aq/b, aq/c, aq/d, aq/bcd; q)_n}, \end{aligned} \tag{2.6.2}$$

when $n = 0, 1, 2, \ldots$. This formula follows directly from (2.5.1), since the ${}_4\phi_3$ series on the right of (2.5.1) becomes a balanced ${}_3\phi_2$ series when (2.6.1) holds, and therefore can be summed by the q-Saalschütz formula. Note that (2.6.2) is a q-analogue of (2.1.5), as can be seen by replacing a, b, c, d, e by q^a, q^b, q^c, q^d, q^e, respectively, and then letting $q \to 1$. It should be observed that the series ${}_8\phi_7$ in (2.6.2) is balanced, while the limiting series ${}_7F_6$ in (2.1.5) is 2-balanced. The reason for this apparent discrepancy is that the appropriate q-analogue of the term $(1 + \frac{1}{2}a)_k/(\frac{1}{2}a)_k = (a + 2k)/a$ in the ${}_7F_6$ series is not $(qa^{\frac{1}{2}}; q)_k/(a^{\frac{1}{2}}; q)_k = (1 - a^{\frac{1}{2}}q^k)/(1 - a^{\frac{1}{2}})$ but $(qa^{\frac{1}{2}}, -qa^{\frac{1}{2}}; q)_k/(a^{\frac{1}{2}}, -a^{\frac{1}{2}}; q)_k$, which introduces an additional q-factor in the ratio of the products of the numerator and denominator parameters.

2.7 Some special and limiting cases of Jackson's and Watson's formulas: the Rogers-Ramanujan identities

Many of the known summation formulas for basic hypergeometric series are special or limiting cases of Jackson's formula (2.6.2). For example, if we take $d \to \infty$ in (2.6.2) we get (2.4.2). On the other hand, taking the limit $a \to 0$ after replacing d by aq/d gives the q-Saalschütz formula (1.7.2). Let us now take the limit $n \to \infty$ in (2.6.2). This gives

$$
{}_6\phi_5\left[\begin{matrix} a, qa^{\frac{1}{2}}, -qa^{\frac{1}{2}}, b, c, d \\ a^{\frac{1}{2}}, -a^{\frac{1}{2}}, aq/b, aq/c, aq/d \end{matrix}; q, \frac{aq}{bcd}\right]
= \frac{(aq, aq/bc, aq/bd, aq/cd; q)_\infty}{(aq/b, aq/c, aq/d, aq/bcd; q)_\infty}, \tag{2.7.1}
$$

provided $|aq/bcd| < 1$. This is clearly a q-analogue of Dougall's formula (2.1.6). Setting $d = a^{\frac{1}{2}}$ in (2.7.1), we get a q-analogue of Dixon's formula (2.1.4)

$$
{}_4\phi_3\left[\begin{matrix} a, -qa^{\frac{1}{2}}, b, c \\ -a^{\frac{1}{2}}, aq/b, aq/c \end{matrix}; q, \frac{qa^{\frac{1}{2}}}{bc}\right]
= \frac{(aq, aq/bc, qa^{\frac{1}{2}}/b, qa^{\frac{1}{2}}/c; q)_\infty}{(aq/b, aq/c, qa^{\frac{1}{2}}, qa^{\frac{1}{2}}/bc; q)_\infty}, \tag{2.7.2}
$$

provided $|qa^{\frac{1}{2}}/bc| < 1$.

Watson [1929a] used his transformation formula (2.5.1) to give a simple proof of the famous Rogers-Ramanujan identities (Hardy [1937]):

$$
\sum_{n=0}^{\infty} \frac{q^{n^2}}{(q; q)_n} = \frac{(q^2, q^3, q^5; q^5)_\infty}{(q; q)_\infty}, \tag{2.7.3}
$$

$$
\sum_{n=0}^{\infty} \frac{q^{n(n+1)}}{(q; q)_n} = \frac{(q, q^4, q^5; q^5)_\infty}{(q; q)_\infty}, \tag{2.7.4}
$$

where $|q| < 1$. First let $b, c, d, e \to \infty$ in (2.5.1) to obtain

$$
\sum_{k=0}^{n} \frac{(a; q)_k\,(1 - aq^{2k})\,(q^{-n}; q)_k}{(q; q)_k\,(1 - a)\,(aq^{n+1}; q)_k} q^{2k^2}(a^2q^n)^k
$$

$$= (aq;q)_n \sum_{k=0}^{n} \frac{(q^{-n};q)_k}{(q;q)_k} \left(aq^{n+1}\right)^k q^{k(k-1)/2}. \tag{2.7.5}$$

Since the series on both sides are finite this limiting procedure is justified as long as the term-by-term limits are assumed to exist. However, our next step is to take the limit $n \to \infty$ on both sides of (2.7.5), which we can justify by applying the dominated convergence theorem. Thus we have

$$1 + \sum_{k=1}^{\infty} \frac{(aq;q)_{k-1}(1-aq^{2k})}{(q;q)_k} (-1)^k a^{2k} q^{k(5k-1)/2}$$
$$= (aq;q)_\infty \sum_{k=0}^{\infty} \frac{a^k q^{k^2}}{(q;q)_k}. \tag{2.7.6}$$

In view of Jacobi's triple product identity (1.6.1) the series on the left side of (2.7.6) can be summed in the cases $a = 1$ and $a = q$ by observing that

$$1 + \sum_{k=1}^{\infty} \frac{(q;q)_{k-1}(1-q^{2k})}{(q;q)_k} (-1)^k q^{k(5k-1)/2}$$
$$= \sum_{n=-\infty}^{\infty} (-1)^n q^{n(5n+1)/2} = \left(q^2, q^3, q^5; q^5\right) \tag{2.7.7}$$

and

$$\sum_{k=0}^{\infty} (1-q^{2k+1})(-1)^k q^{k(5k+3)/2}$$
$$= \sum_{n=-\infty}^{\infty} (-1)^n q^{n(5n+3)/2} = \left(q, q^4, q^5; q^5\right)_\infty. \tag{2.7.8}$$

The identities (2.7.3) and (2.7.4) now follow immediately by using (2.7.6). For an early history of these identities see Hardy [1940, pp. 90-99].

2.8 Bailey's transformation formulas for terminating ${}_5\phi_4$ and ${}_7\phi_6$ series

Using Jackson's formula (2.6.2), it can be easily shown that

$$\frac{(a,b,c;q)_k}{(q,aq/b,aq/c;q)_k} = \frac{(\lambda bc/a;q)_k}{(qa^2/\lambda bc;q)_k}$$
$$\cdot \sum_{j=0}^{k} \frac{(\lambda;q)_j (1-\lambda q^{2j})(\lambda b/a, \lambda c/a, aq/bc;q)_j}{(q;q)_j (1-\lambda)(aq/b, aq/c, \lambda bc/a;q)_j}$$
$$\cdot \frac{(a;q)_{k+j} (a/\lambda;q)_{k-j}}{(\lambda q;q)_{k+j}\ (q;q)_{k-j}} \left(\frac{a}{\lambda}\right)^j, \tag{2.8.1}$$

where λ is an arbitrary parameter. If $\{A_k\}$ is an arbitrary sequence, it follows that

$$\sum_{k=0}^{\infty} \frac{(a,b,c;q)_k}{(q,aq/b,aq/c;q)_k} A_k$$

$$= \sum_{k=0}^{\infty} \frac{(\lambda bc/a;q)_k}{(qa^2/\lambda bc;q)_k} A_k \sum_{j=0}^{k} \frac{(\lambda;q)_j(1-\lambda q^{2j})\,(\lambda b/a, \lambda c/a, aq/bc;q)_j}{(q;q)_j(1-\lambda)(aq/b, aq/c, \lambda bc/a;q)_j}$$
$$\cdot \frac{(a;q)_{k+j}(a/\lambda;q)_{k-j}}{(\lambda q;q)_{k+j}(q;q)_{k-j}} \left(\frac{a}{\lambda}\right)^j$$
$$= \sum_{j=0}^{\infty} \frac{(\lambda;q)_j(1-\lambda q^{2j})(\lambda b/a, \lambda c/a, aq/bc;q)_j (a;q)_{2j}}{(q;q)_j(1-\lambda)(aq/b, aq/c, qa^2/\lambda bc;q)_j (\lambda q;q)_{2j}} \left(\frac{a}{\lambda}\right)^j$$
$$\cdot \sum_{k=0}^{\infty} \frac{(aq^{2j}, a/\lambda, \lambda bcq^j/a;q)_k}{(q, \lambda q^{2j+1}, a^2q^{j+1}/\lambda bc;q)_k} A_{j+k}, \tag{2.8.2}$$

provided the change in order of summation is justified (e.g., if all of the series terminate or are absolutely convergent).

It is clear that appropriate choices of λ and A_k will lead to transformation formulas for basic hypergometric series which have at least a partial well-poised structure.

First, let us take $A_k = q^k(d, q^{-n};q)_k/(aq/d, a^2q^{-n}/\lambda^2;q)_k$ and $\lambda = qa^2/bcd$, so that the inner series on the right of (2.8.2) becomes a balanced and terminating ${}_3\phi_2$. Summing this ${}_3\phi_2$ series and simplifying the coefficients, we obtain Bailey's [1947a,c] formula

$${}_5\phi_4 \left[\begin{matrix} a, b, c, d, q^{-n} \\ aq/b, aq/c, aq/d, a^2q^{-n}/\lambda^2 \end{matrix} ; q, q \right]$$
$$= \frac{(\lambda q/a, \lambda^2 q/a;q)_n}{(\lambda q, \lambda^2 q/a^2;q)_n} {}_{12}\phi_{11} \left[\begin{matrix} \lambda, q\lambda^{\frac12}, -q\lambda^{\frac12}, b\lambda/a, c\lambda/a, d\lambda/a, \\ \lambda^{\frac12}, -\lambda^{\frac12}, aq/b, aq/c, aq/d, \end{matrix} \right.$$
$$\left. \begin{matrix} a^{\frac12}, -a^{\frac12}, (aq)^{\frac12}, -(aq)^{\frac12}, \lambda^2 q^{n+1}/a, q^{-n} \\ \lambda q/a^{\frac12}, -\lambda q/a^{\frac12}, \lambda(q/a)^{\frac12}, -\lambda(q/a)^{\frac12}, aq^{-n}/\lambda, \lambda q^{n+1} \end{matrix} ; q, q \right], \tag{2.8.3}$$

where $\lambda = qa^2/bcd$.

Note that the ${}_5\phi_4$ series on the left is balanced and nearly-poised of the second kind, while the ${}_{12}\phi_{11}$ series on the right is balanced and very-well-poised. Note also that a terminating nearly-poised series of the second kind can be expressed as a multiple of a nearly-poised series of the first kind by simply reversing the series.

By proceeding as in the proof of (2.8.3), one can obtain the following variation of (2.8.3)

$${}_5\phi_4 \left[\begin{matrix} q^{-n}, b, c, d, e \\ q^{1-n}/b, q^{1-n}/c, q^{1-n}/d, eq^{-2n}/\lambda^2 \end{matrix} ; q, q \right]$$
$$= \frac{(\lambda^2 q^{n+1}, \lambda q/e;q)_n}{(\lambda^2 q^{n+1}/e, \lambda q;q)_n} {}_{12}\phi_{11} \left[\begin{matrix} \lambda, q\lambda^{\frac12}, -q\lambda^{\frac12}, \lambda bq^n, \lambda cq^n, \lambda dq^n, \\ \lambda^{\frac12}, -\lambda^{\frac12}, q^{1-n}/b, q^{1-n}/c, q^{1-n}/d, \end{matrix} \right.$$
$$\left. \begin{matrix} q^{-\frac{n}{2}}, -q^{-\frac{n}{2}}, q^{\frac12-\frac{n}{2}}, -q^{\frac12-\frac{n}{2}}, e, \lambda^2 q^{n+1}/e \\ \lambda q^{\frac{n}{2}+1}, -\lambda q^{\frac{n}{2}+1}, \lambda q^{\frac{n}{2}+\frac12}, -\lambda q^{\frac{n}{2}+\frac12}, \lambda q/e, eq^{-n}/\lambda \end{matrix} ; q, q \right], \tag{2.8.4}$$

where $\lambda = q^{1-2n}/bcd$.

Next, let us choose $A_k = q^k(1-aq^{2k})(d,q^{-n};q)_k/(1-a)(aq/d, a^2q^{2-n}/\lambda^2;q)_k$ and $\lambda = qa^2/bcd$ in (2.8.2) so that the inner series on the right side takes the form

$$q^j \frac{1-aq^{2j}}{1-a} \frac{(d,q^{-n};q)_j}{(aq/d, a^2q^{2-n}/\lambda^2;q)_j} \cdot {}_5\phi_4 \left[\begin{matrix} aq^{2j}, q^{j+1}a^{\frac{1}{2}}, -q^{j+1}a^{\frac{1}{2}}, a/\lambda, q^{j-n} \\ q^j a^{\frac{1}{2}}, -q^j a^{\frac{1}{2}}, \lambda q^{2j+1}, a^2q^{j-n+2}/\lambda^2 \end{matrix} ; q,q \right].$$

This ${}_5\phi_4$ series is a special case of the ${}_5\phi_4$ series on the left side of (2.8.3); in fact, the ${}_{12}\phi_{11}$ series on the right side of (2.8.3) in this case reduces to a terminating balanced very-well-poised ${}_8\phi_7$ series which we can sum by Jackson's formula (2.6.2). Carrying out the straightforward calculations, we get Bailey's [1947c] second transformation formula

$$\begin{aligned} &{}_7\phi_6 \left[\begin{matrix} a, qa^{\frac{1}{2}}, -qa^{\frac{1}{2}}, b, c, d, q^{-n} \\ a^{\frac{1}{2}}, -a^{\frac{1}{2}}, aq/b, aq/c, aq/d, a^2q^{2-n}/\lambda^2 \end{matrix} ; q,q \right] \\ &= \frac{(\lambda/aq, \lambda^2/aq;q)_n}{(\lambda q, \lambda^2/a^2q;q)_n} \frac{1-\lambda^2q^{2n-1}/a}{1-\lambda^2/aq} \\ &\quad \cdot {}_{12}\phi_{11} \left[\begin{matrix} \lambda, q\lambda^{\frac{1}{2}}, -q\lambda^{\frac{1}{2}}, b\lambda/a, c\lambda/a, d\lambda/a, (aq)^{\frac{1}{2}}, -(aq)^{\frac{1}{2}}, \\ \lambda^{\frac{1}{2}}, -\lambda^{\frac{1}{2}}, aq/b, aq/c, aq/d, \lambda(q/a)^{\frac{1}{2}}, -\lambda(q/a)^{\frac{1}{2}}, \end{matrix} \right. \\ &\qquad \left. \begin{matrix} qa^{\frac{1}{2}}, -qa^{\frac{1}{2}}, \lambda^2q^{n-1}/a, q^{-n} \\ \lambda/a^{\frac{1}{2}}, -\lambda/a^{\frac{1}{2}}, aq^{2-n}/\lambda, \lambda q^{n+1} \end{matrix} ; q,q \right], \end{aligned} \tag{2.8.5}$$

where $\lambda = qa^2/bcd$.

2.9 Bailey's transformation formula for a terminating ${}_{10}\phi_9$ series

One of the most important transformation formulas for basic hypergeometric series is Bailey's [1929] formula transforming a terminating ${}_{10}\phi_9$ series, which is both balanced and very-well-poised, into a series of the same type:

$$\begin{aligned} &{}_{10}\phi_9 \left[\begin{matrix} a, qa^{\frac{1}{2}}, -qa^{\frac{1}{2}}, b, c, d, e, f, \lambda aq^{n+1}/ef, q^{-n} \\ a^{\frac{1}{2}}, -a^{\frac{1}{2}}, aq/b, aq/c, aq/d, aq/e, aq/f, efq^{-n}/\lambda, aq^{n+1} \end{matrix} ; q,q \right] \\ &= \frac{(aq, aq/ef, \lambda q/e, \lambda q/f;q)_n}{(aq/e, aq/f, \lambda q/ef, \lambda q;q)_n} {}_{10}\phi_9 \left[\begin{matrix} \lambda, q\lambda^{\frac{1}{2}}, -q\lambda^{\frac{1}{2}}, \lambda b/a, \lambda c/a, \lambda d/a, \\ \lambda^{\frac{1}{2}}, -\lambda^{\frac{1}{2}}, aq/b, aq/c, aq/d, \end{matrix} \right. \\ &\qquad \left. \begin{matrix} e, f, \lambda aq^{n+1}/ef, q^{-n}, \\ \lambda q/e, \lambda q/f, efq^{-n}/a, \lambda q^{n+1} \end{matrix} ; q,q \right], \end{aligned} \tag{2.9.1}$$

where $\lambda = qa^2/bcd$.

To derive this formula, first observe that by (2.6.2)

$$\begin{aligned} &{}_8\phi_7 \left[\begin{matrix} \lambda, q\lambda^{\frac{1}{2}}, -q\lambda^{\frac{1}{2}}, \lambda b/a, \lambda c/a, \lambda d/a, aq^m, q^{-m} \\ \lambda^{\frac{1}{2}}, -\lambda^{\frac{1}{2}}, aq/b, aq/c, aq/d, \lambda q^{1-m}/a, \lambda q^{m+1} \end{matrix} ; q,q \right] \\ &= \frac{(b,c,d,\lambda q;q)_m}{(aq/b, aq/c, aq/d, a/\lambda;q)_m}, \end{aligned} \tag{2.9.2}$$

and hence the left side of (2.9.1) equals

$$\sum_{m=0}^{n} \frac{(a;q)_m(1-aq^{2m})(e,f,\lambda aq^{n+1}/ef,q^{-n};q)_m(a/\lambda;q)_m}{(q;q)_m(1-a)(aq/e,aq/f,efq^{-n}/\lambda,aq^{n+1};q)_m(\lambda q;q)_m} q^m$$
$$\cdot \sum_{j=0}^{m} \frac{(\lambda;q)_j(1-\lambda q^{2j})(\lambda b/a,\lambda c/a,\lambda d/a,aq^m,q^{-m};q)_j}{(q;q)_j(1-\lambda)(aq/b,aq/c,aq/d,\lambda q^{1-m}/a,\lambda q^{m+1};q)_j} q^j$$
$$= \sum_{m=0}^{n}\sum_{j=0}^{m} \frac{(a;q)_{m+j}(1-aq^{2m})(e,f,\lambda aq^{n+1}/ef,q^{-n};q)_m}{(q;q)_{m-j}(1-a)(aq/e,aq/f,efq^{-n}/\lambda,aq^{n+1};q)_m} q^m$$
$$\cdot \frac{(a/\lambda;q)_{m-j}(\lambda;q)_j(1-\lambda q^{2j})(\lambda b/a,\lambda c/a,\lambda d/a;q)_j}{(\lambda q;q)_{m+j}(q;q)_j(1-\lambda)(aq/b,aq/c,aq/d;q)_j} \left(\frac{a}{\lambda}\right)^j$$
$$= \sum_{j=0}^{n} \frac{(\lambda;q)_j(1-\lambda q^{2j})(\lambda b/a,\lambda c/a,\lambda d/a,e,f,\lambda aq^{n+1}/ef,q^{-n};q)_j}{(q;q)_j(1-\lambda)(aq/b,aq/c,aq/d,aq/e,aq/f,efq^{-n}/\lambda,aq^{n+1};q)_j}$$
$$\cdot \left(\frac{aq}{\lambda}\right)^j \frac{(aq;q)_{2j}}{(\lambda q;q)_{2j}} \, {}_8W_7\left(aq^{2j};eq^j,fq^j,a/\lambda,\lambda aq^{n+j+1}/ef,q^{j-n};q,q\right), \tag{2.9.3}$$

where the ${}_8W_7$ series is defined as in §2.1. Summing the above ${}_8W_7$ series by means of (2.6.2) and simplifying the coefficients, we obtain (2.9.1). It is sometimes helpful to rewrite (2.9.1) in a somewhat more symmetrical form:

$$\begin{aligned} &{}_{10}W_9\left(a;b,c,d,e,f,g,h;q,q\right) \\ &= \frac{(aq,aq/ef,aq/eg,aq/eh,aq/fg,aq/fh,aq/gh,aq/efgh;q)_\infty}{(aq/e,aq/f,aq/g,aq/h,aq/efg,aq/efh,aq/egh,aq/fgh;q)_\infty} \\ &\quad \cdot {}_{10}W_9\left(qa^2/bcd;aq/bc,aq/bd,aq/cd,e,f,g,h;q,q\right), \end{aligned} \tag{2.9.4}$$

where at least one of the parameters e,f,g,h is of form $q^{-n}, n=0,1,2,...,$ and

$$q^2a^3 = bcdefgh. \tag{2.9.5}$$

2.10 Limiting cases of Bailey's ${}_{10}\phi_9$ transformation formula

A number of the known transformation formulas for basic hypergeometric series follow as limiting cases of the transformation formula (2.9.1). If we let $b,c,$ or $d \to \infty$ in (2.9.1), we obtain Watson's formula (2.5.1). On the other hand, if we take the limit $n \to \infty$, we get the transformation formula for a nonterminating ${}_8\phi_7$ series

$$\begin{aligned} &{}_8\phi_7\left[\begin{matrix} a, qa^{\frac12}, -qa^{\frac12}, b, c, d, e, f \\ a^{\frac12}, -a^{\frac12}, aq/b, aq/c, aq/d, aq/e, aq/f \end{matrix}; q, \frac{a^2q^2}{bcdef}\right] \\ &= \frac{(aq,aq/ef,\lambda q/e,\lambda q/f;q)_\infty}{(aq/e,aq/f,\lambda q,\lambda q/ef;q)_\infty} \\ &\quad \cdot {}_8\phi_7\left[\begin{matrix} \lambda, q\lambda^{\frac12}, -q\lambda^{\frac12}, \lambda b/a, \lambda c/a, \lambda d/a, e, f \\ \lambda^{\frac12}, -\lambda^{\frac12}, aq/b, aq/c, aq/d, \lambda q/e, \lambda q/f \end{matrix}; q, \frac{aq}{ef}\right], \end{aligned} \tag{2.10.1}$$

where $\lambda = qa^2/bcd$ and

$$\max(|aq/ef|,\ |\lambda q/ef|) < 1. \tag{2.10.2}$$

The convergence of the two series in (2.10.1) is ensured by the inequalities (2.10.2) which, of course, are not required if both series terminate. For example, if $f = q^{-n}, n = 0, 1, 2, \ldots$, then (2.10.1) becomes

$$\begin{aligned} &{}_8\phi_7\left[\begin{matrix} a, qa^{\frac12}, -qa^{\frac12}, b, c, d, e, q^{-n} \\ a^{\frac12}, -a^{\frac12}, aq/b, aq/c, aq/d, aq/e, aq^{n+1} \end{matrix}; q, \frac{a^2q^{n+2}}{bcde}\right] \\ &= \frac{(aq, \lambda q/e; q)_n}{(aq/e, \lambda q; q)_n} \\ &\quad \cdot {}_8\phi_7\left[\begin{matrix} \lambda, q\lambda^{\frac12}, -q\lambda^{\frac12}, \lambda b/a, \lambda c/a, \lambda d/a, e, q^{-n} \\ \lambda^{\frac12}, -\lambda^{\frac12}, aq/b, aq/c, aq/d, \lambda q/e, \lambda q^{n+1} \end{matrix}; q, \frac{aq^{n+1}}{e}\right]. \end{aligned} \tag{2.10.3}$$

This identity expresses one terminating very-well-poised ${}_8\phi_7$ series in terms of another. However, it should be noticed that in both terminating and nonterminating cases of the ${}_8\phi_7$ transformation derived above the arguments of the series are related to the parameters in the same way as described in §2.4.

Using (2.5.1) we can now express (2.10.3) as a transformation formula between two terminating balanced ${}_4\phi_3$ series:

$$\begin{aligned} &{}_4\phi_3\left[\begin{matrix} q^{-n}, a, b, c \\ d, e, f \end{matrix}; q, q\right] \\ &= \frac{(e/a, f/a; q)_n}{(e, f; q)_n}\, a^n\, {}_4\phi_3\left[\begin{matrix} q^{-n}, a, d/b, d/c \\ d, aq^{1-n}/e, aq^{1-n}/f \end{matrix}; q, q\right], \end{aligned} \tag{2.10.4}$$

where $abc = defq^{n-1}$. This is a very useful formula which was first derived by Sears [1951c], and hence is called the *Sears' ${}_4\phi_3$ transformation formula.* It is a q-analogue of Whipple's [1926b] formula

$$\begin{aligned} &{}_4F_3\left[\begin{matrix} -n, & a, & b, & c \\ & d, & e, & f \end{matrix}; 1\right] \\ &= \frac{(e-a)_n(f-a)_n}{(e)_n(f)_n}\, {}_4F_3\left[\begin{matrix} -n, a, d-b, d-c \\ d, 1+a-e-n, 1+a-f-n \end{matrix}; 1\right], \end{aligned} \tag{2.10.5}$$

where $a + b + c + 1 = d + e + f + n$.

Use of (2.5.1) and (2.10.1) also enables us to express a terminating balanced ${}_4\phi_3$ series in terms of a nonterminating ${}_8\phi_7$ series. For example, if b, c, or d in (2.10.1) is of the form $q^{-n}, n = 0, 1, 2, \ldots$, then the series on the left side of (2.10.1) terminates, but that on the right side does not. In particular, setting $d = q^{-n}$, and then replacing e and f by d and e, respectivey, we obtain

$$\begin{aligned} &{}_8W_7\left(a; b, c, d, e, q^{-n}; q, a^2q^{n+2}/bcde\right) \\ &= \frac{(aq, aq/de, a^2q^{n+2}/bcd, a^2q^{n+2}/bce; q)_\infty}{(aq/d, aq/e, a^2q^{n+2}/bc, a^2q^{n+2}/bcde; q)_\infty} \\ &\quad \cdot {}_8W_7\left(a^2q^{n+1}/bc; aq^{n+1}/b, aq^{n+1}/c, aq/bc, d, e; q, aq/de\right), \end{aligned} \tag{2.10.6}$$

provided $|aq/de| < 1$ to ensure that the nonterminating series on the right side converges. Use of (2.5.1) then leads to the formula

$$\begin{aligned}{}_4\phi_3\left[\begin{matrix}q^{-n}, a, b, c\\ d, e, f\end{matrix}; q, q\right] &= \frac{(deq^n/a, deq^n/b, deq^n/c, deq^n/abc; q)_\infty}{(deq^n, deq^n/ab, deq^n/ac, deq^n/bc; q)_\infty}\\ &\cdot {}_8\phi_7\left[\begin{matrix}deq^{n-1}, & q\,(deq^{n-1})^{\frac{1}{2}}, & -q\,(deq^{n-1})^{\frac{1}{2}},\\ & (deq^{n-1})^{\frac{1}{2}}, & -(deq^{n-1})^{\frac{1}{2}},\end{matrix}\right.\\ &\left.\begin{matrix}a, b, c, dq^n, eq^n\\ deq^n/a, deq^n/b, deq^n/c, e, d\end{matrix}; q, \frac{de}{abc}\right], \qquad (2.10.7)\end{aligned}$$

provided $def = q^{1-n}abc$ and $|de/abc| < 1$.

As another limiting case of (2.9.1) Bailey [1935, 8.5 (3)] found a nonterminating extension of (2.5.1) that expresses a very-well-poised ${}_8\phi_7$ series in terms of two balanced ${}_4\phi_3$ series. First use (2.9.1) to get

$$\begin{aligned}&{}_{10}W_9\left(a; b, c, d, e, f, a^3q^{n+2}/bcdef, q^{-n}; q, q\right)\\ &= \frac{(aq, aq/de, aq/df, aq/ef; q)_n}{(aq/d, aq/e, aq/f, aq/def; q)_n}\\ &\cdot {}_{10}W_9\left(defq^{-n-1}/a; aq/bc, d, e, f, bdefq^{-n-1}/a^2, cdefq^{-n-1}/a^2, q^{-n}; q, q\right). \qquad (2.10.8)\end{aligned}$$

Clearly, the ${}_{10}W_9$ on the left side of (2.10.8) tends to the ${}_8\phi_7$ series on the left side of (2.10.1) as $n \to \infty$. However, the terms near both ends of the series on the right side of (2.10.8) are large compared to those in the middle for large n, which prevents us from taking the term-by-term limit directly. To circumvent this difficulty, Bailey chose n to be an odd integer, say $n = 2m + 1$ (this is not necessary, but it makes the analysis simpler), and divided the series on the right into two halves, each containing $m + 1$ terms, and then reversed the order of the second series. The procedure is schematized as follows:

$$\begin{aligned}\sum_{k=0}^{2m+1} \lambda_k &= \sum_{k=0}^{m} \lambda_k + \sum_{k=m+1}^{2m+1} \lambda_k\\ &= \sum_{k=0}^{m} \lambda_k + \sum_{k=0}^{m} \lambda_{2m+1-k}, \qquad (2.10.9)\end{aligned}$$

where $\{\lambda_k\}$ is an arbitrary sequence. Letting $m \to \infty$ (and hence $n \to \infty$), it follows from (2.10.8) that

$$\begin{aligned}&{}_8\phi_7\left[\begin{matrix}a, qa^{\frac{1}{2}}, -qa^{\frac{1}{2}}, b, c, d, e, f\\ a^{\frac{1}{2}}, -a^{\frac{1}{2}}, aq/b, aq/c, aq/d, aq/e, aq/f\end{matrix}; q, \frac{a^2q^2}{bcdef}\right]\\ &= \frac{(aq, aq/de, aq/df, aq/ef; q)_\infty}{(aq/d, aq/e, aq/f, aq/def; q)_\infty}\,{}_4\phi_3\left[\begin{matrix}aq/bc, d, e, f\\ aq/b, aq/c, def/a\end{matrix}; q, q\right]\\ &+ \frac{(aq, aq/bc, d, e, f, a^2q^2/bdef,\ a^2q^2/cdef; q)_\infty}{(aq/b, aq/c, aq/d, aq/e, aq/f, a^2q^2/bcdef, def/aq; q)_\infty}\\ &\cdot {}_4\phi_3\left[\begin{matrix}aq/de, aq/df, aq/ef, a^2q^2/bcdef\\ a^2q^2/bdef, a^2q^2/cdef, aq^2/def\end{matrix}; q, q\right], \qquad (2.10.10)\end{aligned}$$

where $|a^2q^2/bcdef| < 1$, if the ${}_8\phi_7$ series does not terminate. Note that if either b or c is of the form $q^{-n}, n = 0, 1, 2, ...$, then the ${}_8\phi_7$ series on the left side terminates but the series on the right side do not necessarily terminate. On the other hand if one of the numerator parameters (except $a^2q^2/bcdef$) in either of the two ${}_4\phi_3$ series in (2.10.10) is of the form q^{-n}, then the coefficient of the other ${}_4\phi_3$ series vanishes and we get either (2.5.1) or (2.10.7).

If $aq/bc, aq/de, aq/df$ or aq/ef equals 1, then (2.10.10) reduces to the ${}_6\phi_5$ summation formula (2.7.1). If , on the other hand, $aq/cd = 1$ then the ${}_8\phi_7$ series in (2.10.10) reduces to a ${}_6\phi_5$ which, via (2.7.1), leads to the summation formula

$$\begin{aligned}
&\frac{(aq, aq/be, aq/bf, aq/ef;q)_\infty}{(aq/b, aq/e, aq/f, aq/bef;q)_\infty} \\
&= \frac{(aq, c/e, c/f, aq/ef;q)_\infty}{(c, aq/e, aq/f, c/ef;q)_\infty} {}_3\phi_2 \left[\begin{matrix} aq/bc, e, f \\ aq/b, efq/c \end{matrix} ; q, q \right] \\
&\quad + \frac{(aq, aq/ef, e, f, aq/bc, acq/bef;q)_\infty}{(aq/e, aq/f, ef/c, c, aq/b, aq/bef;q)_\infty} \\
&\quad \cdot {}_3\phi_2 \left[\begin{matrix} c/e, c/f, aq/bef \\ cq/ef, acq/bef \end{matrix} ; q, q \right]. \qquad (2.10.11)
\end{aligned}$$

Solving for the first ${}_3\phi_2$ series on the right and relabelling the parameters we get the following nonterminating extension of the q-Saalschütz formula

$$\begin{aligned}
{}_3\phi_2 \left[\begin{matrix} a, b, c \\ e, f \end{matrix} ; q, q \right] &= \frac{(q/e, f/a, f/b, f/c;q)}{(aq/e, bq/e, cq/e, f;q)_\infty} \\
&- \frac{(q/e, a, b, c, qf/e;q)_\infty}{(e/q, aq/e, bq/e, cq/e, f;q)_\infty} \\
&\cdot {}_3\phi_2 \left[\begin{matrix} aq/e, bq/e, cq/e \\ q^2/e, qf/e \end{matrix} ; q, q \right], \qquad (2.10.12)
\end{aligned}$$

where $ef = abcq$. Sears [1951a, (5.2)] derived this formula by a different method. If a, b, or c is of the form $q^{-n}, n = 0, 1, 2, \ldots$, then (2.10.12) reduces to (1.7.2).

A special case of (2.10.12) which is worth mentioning is obtained by setting $c = 0, f = 0$, and then replacing e by c to get

$$\begin{aligned}
{}_2\phi_1(a, b; c; q, q) &= \frac{(q/c, abq/c;q)_\infty}{(aq/c, bq/c;q)_\infty} \\
&- \frac{(q/c, a, b;q)_\infty}{(c/q, aq/c, bq/c;q)_\infty} {}_2\phi_1(aq/c, bq/c; q^2/c; q, q). \qquad (2.10.13)
\end{aligned}$$

If a or b is of the form $q^{-n}, n = 0, 1, 2, ...$, then (2.10.13) reduces to (1.5.3). In general, a ${}_2\phi_1(a, b; c; q, q)$ series does not have a sum which can be written as a ratio of infinite products. However, we can still express (2.10.13) as the summation formula for a single bilateral infinite series in the following way.

First, use Heine's transformation formula (1.4.1) to transform both ${}_2\phi_1$ series in (2.10.13):

$${}_2\phi_1(a, b; c; q, q) = \frac{(b, aq;q)_\infty}{(c, q;q)_\infty} {}_2\phi_1(c/b, q; aq; q, b)$$

$$= \frac{(b, aq; q)_\infty}{(c, q; q)_\infty} \sum_{n=0}^{\infty} \frac{(c/b; q)_n}{(aq; q)_n} b^n, \tag{2.10.14}$$

$${}_2\phi_1(aq/c, bq/c; q^2/c; q, q) = \frac{(aq/c, bq^2/c; q)_\infty}{(q^2/c, q; q)_\infty} \sum_{n=0}^{\infty} \frac{(q/a; q)_n}{(bq^2/c; q)_n} \left(\frac{aq}{c}\right)^n. \tag{2.10.15}$$

Next, note that

$$\sum_{n=0}^{\infty} \frac{(q/a; q)_n}{(bq^2/c; q)_n} \left(\frac{aq}{c}\right)^n = -\frac{c}{q} \frac{1 - bq/c}{1 - a} \sum_{n=1}^{\infty} \frac{(1/a; q)_n}{(bq/c; q)_n} \left(\frac{aq}{c}\right)^n. \tag{2.10.16}$$

Using (2.10.14) - (2.10.16) and the identity (1.2.28) in (2.10.13) we obtain

$$\sum_{n=-\infty}^{\infty} \frac{(c/b; q)_n}{(aq; q)_n} b^n = \frac{(c, q/c, abq/c, q; q)_\infty}{(b, aq, aq/c, bq/c; q)_\infty}, \tag{2.10.17}$$

which is Ramanujan's sum (see Chapter 5 and Andrews and Askey [1978]). However, the conditions under which (2.10.17) is valid, namely, $|q| < 1, |b| < 1$ and $|aq/c| < 1$, are more restrictive then those for (2.10.13). Note that (2.10.17) tends to Jacobi's triple product identity (1.6.1) when $a = 0$ and $b \to 0$. We shall give an alternative derivation of this important sum in Chapter 5 where bilateral basic series are considered.

As was pointed out by Al-Salam and Verma [1982a], both (2.10.10) and (2.10.12) can be conveniently expressed as q-integrals. Thus (2.10.12) is equivalent to

$$\int_a^b \frac{(qt/a, qt/b, ct; q)_\infty}{(dt, et, ft; q)_\infty} d_q t$$
$$= b(1-q) \frac{(q, bq/a, a/b, c/d, c/e, c/f; q)_\infty}{(ad, ae, af, bd, be, bf; q)_\infty}, \tag{2.10.18}$$

where $c = abdef$, while (2.10.10) is equivalent to

$$\int_a^b \frac{(qt/a, qt/b, ct, dt; q)_\infty}{(et, ft, gt, ht; q)_\infty} d_q t$$
$$= b(1-q) \frac{(q, bq/a, a/b, cd/eh, cd/fh, cd/gh, bc, bd; q)_\infty}{(ae, af, ag, be, bf, bg, bh, bcd/h; q)_\infty}$$
$$\cdot {}_8W_7(bcd/hq; be, bf, bg, c/h, d/h; q, ah), \tag{2.10.19}$$

provided $cd = abefgh$ and $|ah| < 1$.

By substituting $c = abdef$ into (2.10.18), letting $f \to 0$ and then replacing a, d, e by $-a, -c/a, d/b$, respectively, we obtain Andrews and Askey's [1981] formula

$$\int_{-a}^b \frac{(-qt/a, qt/b; q)_\infty}{(-ct/a, dt/b; q)_\infty} d_q t$$
$$= \frac{b(1-q)(q, -a/b, -bq/a, cd; q)_\infty}{(c, d, -bc/a, -ad/b; q)_\infty}. \tag{2.10.20}$$

In view of (1.3.18), (2.10.20) is a q-extension of the beta-type integral

$$\int_{-c}^{d} (1+t/c)^{a-1}(1-t/d)^{b-1}\, dt = B(a,b)\frac{(c+d)^{a+b-1}}{c^{a-1}d^{b-1}}, \tag{2.10.21}$$

which follows from (1.11.8) by a change of variable.

The notational compactness of (2.10.18) and (2.10.19) is advantageous in many applications (see, e.g., the next section). In addition, the symmetry of the parameters in the q-integral on the left side of (2.10.19) implies the transformation formula (2.10.1).

2.11 Bailey's three-term transformation formula for very-well-poised ${}_8\phi_7$ series

The q-integral representation (2.10.19) of an ${}_8\phi_7$ series can be put to advantage in deriving Bailey's [1936] 3-term transformation formula for an ${}_8\phi_7$ series:

$$\begin{aligned}
&{}_8\phi_7\left[\begin{matrix} a, qa^{\frac12}, -qa^{\frac12}, b, c, d, e, f \\ a^{\frac12}, -a^{\frac12}, aq/b, aq/c, aq/d, aq/e, aq/f \end{matrix}; q, \frac{a^2q^2}{bcdef}\right] \\
&= \frac{(aq, aq/de, aq/df, aq/ef, eq/c, fq/c, b/a, bef/a; q)_\infty}{(aq/d, aq/e, aq/f, aq/def, q/c, efq/c, be/a, bf/a; q)_\infty} \\
&\cdot {}_8\phi_7\left[\begin{matrix} ef/c, q(ef/c)^{\frac12}, -q(ef/c)^{\frac12}, aq/bc, aq/cd, ef/a, e, f \\ (ef/c)^{\frac12}, -(ef/c)^{\frac12}, bef/a, def/a, aq/c, fq/c, eq/c \end{matrix}; q, \frac{bd}{a}\right] \\
&+ \frac{b}{a}\frac{(aq, bq/a, bq/c, bq/d, bq/e, bq/f, d, e, f; q)_\infty}{(aq/b, aq/c, aq/d, aq/e, aq/f, bd/a, be/a, bf/a, def/a; q)_\infty} \\
&\cdot \frac{(aq/bc, bdef/a^2, a^2q/bdef; q)_\infty}{(aq/def, q/c, b^2q/a; q)_\infty} \\
&\cdot {}_8\phi_7\left[\begin{matrix} b^2/a, qba^{-\frac12}, -qba^{-\frac12}, b, bc/a, bd/a, be/a, bf/a \\ ba^{-\frac12}, -ba^{-\frac12}, bq/a, bq/c, bq/d, bq/e, bq/f \end{matrix}; q, \frac{a^2q^2}{bcdef}\right],
\end{aligned} \tag{2.11.1}$$

where $|bd/a| < 1$ and $|a^2q^2/bcdef| < 1$.

To prove this formula, first note that by (2.10.19)

$$\begin{aligned}
&{}_8W_7\left(a; b, c, d, e, f; q, a^2q^2/bcdef\right) \\
&= \frac{aq-def}{adefq(1-q)}\frac{(aq, d, e, f, aq/bc, aq/de, aq/df, aq/ef; q)_\infty}{(q, aq/b, aq/c, aq/d, aq/e, aq/f, def/aq, aq/def; q)_\infty} \\
&\cdot \int_{aq}^{def} \frac{(t/a, qt/def, aqt/bdef, aqt/cdef; q)_\infty}{(t/de, t/df, t/ef, aqt/bcdef; q)_\infty}\, d_qt.
\end{aligned} \tag{2.11.2}$$

Since

$$\int_{aq}^{bdef/a} f(t)\, d_qt + \int_{bdef/a}^{def} f(t)\, d_qt = \int_{aq}^{def} f(t)\, d_qt, \tag{2.11.3}$$

where

$$f(t)=\frac{(t/a,qt/def,aqt/bdef,aqt/cdef;q)_\infty}{(t/de,t/df,t/ef,aqt/bcdef;q)_\infty}, \tag{2.11.4}$$

and

$$\begin{aligned}&\int_{aq}^{bdef/a} f(t)\,d_qt\\ &=\frac{bdef(1-q)(q,bdef/a^2,a^2q/bdef,bq/d,bq/e,bq/f,bq/a,bq/c;q)_\infty}{a(aq/de,aq/df,aq/ef,bd/a,be/a,bf/a,q/c,b^2q/a;q)_\infty}\\ &\quad\cdot {}_8W_7\left(b^2/a;\ b,\ bc/a,\ bd/a,\ be/a,\ bf/a;\ q,\ a^2q^2/bcdef\right),\end{aligned} \tag{2.11.5}$$

$$\begin{aligned}\int_{bdef/a}^{def} f(t)\,d_qt&=\frac{def(1-q)(q,aq/b,aq/c,eq/c,fq/c,bef/a,def/a;q)_\infty}{(d,e,f,be/a,bf/a,aq/bc,q/c,efq/c;q)_\infty}\\ &\quad\cdot {}_8W_7\left(ef/c;aq/bc,aq/cd,ef/a,e,f;q,bd/a\right),\end{aligned} \tag{2.11.6}$$

we immediately get (2.11.1) by using (2.11.5) and (2.11.6) in (2.11.3). The advantage of our use of the q-integral notation can be seen by comparing the above proof with that given in Bailey [1936].

The special case $qa^2=bcdef$ is particularly important since the series on the left side of (2.11.1) and the second series on the right become balanced, while the first series on the right becomes a ${}_6\phi_5$ series with sum

$$\frac{(aq/ce,aq/cf,efq/c,q/c;q)_\infty}{(aq/c,eq/c,fq/c,aq/cef;q)_\infty},$$

provided $|aq/cef|<1$. This gives Bailey's summation formula:

$$\begin{aligned}&{}_8\phi_7\left[\begin{matrix}a,qa^{\frac12},-qa^{\frac12},b,c,d,e,f\\ a^{\frac12},-a^{\frac12},aq/b,aq/c,aq/d,aq/e,aq/f\end{matrix};q,q\right]\\ &-\frac{b}{a}\frac{(aq,c,d,e,f,bq/a,bq/c,bq/d,bq/e,bq/f;q)_\infty}{(aq/b,aq/c,aq/d,aq/e,aq/f,bc/a,bd/a,be/a,bf/a,b^2q/a;q)_\infty}\\ &\cdot{}_8\phi_7\left[\begin{matrix}b^2/a,qba^{-\frac12},-qba^{-\frac12},b,bc/a,bd/a,be/a,bf/a\\ ba^{-\frac12},-ba^{-\frac12},bq/a,bq/c,bq/d,bq/e,bq/f\end{matrix};q,q\right]\\ &=\frac{(aq,b/a,aq/cd,aq/ce,aq/cf,aq/de,aq/df,aq/ef;q)_\infty}{(aq/c,aq/d,aq/e,aq/f,bc/a,bd/a,be/a,bf/a;q)_\infty},\end{aligned} \tag{2.11.7}$$

where $qa^2=bcdef$, which is a nonterminating extension of Jackson's formula (2.6.2). This can also be written in the following equivalent form

$$\begin{aligned}&\int_a^b \frac{(qt/a,qt/b,t/a^{\frac12},-t/a^{\frac12},qt/c,qt/d,qt/e,qt/f;q)_\infty}{(t,bt/a,qt/a^{\frac12},-qt/a^{\frac12},ct/a,dt/a,et/a,ft/a;q)_\infty}\,d_qt\\ &=\frac{b(1-q)(q,a/b,bq/a,aq/cd,aq/ce,aq/cf,aq/de,aq/df,aq/ef;q)_\infty}{(b,c,d,e,f,bc/a,bd/a,be/a,bf/a;q)_\infty}.\end{aligned} \tag{2.11.8}$$

2.12 Bailey's four-term transformation formula for balanced ${}_{10}\phi_9$ series

Let us start by replacing a, b, c, d, e, and f in (2.11.8) by $\lambda, bq^n, \lambda c/a, \lambda d/a, a/bq^n$ and a, respectively, to obtain

$$\int_{\lambda}^{bq^n} \frac{(qt/\lambda, tq^{1-n}/b, t/\lambda^{\frac{1}{2}}, -t/\lambda^{\frac{1}{2}}, aqt/c\lambda, aqt/d\lambda, aqt/e\lambda, btq^{n+1}/a;q)_\infty}{(t, btq^n/\lambda, qt/\lambda^{\frac{1}{2}}, -qt/\lambda^{\frac{1}{2}}, ct/a, dt/a, et/a, atq^{-n}/b\lambda;q)_\infty} d_q t$$
$$= \frac{b(1-q)(q, \lambda/b, bq/\lambda, bq/c, bq/d, bq/e, c, d, e;q)_\infty}{(a/\lambda, b, \lambda c/a, \lambda d/a, \lambda e/a, a/b, bc/a, bd/a, be/a;q)_\infty}$$
$$\cdot \frac{(b, bc/a, bd/a, be/a;q)_n}{(bq/a, bq/c, bq/d, bq/e;q)_n} \left(\frac{\lambda q}{a}\right)^n, \tag{2.12.1}$$

where $n = 0, 1, 2, \ldots$. Let f, g, h be arbitrary complex numbers such that $|a^3q^3/bcdefgh| < 1$. Set $\rho = a^3q^2/bcdefgh$, multiply both sides of (2.12.1) by

$$\frac{(b^2/a;q)_n(1-b^2q^{2n}/a)(bf/a, bg/a, bh/a;q)_n}{(q;q)_n(1-b^2/a)(bq/f, bq/g, bq/h;q)_n} \left(\frac{a\rho}{\lambda}\right)^n$$

and sum over n from 0 to ∞. Then the right side of (2.12.1) leads to

$$\frac{b(1-q)(q, \lambda/b, bq/\lambda, bq/c, bq/d, bq/e, c, d, e;q)_\infty}{(a/\lambda, b, \lambda c/a, \lambda d/a, \lambda e/a, a/b, bc/a, bd/a, be/a;q)_\infty}$$
$$\cdot {}_{10}W_9\left(b^2/a; b, bc/a, bd/a, be/a, bf/a, bg/a, bh/a; q, \rho q\right). \tag{2.12.2}$$

The left side leads to two double sums, one from each of the two limits of the q-integral. From the upper limit, bq^n, we get

$$\frac{b(1-q)(q, bq/\lambda, abq/c\lambda, abq/d\lambda, abq/e\lambda, b^2q/a;q)_\infty}{(b, a/\lambda, b^2q/\lambda, bc/a, bd/a, be/a;q)_\infty}$$
$$\cdot \sum_{n=0}^{\infty} \frac{(b^2/a;q)_n(1-b^2q^{2n}/a)(bf/a, bg/a, bh/a;q)_n}{(q;q)_n(1-b^2/a)(bq/f, bq/g, bq/h;q)_n} \left(\frac{a\rho}{\lambda}\right)^n$$
$$\cdot \sum_{j=0}^{\infty} \frac{(b^2/\lambda;q)_{2n+j}(1-b^2q^{2n+2j}/\lambda)(b, bc/a, bd/a, be/a;q)_{n+j}(a/\lambda;q)_j}{(q;q)_j(1-b^2/\lambda)(bq/\lambda, abq/c\lambda, abq/d\lambda, abq/e\lambda;q)_{n+j}(b^2q/a;q)_{2n+j}} q^j$$
$$= \frac{b(1-q)(q, bq/\lambda, abq/c\lambda, abq/d\lambda, abq/e\lambda, b^2q/a;q)_\infty}{(b, a/\lambda, b^2q/\lambda, bc/a, bd/a, be/a;q)_\infty}$$
$$\cdot \sum_{m=0}^{\infty} \frac{(b^2/\lambda;q)_m(1-b^2q^{2m}/\lambda)(b, bc/a, bd/a, be/a, a/\lambda;q)_m}{(q;q)_m(1-b^2/\lambda)(bq/\lambda, abq/c\lambda, abq/d\lambda, abq/e\lambda, b^2q/a;q)_m} q^m$$
$$\cdot {}_8W_7\left(b^2/a; bf/a, bg/a, bh/a, b^2q^m/\lambda, q^{-m}; q, \rho q\right). \tag{2.12.3}$$

Let us now assume that $\rho = 1$, that is,

$$a^3q^2 = bcdefgh. \tag{2.12.4}$$

Then, by Jackson's formula (2.6.2), the ${}_8\phi_7$ series in (2.12.3) has the sum

$$\frac{(b^2q/a, aq/fg, aq/fh, aq/gh;q)_m}{(bq/f, bq/g, bq/h, aq/fgh;q)_m}$$

and the expression in (2.12.3) simplifies to

$$\frac{b(1-q)(q,bq/\lambda,abq/c\lambda,abq/d\lambda,abq/e\lambda,b^2q/a;q)_\infty}{(b,a/\lambda,b^2q/\lambda,bc/a,bd/a,be/a;q)_\infty} \cdot {}_{10}W_9\left(b^2/\lambda;b,bc/a,bd/a,be/a,bf/\lambda,bg/\lambda,bh/\lambda;q,q\right), \quad (2.12.5)$$

since, by (2.12.4), $aq/fg = bh/\lambda$, $aq/fh = bg/\lambda$, and $aq/gh = bf/\lambda$.

We now turn to the double sum that corresponds to the lower limit, λ, in the q-integral (2.12.1). This leads to the series

$$\begin{aligned}
&\frac{-\lambda(1-q)(q,aq/c,aq/d,aq/e,\lambda q/b,b\lambda q/a;q)_\infty}{(b,\lambda q,a/b,c\lambda/a,d\lambda/a,e\lambda/a;q)_\infty} \\
&\cdot\sum_{n=0}^{\infty}\frac{(b^2/a;q)_n(1-b^2q^{2n}/a)(b,bf/a,bg/a,bh/a,b/\lambda;q)_n}{(q;q)_n(1-b^2/a)(bq/a,bq/f,bq/g,bq/h,b\lambda q/a;q)_n}q^n \\
&\cdot {}_8W_7\left(\lambda;bq^n,c\lambda/a,d\lambda/a,e\lambda/a,aq^{-n}/b;q,q\right) \\
&= \frac{-\lambda(1-q)(q,aq/c,aq/d,aq/e,\lambda q/b,b\lambda q/a;q)_\infty}{(b,\lambda q,a/b,c\lambda/a,d\lambda/a,e\lambda/a;q)_\infty} \\
&\cdot\sum_{m=0}^{\infty}\frac{(\lambda;q)_m(1-\lambda q^{2m})(b,c\lambda/a,d\lambda/a,e\lambda/a,a/b;q)_m}{(q;q)_m(1-\lambda)(\lambda q/b,aq/c,aq/d,aq/e,b\lambda q/a;q)_m}q^m \\
&\cdot {}_8W_7\left(b^2/a;bq^m,bq^{-m}/\lambda,bf/a,bg/a,bh/a;q,q\right). \qquad (2.12.6)
\end{aligned}$$

The last ${}_8\phi_7$ series in (2.12.6) is balanced and nonterminating, so we may use (2.11.7) to get

$$\begin{aligned}
&{}_8W_7\left(b^2/a;bq^m,bq^{-m}/\lambda,bf/a,bg/a,bh/a;q,q\right) \\
&= \frac{(b^2q/a,\lambda q^{m+1}/f,\lambda q^{m+1}/g,\lambda q^{m+1}/h,aq/fg,aq/fh,aq/gh,aq^m/b;q)_\infty}{(b\lambda q^{m+1}/a,bq/f,bq/g,bq/h,a/\lambda,fq^m,gq^m,hq^m;q)_\infty} \\
&+\frac{aq^m}{b}\frac{(b^2q/a,bq^{-m}/\lambda,bf/a,bg/a,bh/a,aq^{m+1}/b,aq^{m+1}/f;q)_\infty}{(bq^{1-m}/a,b\lambda q^{m+1}/a,bq/f,bq/g,bq/h,a/\lambda,fq^m,gq^m;q)_\infty} \\
&\cdot\frac{(aq^{m+1}/g,aq^{m+1}/h,\lambda q^{2m+1};q)_\infty}{(hq^m,aq^{2m+1};q)_\infty} \\
&\cdot {}_8W_7\left(aq^{2m};bq^m,fq^m,gq^m,hq^m,a/\lambda;q,q\right). \qquad (2.12.7)
\end{aligned}$$

Use of this breaks up the double series in (2.12.6) into two parts:

$$\begin{aligned}
&\frac{-\lambda(1-q)(q,b^2q/a,\lambda q/b,\lambda q/f,\lambda q/g,\lambda q/h,bf/\lambda,bg/\lambda;q)_\infty}{(b,f,g,h,bq/f,bq/g,bq/h,\lambda q;q)_\infty} \\
&\cdot\frac{(bh/\lambda,aq/c,aq/d,aq/e;q)_\infty}{(a/\lambda,c\lambda/a,d\lambda/a,e\lambda/a;q)_\infty} \\
&\cdot {}_{10}W_9\left(\lambda;b,c\lambda/a,d\lambda/a,e\lambda/a,f,g,h;q,q\right) \\
&-\frac{a\lambda(1-q)(q,b/\lambda,\lambda q/b,b^2q/a,aq/b,aq/c,aq/d,aq/e,aq/f,aq/g;q)_\infty}{(b,a/b,aq,c\lambda/a,d\lambda/a,e\lambda/a,bq/a,bq/f,bq/g,bq/h;q)_\infty} \\
&\cdot\frac{(aq/h,bf/a,bg/a,bh/a;q)_\infty}{(a/\lambda,f,g,h;q)_\infty}
\end{aligned}$$

$$\cdot \sum_{n=0}^{\infty} \frac{(a;q)_n(1-aq^{2n})(b,f,g,h,a/\lambda;q)_n}{(q;q)_n(1-a)(aq/b,aq/f,aq/g,aq/h,\lambda q;q)_n} q^n$$
$$\cdot {}_8W_7\,(\lambda;c\lambda/a,d\lambda/a,e\lambda/a,aq^n,q^{-n};q,q)\,. \tag{2.12.8}$$

Summing the last ${}_8\phi_7$ series by (2.6.2) we find that the sum over n in (2.12.8) equals ${}_{10}W_9\,(a;b,c,d,e,f,g,h;q,q)$ which is, of course, balanced by virtue of (2.12.4). Equating the expression in (2.12.2) with the sum of those in (2.12.5) and (2.12.8), and simplifying the coefficients, we finally obtain Bailey's [1947b] four-term transformation formula

$${}_{10}\phi_9\left[\begin{matrix}a,qa^{\frac12},-qa^{\frac12},b,c,d,e,f,g,h\\ a^{\frac12},-a^{\frac12},aq/b,aq/c,aq/d,aq/e,aq/f,aq/g,aq/h\end{matrix};q,q\right]$$
$$+\frac{(aq,b/a,c,d,e,f,g,h,bq/c,bq/d;q)_\infty}{(b^2q/a,a/b,aq/c,aq/d,aq/e,aq/f,aq/g,aq/h,bc/a,bd/a;q)_\infty}$$
$$\cdot\frac{(bq/e,bq/f,bq/g,bq/h;q)_\infty}{(be/a,bf/a,bg/a,bh/a;q)_\infty}$$
$$\cdot\,{}_{10}\phi_9\left[\begin{matrix}b^2/a,qba^{-\frac12},-qba^{-\frac12},b,bc/a,bd/a,be/a,bf/a,bg/a,bh/a\\ ba^{-\frac12},-ba^{-\frac12},bq/a,bq/c,bq/d,bq/e,bq/f,bq/g,bq/h\end{matrix};q,q\right]$$
$$=\frac{(aq,b/a,\lambda q/f,\lambda q/g,\lambda q/h,bf/\lambda,bg/\lambda,bh/\lambda;q)_\infty}{(\lambda q,b/\lambda,aq/f,aq/g,aq/h,bf/a,bg/a,bh/a;q)_\infty}$$
$$\cdot\,{}_{10}\phi_9\left[\begin{matrix}\lambda,q\lambda^{\frac12},-q\lambda^{\frac12},b,\lambda c/a,\lambda d/a,\lambda e/a,f,g,h\\ \lambda^{\frac12},-\lambda^{\frac12},\lambda q/b,aq/c,aq/d,aq/e,\lambda q/f,\lambda q/g,\lambda q/h\end{matrix};q,q\right]$$
$$+\frac{(aq,b/a,f,g,h,bq/f,bq/g,bq/h,\lambda c/a,\lambda d/a;q)_\infty}{(b^2q/\lambda,\lambda/b,aq/c,aq/d,aq/e,aq/f,aq/g,aq/h,bc/a,bd/a;q)_\infty}$$
$$\cdot\frac{(\lambda e/a,abq/\lambda c,abq/\lambda d,abq/\lambda e;q)_\infty}{(be/a,bf/a,bg/a,bh/a;q)_\infty}$$
$$\cdot\,{}_{10}\phi_9\left[\begin{matrix}b^2/\lambda,qb\lambda^{-\frac12},-qb\lambda^{-\frac12},b,bc/a,bd/a,be/a,bf/\lambda,bg/\lambda,bh/\lambda\\ b\lambda^{-\frac12},-b\lambda^{-\frac12},bq/\lambda,abq/c\lambda,abq/d\lambda,abq/e\lambda,bq/f,bq/g,bq/h\end{matrix};q,q\right]. \tag{2.12.9}$$

In terms of the q-integrals this can be written in a more compact form:

$$\int_a^b \frac{(qt/a,qt/b,ta^{-\frac12},-ta^{-\frac12},qt/c,qt/d,qt/e,qt/f,qt/g,qt/h;q)_\infty}{(t,bt/a,qta^{-\frac12},-qta^{-\frac12},ct/a,dt/a,et/a,ft/a,gt/a,ht/a;q)_\infty}\,d_qt$$
$$=\frac{a}{\lambda}\frac{(b/a,aq/b,\lambda c/a,\lambda d/a,\lambda e/a,bf/\lambda,bg/\lambda,bh/\lambda;q)_\infty}{(b/\lambda,\lambda q/b,c,d,e,bf/a,bg/a,bh/a;q)_\infty}$$
$$\cdot\int_\lambda^b \frac{(qt/\lambda,qt/b,t\lambda^{-\frac12},-t\lambda^{-\frac12},aqt/c\lambda,aqt/d\lambda,aqt/e\lambda,qt/f,qt/g,qt/h;q)_\infty}{(t,bt/\lambda,qt\lambda^{-\frac12},-qt\lambda^{-\frac12},ct/a,dt/a,et/a,ft/\lambda,gt/\lambda,ht/\lambda;q)_\infty}\,d_qt, \tag{2.12.10}$$

where $\lambda=qa^2/cde$ and $a^3q^2=bcdefgh$.

Exercises 2

2.1 Show that

$$ {}_3\phi_2\left[\begin{matrix} a, qa^{\frac{1}{2}}, -qa^{\frac{1}{2}} \\ a^{\frac{1}{2}}, -a^{\frac{1}{2}} \end{matrix}; q, t\right] = (1-aqt^2)\frac{(atq^2;q)_\infty}{(t;q)_\infty}, \quad |t|<1. $$

2.2 Show that, for $\max(|t|, |aq|) < 1$,

$$ {}_4\phi_3\left[\begin{matrix} a, qa^{\frac{1}{2}}, -qa^{\frac{1}{2}}, b \\ a^{\frac{1}{2}}, -a^{\frac{1}{2}}, aq/b \end{matrix}; q, t\right] = \frac{(aq, bqt;q)_\infty}{(t, aq/b;q)_\infty}\, {}_2\phi_1\left(b^{-1}, t; bqt; q, aq\right). $$

2.3 Give an alternate proof of the ${}_6\phi_5$ summation formula (2.4.2) by first using (2.2.4) to derive a terminating form of the q-Dixon formula (2.7.2) and then using it along with the q-Saalschütz formula (1.7.2).

2.4 Prove Sears' identity (2.10.4) by using (1.4.3) and the coefficients in the power series expansion of the product

$$ {}_2\phi_1(a,b;c;q,z)\ {}_2\phi_1(d,e;abde/c;q,abz/c). $$

2.5 Prove that the sum of the first $n+1$ terms of the series

$$ \sum_{k=0}^\infty \frac{(a;q)_k(1-aq^{2k})(b,c,a/bc;q)_k}{(q;q)_k(1-a)(aq/b,aq/c,bcq;q)_k} q^k $$

is

$$ \frac{(aq,bq,cq,aq/bc;q)_n}{(aq/b,aq/c,bcq,q;q)_n}. $$

2.6 Show that

$$ {}_4\phi_3\left[\begin{matrix} q^{-n}, b, c, -q^{1-n}/bc \\ q^{1-n}/b, q^{1-n}/c, -bc \end{matrix}; q, q\right] = \begin{cases} 0, & n = 2m+1, \\ \dfrac{(q,b^2,c^2;q^2)_m\ (bc;q)_{2m}}{(b,c;q)_{2m}\ (b^2c^2;q^2)_m}, & n = 2m, \end{cases} $$

where $m = 0, 1, 2, \ldots$. (Bailey [1941], Carlitz [1969a])

2.7 Derive Jackson's terminating q-analogue of Dixon's sum:

$$ {}_3\phi_2\left[\begin{matrix} q^{-2n}, & b, & c \\ & q^{1-2n}/b, & q^{1-2n}/c \end{matrix}; q, \frac{q^{2-n}}{bc}\right] = \frac{(b,c;q)_n(q,bc;q)_{2n}}{(q,bc;q)_n(b,c;q)_{2n}}, $$

where $n = 0, 1, 2, \ldots$. (See Jackson [1921, 1941], Bailey [1941], and Carlitz [1969a])

2.8 If $b = q^{-n}, n = 0, 1, 2, \ldots$, show that

$$ {}_4\phi_3\left[\begin{matrix} a, b, c^{\frac{1}{2}}, -c^{\frac{1}{2}} \\ (abq)^{\frac{1}{2}}, -(abq)^{\frac{1}{2}}, c \end{matrix}; q, q\right] = \frac{(aq, bq, cq/a, cq/b; q^2)_\infty}{(q, abq, cq, cq/ab; q^2)_\infty} a^{n/2}. \qquad \text{(Andrews [1976a])} $$

2.9 Prove that

$$
{}_4\phi_3\left[\begin{matrix} a, b, -b, aq/c^2 \\ aq/c, -aq/c, b^2 \end{matrix}; q, q\right]
+ \frac{(q/b^2, a, b, -b, aq/c^2, aq^2/b^2c, -aq^2/b^2c; q)_\infty}{(b^2/q, q/b, -q/b, aq/b^2, aq/c, -aq/c, aq^2/b^2c^2; q)_\infty}
\cdot {}_4\phi_3\left[\begin{matrix} q/b, -q/b, aq/b^2, aq^2/b^2c^2 \\ aq^2/b^2c, -aq^2/b^2c, q^2/b^2 \end{matrix}; q, q\right]
= \frac{(q/b^2, -q; q)_\infty (a^2q^2, aq^2/b^2, aq^2/c^2, a^2q^2/b^2c^2; q^2)_\infty}{(aq/b^2, -aq; q)_\infty (q^2/b^2, a^2q^2/c^2, aq^2, aq^2/b^2c^2; q^2)_\infty}.
$$

2.10 The *q-Racah polynomials*, which were introduced by Askey and Wilson [1979], are defined by

$$
W_n(x; a, b, c, N; q) = {}_4\phi_3\left[\begin{matrix} q^{-n}, abq^{n+1}, q^{-x}, cq^{x-N} \\ aq, bcq, q^{-N} \end{matrix}; q, q\right],
$$

where $n = 0, 1, 2, ..., N$. Show that

$$
W_n(x; a, b, c, N; q) = \frac{(aq/c, bq; q)_n}{(aq, bcq; q)_n} c^n\, W_n(N - x; b, a, c^{-1}, N; q).
$$

2.11 The *Askey-Wilson polynomials* are defined in Askey and Wilson [1985] by

$$
p_n(x; a, b, c, d \mid q)
= a^{-n}(ab, ac, ad; q)_n\, {}_4\phi_3\left[\begin{matrix} q^{-n}, abcdq^{n-1}, ae^{i\theta}, ae^{-i\theta} \\ ab, ac, ad \end{matrix}; q, q\right],
$$

where $x = \cos\theta$. Show that

(i) $p_n(x; a, b, c, d \mid q) = p_n(x; b, a, c, d \mid q)$,

(ii) $p_n(-x; a, b, c, d \mid q) = (-1)^n p_n(x; -a, -b, -c, -d \mid q)$.

2.12 Show that

$$
{}_{10}W_9(a; b^{\frac{1}{2}}, -b^{\frac{1}{2}}, (bq)^{\frac{1}{2}}, -(bq)^{\frac{1}{2}}, a/b, a^2q^{n+1}/b, q^{-n}; q, q)
= \frac{(aq, a^2q/b^2; q)_n}{(aq/b, a^2q/b; q)_n},\quad n = 0, 1, 2, \ldots .
$$

2.13 If $\lambda = qa^2/bcd$ and $|\lambda/a| < 1$, prove that

$$
(i)\quad {}_4\phi_3\left[\begin{matrix} a, b, c, d \\ aq/b, aq/c, aq/d \end{matrix}; q, \frac{q\lambda^2}{a^2}\right] = \frac{(\lambda q/a, \lambda^2 q/a; q)_\infty}{(\lambda q, \lambda^2 q/a^2; q)_\infty}
\cdot {}_{10}W_9\left(\lambda; a^{\frac{1}{2}}, -a^{\frac{1}{2}}, (aq)^{\frac{1}{2}}, -(aq)^{\frac{1}{2}}, \lambda b/a, \lambda c/a, \lambda d/a, ; q, \lambda/a\right),
$$

$$
(ii)\quad {}_4\phi_3\left[\begin{matrix} a, b, c, d \\ aq/b, aq/c, aq/d \end{matrix}; q, \frac{-\lambda q}{a}\right]
= \frac{(aq, -q, \lambda q a^{-\frac{1}{2}}, -\lambda q a^{-\frac{1}{2}}; q)_\infty}{(\lambda q, -\lambda q/a, q a^{\frac{1}{2}}, -q a^{\frac{1}{2}}; q)_\infty}
\cdot {}_8W_7\left(\lambda; a^{\frac{1}{2}}, -a^{\frac{1}{2}}, \lambda b/a, \lambda c/a, \lambda d/a; q, -q\right).
$$

2.14 (i) Show that

$$
{}_4\phi_3\left[\begin{matrix} a, qa^{\frac{1}{2}}, b, q^{-n} \\ a^{\frac{1}{2}}, aq/b, b^2q^{1-n} \end{matrix}; q, q\right]
= \frac{(ab^{-2}, b^{-1}, -qb^{-1}a^{\frac{1}{2}}; q)_n}{(b^{-2}, aq/b, -b^{-1}a^{\frac{1}{2}}; q)_n},
$$

which is a q-analogue of Bailey [1935, 4.5(1.3)].

(ii) Using (i) in the formula (2.8.2) prove the following q-analogue of Bailey [1935, 4.5(4)]:

$$
\begin{aligned}
&{}_6\phi_5\left[\begin{matrix} a, qa^{\frac{1}{2}}, b, c, d, q^{-n} \\ a^{\frac{1}{2}}, aq/b, aq/c, aq/d, a^2q^{1-n}/\lambda^2 \end{matrix}; q, q\right] \\
&= \frac{(\lambda/a, \lambda^2/a, -\lambda qa^{-\frac{1}{2}}; q)_n}{(\lambda q, \lambda^2/a^2, -\lambda a^{-\frac{1}{2}}; q)_n} \\
&\quad \cdot {}_{12}\phi_{11}\left[\begin{matrix} \lambda, q\lambda^{\frac{1}{2}}, -q\lambda^{-\frac{1}{2}}, \lambda b/a, \lambda c/a, \lambda d/a, \\ \lambda^{\frac{1}{2}}, -\lambda^{\frac{1}{2}}, aq/b, aq/c, aq/d, \end{matrix}\right. \\
&\qquad \left.\begin{matrix} qa^{\frac{1}{2}}, -a^{\frac{1}{2}}, (aq)^{\frac{1}{2}}, -(aq)^{\frac{1}{2}}, \lambda^2q^n/a, q^{-n} \\ \lambda a^{-\frac{1}{2}}, -\lambda qa^{-\frac{1}{2}}, \lambda(q/a)^{\frac{1}{2}}, -\lambda(q/a)^{\frac{1}{2}}, aq^{1-n}/\lambda, \lambda q^{n+1} \end{matrix}; q, q\right].
\end{aligned}
$$

This formula is equivalent to Jain's [1982, (4.6)] transformation formula.

2.15 By taking suitable q-integrals of the function

$$
f(t) = \frac{(qt/b, qt/c, aqt/bc, tq^2/bcdef; q)_\infty}{(at, qt/bcd, qt/bce, qt/bcf; q)_\infty},
$$

prove Bailey's [1936, (4.6)] identity

$$
\begin{aligned}
&a^{-1}\frac{(aq/d, aq/e, aq/f, q/ad, q/ae, q/af; q)_\infty}{(qa^2, ab, ac, b/a, c/a; q)_\infty} \\
&\cdot {}_8W_7\left(a^2; ab, ac, ad, ae, af; q, q^2/abcdef\right) \\
&+ b^{-1}\frac{(bq/d, bq/e, bq/f, q/bd, q/be, q/bf; q)_\infty}{(qb^2, ba, bc, a/b, c/b; q)_\infty} \\
&\cdot {}_8W_7\left(b^2; ba, bc, bd, be, bf; q, q^2/abcdef\right) \\
&+ c^{-1}\frac{(cq/d, cq/e, cq/f, q/cd, q/ce, q/cf; q)_\infty}{(qc^2, ca, cb, a/c, b/c; q)_\infty} \\
&\cdot {}_8W_7\left(c^2; ca, cb, cd, ce, cf; q, q^2/abcdef\right) \\
&= 0,
\end{aligned}
$$

provided $|q^2/abcdef| < 1$.

2.16 Let $S(\lambda, \mu, \nu, \rho) = (\lambda, q/\lambda, \mu, q/\mu, \nu, q/\nu, \rho, q/\rho; q)_\infty$. Using Ex. 2.15 prove that

$$
\begin{aligned}
&S(x\lambda, x/\lambda, \mu\nu, \mu/\nu) - S(x\nu, x/\nu, \lambda\mu, \mu/\lambda) \\
&= \frac{\mu}{\lambda}S(x\mu, x/\mu, \lambda\nu, \lambda/\nu),
\end{aligned}
$$

where x, λ, μ, ν are non-zero complex numbers. (Sears [1951c,d], Bailey [1936])

2.17 Show that

$$(i) \quad {}_8\phi_7\left[\begin{matrix} \lambda, q\lambda^{\frac{1}{2}}, -q\lambda^{\frac{1}{2}}, a, b, c, -c, \lambda q/c^2 \\ \lambda^{\frac{1}{2}}, -\lambda^{\frac{1}{2}}, \lambda q/a, \lambda q/b, \lambda q/c, -\lambda q/c, c^2 \end{matrix}; q, -\frac{\lambda q}{ab}\right]$$
$$= \frac{(\lambda q, c^2/\lambda; q)_\infty (aq, bq, c^2q/a, c^2q/b; q^2)_\infty}{(\lambda q/a, \lambda q/b; q)_\infty (q, abq, c^2q, c^2q/ab; q^2)_\infty}$$

where $\lambda = -c(ab/q)^{\frac{1}{2}}$ and $|\lambda q/ab| < 1$;

$$(ii) \quad {}_8\phi_7\left[\begin{matrix} -c, q(-c)^{\frac{1}{2}}, -q(-c)^{\frac{1}{2}}, a, q/a, c, -d, -q/d \\ (-c)^{\frac{1}{2}}, -(-c)^{\frac{1}{2}}, -cq/a, -ac, -q, cq/d, cd \end{matrix}; q, c\right]$$
$$= \frac{(-c, -cq; q)_\infty (acd, acq/d, cdq/a, cq^2/ad; q^2)_\infty}{(cd, cq/d, -ac, -cq/a; q)_\infty}, \quad |c| < 1.$$

Verify that (i) is a q-analogue of Watson's summation formula (Bailey [1935, 3.3(1)]) while (ii) is a q-analogue of Whipple's formula (Bailey [1935, 3.4(1)]). (See Jain and Verma [1985] and Gasper and Rahman [1986]).

2.18 In the ${}_6\phi_5$ summation formula (2.7.1) let $b, c, d \to \infty$. Then set $a = 1$ to prove Euler's [1748] identity

$$1 + \sum_{n=1}^{\infty} (-1)^n \left\{q^{n(3n-1)/2} + q^{n(3n+1)/2}\right\} = (q; q)_\infty.$$

2.19 Show that

$$\begin{aligned} &{}_{10}W_9\left(a; b, c, d, e, f, g, q^{-n}; q, q\right) \\ &= \frac{(aq, aq/ce, aq/de, aq/ef, aq/eg, b; q)_n}{(aq/c, aq/d, aq/e, aq/f, aq/g, b/e; q)_n} e^n \\ &\quad \cdot {}_{10}W_9\left(eq^{-n}/b; e, aq/bc, aq/bd, aq/bf, aq/bg, eq^{-n}/a, q^{-n}; q, q\right), \end{aligned}$$

where $a^3q^{n+2} = bcdefg$ and $n = 0, 1, 2, \ldots$.

2.20 Prove that

$$\begin{aligned} &{}_{10}W_9\left(a; b, c, d, e, f, g, q^{-n}; q, a^3q^{n+3}/bcdefg\right) \\ &= \frac{(aq, aq/fg; q)_n}{(aq/f, aq/g; q)_n} \sum_{j=0}^{n} \frac{(q^{-n}, f, g, aq/de; q)_j q^j}{(q, aq/d, aq/e, fgq^{-n}/a; q)_j} \\ &\quad \cdot {}_4\phi_3\left[\begin{matrix} q^{-j}, d, e, aq/bc \\ aq/b, aq/c, deq^{-j}/a \end{matrix}; q, q\right], \end{aligned}$$

for $n = 0, 1, 2, \ldots$.

2.21 Show that

$$
\begin{aligned}
&{}_{10}W_9\left(a;b,c,d,e,f,g,h;q,a^3q^3/bcdefgh\right)\\
&=\sum_{n=0}^{\infty}\frac{(\lambda;q)_n(1-\lambda q^{2n})(\lambda b/a,\lambda c/a,\lambda d/a,e,f,g,h;q)_n\,(aq;q)_{2n}}{(q;q)_n(1-\lambda)(aq/b,aq/c,aq/d,aq/e,aq/f,aq/g,aq/h;q)_n\,(\lambda q;q)_{2n}}\\
&\cdot\left(\frac{aq}{\lambda}\right)^n\ {}_8W_7\left(aq^{2n};a/\lambda,eq^n,fq^n,gq^n,hq^n;q,a^3q^3/bcdefgh\right),
\end{aligned}
$$

where $\lambda=qa^2/bcd$ and $|\ a^3q^3/bcdefgh\ |<1$.

2.22 Prove that

$$
\begin{aligned}
&\sum_{n=0}^{\infty}\frac{(a;q)_n(1-aq^{2n})(b,c,d,e;q)_n}{(q;q)_n(1-a)(aq/b,aq/c,aq/d,aq/e;q)_n}\left(-\frac{a^2q^2}{bcde}\right)^n q^{\binom{n}{2}}\\
&=\frac{(aq,aq/de;q)_\infty}{(aq/d,aq/e;q)_\infty}\ {}_3\phi_2\left[\begin{matrix}aq/bc,d,e\\ aq/b,aq/c\end{matrix};q,\frac{aq}{de}\right],\quad |aq/de|<1.
\end{aligned}
$$

Deduce that

$$
\begin{aligned}
&\sum_{n=0}^{\infty}\frac{(a;q)_n(1-aq^{2n})(d,e;q)_n}{(q;q)_n(1-a)(aq/d,aq/e;q)_n}\left(-\frac{aq}{de}\right)^n q^{\binom{n}{2}}\\
&=\frac{(aq,aq/de;q)_\infty}{(aq/d,aq/e;q)_\infty}.
\end{aligned}
$$

2.23 Prove that

$$
\begin{aligned}
&\sum_{j=0}^{n}\frac{(ab,ac,ad;q)_j}{(abcd,aqz,aq/z;q)_j}\,q^j\\
&=\frac{(1-z/a)(1-abcz)}{(1-bz)(1-cz)}\ {}_8W_7\left(abcz;ab,ac,bc,qz/d,q;q,dz\right)\\
&-\frac{(ab,ac,ad;q)_{n+1}}{(abcd,aqz,aq/z;q)_{n+1}}\ \frac{(1-aq^{n+1}/z)(1-abczq^{n+1})}{(1-a/z)(1-abcz)}\\
&\cdot{}_8W_7\left(abczq^{n+1};abq^{n+1},acq^{n+1},bc,qz/d,q;q,dz\right).
\end{aligned}
$$

2.24 Show that

$$
\begin{aligned}
&{}_5\phi_4\left[\begin{matrix}a,b,c,d,e\\ aq/b,aq/c,aq/d,f\end{matrix};q,q\right]=\frac{(\lambda q/a,\lambda q/e,q\lambda^2/a,q/f;q)_\infty}{(\lambda q,aq/f,eq/f,aq/\lambda f;q)_\infty}\\
&\cdot\ {}_{12}W_{11}\left(\lambda;a^{\frac12},-a^{\frac12},(aq)^{\frac12},-(aq)^{\frac12},\lambda b/a,\lambda c/a,\lambda d/a,e,aq/f;q,q\right)\\
&-\frac{(a,e,a/\lambda,q/f,\lambda q^2/f;q)_\infty}{(f/q,\lambda q,aq/f,aq/\lambda f,eq/f;q)_\infty}\\
&\cdot\sum_{j=0}^{\infty}\sum_{k=0}^{\infty}\frac{(\lambda;q)_j(1-\lambda q^{2j})(\lambda b/a,\lambda c/a,\lambda d/a;q)_j}{(q;q)_j(1-\lambda)(aq/b,aq/c,aq/d;q)_j}\\
&\cdot\frac{(f/q,aq/f;q)_j\,(aq^{j+1}/f,aq^{1-j}/\lambda f,eq/f;q)_k}{(\lambda q^2/f,\lambda f/a;q)_j\,(q,q^{2-j}/f,\lambda q^{2+j}/f;q)_k}\,q^{j+k},
\end{aligned}
$$

where $\lambda = qa^2/bcd$ and $f = ea^2/\lambda^2$. Note that this reduces to (2.8.3) when $e = q^{-n}$, $n = 0, 1, 2, \ldots$.

2.25 By interchanging the order of summation in the double sum in Ex. 2.24 and using Bailey's summation formula (2.11.7), prove Jain and Verma's [1982, (7.1)] transformation formula

$$
\begin{aligned}
&{}_5\phi_4\left[\begin{matrix} a,b,c,d,e \\ aq/b,aq/c,aq/d,f \end{matrix};q,q\right] \\
&\quad + \frac{(a,b,c,d,e,q/f,aq^2/bf,aq^2/cf,aq^2/df;q)_\infty}{(aq/b,aq/c,aq/d,f/q,aq/f,bq/f,cq/f,dq/f,eq/f;q)_\infty} \\
&\quad \cdot {}_5\phi_4\left[\begin{matrix} eq/f,aq/f,bq/f,cq/f,dq/f \\ q^2/f,aq^2/bf,aq^2/cf,aq^2/df \end{matrix};q,q\right] \\
&= \frac{(\lambda q/a,\lambda q/e,q\lambda^2/a,q/f;q)_\infty}{(\lambda q,aq/f,eq/f,aq/\lambda f;q)_\infty} \\
&\quad \cdot {}_{12}W_{11}\left(\lambda;a^{\frac{1}{2}},-a^{\frac{1}{2}},(aq)^{\frac{1}{2}},-(aq)^{\frac{1}{2}},\lambda b/a,\lambda c/a,\lambda d/a,e,aq/f;q,q\right) \\
&\quad + \frac{(a,e,\lambda b/a,\lambda c/a,\lambda d/a,q/f,a^2q^2/\lambda bf,a^2q^2/\lambda cf,a^2q^2/\lambda df,aq^3/f^2;q)_\infty}{(aq/b,aq/c,aq/d,aq/f,bq/f,cq/f,dq/f,eq/f,\lambda f/aq,a^2q^3/\lambda f^2;q)_\infty} \\
&\quad \cdot {}_{12}W_{11}(a^2q^2/\lambda f^2;qa^{\frac{3}{2}}/\lambda f,-qa^{\frac{3}{2}}/\lambda f,(qa)^{\frac{3}{2}}/\lambda f,-(qa)^{\frac{3}{2}}/\lambda f, \\
&\quad \lambda q/a,aq/f,bq/f,cq/f,dq/f;q,q),
\end{aligned}
$$

where the parameters are related in the same way as in Ex. 2.24. Note that this is a nonterminating extension of (2.8.3) and that the first ${}_5\phi_4$ series on the left is a nearly-poised series of the second kind while the second ${}_5\phi_4$ series is a nearly-poised series of the first kind.

2.26 If $a = q^{-n}, n = 0, 1, 2, \ldots$, prove that

$$
\begin{aligned}
{}_3\phi_2\left[\begin{matrix} a,b,c \\ aq/b,aq/c \end{matrix};q,\frac{aqx}{bc}\right] &= \frac{(ax;q)_\infty}{(x;q)_\infty} \\
&\cdot {}_5\phi_4\left[\begin{matrix} a^{\frac{1}{2}},-a^{\frac{1}{2}},(aq)^{\frac{1}{2}},-(aq)^{\frac{1}{2}},aq/bc \\ aq/b,aq/c,ax,q/x \end{matrix};q,q\right].
\end{aligned}
$$

(Sears [1951a, (4.1)], Carlitz [1969a, (2.4)])

2.27 Show that

$$
\begin{aligned}
{}_{r+3}\phi_{r+2}\left[\begin{matrix} q^{-n},c,ab/c,a_1,\ldots,a_r \\ a,b,b_1,\ldots,b_r \end{matrix};q,z\right] &= \frac{(c,ab/c;q)_n}{(a,b;q)_n} \\
\cdot \sum_{k=0}^{n} \frac{(q^{-n},c/a,c/b;q)_k}{(q,c,ca^{-1}b^{-1}q^{1-n};q)_k}\, q^k \, {}_{r+2}\phi_{r+1}&\left[\begin{matrix} q^{k-n},c,a_1,\ldots,a_r \\ cq^k,b_1,\ldots,b_r \end{matrix};q,z\right].
\end{aligned}
$$

2.28 Show that

$$ {}_{r+3}\phi_{r+2}\left[\begin{matrix} a, b_1, ..., b_r, c, q^{-n} \\ aq/b_1, ..., aq/b_r, aq/c, aq^{n+1} \end{matrix}; q, qz\right] = \frac{(aq/cz, q^{1-n}/c; q)_n}{(aq/c, q^{1-n}/cz; q)_n} $$

$$ \sum_{k=0}^{n} \frac{(q^{-n}/cz, 1/z, q^{-n}/a, q^{-n}; q)_k \; (1-q^{2k-n}/cz)}{(q, aq/cz, q/cz, q^{1-n}/c; q)_k \; (1-q^{-n}/cz)} \left(\frac{aq^{n+1}}{c}\right)^k $$

$$ \cdot {}_{r+3}\phi_{r+2}\left[\begin{matrix} a, b_1, ..., b_r, czq^{-k}, q^{k-n} \\ aq/b_1, ..., aq/b_r, aq^{k+1}/cz, aq^{n+1-k} \end{matrix}; q, q\right]. $$

2.29 Show that

$$ {}_{r+2}\phi_{r+1}\left[\begin{matrix} q^{-n}, cdq^{n+1}, a_1, ..., a_r \\ cq, b_1, ..., b_r \end{matrix}; q, z\right] $$

$$ = \sum_{k=0}^{n} \frac{(-1)^k q^{k(k+1)/2} (aq, cdq^{n+1}, q^{-n}; q)_k}{(q, cq, abq^{k+1}; q)_k} $$

$$ \cdot {}_3\phi_2\left[\begin{matrix} q^{k-n}, cdq^{n+k+1}, aq^{k+1} \\ cq^{k+1}, abq^{2k+2} \end{matrix}; q, q\right] {}_{r+2}\phi_{r+1}\left[\begin{matrix} q^{-k}, abq^{k+1}, a_1, ..., a_r \\ aq, b_1, ..., b_r \end{matrix}; q, z\right]. $$

2.30 Iterate (2.12.9) to prove Bailey's [1947b, (8.1)] transformation formula:

$$ {}_{10}W_9(a; b, c, d, e, f, g, h; q, q) $$

$$ + \frac{(aq, c, d, e, f, g, h, b/a, bq/c, bq/d, bq/e; q)_\infty}{(qb^2/a, bc/a, bd/a, be/a, bf/a, bg/a, bh/a, a/b, aq/c, aq/d, aq/e; q)_\infty} $$

$$ \cdot \frac{(bq/f, bq/g, bq/h; q)_\infty}{(aq/f, aq/g, ag/h; q)_\infty} {}_{10}W_9(b^2/a; b, bc/a, bd/a, be/a, bf/a, bg/a, bh/a; q, q) $$

$$ = \frac{(aq, b/a, g, bq/g, aq/ch, aq/dh, aq/eh, aq/fh, bch/a, bdh/a; q)_\infty}{(bhq/g, bh/a, g/h, aq/h, aq/c, aq/d, aq/e, aq/f, bc/a, bd/a; q)_\infty} $$

$$ \cdot \frac{(beh/a, bfh/a; q)_\infty}{(be/a, bf/a; q)_\infty} {}_{10}W_9(bh/g; b, aq/cg, aq/dg, aq/eg, aq/fg, bh/a, h; q, q) $$

$$ + \frac{(aq, b/a, h, bq/h, aq/cg, aq/dg, aq/eg, aq/fg, bcg/a, bdg/a; q)_\infty}{(bgq/h, bg/a, h/g, aq/g, aq/c, aq/d, aq/e, aq/f, bc/a, bd/a; q)_\infty} $$

$$ \cdot \frac{(beg/a, bfg/a; q)_\infty}{(be/a, bf/a; q)_\infty} {}_{10}W_9(bg/h; b, aq/ch, aq/dh, aq/eh, aq/fh, bg/a, g; q, q), $$

where $a^3q^2 = bdefgh$.

2.31 By using the q-Dixon formula (2.7.2) prove that the constant term in the Laurent expansion of

$$ (x_1/x_2, x_1/x_3; q)_{a_1} \; (x_2/x_3, qx_2/x_1; q)_{a_2} \; (qx_3/x_1, qx_3/x_2; q)_{a_3} $$

is

$$ (q; q)_{a_1+a_2+a_3}/(q; q)_{a_1}(q; q)_{a_2}(q; q)_{a_3}. $$

where a_1, a_2 and a_3 are nonnegative integers.
(Andrews [1975a])

2.32 Use (2.10.18), the q-binomial theorem, and the generating function in Ex. 1.29 to derive the formula

$$C_n(\cos\theta;\beta|q) = \frac{2i\sin\theta}{1-q}\frac{(\beta,\beta,\beta e^{2i\theta},\beta e^{-2i\theta};q)_\infty}{(q,\beta^2,e^{2i\theta},e^{-2i\theta};q)_\infty}$$
$$\cdot\frac{(\beta^2;q)_n}{(q;q)_n}\int_{e^{i\theta}}^{e^{-i\theta}}\frac{(qte^{i\theta},qte^{-i\theta};q)_\infty}{(\beta te^{i\theta},\beta te^{-i\theta};q)_\infty}t^n\,d_qt,\quad 0<\theta<\pi.$$

(Rahman and Verma [1986a])

Notes 2

§2.5 Some applications of Watson's transformation formula (2.5.1) to mock theta functions are presented in Watson [1936, 1937].

§2.7 For additional proofs of the Rogers-Ramanujan identities, identities of Rogers-Ramanujan type, applications to combinatorics, Lie algebras, statistical mechanics, etc., see Andrews [1970b, 1974b,c, 1976, 1976b, 1979b, 1981a, 1984b,d, 1986, 1987a,b], Andrews, Askey, Berndt *et al.* [1988], Andrews and Baxter [1986, 1987], Andrews, Baxter, Bressoud *et al.* [1987], Andrews, Baxter, and Forrester [1984], Bailey [1947a, 1949, 1951], Baxter [1980–1988], Baxter and Andrews [1986], Baxter and Pearce [1983, 1984], Berndt [1985–1989], Berndt and Joshi [1983], Borwein and Borwein [1988], Bressoud [1980a, 1981a,b, 1983a], Dobbie [1962], Dyson [1988], Fine [1988], Garsia and Milne [1981], Garvan [1988], Lepowsky [1982], Lepowsky and Wilson [1982], Misra [1988], Paule [1985], Ramanujan [1919], Rogers and Ramanujan [1919], Schur [1917], Slater [1952a], and Watson [1931].

§2.9 Agarwal [1953e] showed that Bailey's transformation (2.9.1) gives a transformation formula for truncated ${}_8\phi_7$ series, where the sum of the first N terms of an infinite series is called a truncated series.

§2.10 Many additional transformation formulas for hypergeometric series are derived in Whipple [1926a,b].

§2.11 Additional transformation formulas for ${}_8\phi_7$ series are derived in Agarwal [1953c].

Ex. 2.6 Also see the summation formulas for very-well-poised series in C.M. Joshi and Verma [1979].

Ex. 2.31 This exercise is the $n = 3$ case of the Zeilberger and Bressoud [1985] theorem that if $x_1,\ldots,x_n,q$ are commuting indeterminates and $a_1,\ldots,a_n$ are nonnegative integers, then the constant term in the Laurent expansion of

$$\prod_{1\le i<j\le n}(x_i/x_j;q)_{a_i}(qx_j/x_i;q)_{a_j}$$

is equal to the q-multinomial coefficient

$$\frac{(q;q)_{a_1+\cdots+a_n}}{(q;q)_{a_1}\cdots(q;q)_{a_n}}.$$

This was called the Andrews' q-Dyson conjecture because Andrews [1975a] had conjectured it as a q-analogue of a previously proved conjecture of Dyson

[1962] that the constant term in the Laurent expansion of

$$\prod_{1\leq i\neq j\leq n}\left(1-\frac{x_i}{x_j}\right)^{a_i}$$

is equal to the multinomial coefficient

$$\frac{(a_1+\cdots+a_n)!}{a_1!\cdots a_n!}.$$

The $n=4$ case of the Andrews' q-Dyson conjecture was proved independently by Kadell [1985b]. Additional constant term results are derived in Bressoud and Goulden [1985], Evans, Ismail and Stanton [1982], Kadell [1989a–c], Macdonald [1972–1989], Morris [1982], Stanton [1986b, 1989], and Zeilberger [1987, 1988, 1989a].

3

ADDITIONAL SUMMATION, TRANSFORMATION, AND EXPANSION FORMULAS

3.1 Introduction

In this chapter we shall use the summation and transformation formulas of Chapters 1 and 2 to deduce additional transformation formulas for basic hypergeometric series which are useful in many applications. In §3.2 and §3.3 we shall obtain q-analogues of some of Thomae's [1879] ${}_3F_2$ transformation formulas, typical among which are

$$ {}_3F_2\left[\begin{matrix} -n, a, b \\ c, d \end{matrix}; q\right] = \frac{(d-b)_n}{(d)_n}\, {}_3F_2\left[\begin{matrix} -n, c-a, b \\ c, 1+b-d-n \end{matrix}; 1\right], \tag{3.1.1} $$

$n = 0, 1, 2, \ldots,$

$$ {}_3F_2\left[\begin{matrix} a, b, c \\ d, e \end{matrix}; 1\right] = \frac{\Gamma(d)\Gamma(e)\Gamma(s)}{\Gamma(a)\Gamma(s+b)\Gamma(s+c)}\, {}_3F_2\left[\begin{matrix} d-a, e-a, s \\ s+b, s+c \end{matrix}; 1\right], \tag{3.1.2} $$

$s = d + e - a - b - c$, and

$$ \begin{aligned} {}_3F_2\left[\begin{matrix} a, b, c \\ d, e \end{matrix}; 1\right] &= \frac{\Gamma(1-a)\Gamma(d)\Gamma(e)\Gamma(c-b)}{\Gamma(d-b)\Gamma(e-b)\Gamma(1+b-a)\Gamma(c)} \\ &\quad \cdot {}_3F_2\left[\begin{matrix} b, b-d+1, b-e+1 \\ 1+b-c, 1+b-a \end{matrix}; 1\right] + \text{idem } (b; c), \end{aligned} \tag{3.1.3} $$

where the symbol "idem $(b; c)$" after an expression means that the preceding expression is repeated with b and c interchanged.

The main topic of this chapter, however, will be the q-analogues of a large class of transformations known as quadratic transformations. Two functions $f(z)$ and $g(w)$ are said to satisfy a *quadratic transformation* if z and w identically satisfy a quadratic equation and $f(z) = g(w)$. Among the important examples of quadratic transformation formulas are

$$ (1+z)^a\, {}_2F_1(a, b; 1+a-b; -z) = {}_2F_1\left(\frac{a}{2}, \frac{a+1}{2} - b; 1+a-b; \frac{4z}{(1+z)^2}\right), \tag{3.1.4} $$

$$ (1-z)^a\, {}_2F_1(a, b; 2b; 2z) = {}_2F_1\left(\frac{a}{2}, \frac{a+1}{2}; b+\frac{1}{2}; \frac{z^2}{(1-z)^2}\right), \tag{3.1.5} $$

$$ (1-z)^a\, {}_2F_1(2a, a+b; 2a+2b; z) = {}_2F_1\left(a, b; a+b+\frac{1}{2}; \frac{z^2}{4(z-1)}\right), \tag{3.1.6} $$

$$
{}_2F_1(2a,2b;a+b+\frac{1}{2};z) = {}_2F_1(a,b;a+b+\frac{1}{2};4z(1-z)), \tag{3.1.7}
$$

$$
\begin{aligned}
&(1-z)^a\ {}_3F_2\left[\begin{matrix} a,b,c \\ 1+a-b,1+a-c \end{matrix};z\right] \\
&\qquad = {}_3F_2\left[\begin{matrix} \frac{1}{2}a,\frac{1}{2}(a+1),1+a-b-c \\ 1+a-b,1+a-c \end{matrix};-\frac{4z}{(1-z)^2}\right],
\end{aligned} \tag{3.1.8}
$$

$$
\begin{aligned}
&\frac{(1-z)^{a+1}}{1+z}\ {}_4F_3\left[\begin{matrix} a,1+\frac{1}{2}a,b,c \\ \frac{1}{2}a,1+a-b,1+a-c \end{matrix};z\right] \\
&\qquad = {}_3F_2\left[\begin{matrix} \frac{1}{2}(a+1),1+\frac{1}{2}a,1+a-b-c \\ 1+a-b,1+a-c \end{matrix};-\frac{4z}{(1-z)^2}\right].
\end{aligned} \tag{3.1.9}
$$

The above definition of a quadratric transformation cannot be directly applied to basic hypergeometric series. For example, the $a=q^{-n}$ case of the identity in Ex. 2.26 is a q-analogue of the $a=-n$ case of (3.1.8), but it does not fit into the above definition of a quadratic transformation. So we shall just say that a transformation between basic hypergeometric series is "*quadratic*" if it is a q-analogue of a quadratic transformation for hypergeometric series.

It will be seen that one important feature of the quadratic transformations derived for basic hypergeometric series in the following sections is that the series obtained from an ${}_r\phi_s(a_1,\ldots,a_r;\ b_1,\ldots,b_s;q,z)$ series by a quadratic transformation will have squares or square roots of at least one of $a_1,\ldots,a_r$, $b_1,\ldots,b_s,q,z$ and possibly a square or square root of q as its base.

3.2 Two-term transformation formulas for ${}_3\phi_2$ series

In general, a convergent ${}_3\phi_2(a,b,c;d,e;q,z)$ series cannot be expressed as a multiple of another ${}_3\phi_2$ or of any other ${}_r\phi_s$ series. It is natural to expect that for such a transformation to exist there has to be some relationship among the parameters and the argument z. Sears [1951a,c] found that in the cases $z=q$ and $z=de/abc$ there is a whole family of transformation formulas for ${}_3\phi_2$ series, analogous to Thomae's [1879] transformation formulas for ${}_3F_2$ series. For the sake of convenience we will say that a basic hypergeometric series ${}_r\phi_s(a_1,\ldots,a_r;b_1,\ldots,b_s;q,z)$ is of *type* I if $z=q$, and of *type* II if z is the product of the denominator parameters divided by the product of the numerator parameters. Note that a series is of both types if it is balanced.

In this section we shall consider transformations between two ${}_3\phi_2$ series. Such formulas may be obtained in a very straightforward manner as special and limiting cases of Sears' identity (2.10.4) which, for our present purposes, is rewritten in the form

$$
\begin{aligned}
&{}_4\phi_3\left[\begin{matrix} q^{-n},a,b,c, \\ d,e,abcq^{1-n}/de \end{matrix};q,q\right] \\
&\qquad = \frac{(e/a,de/bc;q)_n}{(e,de/abc;q)_n}\ {}_4\phi_3\left[\begin{matrix} q^{-n},a,d/b,d/c, \\ d,de/bc,aq^{1-n}/e \end{matrix};q,q\right],
\end{aligned} \tag{3.2.1}
$$

with $n = 0, 1, 2, \ldots$.

Keeping n fixed and choosing special or limiting values of one of the other parameters leads to transformation formulas for terminating ${}_3\phi_2$ series. Let us consider this class of formulas first.

Case (i) Letting $c \to 0$ in (3.2.1) we get

$$ {}_3\phi_2\left[\begin{matrix} q^{-n}, a, b \\ d, e \end{matrix}; q, q\right] = \frac{(e/a;q)_n}{(e;q)_n} a^n \, {}_3\phi_2\left[\begin{matrix} q^{-n}, a, d/b \\ d, aq^{1-n}/e \end{matrix}; q, \frac{bq}{e}\right]. \tag{3.2.2} $$

Note that the series on the left is of type I and that on the right is of type II. Formula (3.2.2) is a q-analogue of (3.1.1).

Case (ii) Letting $a \to 0$ in (3.2.1) gives

$$ {}_3\phi_2\left[\begin{matrix} q^{-n}, b, c \\ d, e \end{matrix}; q, q\right] = \frac{(de/bc;q)_n}{(e;q)_n} \left(\frac{bc}{d}\right)^n {}_3\phi_2\left[\begin{matrix} q^{-n}, d/b, d/c \\ d, de/bc \end{matrix}; q, q\right]. \tag{3.2.3} $$

If we let $c \to 0$ in (3.2.3) we obtain

$$ {}_2\phi_1\left(q^{-n}, d/b; d; q, bq/e\right) = (-1)^n q^{-\binom{n}{2}} (e;q)_n e^{-n} \, {}_3\phi_2\left[\begin{matrix} q^{-n}, b, 0 \\ d, e \end{matrix}; q, q\right], $$

which may be written in the form (Ex. 1.15(i))

$$ {}_2\phi_1(a, b; c; q, z) = \frac{(abz/c;q)_\infty}{(bz/c;q)_\infty} \, {}_3\phi_2\left[\begin{matrix} a, c/b, 0 \\ c, cq/bz \end{matrix}; q, q\right], \tag{3.2.4} $$

where $a = q^{-n}$, $n = 0, 1, 2, \ldots$.

Case (iii) Let $c \to \infty$ in (3.2.1). This gives Sears' [1951c, (4.5)] formula

$$ {}_3\phi_2\left[\begin{matrix} q^{-n}, a, b \\ d, e \end{matrix}; q, \frac{deq^n}{ab}\right] = \frac{(e/a;q)_n}{(e;q)_n} \, {}_3\phi_2\left[\begin{matrix} q^{-n}, a, d/b \\ d, aq^{1-n}/e \end{matrix}; q, q\right]. \tag{3.2.5} $$

Note that there is no essential difference between (3.2.2) and (3.2.5) since one can be obtained from the other by a change of parameters.

Case (iv) Replacing a by aq^n in (3.2.1) and simplifying, we get

$$ \begin{aligned} &{}_4\phi_3\left[\begin{matrix} q^{-n}, aq^n, b, c \\ d, e, abcq/de \end{matrix}; q, q\right] \\ &= \frac{(aq/e, de/bc;q)_n}{(e, abcq/de;q)_n} \left(\frac{bc}{d}\right)^n {}_4\phi_3\left[\begin{matrix} q^{-n}, aq^n, d/b, d/c \\ d, de/bc, aq/e \end{matrix}; q, q\right]. \end{aligned} $$

Set $d = \lambda c$ and then let $c \to \infty$. In the resulting formula we replace λ, e and $abq/\lambda e$ by c, d and e, respectively, to get

$$ \begin{aligned} &{}_3\phi_2\left[\begin{matrix} q^{-n}, aq^n, b \\ d, e \end{matrix}; q, \frac{de}{ab}\right] \\ &= \frac{(aq/d, aq/e;q)_n}{(d, e;q)_n} \left(\frac{de}{aq}\right)^n {}_3\phi_2\left[\begin{matrix} q^{-n}, aq^n, abq/de \\ aq/d, aq/e \end{matrix}; q, \frac{q}{b}\right] \end{aligned} \tag{3.2.6} $$

which is a transformation formula between two terminating ${}_3\phi_2$ series of type II.

Let us now consider the class of transformation formulas that connect two nonterminating ${}_3\phi_2$ series.

Case (v) In (3.2.1) let us take $n \to \infty$. A straightforward term-by-term limiting process gives the formula

$$ {}_3\phi_2\left[\begin{matrix} a, b, c \\ d, e \end{matrix}; q, \frac{de}{abc}\right] = \frac{(e/a, de/bc; q)_\infty}{(e, de/abc; q)_\infty} \, {}_3\phi_2\left[\begin{matrix} a, d/b, d/c \\ d, de/bc \end{matrix}; q, \frac{e}{a}\right]. \tag{3.2.7} $$

Apart from the general requirement that no zero shall appear in the denominators of the two ${}_3\phi_2$ series, the parameters must be restricted by the convergence conditions: $|de/abc| < 1$ and $|e/a| < 1$. This formula is a q-analogue of the Kummer-Thomae-Whipple formula

$$ {}_3F_2\left[\begin{matrix} a, b, c \\ d, e \end{matrix}; 1\right] = \frac{\Gamma(e)\Gamma(d+e-a-b-c)}{\Gamma(e-a)\Gamma(d+e-b-c)} \, {}_3F_2\left[\begin{matrix} a, d-b, d-c \\ d, d+e-b-c \end{matrix}; 1\right], \tag{3.2.8} $$

where Re $(e-a) > 0$ and Re $(d+e-a-b-c) > 0$.

Case (vi) Iterating (3.2.1) once gives

$$ \begin{aligned} &{}_4\phi_3\left[\begin{matrix} q^{-n}, a, b, c \\ d, e, abcq^{1-n}/de \end{matrix}; q, q\right] \\ &= \frac{(b, de/ab, de/bc; q)_n}{(d, e, de/abc; q)_n} \, {}_4\phi_3\left[\begin{matrix} q^{-n}, d/b, e/b, de/abc \\ de/ab, de/bc, q^{1-n}/b \end{matrix}; q, q\right]. \end{aligned} \tag{3.2.9} $$

Let us assume that min $(|b|, |de/abc|) < 1$. Then, taking the limit $n \to \infty$, we obtain Hall's [1936] formula

$$ \begin{aligned} &{}_3\phi_2\left[\begin{matrix} a, b, c \\ d, e \end{matrix}; q, \frac{de}{abc}\right] \\ &= \frac{(b, de/ab, de/bc; q)_\infty}{(d, e, de/abc; q)_\infty} \, {}_3\phi_2\left[\begin{matrix} d/b, e/b, de/abc \\ de/ab, de/bc \end{matrix}; q, b\right]. \end{aligned} \tag{3.2.10} $$

Note that this is a q-analogue of formula (3.1.2).

Before leaving this section it is worth mentioning that by taking the limit $n \to \infty$ in Watson's formula (2.5.1), we get another transformation formula:

$$ \begin{aligned} &{}_3\phi_2\left[\begin{matrix} aq/bc, d, e \\ aq/b, aq/c \end{matrix}; q, \frac{aq}{de}\right] = \frac{(aq/d, aq/e; q)_\infty}{(aq, aq/de; q)_\infty} \\ &\cdot \sum_{k=0}^{\infty} \frac{(a;q)_k (1-aq^{2k})(b, c, d, e; q)_k}{(q;q)_k (1-a)(aq/b, aq/c, aq/d, aq/e; q)_k} q^{\binom{k}{2}} \left(-\frac{a^2q^2}{bcde}\right)^k, \end{aligned} \tag{3.2.11} $$

provided $|aq/de| < 1$. This is a q-analogue of the formula

$$ \begin{aligned} &{}_3F_2\left[\begin{matrix} 1+a-b-c, d, e \\ 1+a-b, 1+a-c \end{matrix}; 1\right] = \frac{\Gamma(1+a)\Gamma(1+a-d-e)}{\Gamma(1+a-d)\Gamma(1+a-e)} \\ &\cdot {}_6F_5\left[\begin{matrix} a, 1+\frac{1}{2}a, b, c, d, e \\ \frac{1}{2}a, 1+a-b, 1+a-c, 1+a-d, 1+a-e \end{matrix}; -1\right], \end{aligned} \tag{3.2.12} $$

where Re $(1+a-d-e)>0$; see Bailey [1935, 4.4(2)].

3.3 Three-term transformation formulas for ${}_3\phi_2$ series

In (2.10.10) let us replace a,b,c,d,e,f by $Aq^N, Bq^N, C, D, E, Fq^N$, respectively, and then let $N\to\infty$. In the resulting formula replace $C, D, E, Aq/B$ and Aq/F by a,b,c,d and e, respectively, to obtain

$$ {}_3\phi_2\left[\begin{matrix} a,b,c, \\ d,e \end{matrix}; q, \frac{de}{abc}\right] = \frac{(e/b, e/c; q)_\infty}{(e, e/bc; q)_\infty}\, {}_3\phi_2\left[\begin{matrix} d/a, b, c \\ d, bcq/e \end{matrix}; q, q\right] $$
$$ + \frac{(d/a, b, c, de/bc; q)_\infty}{(d, e, bc/e, de/abc; q)_\infty}\, {}_3\phi_2\left[\begin{matrix} e/b, e/c, de/abc \\ de/bc, eq/bc \end{matrix}; q, q\right], \tag{3.3.1} $$

where $|de/abc|<1$, and bc/e is not an integer power of q. This expresses a ${}_3\phi_2$ series of type II in terms of a ${}_3\phi_2$ series of type I. As a special case of (3.3.1), let $a=q^{-n}$ with $n=0,1,2,\ldots$. Then

$$ {}_3\phi_2\left[\begin{matrix} q^{-n}, b, c, \\ d, e \end{matrix}; q, \frac{deq^n}{bc}\right] $$
$$ = \frac{(e/b, e/c; q)_\infty}{(e, e/bc; q)_\infty}\, {}_3\phi_2\left[\begin{matrix} b, c, dq^n \\ bcq/e, d \end{matrix}; q, q\right] $$
$$ + \frac{(b, c; q)_\infty}{(e, bc/e; q)_\infty}\frac{(de/bc; q)_n}{(d; q)_n}\, {}_3\phi_2\left[\begin{matrix} e/b, e/c, deq^n/bc \\ eq/bc, de/bc \end{matrix}; q, q\right]. \tag{3.3.2} $$

Setting $n=0$ in (3.3.2) gives the summation formula (2.10.13).

We shall now obtain a transformation formula involving three ${}_3\phi_2$ series of type II. We start by replacing a,b,c,d,e,f in (2.11.1) by $Aq^N, Bq^N, C, D, E, Fq^N$, respectively, and then taking the limit $N\to\infty$. In the resulting formula we replace $C, D, E, Aq/B, Aq/F$ by a,b,c,d,e, respectively, and obtain

$$ {}_3\phi_2\left[\begin{matrix} a, b, c \\ d, e \end{matrix}; q, \frac{de}{abc}\right] $$
$$ = \frac{(e/b, e/c, cq/a, q/d; q)_\infty}{(e, cq/d, q/a, e/bc; q)_\infty}\, {}_3\phi_2\left[\begin{matrix} c, d/a, cq/e \\ cq/a, bcq/e \end{matrix}; q, \frac{bq}{d}\right] $$
$$ - \frac{(q/d, eq/d, b, c, d/a, de/bcq, bcq^2/de; q)_\infty}{(d/q, e, bq/d, cq/d, q/a, e/bc, bcq/e; q)_\infty}\, {}_3\phi_2\left[\begin{matrix} aq/d, bq/d, cq/d \\ q^2/d, eq/d \end{matrix}; q, \frac{de}{abc}\right], \tag{3.3.3} $$

provided $|bq/d|<1$, $|de/abc|<1$ and none of the denominator parameters on either side produces a zero factor. If $|q|<|de/abc|<1$, then

$$ {}_3\phi_2\left[\begin{matrix} a, b, c \\ d, e \end{matrix}; q, \frac{de}{abc}\right] $$
$$ = \frac{(e/b, e/c, q/d, bq/a, cq/a, abcq/de; q)_\infty}{(e, e/bc, q/a, bq/d, cq/d, bcq/e; q)_\infty}\, {}_3\phi_2\left[\begin{matrix} q/a, d/a, e/a \\ bq/a, cq/a \end{matrix}; q, \frac{abcq}{de}\right] $$

$$-\frac{(b,c,q/d,d/a,eq/d,de/bcq,bcq^2/de;q)_\infty}{(e,e/bc,q/a,bq/d,cq/d,bcq/e,d/q;q)_\infty}\,{}_3\phi_2\left[\begin{matrix}aq/d,bq/d,cq/d\\ q^2/d,eq/d\end{matrix};q,\frac{de}{abc}\right], \tag{3.3.4}$$

by observing that from (3.2.7)

$${}_3\phi_2\left[\begin{matrix}c,d/a,cq/e\\ cq/a,bcq/e\end{matrix};q,\frac{bq}{d}\right]$$
$$=\frac{(abcq/de,bq/a;q)_\infty}{(bcq/e,bq/d;q)_\infty}\,{}_3\phi_2\left[\begin{matrix}q/a,d/a,e/a\\ bq/a,cq/a\end{matrix};q,\frac{abcq}{de}\right].$$

If we set $e=\lambda c$ in (3.3.3), let $c\to 0$ and then replace d and λ by c and abz/c, respectively, where $|z|<1, |bq/c|<1$, then we obtain

$${}_2\phi_1(a,b;c;q,z)=\frac{(abz/c,q/c;q)_\infty}{(az/c,q/a;q)_\infty}\,{}_2\phi_1(c/a,cq/abz;cq/az;q,bq/c)$$
$$-\frac{(b,q/c,c/a,az/q,q^2/az;q)_\infty}{(c/q,bq/c,q/a,az/c,cq/az;q)_\infty}\,{}_2\phi_1(aq/c,bq/c;q^2/c;q,z). \tag{3.3.5}$$

Sears' [1951c, p.173] four-term transformation formulas involving ${}_3\phi_2$ series of types I and II can also be derived by a combination of the formulas obtained in this and the previous section. Some of these transformation formulas also arise as special cases of the more general formulas that we shall obtain in the next chapter by using contour integrals.

3.4 Transformation formulas for well-poised ${}_3\phi_2$ and very-well-poised ${}_5\phi_4$ series with arbitrary arguments

Gasper and Rahman [1986] found the following formula connecting a well-poised ${}_3\phi_2$ series with two balanced ${}_5\phi_4$ series:

$${}_3\phi_2\left[\begin{matrix}a,b,c\\ aq/b,aq/c\end{matrix};q,\frac{aqx}{bc}\right]$$
$$=\frac{(ax;q)_\infty}{(x;q)_\infty}\,{}_5\phi_4\left[\begin{matrix}a^{\frac12},-a^{\frac12},(aq)^{\frac12},-(aq)^{\frac12},aq/bc\\ aq/b,aq/c,ax,q/x\end{matrix};q,q\right]$$
$$+\frac{(a,aq/bc,aqx/b,aqx/c;q)_\infty}{(aq/b,aq/c,aqx/bc,x^{-1};q)_\infty}$$
$$\cdot\,{}_5\phi_4\left[\begin{matrix}xa^{\frac12},-xa^{\frac12},x(aq)^{\frac12},-x(aq)^{\frac12},aqx/bc\\ aqx/b,aqx/c,xq,ax^2\end{matrix};q,q\right]. \tag{3.4.1}$$

Convergence of the ${}_3\phi_2$ series on the left requires that $|aqx/bc|<1$. It is also essential to assume that x does not equal $q^{\pm j}, j=0,1,2,...,$ because of the factors $(x;q)_\infty$ and $(x^{-1};q)_\infty$ appearing in the denominators on the right side of (3.4.1). Note that if either a or aq/bc is 1 or a negative integer power of q, then the coefficient of the second ${}_5\phi_4$ series on the right vanishes, so that (3.4.1) reduces to the Sears-Carlitz formula (Ex. 2.26). An important application of (3.4.1) is given in §8.8.

To prove (3.4.1) we replace d by dq^n in (2.8.3) and then let $n \to \infty$. This gives

$$
{}_3\phi_2\left[\begin{matrix} a, b, c \\ aq/b, aq/c \end{matrix}; q, \frac{d}{a}\right] = \frac{(bcd/aq;q)_\infty}{(bcd/qa^2;q)_\infty}
$$
$$
\cdot \lim_{n\to\infty} {}_{12}W_{11}\left(a^2q^{1-n}/bcd;\ a^{\frac{1}{2}},\ -a^{\frac{1}{2}},\ (aq)^{\frac{1}{2}},\ -(aq)^{\frac{1}{2}}, aq^{1-n}/bc,\right.
$$
$$
\left. aq^{1-n}/bd,\ aq^{1-n}/cd,\ a^3q^{3-n}/b^2c^2d^2,\ q^{-n}; q, q\right). \tag{3.4.2}
$$

To take the limit on the right side of (3.4.2) it suffices to proceed as in (2.10.9) to obtain

$$
\lim_{n\to\infty} {}_{12}W_{11}[\ \ ]
$$
$$
= {}_5\phi_4\left[\begin{matrix} a^{\frac{1}{2}}, -a^{\frac{1}{2}}, (aq)^{\frac{1}{2}}, -(aq)^{\frac{1}{2}}, aq/bc \\ aq/b, aq/c, bcd/aq, a^2q^2/bcd \end{matrix}; q, q\right]
$$
$$
- \frac{bcd}{qa^2}\frac{(bcd/a^2, bd/a, cd/a, aq/bc, a;q)_\infty}{(d/a, aq/b, aq/c, bcd/a, a^2q^2/bcd;q)_\infty}
$$
$$
\cdot {}_5\phi_4\left[\begin{matrix} d/a, bcd/qa^{\frac{3}{2}}, -bcd/qa^{\frac{3}{2}}, bcd/q^{\frac{1}{2}}a^{\frac{3}{2}}, -bcd/q^{\frac{1}{2}}a^{\frac{3}{2}} \\ bd/a, cd/a, bcd/a^2, b^2c^2d^2/qa^2 \end{matrix}; q, q\right]. \tag{3.4.3}
$$

Using this in (3.4.2) and replacing d by qxa^2/bc, we get (3.4.1).

If we now replace d by dq^n in (2.8.5) and then let $n \to \infty$, we obtain the transformation formula

$$
{}_5\phi_4\left[\begin{matrix} a, qa^{\frac{1}{2}}, -qa^{\frac{1}{2}}, b, c \\ a^{\frac{1}{2}}, -a^{\frac{1}{2}}, aq/b, aq/c \end{matrix}; q, \frac{x(aq)^{\frac{1}{2}}}{bc}\right]
$$
$$
= \frac{(1-x^2)(xq(aq)^{\frac{1}{2}};q)_\infty}{(x(aq)^{-\frac{1}{2}};q)_\infty}
$$
$$
\cdot {}_5\phi_4\left[\begin{matrix} (aq)^{\frac{1}{2}}, -(aq)^{\frac{1}{2}}, qa^{\frac{1}{2}}, -qa^{\frac{1}{2}}, aq/bc \\ aq/b, aq/c, xq(aq)^{\frac{1}{2}}, q(aq)^{\frac{1}{2}}/x \end{matrix}; q, q\right]
$$
$$
+ \frac{(aq, aq/bc, x(aq)^{\frac{1}{2}}/b, x(aq)^{\frac{1}{2}}/c;q)_\infty}{(aq/b, aq/c, x(aq)^{\frac{1}{2}}/bc, (aq)^{\frac{1}{2}}/x;q)_\infty}
$$
$$
\cdot {}_5\phi_4\left[\begin{matrix} x, -x, xq^{\frac{1}{2}}, -xq^{\frac{1}{2}}, x(aq)^{\frac{1}{2}}/bc \\ x(aq)^{\frac{1}{2}}/b, x(aq)^{\frac{1}{2}}/c, x(q/a)^{\frac{1}{2}}, qx^2 \end{matrix}; q, q\right]. \tag{3.4.4}
$$

In terms of q-integrals formulas (3.4.1) and (3.4.4) are equivalent to

$$
{}_3\phi_2\left[\begin{matrix} a, b, c \\ aq/b, aq/c \end{matrix}; q, \frac{aqx}{bc}\right]
$$
$$
= \frac{(a, aq/bc;q)_\infty}{s(1-q)(q, aq/b, aq/c, q/x, x;q)_\infty}
$$

$$\cdot \int_{sx}^{s} \frac{(qu/xs, qu/s, aqu/bs, aqu/cs, axu/s; q)_\infty}{(ua^{\frac{1}{2}}/s, -ua^{\frac{1}{2}}/s, u(aq)^{\frac{1}{2}}/s, -u(aq)^{\frac{1}{2}}/s, aqu/bcs; q)_\infty} d_q u, \tag{3.4.5}$$

and

$$ {}_5\phi_4 \left[\begin{matrix} a, qa^{\frac{1}{2}}, -qa^{\frac{1}{2}}, b, c \\ a^{\frac{1}{2}}, -a^{\frac{1}{2}}, aq/b, aq/c \end{matrix} ; q, \frac{x(aq)^{\frac{1}{2}}}{bc} \right]$$

$$= \frac{(1-x^2)(aq, aq/bc; q)_\infty}{s(1-q)(q, aq/b, aq/c, x(aq)^{-\frac{1}{2}}, q(aq)^{\frac{1}{2}}/x; q)_\infty}$$

$$\cdot \int_{sx(aq)^{-\frac{1}{2}}}^{s} \frac{(uq(aq)^{\frac{1}{2}}/sx, qu/s, aqu/bs, aqu/cs, uxq(aq)^{\frac{1}{2}}/s; q)_\infty}{(u(aq)^{\frac{1}{2}}/s, -u(aq)^{\frac{1}{2}}/s, uqa^{\frac{1}{2}}/s, -uqa^{\frac{1}{2}}/s, aqu/bsc; q)_\infty} d_q u, \tag{3.4.6}$$

respectively, where $s \neq 0$ is an arbitrary parameter.

If we now set $c = (aq)^{\frac{1}{2}}$ in (3.4.5), replace x by $x/b(aq)^{\frac{1}{2}}$, and use (2.10.19), then we get

$$ {}_2\phi_1(a, b; aq/b; q, qx/b^2) = \frac{(xq/b, aqx^2/b^2; q)_\infty}{(aqx/b, qx^2/b^2; q)_\infty}$$

$$\cdot {}_8\phi_7 \left[\begin{matrix} ax/b, q(ax/b)^{\frac{1}{2}}, -q(ax/b)^{\frac{1}{2}}, x, a^{\frac{1}{2}}, -a^{\frac{1}{2}}, (aq)^{\frac{1}{2}}, -(aq)^{\frac{1}{2}} \\ (ax/b)^{\frac{1}{2}}, -(ax/b)^{\frac{1}{2}}, aq/b, xqa^{\frac{1}{2}}/b, -xqa^{\frac{1}{2}}/b, x(aq)^{\frac{1}{2}}/b, -x(aq)^{\frac{1}{2}}/b \end{matrix} ; q, \frac{qx}{b^2} \right] \tag{3.4.7}$$

provided $|qx/b^2| < 1$ when the two series do not terminate.

Similarly, setting $c = (aq)^{\frac{1}{2}}$ and replacing x by x/aq in (3.4.6) we obtain

$$ {}_4\phi_3 \left[\begin{matrix} a, qa^{\frac{1}{2}}, -qa^{\frac{1}{2}}, b \\ a^{\frac{1}{2}}, -a^{\frac{1}{2}}, aq/b \end{matrix} ; q, \frac{x}{qb^2} \right] = \frac{(ax^2/b^2, x/qb; q)_\infty}{(aqx/b, x^2/qb^2; q)_\infty}$$

$$\cdot {}_8\phi_7 \left[\begin{matrix} ax/b, q(ax/b)^{\frac{1}{2}}, -q(ax/b)^{\frac{1}{2}}, (aq)^{\frac{1}{2}}, -(aq)^{\frac{1}{2}}, qa^{\frac{1}{2}}, -qa^{\frac{1}{2}}, x \\ (ax/b)^{\frac{1}{2}}, -(ax/b)^{\frac{1}{2}}, x(aq)^{\frac{1}{2}}/b, -x(aq)^{\frac{1}{2}}/b, xa^{\frac{1}{2}}/b, -xa^{\frac{1}{2}}/b, aq/b \end{matrix} ; q, \frac{x}{qb^2} \right], \tag{3.4.8}$$

provided $|x/qb^2| < 1$ when the series do not terminate.

3.5 Transformations of series with base q^2 to series with base q

If in Sears' summation formula (2.10.12) we set $b = -c,\ e = -q$, replace a by $aq^r, r = 0, 1, 2, \ldots$, multiply both sides by

$$\frac{(x^2, y^2; q^2)_r}{(-q;q)_r (x^2y^2b^2;q^2)_r} b^{2r} q^r$$

and then sum over r from 0 to ∞, we get

$$\frac{(-1,-q,ab,-ab,b^2;q)_\infty}{(a,b,-a,-b,b,-b;q)_\infty}\ {}_3\phi_2\left[\begin{matrix}a^2, x^2, y^2\\ a^2b^2, x^2y^2b^2\end{matrix}; q^2, qb^2\right]$$

$$= \frac{(-q,ab^2;q)_\infty}{(a,b,-b;q)_\infty}\sum_{j=0}^{\infty}\frac{(a,b,-b;q)_j}{(q,-q,ab^2;q)_j} q^j\ {}_3\phi_2\left[\begin{matrix}q^{-2j}, x^2, y^2\\ x^2y^2b^2, q^{2-2j}/b^2\end{matrix}; q^2, q^2\right]$$

$$+ \frac{(-q,-ab^2;q)_\infty}{(-a,-b,b;q)_\infty}\sum_{j=0}^{\infty}\frac{(-a,-b,b;q)_j}{(q,-q,-ab^2;q)_j} q^j\ {}_3\phi_2\left[\begin{matrix}q^{-2j}, x^2, y^2\\ x^2y^2b^2, q^{2-2j}/b^2\end{matrix}; q^2, q^2\right] \tag{3.5.1}$$

assuming that $|qb^2| < 1$ when the series on the left is nonterminating.

Since the two ${}_3\phi_2$ series on the right side can be summed by the q-Saalschütz formula (1.7.2) with the base q replaced by q^2, it follows from (3.5.1) that

$$\frac{(a^2b^2;q^2)_\infty}{(b^2;q^2)_\infty}\ {}_3\phi_2\left[\begin{matrix}a^2, x^2, y^2\\ a^2b^2, x^2y^2b^2\end{matrix}; q^2, qb^2\right]$$

$$= \frac{(-a,ab^2;q)_\infty}{(-1,b^2;q)_\infty}\ {}_5\phi_4\left[\begin{matrix}a, bx, -bx, by, -by\\ -q, ab^2, bxy, -bxy\end{matrix}; q, q\right]$$

$$+ \frac{(a,-ab^2;q)_\infty}{(-1,b^2;q)_\infty}\ {}_5\phi_4\left[\begin{matrix}-a, -bx, bx, -by, by\\ -q, -ab^2, -bxy, bxy\end{matrix}; q, q\right]. \tag{3.5.2}$$

Note that one of the terms on the right side of (3.5.2) drops out when $a = \pm q^{-n}, n = 0, 1, 2, \ldots$. Setting $y = ab$ and using (2.10.10) gives

$${}_2\phi_1(a^2, x^2; a^2x^2b^4; q^2, qb^2) = \frac{(b^2, a^2b^2x^2;q^2)_\infty (ab^2, b^2x^2;q)_\infty}{(a^2b^2, b^2x^2;q^2)_\infty (b^2, ab^2x^2;q)_\infty}$$

$$\cdot\ {}_8W_7(ab^2x^2/q; a, x, -x, bx, -bx; q, ab^2), \tag{3.5.3}$$

where $|qb^2| < 1$ and $|ab^2| < 1$ when the series do not terminate. By applying Heine's transformation formula (1.4.1) twice to the ${}_2\phi_1$ series above and replacing b by $q^{\frac{1}{2}}/b$ we find that

$${}_2\phi_1(a^2, b^2; a^2q^2/b^2; q^2, x^2q^2/b^4)$$

$$= \frac{(qa^2x^2/b^2, q^2a^2x^2/b^4; q^2)_\infty (aq/b^2, qx^2/b^2; q)_\infty}{(qx^2/b^2, q^2x^2/b^4; q^2)_\infty (qa^2/b^2, aqx^2/b^2; q)_\infty}$$

$$\cdot\, {}_8\phi_7\left[\begin{matrix} \frac{ax^2}{b^2}, q\left(\frac{ax^2}{b^2}\right)^{\frac12}, -q\left(\frac{ax^2}{b^2}\right)^{\frac12}, a, x, -x, \frac{xq^{\frac12}}{b}, \frac{-xq^{\frac12}}{b} \\ \left(\frac{ax^2}{b^2}\right)^{\frac12}, -\left(\frac{ax^2}{b^2}\right)^{\frac12}, \frac{qx^2}{b^2}, \frac{aqx}{b^2}, \frac{-aqx}{b^2}, \frac{axq^{\frac12}}{b}, \frac{-axq^{\frac12}}{b} \end{matrix}; q, \frac{aq}{b^2}\right], \tag{3.5.4}$$

provided $|aq/b^2| < 1$ and $|xq/b^2| < 1$ when the series do not terminate. This formula was derived by Gasper and Rahman [1986], and a terminating version of it was given earlier by Verma [1980]. Application of the transformation formula (2.10.1) to the ${}_8\phi_7$ series on the right of (3.5.4) yields an equivalent formula

$$\begin{aligned} &{}_2\phi_1(a^2, b^2; a^2q^2/b^2; q^2; x^2q^2/b^4) \\ &= \frac{(qa^2x^2/b^2, q^2a^2x^2/b^4; q^2)_\infty (-xq^{\frac12}/b, -axq^{\frac32}/b^3; q)_\infty}{(qx^2/b^2, q^2x^2/b^4; q^2)_\infty (-axq^{\frac12}/b, -a^2xq^{\frac32}/b^3; q)_\infty} \\ &\cdot\, {}_8\phi_7\left[\begin{matrix} -xa^2q^{\frac12}/b^3, & q(-xa^2q^{\frac12}/b^3)^{\frac12}, & -q(-xa^2q^{\frac12}/b^3)^{\frac12}, & a, \\ & (-xa^2q^{\frac12}/b^3)^{\frac12}, & -(-xa^2q^{\frac12}/b^3)^{\frac12}, & -axq^{\frac32}/b^3, \end{matrix}\right. \\ &\qquad \left.\begin{matrix} aq^{\frac12}/b, & -aq^{\frac12}/b, & -aq/b^2, & -xq^{\frac12}/b \\ -aqx/b^2, & aqx/b^2, & axq^{\frac12}/b, & qa^2/b^2 \end{matrix}; q, -\frac{xq^{\frac12}}{b}\right], \end{aligned} \tag{3.5.5}$$

where $|xq/b^2| < 1$ and $|xq^{\frac12}/b| < 1$ when the series do not terminate. It is clear that formula (3.5.5) is a q-analogue of the quadratic transformation formula

$${}_2F_1(a, b; 1 + a - b; x^2) = (1 - x)^{-2a}\, {}_2F_1(a, a + \frac12 - b; 2a + 1 - 2b; \frac{-4x}{(1 - x)^2}). \tag{3.5.6}$$

We shall now prove the following transformation formula due to Jain and Verma [1982]:

$$\begin{aligned} &{}_{10}\phi_9\left[\begin{matrix} a, q^2a^{\frac12}, -q^2a^{\frac12}, b, c, cq, d, dq, e, eq \\ a^{\frac12}, -a^{\frac12}, aq^2/b, aq^2/c, aq/c, aq^2/d, aq/d, aq^2/e, aq/e \end{matrix}; q^2, \frac{a^3q^3}{bc^2d^2e^2}\right] \\ &= \frac{(aq, aq/cd, aq/ce, aq/de; q)_\infty}{(aq/c, aq/d, aq/e, aq/cde; q)_\infty}\, {}_5\phi_4\left[\begin{matrix} (aq/b)^{\frac12}, -(aq/b)^{\frac12}, c, d, e \\ (aq)^{\frac12}, -(aq)^{\frac12}, aq/b, cde/a \end{matrix}; q, q\right] \\ &+ \frac{(aq^2, a^3q^3/c^2d^2e^2; q^2)_\infty (c, d, e, a^2q^2/bcde; q)_\infty}{(aq^2/b, a^3q^3/bc^2d^2e^2; q^2)_\infty (aq/c, aq/d, aq/e, cde/aq; q)_\infty} \\ &\cdot\, {}_5\phi_4\left[\begin{matrix} (a^3q^3/bc^2d^2e^2)^{\frac12}, -(a^3q^3/bc^2d^2e^2)^{\frac12}, aq/cd, aq/ce, aq/de \\ (a^3q^3/c^2d^2e^2)^{\frac12}, -(a^3q^3/c^2d^2e^2)^{\frac12}, a^2q^2/bcde, aq^2/cde \end{matrix}; q, q\right], \end{aligned} \tag{3.5.7}$$

with the usual understanding that if the ${}_{10}\phi_9$ series on the left does not terminate then the convergence condition $|a^3q^3/bc^2d^2e^2| < 1$ must be assumed to hold.

First we rewrite (2.10.12) in the form

$$\frac{(aq^{4n+1}, aq/cd, aq/ce, aq/de; q)_\infty}{(cq^{2n}, dq^{2n}, eq^{2n}, aq^{1-2n}/cde; q)_\infty} \sum_{r=0}^{\infty} \frac{(cq^{2n}, dq^{2n}, eq^{2n}; q)_r}{(q, aq^{4n+1}, cdeq^{2n}/a; q)_r} q^r$$

$$+ \frac{(a^2q^{2n+2}/cde; q)_\infty}{(cdeq^{2n-1}/a; q)_\infty} \sum_{r=0}^{\infty} \frac{(aq/cd, aq/ce, aq/de; q)_r}{(q, a^2q^{2n+2}/cde, aq^{2-2n}/cde; q)_r} q^r$$

$$= \frac{(aq/c, aq/d, aq/e; q)_\infty}{(c, d, e; q)_\infty} \frac{(c, d, e; q)_{2n}}{(aq/c, aq/d, aq/e; q)_{2n}}, \tag{3.5.8}$$

where n is a nonnegative integer. Using (1.2.39) and (1.2.40), multiplying both sides of (3.5.8) by

$$\frac{(a, b; q^2)_n(1 - aq^{4n})}{(q^2, aq^2/b; q^2)_n(1 - a)} \left(\frac{a^3q^3}{bc^2d^2e^2}\right)^n,$$

and summing over n from 0 to ∞, we get

$$
\begin{aligned}
&{}_{10}W_9(a; b, c, cq, d, dq, e, eq; q^2, a^3q^3/bc^2d^2e^2) \\
&= \frac{(aq, aq/cd, aq/ce, aq/de; q)_\infty}{(aq/c, aq/d, aq/e, aq/cde; q)_\infty} \sum_{n=0}^{\infty} \frac{(a, b; q^2)_n(1 - aq^{4n})}{(q^2, aq^2/b; q^2)_n(1 - a)} \\
&\quad \cdot \frac{(c, d, e; q)_{2n}}{(cde/a; q)_{2n}(aq; q)_{4n}} q^{n(2n-1)} \left(\frac{aq^3}{b}\right)^n {}_3\phi_2\left[\begin{matrix} cq^{2n}, dq^{2n}, eq^{2n} \\ aq^{4n+1}, cdeq^{2n}/a \end{matrix}; q, q\right] \\
&\quad + \frac{(c, d, e, a^2q^2/cde; q)_\infty}{(aq/c, aq/d, aq/e, cde/aq; q)_\infty} \sum_{n=0}^{\infty} \frac{(a, b; q^2)_n(1 - aq^{4n})}{(q^2, aq^2/b; q^2)_n(1 - a)} \\
&\quad \cdot \frac{(cde/aq; q)_{2n}}{(a^2q^2/cde; q)_{2n}} \left(\frac{a^3q^3}{bc^2d^2e^2}\right)^n {}_3\phi_2\left[\begin{matrix} aq/cd, aq/ce, aq/de \\ a^2q^{2n+2}/cde, aq^{2-2n}/cde \end{matrix}; q, q\right].
\end{aligned} \tag{3.5.9}
$$

The first double series on the right side of (3.5.9) easily transforms to

$$\sum_{m=0}^{\infty} \frac{(c, d, e; q)_m}{(q, aq, cde/a; q)_m} q^m \, {}_6W_5\left(a; b, q^{1-m}, q^{-m}; q^2, aq^{2m+1}/b\right),$$

which, by (2.4.2), equals

$${}_5\phi_4\left[\begin{matrix} c, d, e, (aq/b)^{\frac{1}{2}}, -(aq/b)^{\frac{1}{2}} \\ aq/b, cde/a, (aq)^{\frac{1}{2}}, -(aq)^{\frac{1}{2}} \end{matrix}; q, q\right].$$

Similarly we can express the second double series on the right side of (3.5.9) as a single balanced ${}_5\phi_4$ series. Combining the two we get (3.5.7).

The special case of (3.5.7) that results from setting $e = (aq)^{\frac{1}{2}}$ is particularly interesting because both ${}_5\phi_4$ series on the right side become balanced ${}_4\phi_3$

series which, via (2.10.10), combine into a single ${}_8\phi_7$ series with base q. Thus we have the formula

$$
{}_8\phi_7\left[\begin{matrix} a, q^2a^{\frac{1}{2}}, -q^2a^{\frac{1}{2}}, b, c, cq, d, dq \\ a^{\frac{1}{2}}, -a^{\frac{1}{2}}, aq^2/b, aq^2/c, aq/c, aq^2/d, aq/d \end{matrix}; q^2, \frac{a^2q^2}{bc^2d^2}\right]
$$

$$
= \frac{(aq, aq/bc, aq/cd, -aq/cd, aq/db^{\frac{1}{2}}, -aq/db^{\frac{1}{2}}; q)_\infty}{(aq/b, aq/c, aq/d, -aq/d, aq/cdb^{\frac{1}{2}}, -aq/cdb^{\frac{1}{2}}; q)_\infty}
$$

$$
\cdot {}_8\phi_7\left[\begin{matrix} -a/d, q(-a/d)^{\frac{1}{2}}, -q(-a/d)^{\frac{1}{2}}, c, b^{\frac{1}{2}}, -b^{\frac{1}{2}}, (aq)^{\frac{1}{2}}/d, -(aq)^{\frac{1}{2}}/d \\ (-a/d)^{\frac{1}{2}}, -(-a/d)^{\frac{1}{2}}, -aq/cd, -aq/db^{\frac{1}{2}}, aq/db^{\frac{1}{2}}, -(aq)^{\frac{1}{2}}, (aq)^{\frac{1}{2}} \end{matrix}; q, \frac{aq}{bc}\right], \tag{3.5.10}
$$

where $|a^2q^2/bc^2d^2| < 1$ and $|aq/bc| < 1$ when the series do not terminate.

3.6 Bibasic summation formulas

Our main objective in this section is to derive summation formulas containing two independent bases. Let us start by observing that when $d = a/bc$ Jackson's ${}_8\phi_7$ summation formula (2.6.2) reduces to the following sum of a truncated series

$$
\sum_{k=0}^{n} \frac{1-aq^{2k}}{1-a} \frac{(a, b, c, a/bc; q)_k}{(q, aq/b, aq/c, bcq; q)_k} q^k = \frac{(aq, bq, cq, aq/bc; q)_n}{(q, aq/b, aq/c, bcq; q)_n}, \tag{3.6.1}
$$

where $n = 0, 1, \ldots$. Notice that this series telescopes, for if we set $\sigma_{-1} = 0$ and

$$
\sigma_k = \frac{(aq, bq, cq, aq/bc; q)_k}{(q, aq/b, aq/c, bcq; q)_k} \tag{3.6.2}
$$

for $k = 0, 1, \ldots$, and apply the difference operator Δ defined by $\Delta u_k = u_k - u_{k-1}$ to σ_k, then we get

$$
\Delta\sigma_k = \frac{(1-aq^{2k})(a, b, c, a/bc; q)_k}{(1-a)(q, aq/b, aq/c, bcq; q)_k} q^k, \tag{3.6.3}
$$

which gives (3.6.1), since

$$
\sum_{k=0}^{n} \Delta u_k = u_n - u_{-1} \tag{3.6.4}
$$

for any sequence $\{u_k\}$.

These observations and the bibasic extension

$$
\tau_k = \frac{(ap, bp; p)_k (cq, aq/bc; q)_k}{(q, aq/b; q)_k (ap/c, bcp; p)_k} \tag{3.6.5}
$$

of σ_k were used in Gasper [1989a] to show that

$$
\Delta\tau_k = \frac{(1-ap^kq^k)(1-bp^kq^{-k})}{(1-a)(1-b)} \frac{(a, b; p)_k (c, a/bc; q)_k}{(q, aq/b; q)_k (ap/c, bcp; p)_k} q^k, \tag{3.6.6}
$$

which, by (3.6.4), gave the indefinite bibasic summation formula

$$\sum_{k=0}^{n} \frac{(1-ap^kq^k)(1-bp^kq^{-k})}{(1-a)(1-b)} \frac{(a,b;p)_k(c,a/bc;q)_k}{(q,aq/b;q)_k(ap/c,bcp;p)_k} q^k$$
$$= \frac{(ap,bp;p)_n(cq,aq/bc;q)_n}{(q,aq/b;q)_n(ap/c,bcp;p)_n} \tag{3.6.7}$$

for $n = 0, 1, \ldots$. Notice that the part of the series on the left side of (3.6.7) containing the q-shifted factorials is split-poised in the sense that $aq = b(aq/b)$ and $c(ap/c) = (a/bc)(bcp) = ap$, while the expression on the right side is balanced and well-poised since

$$(ap)(bp)(cq)(aq/bc) = q(aq/b)(ap/c)(bcp)$$

and

$$(ap)q = (bp)(aq/b) = (cq)(ap/c) = (aq/bc)(bcp).$$

The $b \to 0$ case of (3.6.7)

$$\sum_{k=0}^{n} \frac{1-ap^kq^k}{1-a} \frac{(a;p)_k(c;q)_k}{(q;q)_k(ap/c;p)_k} c^{-k} = \frac{(ap;p)_n(cq;q)_n}{(q;q)_n(ap/c;p)_n} c^{-n} \tag{3.6.8}$$

is due to Gosper.

To derive a useful extension of (3.6.7), Gasper and Rahman [1989a] set

$$s_k = \frac{(ap,bp;p)_k(cq,ad^2q/bc;q)_k}{(dq,adq/b;q)_k(adp/c,bcp/d;p)_k} \tag{3.6.9}$$

for $k = 0, \pm 1, \pm 2, \ldots$, and observed that

$$\begin{aligned} \Delta s_k &= s_k - s_{k-1} \\ &= \frac{(ap,bp;p)_{k-1}(cq,ad^2q/bc;q)_{k-1}}{(dq,adq/b;q)_k(adp/c,bcp/d;p)_k} \\ &\quad \cdot \{(1-ap^k)(1-bp^k)(1-cq^k)(1-ad^2q^k/bc) \\ &\quad -(1-dq^k)(1-adq^k/b)(1-adp^k/c)(1-bcp^k/d)\} \\ &= \frac{d(1-c/d)(1-ad/bc)(1-adp^kq^k)(1-bp^k/dq^k)}{(1-a)(1-b)(1-c)(1-ad^2/bc)} \\ &\quad \cdot \frac{(a,b;p)_k(c,ad^2/bc;q)_kq^k}{(dq,adq/b;q)_k(adp/c,bcp/d;p)_k}. \end{aligned} \tag{3.6.10}$$

Since (3.6.4) extends to

$$\sum_{k=-m}^{n} \Delta u_k = u_n - u_{-m-1}, \tag{3.6.11}$$

where we employed the standard convention of defining

$$\sum_{k=m}^{n} a_k = \begin{cases} a_m + a_{m+1} + \cdots + a_n, & m \le n, \\ 0, & m = n+1, \\ -(a_{n+1} + a_{n+2} + \cdots + a_{m-1}), & m \ge n+2, \end{cases} \tag{3.6.12}$$

for $n, m = 0, \pm 1, \pm 2, \ldots$, it follows from (3.6.10) that (3.6.7) extends to the indefinite bibasic summation formula

$$\sum_{k=-m}^{n} \frac{(1-adp^k q^k)(1-bp^k/dq^k)}{(1-ad)(1-b/d)} \frac{(a,b;p)_k (c,ad^2/bc;q)_k}{(dq,adq/b;q)_k (adp/c,bcp/d;p)_k} q^k$$
$$= \frac{(1-a)(1-b)(1-c)(1-ad^2/bc)}{d(1-ad)(1-b/d)(1-c/d)(1-ad/bc)}$$
$$\cdot \left\{ \frac{(ap,bp;p)_n (cq,ad^2q/bc;q)_n}{(dq,adq/b;q)_n (adp/c,bcp/d;p)_n} - \frac{(c/ad,d/bc;p)_{m+1} (1/d,b/ad;q)_{m+1}}{(1/c,bc/ad^2;q)_{m+1} (1/a,1/b;p)_{m+1}} \right\} \tag{3.6.13}$$

for $n, m = 0, \pm 1, \pm 2, \ldots$, by applying the identity (1.2.28). Observe that (3.6.7) is the case $d = 1$ of (3.6.13) and that the right side of (3.6.9) is balanced and well-poised since

$$(ap)(bp)(cq)(ad^2q/bc) = (dq)(adq/b)(adp/c)(bcp/d)$$

and

$$(ap)(dq) = (bp)(adq/b) = (cq)(adp/c) = (ad^2q/bc)(bcp/d).$$

It is these observations and the factorization that occurred in (3.6.10) which motivated the choice of s_k in (3.6.9).

If $|p| < 1$ and $|q| < 1$, then by letting n or m tend to infinity in (3.6.13) we find that (3.6.13) also holds with n or m replaced by ∞. In particular, this yields the following evaluation of a bilateral bibasic series

$$\sum_{k=-\infty}^{\infty} \frac{(1-adp^k q^k)(1-bp^k/dq^k)}{(1-ad)(1-b/d)} \frac{(a,b;p)_k (c,ad^2/bc;q)_k}{(dq,adq/b;q)_k (adp/c,bcp/d;p)_k} q^k$$
$$= \frac{(1-a)(1-b)(1-c)(1-ad^2/bc)}{d(1-ad)(1-b/d)(1-c/d)(1-ad/bc)}$$
$$\cdot \left\{ \frac{(ap,bp;p)_\infty (cq,ad^2q/bc;q)_\infty}{(dq,adq/b;q)_\infty (adp/c,bcp/d;p)_\infty} - \frac{(c/ad,d/bc;p)_\infty (1/d,b/ad;q)_\infty}{(1/c,bc/ad^2;q)_\infty (1/a,1/b;p)_\infty} \right\}, \tag{3.6.14}$$

where $|p| < 1$ and $|q| < 1$.

In §3.8 we shall use the $m = 0$ case of (3.6.13) in the form

$$\sum_{k=0}^{n} \frac{(1-adp^k q^k)(1-bp^k/dq^k)}{(1-ad)(1-b/d)} \frac{(a,b;p)_k (c,ad^2/bc;q)_k}{(dq,adq/b;q)_k (adp/c,bcp/d;p)_k} q^k$$
$$= \frac{(1-a)(1-b)(1-c)(1-ad^2/bc)}{d(1-ad)(1-b/d)(1-c/d)(1-ad/bc)}$$
$$\cdot \frac{(ap,bp;p)_n (cq,ad^2q/bc;q)_n}{(dq,adq/b;q)_n (adp/c,bcp/d;p)_n}$$
$$- \frac{(1-d)(1-ad/b)(1-ad/c)(1-bc/d)}{d(1-ad)(1-b/d)(1-c/d)(1-ad/bc)}. \tag{3.6.15}$$

There is no loss in generality since, by setting $k = j - m$ in (3.6.13), it is seen that (3.6.13) is equivalent to (3.6.15) with n, a, b, c, d replaced by $n +$

$m, ap^{-m}, bp^{-m}, cq^{-m}, dq^{-m}$, respectively. We shall also use the special case $c = q^{-n}$ of (3.6.15) in the form

$$\sum_{k=0}^{n} \frac{(1-adp^kq^k)(1-bp^k/dq^k)}{(1-ad)(1-b/d)} \frac{(a,b;p)_k(q^{-n},ad^2q^n/b;q)_k}{(dq,adq/b;q)_k(adpq^n,bp/dq^n;p)_k} q^k$$
$$= \frac{(1-d)(1-ad/b)(1-adq^n)(1-dq^n/b)}{(1-ad)(1-d/b)(1-dq^n)(1-adq^n/b)}, \tag{3.6.16}$$

where $n = 0, 1, \ldots$. The $d \to 1$ limit case of (3.6.16)

$$\sum_{k=0}^{n} \frac{(1-ap^kq^k)(1-bp^kq^{-k})}{(1-a)(1-b)} \frac{(a,b;p)_k(q^{-n},aq^n/b;q)_k}{(q,aq/b;q)_k(apq^n,bpq^{-n};p)_k} q^k = \delta_{n,0}, \tag{3.6.17}$$

where $\delta_{n,m}$ is the Kronecker delta function and $n = 0, 1, \ldots$, was derived independently by Bressoud [1988], Gasper [1989a], and Krattenthaler [1989b].

If we replace n, a, b and k in (3.6.17) by $n-m, ap^mq^m, bp^mq^{-m}$ and $j-m$, respectively, we obtain the orthogonality relation

$$\sum_{j=m}^{n} a_{nj} b_{jm} = \delta_{n,m} \tag{3.6.18}$$

with

$$a_{nj} = \frac{(-1)^{n+j}(1-ap^jq^j)(1-bp^jq^{-j})(apq^n,bpq^{-n};q)_{n-1}}{(q;q)_{n-j}(apq^n,bpq^{-n};p)_j(bq^{1-2n}/a;q)_{n-j}}, \tag{3.6.19}$$

$$b_{jm} = \frac{(ap^mq^m,bp^mq^{-m};p)_{j-m}}{(q,aq^{1+2m}/b;q)_{j-m}} \left(-\frac{a}{b}q^{1+2m}\right)^{j-m} q^{2\binom{j-m}{2}}. \tag{3.6.20}$$

This shows that the triangular matrix $A = (a_{nj})$ is inverse to the triangular matrix $B = (b_{jm})$. Since inverse matrices commute, by computing the jk^{th} term of BA, we obtain the orthogonality relation

$$\sum_{n=0}^{j-k} \frac{(1-ap^kq^k)(1-bp^kq^{-k})(ap^{k+1}q^{k+n},bp^{k+1}q^{-k-n};p)_{j-k-1}}{(q;q)_n(q;q)_{j-k-n}(aq^{2k+n}/b;q)_{j-k-1}}$$
$$\cdot \left(1-\frac{a}{b}q^{2k+2n}\right)(-1)^n q^{n(j-k-1)+\binom{j-k-n}{2}} = \delta_{j,k}, \tag{3.6.21}$$

which, by replacing j, n, a, b by $n+k, k, ap^{-k-1}q^{-k}, bp^{-k-1}q^k$, respectively, yields the bibasic summation formula

$$\left(1-\frac{a}{p}\right)\left(1-\frac{b}{p}\right) \sum_{k=0}^{n} \frac{(aq^k,bq^{-k};p)_{n-1}(1-aq^{2k}/b)}{(q;q)_k(q;q)_{n-k}(aq^k/b;q)_{n+1}} (-1)^k q^{\binom{k}{2}} = \delta_{n,0} \tag{3.6.22}$$

for $n = 0, 1, \ldots$. The $b \to 0$ limit case of (3.6.22) was derived in Al-Salam and Verma [1984] by using the fact that the nth q-difference of a polynomial in q of degree less than n is equal to zero. For applications to q-analogues of Lagrange inversion, see Gessel and Stanton [1983, 1986] and Gasper [1989a]. Formulas (3.6.17) and (3.6.22) will be used in §3.7 to derive some useful general expansion formulas.

3.7 Bibasic expansion formulas

One of the most important general expansion formulas for hypergeometric series is the Fields and Wimp [1961] expansion

$$
\begin{aligned}
{}_{r+t}F_{s+u}\left[\begin{matrix} a_R, c_T \\ b_S, d_U \end{matrix}; xw\right] &= \sum_{n=0}^{\infty} \frac{(a_R)_n(\alpha)_n(\beta)_n}{(b_S)_n(\gamma+n)_n} \frac{(-x)^n}{n!} \\
&\quad \cdot {}_{r+2}F_{s+1}\left[\begin{matrix} n+\alpha, n+\beta, n+a_R \\ 1+2n+\gamma, n+b_S \end{matrix}; x\right] \\
&\quad \cdot {}_{t+2}F_{u+2}\left[\begin{matrix} -n, n+\gamma, c_T \\ \alpha, \beta, d_U \end{matrix}; w\right], \qquad (3.7.1)
\end{aligned}
$$

where we employed the contracted notation of representing $a_1, \ldots, a_r$ by a_R, $(a_1)_n \cdots (a_r)_n$ by $(a_R)_n$, and $n+a_1, \ldots, n+a_r$ by $n+a_R$. In (3.7.1), as elsewhere, either the parameters and variables are assumed to be such that the (multiple) series converge absolutely or the series are considered to be formal power series in the variables x and w. Special cases of (3.7.1) were employed, e.g., in Gasper [1975a] to prove the nonnegativity of certain sums (kernels) of Jacobi polynomials and to give additional proofs of the Askey and Gasper [1976] inequalities that de Branges [1985] used at the last step in his proof of the Bieberbach conjecture.

Verma [1972] showed that (3.7.1) is a special case of the expansion

$$
\begin{aligned}
\sum_{n=0}^{\infty} A_n B_n \frac{(xw)^n}{n!} &= \sum_{n=0}^{\infty} \frac{(-x)^n}{n!(\gamma+n)_n} \sum_{n=0}^{\infty} \frac{(\alpha)_{n+k}(\beta)_{n+k}}{k!(\gamma+2n+1)_k} B_{n+k} x^k \\
&\quad \cdot \sum_{j=0}^{n} \frac{(-n)_j(n+\gamma)_j}{j!(\alpha)_j(\beta)_j} A_j w^j \qquad (3.7.2)
\end{aligned}
$$

and derived the q-analogue

$$
\begin{aligned}
&\sum_{n=0}^{\infty} A_n B_n \frac{(xw)^n}{(q;q)_n} \\
&= \sum_{n=0}^{\infty} \frac{(-x)^n}{(q,\gamma q^n;q)_n} q^{\binom{n}{2}} \sum_{k=0}^{\infty} \frac{(\alpha,\beta;q)_{n+k}}{(q,\gamma q^{2n+1};q)_k} B_{n+k} x^k \\
&\quad \cdot \sum_{j=0}^{n} \frac{(q^{-n},\gamma q^n;q)_j}{(q,\alpha,\beta;q)_j} A_j (wq)^j. \qquad (3.7.3)
\end{aligned}
$$

To derive a bibasic extension of (3.7.3) we first observe that, by (3.6.17),

$$
\begin{aligned}
&\sum_{j=0}^{m} \frac{(1-\gamma p^{r+j} q^{r+j})(1-\sigma p^{r+j} q^{-r-j})}{(1-\gamma p^r q^r)(1-\sigma p^r q^{-r})} \frac{(\gamma p^r q^r, \sigma p^r q^{-r}; p)_j}{(q, \gamma\sigma^{-1} q^{2r+1}; q)_j} \\
&\quad \cdot \frac{(q^{-m}, \gamma\sigma^{-1} q^{2r+m}; q)_j}{(\gamma p^{r+1} q^{r+m}, \sigma p^{r+1} q^{-r-m}; p)_j} q^j = \delta_{m,0} \qquad (3.7.4)
\end{aligned}
$$

for $m = 0, 1, \ldots$. Hence, if $C_{r,m}$ are complex numbers such that $C_{r,0} = 1$ for $r = 0, 1, \ldots$, then

$$B_r x^r = \sum_{m=0}^{\infty} \frac{1-\gamma\sigma^{-1}q^{2r+2m}}{1-\gamma\sigma^{-1}q^{2r}} \frac{(\gamma\sigma^{-1}q^{2r};q)_m(\gamma pq^r, \sigma pq^{-r};p)_r}{(q;q)_m(\gamma pq^{r+m}, \sigma pq^{-r-m};p)_r}$$

$$\cdot q^{-mr} B_{r+m} C_{r,m} x^{r+m} \delta_{m,0}$$

$$= \sum_{k=0}^{\infty}\sum_{n=r}^{\infty} \frac{(1-\gamma p^n q^n)(1-\sigma p^n q^{-n})(1-\gamma\sigma^{-1}q^{2n+2k})}{(q;q)_k(q;q)_n(\gamma pq^{n+k}, \sigma pq^{-n-k};p)_n}$$

$$\cdot (\gamma\sigma^{-1}q^{n+r+1};q)_{n+k-r-1}(\gamma pq^r, \sigma pq^{-r};p)_{n-1}(q^{-n};q)_r$$

$$\cdot (-1)^n B_{n+k} C_{r,n+k-r} x^{n+k} q^{n(1+r-n-k)+\binom{n}{2}} \tag{3.7.5}$$

by setting $j = n - r$ and $m = n + k - r$. Then by multiplying both sides of (3.7.5) by $A_r w^r/(q;q)_r$ and summing from $r = 0$ to ∞ we obtain Gasper's [1989a] bibasic expansion formula

$$\sum_{n=0}^{\infty} A_n B_n \frac{(xw)^n}{(q;q)_n} = \sum_{n=0}^{\infty} \frac{(1-\gamma p^n q^n)(1-\sigma p^n q^n)}{(q;q)_n} (-x)^n q^{n+\binom{n}{2}}$$

$$\cdot \sum_{k=0}^{\infty} \frac{1-\gamma\sigma^{-1}q^{2n+2k}}{(q;q)_k(\gamma pq^{n+k}, \sigma pq^{-n-k};p)_n} B_{n+k} x^k$$

$$\cdot \sum_{j=0}^{n} \frac{(q^{-n};q)_j(\gamma\sigma^{-1}q^{n+j+1};q)_{n+k-j-1}}{(q;q)_j}$$

$$\cdot (\gamma pq^j, \sigma pq^{-j};p)_{n-1} A_j C_{j,n+k-j} w^j q^{n(j-n-k)}, \tag{3.7.6}$$

where $C_{j,0} = 1$, for $j = 0, 1, \ldots$.

Note that if $p = q$ and $C_{j,m} \equiv 1$, then (3.7.6) reduces to an expansion which is equivalent to

$$\sum_{n=0}^{\infty} A_n B_n \frac{(xw)^n}{(q;q)_n} = \sum_{n=0}^{\infty} \frac{(\sigma, \gamma q^{n+1}/\sigma, \alpha, \beta;q)_n}{(q, \gamma q^n;q)_n} \left(\frac{x}{\sigma}\right)^n$$

$$\cdot \sum_{k=0}^{\infty} \frac{(\gamma q^{2n}/\sigma, q^{n+1}\sqrt{\gamma/\sigma}, -q^{n+1}\sqrt{\gamma/\sigma}, 1/\sigma, \alpha q^n, \beta q^n;q)_k}{(q, q^n\sqrt{\gamma/\sigma}, -q^n\sqrt{\gamma/\sigma}, \gamma q^{2n+1};q)_k} B_{n+k} x^k$$

$$\cdot \sum_{j=0}^{n} \frac{(q^{-n}, \gamma q^n;q)_j}{(q, \gamma q^{n+1}/\sigma, q^{1-n}/\sigma, \alpha, \beta;q)_j} A_j (wq)^j. \tag{3.7.7}$$

Verma's expansion (3.7.3) is the $\sigma \to \infty$ limit case of (3.7.7). For basic hypergeometric series, (3.7.7) gives the following q-extension of (3.7.1)

$${}_{r+t}\phi_{s+u}\left[\begin{matrix} a_R, c_T \\ b_S, d_U \end{matrix}; q, xw\right]$$

$$= \sum_{j=0}^{\infty} \frac{(c_T, e_K, \sigma, \gamma q^{j+1}/\sigma; q)_j}{(q, d_U, f_M, \gamma q^j; q)_j} \left(\frac{x}{\sigma}\right)^j [(-1)^j q^{\binom{j}{2}}]^{u+m-t-k}$$

$$\cdot {}_{t+k+4}\phi_{u+m+3} \left[\begin{matrix} \gamma q^{2j}/\sigma, q^{j+1}\sqrt{\gamma/\sigma}, -q^{j+1}\sqrt{\gamma/\sigma}, \sigma^{-1}, \\ q^j\sqrt{\gamma/\sigma}, -q^j\sqrt{\gamma/\sigma}, \gamma q^{2j+1}, d_U q^j, \end{matrix} \right.$$

$$\left. \begin{matrix} c_T q^j, e_K q^j \\ f_M q^j \end{matrix} ; q, x q^{j(u+m-t-k)} \right]$$

$$\cdot {}_{r+m+2}\phi_{s+k+2} \left[\begin{matrix} q^{-j}, \gamma q^j, a_R, f_M \\ \gamma q^{j+1}/\sigma, q^{1-j}/\sigma, b_S, e_K \end{matrix} ; q, wq \right], \tag{3.7.8}$$

where we used a contracted notation analogous to that used in (3.7.1).

Note that by letting $\sigma \to \infty$ in (3.7.8) and setting $m = 2, f_1 = f_2 = 0$ we get the expansion

$${}_{r+t}\phi_{s+u} \left[\begin{matrix} a_R, c_T \\ b_S, d_U \end{matrix} ; q, xw \right]$$

$$= \sum_{j=0}^{\infty} \frac{(c_T, e_K; q)_j}{(q, d_U, \gamma q^j; q)_j} x^j [(-1)^j q^{\binom{j}{2}}]^{u+3-t-k}$$

$$\cdot {}_{t+k}\phi_{u+1} \left[\begin{matrix} c_T q^j, e_K q^j \\ \gamma q^{2j+1}, d_U q^j \end{matrix} ; q, x q^{j(u+2-t-k)} \right]$$

$$\cdot {}_{r+2}\phi_{s+k} \left[\begin{matrix} q^{-j}, \gamma q^j, a_R \\ b_S, e_K \end{matrix} ; q, wq \right], \tag{3.7.9}$$

which is equivalent to Verma's [1966] q-extension of the Fields and Wimp expansion (3.7.1). Other types of expansions are given in Fields and Ismail [1975].

Al-Salam and Verma [1984] used the $b \to 0$ limit case of the summation formula (3.6.22) to show that Euler's transformation formula

$$\sum_{n=0}^{\infty} a_n b_n x^n = \sum_{k=0}^{\infty} (-1)^k \frac{x^k}{k!} f^{(k)}(x) \Delta^k a_0, \tag{3.7.10}$$

where

$$f(x) = b_0 + b_1 x + b_2 x^2 + \cdots$$

and

$$\Delta^k a_0 = \sum_{j=0}^{k} (-1)^j \binom{k}{j} a_{k-j},$$

has the bibasic extension

$$\sum_{n=0}^{\infty} A_n B_n (xw)^n = \sum_{k=0}^{\infty} (apq^k; p)_{k-1} x^k \sum_{n=0}^{k} \frac{(1 - ap^n q^n) w^n A_n}{(q; q)_{k-n} (apq^k; p)_n}$$

$$\cdot \sum_{j=0}^{\infty} \frac{(ap^k q^k;p)_j}{(q;q)_j} B_{j+k}(-x)^j q^{\binom{j}{2}}. \tag{3.7.11}$$

The $p = q$ case of (3.7.11) is due to Jackson [1910a].

In order to employ (3.6.22) to extend (3.7.10), replace n in (3.6.22) by j, multiply both sides by $B_{n+j}x^{n+j}(a/b)^j q^{j^2}$, sum from $j = 0$ to ∞, change the order of summation and then replace k by $k - n$ and j by $j + k - n$ to obtain

$$B_n x^n = \left(1 - \frac{a}{p}\right)\left(1 - \frac{b}{p}\right) \sum_{k=n}^{\infty} \frac{1 - aq^{2k-2n}/b}{(q;q)_{k-n}} x^k$$
$$\cdot \sum_{j=0}^{\infty} \frac{(aq^{k-n}, bq^{n-k};p)_{j+k-n-1}}{(q;q)_j (aq^{k-n}/b;q)_{j+k-n+1}} \left(-\frac{a}{b}\right)^{j+k-n}$$
$$\cdot (-x)^j B_{j+k} q^{(k-n)(j+k-n-1)+\binom{j}{2}+\binom{j+k-n+1}{2}}. \tag{3.7.12}$$

Next we replace a by $ap^{n+1}q^n$, b by $bp^{n+1}q^{-n}$, multiply both sides by $A_n w^n$ and then sum from $n = 0$ to ∞ to get

$$\sum_{n=0}^{\infty} A_n B_n (xw)^n = \sum_{k=0}^{\infty} \frac{(apq^k, bpq^{-k};p)_{k-1}}{(aq^k/b;q)_k} x^k$$
$$\cdot \sum_{n=0}^{k} \frac{(1 - ap^n q^n)(1 - bp^n q^{-n})(aq^k/b;q)_n}{(q;q)_{k-n}(apq^k, bpq^{-k};p)_n} A_n w^n$$
$$\cdot \sum_{j=0}^{\infty} \frac{(ap^k q^k, bp^k q^{-k};p)_j}{(q;q)_j (aq^{2k+1}/b;q)_j} \left(-\frac{a}{b} q^{2n}\right)^{j+k-n}$$
$$\cdot B_{j+k}(-x)^j q^{(k-n)(j+k-n-1)+\binom{j}{2}+\binom{j+k-n+1}{2}}. \tag{3.7.13}$$

This formula tends directly to (3.7.11) as $b \to 0$. By replacing A_n, B_n, x, w by suitable multiples, we may change (3.7.13) to an equivalent form which tends to (3.7.11) as $b \to \infty$. In addition, by replacing A_n, B_n, x, w by $A_n q^{2\binom{n}{2}}$, $B_n q^{-2\binom{n}{2}}$, bx/a, aw/b, respectively, we can write (3.7.13) in the simpler looking equivalent form

$$\sum_{n=0}^{\infty} A_n B_n x^n w^n = \sum_{k=0}^{\infty} \frac{(apq^k, bpq^{-k};p)_{k-1}}{(q, aq^k/b;q)_k} (-x)^k q^{\binom{k+1}{2}}$$
$$\cdot \sum_{n=0}^{k} \frac{(1 - ap^n q^n)(1 - bp^n q^{-n})(q^{-k}, aq^k/b;q)_n}{(apq^k, bpq^{-k};p)_n} A_n w^n$$
$$\cdot \sum_{j=0}^{\infty} \frac{(ap^k q^k, bp^k q^{-k};p)_j}{(q, aq^{2k+1}/b;q)_j} B_{j+k} x^j q^j. \tag{3.7.14}$$

As in the derivation of (3.7.6), one may extend (3.7.14) by replacing B_{j+k} by $B_{j+k}C_{n,j+k-n}$ with $C_{n,0} = 1$ for $n = 0, 1, \ldots$. Multivariable expansions, which are really special cases of (3.7.6) and (3.7.14), may be obtained by replacing A_n and B_n in (3.7.6) and (3.7.14) by multiple power series, see, e.g.

Gasper [1989a], Ex. 3.22 and, in the hypergeometric limit case, Luke [1969]. For a multivariable special case of the Al-Salam and Verma expansion (3.7.11), see Srivastava [1984].

3.8 Quadratic, cubic, and quartic summation and transformation formulas

By setting $p = q^j$ or $q = p^j, j = 2, 3, \ldots$, in the bibasic summation formulas of §3.7 and using summation and transformation formulas for basic hypergeometric series, one can derive families of quadratic, cubic, etc. summation, transformation and expansion formulas. To illustrate this we shall derive a quadratic transformation formula containing five arbitrary parameters by starting with the $q = p^2$ case of (3.6.16)

$$\sum_{k=0}^{n} \frac{(1-adp^{3k})(1-b/dp^k)}{(1-ad)(1-b/d)} \frac{(a,b;p)_k(p^{-2n}, ad^2p^{2n}/b;p^2)_k}{(dp^2, adp^2/b;p^2)_k(adp^{2n+1}, bp^{1-2n}/d;p)_k} p^k$$

$$= \frac{(1-d)(1-ad/b)(1-adp^{2n})(1-dp^{2n}/b)}{(1-ad)(1-d/b)(1-dp^{2n})(1-adp^{2n}/b)}, \tag{3.8.1}$$

where $n = 0, 1, \ldots$.

Change p to q and d to c in (3.8.1), multiply both sides by

$$\frac{(ac^2/b;q^2)_n(c/b;q)_{2n}}{(q^2;q^2)_n(acq;q)_{2n}} C_n$$

and sum over n to get

$$\sum_{n=0}^{\infty} \frac{(ac^2/b;q^2)_n(cq/b;q)_{2n}(1-c)(1-ac/b)}{(q^2;q^2)_n(ac;q)_{2n}(1-cq^{2n})(1-acq^{2n}/b)} C_n$$

$$= \sum_{n=0}^{\infty}\sum_{k=0}^{n} \frac{(1-acq^{3k})(1-b/cq^k)}{(1-ac)(1-b/c)}$$

$$\cdot \frac{(a,b;q)_k(ac^2/b;q^2)_{n+k}(c/b;q)_{2n-k}}{(cq^2, acq^2/b;q^2)_k(q^2;q^2)_{n-k}(acq;q)_{2n+k}} \left(\frac{cq}{b}\right)^k q^{\binom{k}{2}} C_n$$

$$= \sum_{k=0}^{\infty} \frac{(1-acq^{3k})(1-b/cq^k)}{(1-ac)(1-b/c)} \frac{(a,b;q)_k(ac^2/b;q^2)_{2k}(c/b;q)_k}{(cq^2, acq^2/b;q^2)_k(acq;q)_{3k}} \left(\frac{cq}{b}\right)^k q^{\binom{k}{2}}$$

$$\cdot \sum_{m=0}^{\infty} \frac{(ac^2q^{4k}/b;q^2)_m(cq^k/b;q)_{2m}}{(q^2;q^2)_m(acq^{3k+1};q)_{2m}} C_{k+m}. \tag{3.8.2}$$

Setting

$$C_n = \frac{(1-ac^2q^{4n}/b)(d,e,f;q^2)_n(a^2q^3/def)^n}{(1-ac^2/b)(ac^2q^2/bd, ac^2q^2/be, ac^2q^2/bf;q^2)_n},$$

it follows from (3.8.2) that

$${}_{10}W_9(ac^2/b; ac/b, c, cq/b, cq^2/b, d, e, f; q^2, a^2c^2q^3/def)$$

$$= \sum_{k=0}^{\infty} \frac{(1-acq^{3k})(1-b/cq^k)}{(1-ac)(1-b/c)} \frac{(a,b,c/b;q)_k (ac^2q^2/b;q^2)_{2k}}{(cq^2,acq^2/b;q^2)_k (acq;q)_{3k}}$$

$$\cdot \frac{(d,e,f;q^2)_k (a^2c^3q^4/bdef)^k}{(ac^2q^2/bd,ac^2q^2/be,ac^2q^2/bf;q^2)_k} q^{\binom{k}{2}}$$

$$\cdot {}_8W_7(ac^2q^{4k}/b;cq^k/b,cq^{k+1}/b,dq^{2k},eq^{2k},fq^{2k};q^2,a^2c^2q^3/def). \quad (3.8.3)$$

If we now assume that

$$a^2c^2q = def, \quad (3.8.4)$$

then we can apply (2.11.7) to get

$${}_8W_7(ac^2q^{4k}/b;cq^k/b,cq^{k+1}/b,dq^{2k},eq^{2k},fq^{2k};q^2,q^2)$$

$$= \frac{(ac^2q^{4k+2}/b,bf/ac^2q^{2k},abq^{2k+1},acq^{k+2}/d;q)_\infty}{(acq^{3k+2},acq^{3k+1},ac^2q^{2k+2}/bd,ac^2q^{2k+2}/be;q)_\infty}$$

$$\cdot \frac{(acq^{k+2}/e,acq^{k+1}/d,acq^{k+1}/e,ac^2q^2/bde;q^2)_\infty}{(bef/ac^2,bdf/ac^2,f/acq^k,f/acq^{k-1};q^2)_\infty}$$

$$+ \frac{bfq^{-2k}(ac^2q^{4k+2}/b,cq^k/b,cq^{k+1}/b,dq^{2k},eq^{2k};q)_\infty}{ac^2(ac^2q^{2k+2}/bf,acq^{3k+2},acq^{3k+1},ac^2q^{2k+2}/bd;q)_\infty}$$

$$\cdot \frac{(fq^2/e,fq^2/d,bfq^{2-2k}/ac^2,bfq^{k+1}/c,bfq^{k+2}/c;q^2)_\infty}{(ac^2q^{2k+2}/be,bef/ac^2,bdf/ac^2,bf^2q^2/ac^2,f/acq^k,f/acq^{k-1};q^2)_\infty}$$

$$\cdot {}_8W_7(bf^2/ac^2;fq^{2k},bef/ac^2,bdf/ac^2,f/acq^k,f/acq^{k-1};q^2,q^2), \quad (3.8.5)$$

which, combined with (3.8.3), gives

$${}_{10}W_9(ac^2/b;c,d,e,f,ac/b,cq/b,cq^2/b;q^2,q^2)$$

$$= \frac{(ac^2q^2/b,ac^2q^2/bde,abq,bf/ac^2;q^2)_\infty (acq/d,acq/e;q)_\infty}{(ac^2q^2/bd,ac^2q^2/be,bdf/ac^2,bef/ac^2;q^2)_\infty (acq,f/ac;q)_\infty}$$

$$\cdot \sum_{k=0}^{\infty} \frac{(1-acq^{3k})(1-b/cq^k)(a,b,c/b;q)_k (d,e,f;q^2)_k}{(1-ac)(1-b/c)(cq^2,acq^2/b,abq;q^2)_k (acq/d,acq/e,acq/f;q)_k} q^{2k}$$

$$+ \frac{bf(ac^2q^2/b,d,e,fq^2/d,fq^2/e,bfq^2/ac^2;q^2)_\infty}{ac^2(ac^2q^2/bf,ac^2q^2/be,ac^2q^2/bd,bdf/ac^2,bef/ac^2,bf^2q^2/ac^2;q^2)_\infty}$$

$$\cdot \frac{(bfq/c,c/b;q)_\infty}{(acq,f/ac;q)_\infty}$$

$$\cdot \sum_{k=0}^{\infty} \frac{(1-acq^{3k})(1-b/cq^k)}{(1-ac)(1-b/c)} \frac{(a,b;q)_k (ac^2/bf,f;q^2)_k}{(cq^2,acq^2/b;q^2)_k (bfq/c,acq/f;q)_k} q^{2k}$$

$$\cdot {}_8W_7(bf^2/ac^2;fq^{2k},bdf/ac^2,bef/ac^2,f/acq^k,f/acq^{k-1};q^2,q^2). \quad (3.8.6)$$

The last sum over k in (3.8.6) is

$$\sum_{j=0}^{\infty} \frac{(bf^2/ac^2, f, bdf/ac^2, bef/ac^2, f/ac, fq/ac; q^2)_j (1 - bf^2 q^{4j}/ac^2)}{(q^2, bfq^2/ac^2, fq^2/d, fq^2/e, bfq^2/c, bfq/c; q^2)_j (1 - bf^2/ac^2)} q^{2j}$$

$$\cdot \sum_{k=0}^{\infty} \frac{(1 - acq^{3k})(1 - b/cq^k)}{(1 - ac)(1 - b/c)} \frac{(a, b; q)_k (ac^2/bfq^{2j}, fq^{2j}; q^2)_k}{(cq^2, acq^2/b; q^2)_k (bfq^{2j+1}/c, acq^{1-2j}/f; q)_k} q^{2k}$$

$$= \sum_{j=0}^{\infty} \frac{(bf^2/ac^2, f, bdf/ac^2, bef/ac^2, f/ac, fq/ac; q^2)_j (1 - bf^2 q^{4j}/ac^2)}{(q^2, bfq^2/ac^2, fq^2/d, fq^2/e, bfq^2/c, bfq/c; q^2)_j (1 - bf^2/ac^2)} q^{2j}$$

$$\cdot \frac{(1 - c)(1 - ac/b)(1 - ac/fq^{2j})(1 - bfq^{2j}/c)}{c(1 - ac)(1 - b/c)(1 - fq^{2j}/c)(1 - ac/bfq^{2j})}$$

$$\cdot \left\{ \frac{(a, b; q)_\infty (fq^{2j}, ac^2/bfq^{2j}; q^2)_\infty}{(c, ac/b; q^2)_\infty (ac/fq^{2j}, bfq^{2j}/c; q)_\infty} - 1 \right\}$$

$$= \frac{(1 - c)(1 - ac/b)(1 - f/ac)(1 - bf/c)}{(1 - c/b)(1 - ac)(1 - f/c)(1 - bf/ac)}$$

$$\cdot {}_{10}W_9(bf^2/ac^2; f, bdf/ac^2, bef/ac^2, bf/ac, f/c, fq/ac, fq^2/ac; q^2, q^2)$$

$$+ \frac{f}{ac(1 - ac)(1 - c/b)(1 - f/c)(1 - bf/ac)}$$

$$\cdot \frac{(a, b; q)_\infty (f, ac^2/bf; q^2)_\infty}{(cq^2, acq^2/b; q^2)_\infty (acq/f, bfq/c; q)_\infty}$$

$$\cdot \sum_{j=0}^{\infty} \frac{(bf^2/ac^2, bdf/ac^2, bef/ac^2, f/c, bf/ac; q^2)_j}{(q^2, fq^2/d, fq^2/e, bfq^2/ac, fq^2/c; q^2)_j} \frac{(1 - bf^2 q^{4j}/ac^2)}{(1 - bf^2/ac^2)} \left(-\frac{fq^2}{ab} \right)^j q^{j^2} \tag{3.8.7}$$

by the $n = \infty$ case of (3.6.15). Thus,

$$ {}_{10}W_9(ac^2/b; c, d, e, f, ac/b, cq/b, cq^2/b; q^2, q^2) - \frac{bf(1 - c)(1 - ac/b)}{ac^2(1 - f/c)(1 - bf/ac)}$$

$$\cdot \frac{(ac^2q^2/b, d, e, fq^2/d, fq^2/e, bfq^2/ac^2; q^2)_\infty}{(bf^2q^2/ac^2, ac^2q^2/be, ac^2q^2/bd, bdf/ac^2, bef/ac^2, ac^2q^2/bf; q^2)_\infty}$$

$$\cdot \frac{(bf/c, cq/b; q)_\infty}{(ac, fq/ac; q)_\infty} {}_{10}W_9 \left(\frac{bf^2}{ac^2}; \frac{bdf}{ac^2}, \frac{bef}{ac^2}, \frac{f}{c}, \frac{bf}{ac}, \frac{fq}{ac}, \frac{fq^2}{ac^2}; q^2, q^2 \right)$$

$$= \frac{(acq/d, acq/e; q)_\infty (ac^2q^2/b, abq, bf/ac^2, ac^2q^2/bde; q^2)_\infty}{(acq, f/ac; q)_\infty (ac^2q^2/bd, ac^2q^2/be, bdf/ac^2, bef/ac^2; q^2)_\infty}$$

$$\cdot \sum_{k=0}^{\infty} \frac{1 - acq^{3k}}{1 - ac} \frac{(a, b, cq/b; q)_k (d, e, f; q^2)_k}{(cq^2, acq^2/b, abq; q)_k (acq/d, acq/e, acq/f; q)_k} q^k$$

$$+ \frac{bf^2(a, b, cq/b; q)_\infty}{a^2c^3(1 - f/c)(1 - bf/ac)(ac, f/ac, acq/f; q)_\infty}$$

$$\cdot \frac{(f, ac^2/bf, ac^2q^2/b, d, e, fq^2/d, fq^2/e, bfq^2/ac^2; q^2)_\infty}{(cq^2, acq^2/b, ac^2q^2/bf, ac^2q^2/be, ac^2q^2/bd, bdf/ac^2, bef/ac^2, bf^2q^2/ac^2; q^2)_\infty}$$

$$\cdot \sum_{j=0}^{\infty} \frac{(bf^2/ac^2, bdf/ac^2, bef/ac^2, f/c, bf/ac; q^2)_j}{(q^2, fq^2/d, fq^2/e, bfq^2/ac, fq^2/c; q^2)_j}$$

$$\cdot \frac{(1-bf^2q^{4j}/ac^2)}{(1-bf^2/ac^2)} \left(-\frac{fq^2}{ab}\right)^j q^{j^2} \tag{3.8.8}$$

when (3.8.4) holds.

Now observe that since

$$\sum_{j=0}^{\infty} \frac{(a, q\sqrt{a}, -q\sqrt{a}, c, d, e, f; q)_j}{(q, \sqrt{a}, -\sqrt{a}, aq/c, aq/d, aq/e, aq/f; q)_j} \left(-\frac{a^2q^2}{cdef}\right)^j q^{\binom{j}{2}}$$

$$= \frac{(aq, aq/ef; q)_\infty}{(aq/e, aq/f; q)_\infty} {}_3\phi_2 \left[\begin{matrix} aq/cd, e, f \\ aq/c, aq/d \end{matrix}; q, \frac{aq}{ef} \right] \tag{3.8.9}$$

by the $n \to \infty$ limit case of (2.5.1), the sum over j in (3.8.8) equals

$$\frac{(q^2, bf^2q^2/ac^2; q^2)_\infty}{(bfq^2/ac, fq^2/c; q^2)_\infty} {}_3\phi_2 \left[\begin{matrix} f/c, bf/ac, ac^2q^2/bde \\ fq^2/d, fq^2/e \end{matrix}; q^2, q^2 \right]. \tag{3.8.10}$$

Hence, by setting $e = a^2c^2q/df$ in (3.8.8) we obtain the Gasper and Rahman [1989a] quadratic transformation formula

$${}_{10}W_9(ac^2/b; f, ac/b, c, cq/b, cq^2/b, d, a^2c^2q/df; q^2, q^2)$$

$$+ \frac{(ac^2q^2/b, bf/ac^2, ac/b, c, cq/b, cq^2/b, bfq^2/ac; q^2)_\infty}{(bf^2q^2/ac^2, ac^2/bf, ac^2q^2/bd, dfq/ab, bdf/ac^2, abq/d, cq^2; q^2)_\infty}$$

$$\cdot \frac{(fq^2/c, bf/c, bfq/c, fq^2/d, df^2q/a^2c^2, d, a^2c^2q/df; q^2)_\infty}{(acq^2/b, f/c, bf/ac, ac, acq, fq/ac, fq^2/ac; q^2)_\infty}$$

$$\cdot {}_{10}W_9(bf^2/ac^2; f, bdf/ac^2, abq/d, f/c, bf/ac, fq/ac, fq^2/ac; q^2, q^2)$$

$$- \frac{(a, b, cq/f; q)_\infty}{(ac, ac/f, fq/ac; q)_\infty}$$

$$\cdot \frac{(f, d, a^2c^2q/df, bf/ac^2, ac^2q^2/b, fq^2/d, df^2q/a^2c^2, q^2; q^2)_\infty}{(bf/ac, f/c, cq^2, acq^2/b, ac^2q^2/bf, dfq/ab, bdf/ac^2, abq/d; q^2)_\infty}$$

$$\cdot {}_3\phi_2 \left[\begin{matrix} f/c, bf/ac, fq/ab \\ fq^2/d, df^2q/a^2c^2 \end{matrix}; q^2, q^2 \right]$$

$$= \frac{(acq/d, df/ac; q)_\infty (ac^2q^2/b, abq, bf/ac^2, fq/ab; q^2)_\infty}{(acq, f/ac; q)_\infty (ac^2q^2/bd, dfq/ab, bdf/ac^2, abq/d; q^2)_\infty}$$

$$\cdot \sum_{k=0}^{\infty} \frac{1-acq^{3k}}{1-ac} \frac{(a, b, cq/b; q)_k (d, f, a^2c^2q/df; q^2)_k q^k}{(cq^2, acq^2/b, abq; q^2)_k (acq/d, acq/f, df/ac; q)_k}. \tag{3.8.11}$$

Note that the first two terms on the left side of (3.8.11) containing the ${}_{10}W_9$ series can be transformed to another pair of ${}_{10}W_9$ series by applying the four-term transformation formula (2.12.9). Since the ${}_3\phi_2$ series in (3.8.11) is balanced it can be summed by (1.7.2) whenever it terminates. When $c = 1$ formula (3.8.11) reduces to the quadratic summation formula

$$\sum_{k=0}^{\infty} \frac{1-aq^{3k}}{1-a} \frac{(a, b, q/b; q)_k (d, f, a^2q/df; q^2)_k}{(q^2, aq^2/b, abq; q^2)_k (aq/d, aq/f, df/a; q)_k} q^k$$

$$+ \frac{(aq, f/a, b, f/q; q)_\infty (d, a^2q/bd, a^2q/df, fq^2/d, df^2q/a^2; q^2)_\infty}{(a/f, fq/a, aq/d, df/a; q)_\infty (aq^2/b, abq, fq/ab, bf/a, aq^2/bf; q^2)_\infty}$$

$$\cdot {}_3\phi_2 \left[\begin{matrix} f, bf/a, fq/ab \\ fq^2/d, df^2q/a^2 \end{matrix} ; q^2, q^2 \right]$$

$$= \frac{(aq, f/a; q)_\infty (aq^2/bd, abq/d, bdf/a, dfq/ab; q^2)_\infty}{(aq/d, df/a; q)_\infty (aq^2/b, abq, bf/a, fq/ab; q^2)_\infty}. \tag{3.8.12}$$

By multiplying both sides of (3.8.11) by $(f/ac; q)_\infty$ and then setting $f = ac$ we obtain Rahman's [1989d] quadratic transformation formula

$$\sum_{k=0}^{\infty} \frac{(a; q^2)_k (1-aq^{3k})(d, aq/d; q^2)_k (b, c, aq/bc; q)_k}{(q; q)_k (1-a)(aq/d, d; q)_k (aq^2/b, aq^2/c, bcq; q^2)_k} q^k$$

$$= \frac{(aq^2, bq, cq, aq^2/bc; q^2)_\infty}{(q, aq^2/b, aq^2/c, bcq; q^2)_\infty} {}_3\phi_2 \left[\begin{matrix} b, c, aq/bc \\ dq, aq^2/d \end{matrix} ; q^2, q^2 \right]. \tag{3.8.13}$$

Also, the case $d = q^{-n}$ of (3.8.11) gives

$$\sum_{k=0}^{n} \frac{1-acq^{3k}}{1-ac} \frac{(a, b, cq/b; q)_k (f, a^2c^2q^{2n+1}/f, q^{-2n}; q^2)_k}{(cq^2, acq^2/b, abq; q^2)_k (acq/f, f/acq^{2n}, acq^{2n+1}; q)_k} q^k$$

$$= \frac{(acq; q)_{2n} (ac^2q^2/bf, abq/f; q^2)_n}{(acq/f; q)_{2n} (abq, ac^2q^2/b; q^2)_n}$$

$$\cdot {}_{10}W_9(ac^2/b; f, ac/b, c, cq/b, cq^2/b, a^2c^2q^{2n+1}/f, q^{-2n}; q^2, q^2) \tag{3.8.14}$$

and the case $b = cq^{n+1}$ gives

$$\sum_{k=0}^{n} \frac{1-acq^{3k}}{1-ac} \frac{(d, f, a^2c^2q/df; q^2)_k (a, cq^{n+1}, q^{-n}; q)_k}{(acq/d, acq/f, df/ac; q)_k (cq^2, aq^{1-n}, acq^{n+2}; q^2)_k} q^k$$

$$= \frac{(acq^2/d, q/ac, acq/f, dfq/ac; q^2)_n}{(acq^2, fq/ac, dq/ac, acq/df; q^2)_n}$$

$$\cdot {}_{10}W_9(acq^{-n-1}; c, d, f, a^2c^2q/df, aq^{-n-1}, q^{1-n}, q^{-n}; q^2, q^2) \tag{3.8.15}$$

for $n = 0, 1, \ldots$.

Similarly, the special case

$$\sum_{k=0}^{n} \frac{(1-acq^{4k})(1-b/cq^{2k})}{(1-ac)(1-b/c)} \frac{(a, b; q)_k (q^{-3n}, ac^2q^{3n}/b; q^3)_k}{(cq^3, acq^3/b; q^3)_k (acq^{3n+1}, b/cq^{3n-1}; q)_k} q^{3k}$$

$$= \frac{(1-c)(1-ac/b)(1-acq^{3n})(1-cq^{3n}/b)}{(1-ac)(1-c/b)(1-cq^{3n})(1-acq^{3n}/b)}, \quad n = 0, 1, 2, \ldots, \tag{3.8.16}$$

of (3.6.16) is used in Gasper and Rahman [1989a] to show that Wm. Gosper's sum (see Gessel and Stanton [1982])

$$
{}_7F_6\left[\begin{matrix} a, a+1/2, b, 1-b, c, (2a+1)/3-c, a/2+1 \\ 1/2, (2a-b+3)/3, (2a+b+2)/3, 3c, 2a+1-3c, a/2 \end{matrix}; 1\right]
$$

$$
= \frac{2}{\sqrt{3}} \frac{\Gamma\left(c+\frac{1}{3}\right)\Gamma\left(c+\frac{2}{3}\right)\Gamma\left(\frac{2a-b+3}{3}\right)\Gamma\left(\frac{2a+b+2}{3}\right)}{\Gamma\left(\frac{2a+2}{3}\right)\Gamma\left(\frac{2a+3}{3}\right)\Gamma\left(\frac{3c+b+1}{3}\right)\Gamma\left(\frac{3c+2-b}{3}\right)}
$$

$$
\cdot \frac{\Gamma\left(\frac{2+2a-3c}{3}\right)\Gamma\left(\frac{3+2a-3c}{3}\right)\sin\frac{\pi}{3}(b+1)}{\Gamma\left(\frac{2+2a+b-3c}{3}\right)\Gamma\left(\frac{3+2a-b-3c}{3}\right)} \tag{3.8.17}
$$

has a q-analogue of the form

$$
\sum_{k=0}^{\infty} \frac{1-acq^{4k}}{1-ac} \frac{(a, q/a; q)_k (ac; q)_{2k} (d, acq/d; q^3)_k}{(cq^3, a^2cq^2; q^3)_k (q; q)_{2k} (acq/d, d; q)_k} q^k
$$

$$
= \frac{(acq^2, acq^3, d/ac, dq/ac, adq, aq, q^2/a, dq^2/a; q^3)_\infty}{(q, q^2, dq, dq^2, a^2cq^2, cq^3, dq/a^2c, d/c; q^3)_\infty}
$$

$$
+ \frac{d(a, q/a, acq; q)_\infty (q^3, d, acq/d, d^2q^2/ac; q^3)_\infty}{ac(q, d, acq/d; q)_\infty (cq^3, a^2cq^2, d/c, dq/a^2c; q^3)_\infty}
$$

$$
\cdot {}_2\phi_1\left[\begin{matrix} d/c, dq/a^2c \\ d^2q^2/ac \end{matrix}; q^3, q^3\right] \tag{3.8.18}
$$

and to derive the extension

$$
{}_{10}W_9(ac^2/b; d, c, a^2bc/d, ac/b, cq/b, cq^2/b, cq^3/b; q^3, q^3)
$$

$$
+ \frac{(1-c)(1-ac/b)}{(1-d/c)(1-bd/ac)} \frac{(cq/b, bd/c; q)_\infty (ac^2q^3/b, a^2bc/d, d^2q^3/a^2bc, bd/ac^2; q^3)_\infty}{(ac, dq/ac; q)_\infty (ac^2/bd, cdq^3/ab^2, ab^2/c, bd^2q^3/ac^2; q^3)_\infty}
$$

$$
\cdot {}_{10}W_9(bd^2/ac^2; d, bd/ac, ab^2/c, d/c, dq/ac, dq^2/ac, dq^3/ac; q^3, q^3)
$$

$$
- \frac{(a, b, cq/b; q)_\infty (q^3, d, ac^2q^3/b, a^2bc/d, d^2q^3/a^2bc, bd/ac^2; q^3)_\infty}{(ac, dq/ac, ac/d; q)_\infty (cq^3, acq^3/b, d/c, bd/ac, cdq^3/ab^2, ab^2/c; q^3)_\infty}
$$

$$
\cdot {}_2\phi_1\left[\begin{matrix} d/c, bd/ac \\ d^2q^3/a^2bc \end{matrix}; q^3, q^3\right]
$$

$$
= \frac{(ab, dq/ab; q)_\infty (bd/ac^2, ac^2q^3/b; q^3)_\infty}{(acq, d/ac; q)_\infty (ab^2/c, cdq^3/ab^2; q^3)_\infty}
$$

$$
\cdot \sum_{k=0}^{\infty} \frac{1-acq^{4k}}{1-ac} \frac{(a, b; q)_k (cq/b; q)_{2k} (d, a^2bc/d; q^3)_k}{(cq^3, acq^3/b; q^3)_k (ab; q)_{2k} (acq/d, dq/ab; q)_k} q^k \tag{3.8.19}
$$

and some other cubic transformation formulas.

The special case

$$
\sum_{k=0}^{n} \frac{(1-acq^{5k})(1-b/cq^{3k})}{(1-ac)(1-b/c)} \frac{(a, b; q)_k (q^{-4n}, ac^2q^{4n}/b; q^4)_k}{(cq^4, acq^4/b; q^4)_k (acq^{4n+1}, b/cq^{4n-1}; q)_k} q^{4k}
$$

$$= \frac{(1-c)(1-ac/b)(1-acq^{4n})(1-cq^{4n}/b)}{(1-ac)(1-c/b)(1-cq^{4n})(1-acq^{4n}/b)}, \quad n=0,1,2,\ldots, \tag{3.8.20}$$

of (3.6.16) was used in Gasper and Rahman [1989a] to derive the quartic transformation formula

$$\begin{aligned}
&{}_{10}W_9(ac^2/b;a^2b^2/q^2,ac/b,c,cq/b,cq^2/b,cq^3/b,cq^4/b;q^4,q^4)\\
&+\frac{(1-c)(1-ac/b)(a^2b^3/cq^2,cq/b;q)_\infty(ac^2q^4/b,ab^3/c^2q^2;q^4)_\infty}{(1-a^2b^2/cq^2)(1-ab^3/cq^2)(ab^2/cq,ac;q)_\infty(a^3b^5/c^2,c^2q^2/ab^3;q^4)_\infty}\\
&\cdot{}_{10}W_9\left(\frac{a^3b^5}{c^2q^4};\frac{a^2b^2}{q^2},\frac{a^2b^2}{cq^2},\frac{ab^3}{cq^2},\frac{ab^2}{cq},\frac{ab^2}{c},\frac{ab^2q}{c},\frac{ab^2q^2}{c};q^4,q^4\right)\\
&+\frac{ab^2(a,b,cq/b;q)_\infty(ac^2q^4/b,ab^3/c^2q^2,a^2b^2/q^2;q^4)_\infty}{cq^2(1-a^2b^2/cq^2)(ac,ab^2/cq^2,cq^3/ab^2;q)_\infty(ab^3/cq^2,cq^4,acq^4/b;q^4)_\infty}\\
&\cdot{}_1\phi_1\left[\begin{matrix}a^2b^2/cq^2\\a^2b^2q^2/c\end{matrix};q^4,\frac{ab^3q^2}{c}\right]\\
&=\frac{(ab;q)_\infty(ab/q;q^2)_\infty(ac^2q^4/b,ab^3/c^2q^2;q^4)_\infty}{(acq,ab^2/cq^2;q)_\infty}\\
&\cdot\sum_{k=0}^\infty\frac{1-acq^{5k}}{1-ac}\frac{(a,b;q)_k(cq/b,cq^2/b,cq^3/b;q^3)_k(a^2b^2/q^2;q^4)_k}{(cq^4,acq^4/b;q^4)_k(abq,ab,ab/q;q^2)_k(cq^3/ab^2;q)_k}q^k.
\end{aligned} \tag{3.8.21}$$

When $b=q^2/a$ the sum of the two ${}_{10}W_9$ series in (3.8.21) reduces to a sum of two ${}_8W_7$ series, which can be summed by (2.11.7) to obtain the quartic summation formula

$$\begin{aligned}
&\sum_{k=0}^\infty\frac{1-acq^{5k}}{1-ac}\frac{(a,q^2/a;q)_k(ac/q,ac,acq;q^3)_k(q^2;q^4)_k}{(cq^4,a^2cq^2;q^4)_k(q^3,q^2,q;q^2)_k(ac/q;q)_k}q^k\\
&-\frac{q^2(a,q^2/a,acq;q)_\infty(q^2;q^4)_\infty}{ac(1-q^2/c)(q^2,ac;q)_\infty(q;q^2)_\infty(cq^4,a^2cq^2,q^4/a^2c;q^4)_\infty}\\
&\cdot{}_1\phi_1\left[\begin{matrix}q^2/c\\q^6/c\end{matrix};q^4,\frac{q^8}{a^2c}\right]\\
&=\frac{(acq^2,q^2/ac,aq,q^3/a;q^2)_\infty}{(q,q^3;q^2)_\infty(cq^4,q^2/c,a^2cq^2,q^4/a^2c;q^4)_\infty}.
\end{aligned} \tag{3.8.22}$$

Additional quadratic, cubic and quartic summations and transformation formulas are given in the exercises.

3.9 Multibasic hypergeometric series

In view of the observation in §1.2 that a series $\sum_{n=0}^\infty v_n$ is a basic hypergeometric series in base q if $v_0=1$ and v_{n+1}/v_n is a rational function of q^n, a series $\sum_{n=0}^{\infty} v_n$ will be called a bibasic hypergeometric series in bases p and q if v_{n+1}/v_n is a rational function of p^n and q^n, and p and q are independent. More generally, we shall call a series $\sum_{n=0}^{\infty} v_n$ a multibasic (or m-basic) hypergeometric series in bases $q_1, \ldots, q_m$ if v_{n+1}/v_n is a rational function of $q_1^n, \ldots, q_m^n$, and $q_1, \ldots, q_m$ are independent. Similarly a bilateral series $\sum_{n=-\infty}^{\infty} v_n$ will be called a bilateral multibasic (or m-basic) hypergeometric series in bases $q_1, \ldots, q_m$ if v_{n+1}/v_n is a rational function of $q_1^n, \ldots, q_m^n$, and $q_1, \ldots, q_m$ are independent. Multibasic series are sometimes called polybasic series.

Since a multibasic series in bases $q_1, \ldots, q_m$ may contain products and quotients of q-shifted factorials $(a;q)_n$ with q replaced by $q_1^{k_1} \cdots q_m^{k_m}$ where $k_1, \ldots, k_m$ are arbitrary integers, the form of such a series could be quite complicated. Therefore, in working with multibasic series either the series are displayed explicitly or notations are employed which apply only to the series under consideration. For example, to shorten the displays of many of the formulas derived in §3.8 we employ the notation

$$\Phi\left[\begin{matrix} a_1, \ldots, a_r : c_{1,1}, \ldots, c_{1,r_1} : \ldots : c_{m,1}, \ldots, c_{m,r_m} \\ b_1, \ldots, b_s : d_{1,1}, \ldots, d_{1,s_1} : \ldots : d_{m,1}, \ldots, d_{m,s_m} \end{matrix} ; q, q_1, \ldots, q_m; z\right] \tag{3.9.1}$$

to represent the $(m+1)$-basic hypergeometric series

$$\sum_{n=0}^{\infty} \frac{(a_1, \ldots, a_r; q)_n}{(q, b_1, \ldots, b_s; q)_n} z^n \left[(-1)^n q^{\binom{n}{2}}\right]^{1+s-r}$$
$$\cdot \prod_{j=1}^{m} \frac{(c_{j,1}, \ldots, c_{j,r_j}; q_j)_n}{(d_{j,1}, \ldots, d_{j,s_j}; q_j)_n} \left[(-1)^n q_j^{\binom{n}{2}}\right]^{s_j - r_j}. \tag{3.9.2}$$

The notation in (3.9.2) may be abbreviated by using the vector notations: $\mathbf{a} = (a_1, \ldots, a_r)$, $\mathbf{b} = (b_1, \ldots, b_s)$, $\mathbf{c}_j = (c_{j,1}, \ldots, c_{j,r_j})$, $\mathbf{d}_j = (d_{j,1}, \ldots, d_{j,s_j})$ and letting

$$\Phi\left[\begin{matrix} \mathbf{a} : \mathbf{c}_1 : \ldots : \mathbf{c}_m \\ \mathbf{b} : \mathbf{d}_1 : \ldots : \mathbf{d}_m \end{matrix} ; q, q_1, \ldots, q_m; z\right] \tag{3.9.3}$$

denote the series in (3.9.2).

If in (2.2.2) we set

$$A_k = \frac{(a;q)_k}{(q^{-n};q)_k} z^k \prod_{j=1}^{m} \frac{(c_{j,1}, \ldots, c_{j,r_j}; q_j)_k}{(d_{j,1}, \ldots, d_{j,s_j}; q_j)_k} \left[(-1)^k q_j^{\binom{k}{2}}\right]^{s_j - r_j}, \tag{3.9.4}$$

then we obtain the expansion

$$\Phi\left[\begin{matrix} a, b, c : c_{1,1}, \ldots, c_{1,r_1} : \ldots : c_{m,1}, \ldots, c_{m,r_m} \\ aq/b, aq/c : d_{1,1}, \ldots, d_{1,s_1} : \ldots : d_{m,1}, \ldots, d_{m,s_m} \end{matrix} ; q, q_1, \ldots, q_m; z\right]$$
$$= \sum_{n=0}^{\infty} \frac{(aq/bc;q)_n (a;q)_{2n}}{(q, aq/b, aq/c;q)_n} \left(-\frac{bcz}{aq}\right)^n q^{-\binom{n}{2}}$$
$$\cdot \prod_{j=1}^{m} \frac{(c_{j,1}, \ldots, c_{j,r_j}; q_j)_n}{(d_{j,1}, \ldots, d_{j,s_j}; q_j)_n} \left[(-1)^n q_j^{\binom{n}{2}}\right]^{s_j - r_j}$$

$$\cdot \Phi \left[\begin{matrix} aq^{2n} & : c_{1,1}q_1^n, \ldots, c_{1,r_1}q_1^n : \ldots : \\ & : d_{1,1}q_1^n, \ldots, d_{1,s_1}q_1^n : \ldots : \end{matrix} \right.$$

$$\left. \begin{matrix} c_{m,1}q_m^n, \ldots, c_{m,r_m}q_m^n \\ d_{m,1}q_m^n, \ldots, d_{m,s_m}q_m^n \end{matrix} ; q, q_1, \ldots, q_m; \frac{bcz}{a} q^{-n-1} P_n \right] \tag{3.9.5}$$

where

$$P_n = \prod_{j=1}^{m} q_j^{n(s_j - r_j)}, \tag{3.9.6}$$

provided at least one c_{j,r_j} is a negative integer power of q_j so that the series terminate. Note that this formula is valid even if $a = q^{-k}$, $k = 0, 1, \ldots$, in which case the upper limit of the sum on the right side can be replaced by $[k/2]$, where $[k/2]$ denotes the greatest integer less than or equal to $k/2$.

Similarly, use of the expansion formula (2.8.2) leads to the formula

$$\Phi \left[\begin{matrix} a, b, c, d : c_{1,1}, \ldots, c_{1,r_1} : \ldots : c_{m,1}, \ldots, c_{m,r_m} \\ aq/b, aq/c, aq/d : d_{1,1}, \ldots, d_{1,s_1} : \ldots : d_{m,1}, \ldots, d_{m,s_m} \end{matrix} ; q, q_1, \ldots, q_m; z \right]$$

$$= \sum_{n=0}^{\infty} \frac{(1-\lambda q^{2n})(\lambda, \lambda b/a, \lambda c/a, \lambda d/a; q)_n}{(1-\lambda)(q, aq/b, aq/c, aq/d; q)_n} \frac{(a;q)_{2n}}{(\lambda q; q)_{2n}} \left(\frac{az}{\lambda}\right)^n$$

$$\cdot \prod_{j=1}^{m} \frac{(c_{j,1}, \ldots, c_{j,r_j}; q_j)_n}{(d_{j,1}, \ldots, d_{j,s_j}; q)_n} \left[(-1)^n q_j^{\binom{n}{2}} \right]^{s_j - r_j}$$

$$\cdot \Phi \left[\begin{matrix} aq^{2n}, \dfrac{a}{\lambda} : c_{1,1}q_1^n, \ldots, c_{1,r_1}q_1^n : \ldots : c_{m,1}q_m^n, \ldots, c_{m,r_m}q_m^n \\ \lambda q^{2n+1} : d_{1,1}q_1^n, \ldots, d_{1,s_1}q_1^n : \ldots : d_{m,1}q_m^n, \ldots, d_{m,s_m}q_m^n \end{matrix} ; q, q_1, \ldots, q_m; zP_n \right], \tag{3.9.7}$$

where $\lambda = qa^2/bcd$. In (3.9.7) the series on both sides need not terminate as long as they converge absolutely. Formulas (3.9.5) and (3.9.7) are multibasic extensions of formulas 4.3(1) and 4.3(6), respectively, in Bailey [1935].

3.10 Transformations of series with base q to series with base q^2

Following Nassrallah and Rahman [1981], we shall derive some quadratic transformation formulas for basic hypergeometric series by using the following special cases of (3.9.5) and (3.9.7):

$$\Phi \left[\begin{matrix} a^2, -aq^2, b^2, c^2 : a_1, \ldots, a_r, q^{-n} \\ -a, a^2q^2/b^2, a^2q^2/c^2 : b_1, \ldots, b_r, b_{r+1} \end{matrix} ; q^2, q; z \right]$$

$$= \sum_{j=0}^{n} \frac{(a^2q^2/b^2c^2, -aq^2; q^2)_j (a^2;q^2)_{2j} (a_1, \ldots, a_r, q^{-n}; q)_j}{(q^2, -a, a^2q^2/b^2, a^2q^2/c^2; q^2)_j (b_1, \ldots, b_r, b_{r+1}; q)_j} q^{-2\binom{j}{2}} \left(-\frac{b^2c^2z}{a^2q^2}\right)^j$$

$$\cdot \Phi \left[\begin{matrix} a^2q^{4j}, -aq^{2j+2} : a_1q^j, \ldots, a_rq^j, q^{j-n} \\ -aq^{2j} : b_1q^j, \ldots, b_rq^j, b_{r+1}q^j \end{matrix} ; q^2, q; \frac{b^2c^2z}{a^2q^{2j+2}} \right] \tag{3.10.1}$$

and

$$\Phi \left[\begin{matrix} a^2, -aq^2, b^2, c^2, d^2 : a_1, \ldots, a_r, q^{-n} \\ -a, a^2q^2/b^2, a^2q^2/c^2, a^2q^2/d^2 : b_1, \ldots, b_r, b_{r+1} \end{matrix} ; q^2, q; z \right]$$

$$= \sum_{j=0}^{n} \frac{1-\lambda q^{4j}}{1-\lambda} \frac{(\lambda, \lambda b^2/a^2, \lambda c^2/a^2, \lambda d^2/a^2, -aq^2; q^2)_j (a^2; q^2)_{2j}}{(q^2, a^2q^2/b^2, a^2q^2/c^2, a^2q^2/d^2, -a; q^2)_j (\lambda q^2; q^2)_{2j}}$$

$$\cdot \frac{(a_1, \ldots, a_r, q^{-n}; q)_j}{(b_1, \ldots, b_r, b_{r+1}; q)_j} \left(\frac{a^2z}{\lambda} \right)^j$$

$$\cdot \Phi \left[\begin{matrix} a^2q^{4j}, -aq^{2j+2}, a^2/\lambda : a_1q^j, \ldots, a_rq^j, q^{j-n} \\ -aq^{2j}, \lambda q^{4j+2} : b_1q^j, \ldots, b_rq^j, b_{r+1}q^j \end{matrix} ; q^2, q; z \right], \tag{3.10.2}$$

respectively, where $\lambda = \left(qa^2/bcd\right)^2$.

Let us first consider the $r = 1$ case of (3.10.1). If we set $a_1 = -aq/w$, $b_1 = w, b_2 = -aq^{n+1}$, $z = awq^{n+2}/b^2c^2$ and apply (1.2.40), the series on the right side of (3.10.1) reduces to a very-well-poised ${}_6\phi_5$ series in base q, and hence can be summed by (2.4.2). This gives the transformation formula

$$\Phi \left[\begin{matrix} a^2, -aq^2, b^2, c^2 : -aq/w, q^{-n} \\ -a, a^2q^2/b^2, a^2q^2/c^2 : w, -aq^{n+1} \end{matrix} ; q^2, q; \frac{awq^{n+2}}{b^2c^2} \right]$$

$$= \frac{(w/a, -aq; q)_n}{(w, -q; q)_n} \, {}_5\phi_4 \left[\begin{matrix} a, aq, a^2q^2/w^2, a^2q^2/b^2c^2, q^{-2n} \\ a^2q^2/b^2, a^2q^2/c^2, aq^{1-n}/w, aq^{2-n}/w \end{matrix} ; q^2, q^2 \right]. \tag{3.10.3}$$

Note that the above ${}_5\phi_4$ series is balanced and that the Φ series on the left side of (3.10.3) can be written as

$${}_{10}W_9(-a; a, b, -b, c, -c, -aq/w, q^{-n}; q, awq^{n+2}/b^2c^2).$$

Formula (3.10.2) is a q-analogue of Bailey [1935, 4.5(1)]. By reversing the series on both sides of (3.10.3) and relabelling the parameters, this formula can be written, as in Jain and Verma [1980], in the form

$${}_{10}\phi_9 \left[\begin{matrix} a, qa^{\frac{1}{2}}, -qa^{\frac{1}{2}}, b, x, -x, y, -y, -q^{-n}, q^{-n} \\ a^{\frac{1}{2}}, -a^{\frac{1}{2}}, aq/b, aq/x, -aq/x, aq/y, -aq/y, -aq^{n+1}, aq^{n+1} \end{matrix} ; q, -\frac{a^3q^{2n+3}}{bx^2y^2} \right]$$

$$= \frac{(a^2q^2, a^2q^2/x^2y^2; q^2)_n}{(a^2q^2/x^2, a^2q^2/y^2; q^2)_n} \, {}_5\phi_4 \left[\begin{matrix} q^{-2n}, x^2, y^2, -aq/b, -aq^2/b \\ x^2y^2q^{-2n}/a^2, a^2q^2/b^2, -aq, -aq^2 \end{matrix} ; q^2, q^2 \right]. \tag{3.10.4}$$

For a nonterminating extension of (3.10.4) see Jain and Verma [1982].

Since the ${}_5\phi_4$ series on the right side of (3.10.3) is balanced, it can be summed by (1.7.2) whenever it reduces to a ${}_3\phi_2$ series. Thus, we obtain the summation formulas:

$$\Phi\left[\begin{matrix} a^2, aq^2, -aq^2 : -aq/w, q^{-n} \\ a, -a : w, -aq^{n+1} \end{matrix}; q^2, q; \frac{wq^{n+1}}{a}\right]$$
$$= \frac{(w/a, -aq; q)_n (wq^{-n-1}/a, aq^{2-n}/w; q^2)_n}{(w, -q; q)_n (aq^{1-n}/w, wq^{-n}/a; q^2)_n}, \tag{3.10.5}$$

$$\Phi\left[\begin{matrix} a^2, -aq^2, b^2 : -aq^n/b^2, q^{-n} \\ -a, a^2q^2/b^2 : b^2q^{1-n}, -aq^{n+1} \end{matrix}; q^2, q; q^2\right]$$
$$= \frac{(q^n/b^2, -aq; q)_n (aq/b^2, aq^2/b^2; q^2)_n}{(aq^n/b^2, -q; q)_n (q/b^2, a^2q^2/b^2; q^2)_n}, \tag{3.10.6}$$

$$\Phi\left[\begin{matrix} a^2, aq^2, -aq^2, b^2 : -aq^n/b^2, q^{-n} \\ a, -a, a^2q^2/b^2 : b^2q^{1-n}, -aq^{n+1} \end{matrix}; q^2, q; q\right]$$
$$= \frac{(-aq^2, a/b^2; q)_n (1/b^2; q^2)_n}{(-q, 1/b^2; q)_n (a^2q^2/b^2; q^2)_n} q^n \tag{3.10.7}$$

and

$$\Phi\left[\begin{matrix} a^2, aq^2, -aq^2, b^2 : -aq^{n-1}/b^2, q^{-n} \\ a, -a, a^2q^2/b^2 : b^2q^{2-n}, -aq^{n+1} \end{matrix}; q^2, q; q^2\right]$$
$$= \frac{(-aq, a/q^2; q)_n (aq/b^2, 1/b^2q; q^2)_n}{(-q, 1/qb^2; q)_n (a/qb^2, a^2q^2/b^2; q^2)_n} q^n. \tag{3.10.8}$$

These are q-analogues of formulas 4.5(1.1) - 4.5(1.4) in Bailey [1935]. Since the series on the left sides of (3.10.5) and (3.10.6) can also be written as very-well-poised ${}_8\phi_7$ series in base q, which are transformable to balanced ${}_4\phi_3$ series by Watson's formula (2.5.1), formulas (3.10.5) and (3.10.6) are equivalent to the summation formulas

$${}_4\phi_3\left[\begin{matrix} a, qa^{\frac{1}{2}}, -w/qa^{\frac{1}{2}}, q^{-n} \\ a^{\frac{1}{2}}, w, -a^{\frac{1}{2}}q^{1-n} \end{matrix}; q, q\right]$$
$$= \frac{(w/a, -a^{\frac{1}{2}}; q)_n (wq^{-n-1}/a, aq^{2-n}/w; q^2)_n}{(w, -a^{-\frac{1}{2}}; q)_n (aq^{1-n}/w, wq^{-n}/a; q^2)_n} \tag{3.10.9}$$

and

$${}_4\phi_3\left[\begin{matrix} a, b, -bq^{1-n}, q^{-n} \\ aq/b, b^2q^{1-n}, -bq^{-n} \end{matrix}; q, q\right]$$
$$= \frac{(1+1/b)(1+a^{\frac{1}{2}}q^n/b)(a/b^2, qa^{\frac{1}{2}}/b, 1/b; q)_n}{(1+q^n/b)(1+a^{\frac{1}{2}}/b)(aq/b, a^{\frac{1}{2}}/b, 1/b^2; q)_n}, \tag{3.10.10}$$

respectively, which are closer to what one would expect q-analogues of formulas 4.5(1.1) and 4.5(1.2) in Bailey [1935] to look like.

It is also of interest to note that if we set $c^2 = aq$ in (3.10.3), rewrite the Φ series on the left side as an ${}_8\phi_7$ series in base q and transform it to a balanced

${}_4\phi_3$ series, we obtain

$$
{}_4\phi_3\left[\begin{matrix} a, b, -w/b, q^{-n} \\ aq/b, -bq^{-n}, w \end{matrix}; q, q\right]
$$

$$
= \frac{(w/a, -aq/b; q)_n}{(w, -q/b; q)_n}\, {}_4\phi_3\left[\begin{matrix} a, a^2q^2/w^2, aq/b^2, q^{-2n} \\ a^2q^2/b^2, aq^{1-n}/w, aq^{2-n}/w \end{matrix}; q^2, q^2\right], \quad (3.10.11)
$$

which is a q-analogue of the $c = (1+a)/2$ case of Bailey [1935, 4.5(1)]. Using (2.10.4), the left side of (3.10.11) can be transformed to give

$$
{}_4\phi_3\left[\begin{matrix} q^{-n}, a, aq/b^2, -aq/w \\ aq/b, -aq/b, aq^{1-n}/w \end{matrix}; q, q\right]
$$

$$
= {}_4\phi_3\left[\begin{matrix} q^{-2n}, a, aq/b^2, a^2q^2/w^2 \\ a^2q^2/b^2, aq^{1-n}/w, aq^{2-n}/w \end{matrix}; q^2, q^2\right]. \quad (3.10.12)
$$

This formula was first proved by Singh [1959] and more recently by Askey and Wilson [1985]. The latter authors also wrote it in the form

$$
{}_4\phi_3\left[\begin{matrix} a^2, b^2, c, d \\ abq^{\frac{1}{2}}, -abq^{\frac{1}{2}}, -cd \end{matrix}; q, q\right]
$$

$$
= {}_4\phi_3\left[\begin{matrix} a^2, b^2, c^2, d^2 \\ qa^2b^2, -cd, -qcd \end{matrix}; q^2, q^2\right], \quad (3.10.13)
$$

provided that both series terminate.

Now that we have the summation formulas (3.10.5)–(3.10.8), we can use them to produce additional transformation formulas. Set $r = 3$ and $a_1 = -a_2 = qa^{\frac{1}{2}}, a_3 = -aq/w, b_1 = -b_2 = a^{\frac{1}{2}}$, $b_4 = -aq^{n+1}$ in (3.10.1). The Φ series on the right side can now be summed by (3.10.5) and this leads to the following q-analogue of Bailey [1935, 4.5(2)]

$$
\Phi\left[\begin{matrix} a^2, aq^2, -aq^2, b^2, c^2 : -aq/w, q^{-n} \\ a, -a, a^2q^2/b^2, a^2q^2/c^2 : w, -aq^{n+1} \end{matrix}; q^2, q; \frac{awq^{n+1}}{b^2c^2}\right]
$$

$$
= \frac{(w/a, -aq; q)_n (wq^{-n-1}/a, aq^{2-n}/w; q^2)_n}{(w, -q; q)_n (aq^{1-n}/w, wq^{-n}/a; q^2)_n}
$$

$$
\cdot\, {}_5\phi_4\left[\begin{matrix} aq, aq^2, a^2q^2/b^2c^2, a^2q^2/w^2, q^{-2n} \\ a^2q^2/b^2, a^2q^2/c^2, aq^{2-n}/w, aq^{3-n}/w \end{matrix}; q^2, q^2\right]. \quad (3.10.14)
$$

Let us now turn to applications of (3.10.2). If we set $r = 1$, $a_1 = -\lambda q^{n+1}/a$, $b_1 = a^2q^{-n}/\lambda$, $b_2 = -aq^{n+1}$, $z = q$ in (3.10.2), where $\lambda = (qa^2/bcd)^2$, then the Φ series on the right side reduces to a balanced very-well-poised ${}_8\phi_7$ series in base q which can be summed by Jackson's formula (2.6.2). Thus, we obtain the transformation formula

$$
\Phi\left[\begin{matrix} a^2, -aq^2, b^2, c^2, d^2 : -\lambda q^{n+1}, q^{-n} \\ -a, a^2q^2/b^2, a^2q^2/c^2, a^2q^2/d^2 : a^2q^{-n}/\lambda, -aq^{n+1} \end{matrix}; q^2, q; q\right]
$$

$$= \frac{(-aq, \lambda q/a; q)_n (\lambda q^2/a^2; q^2)_n}{(-q, \lambda q/a^2; q)_n (\lambda q^2; q^2)_n}$$

$$\cdot {}_{10}\phi_9 \left[\begin{matrix} \lambda, q^2\lambda^{\frac{1}{2}}, -q^2\lambda^{\frac{1}{2}}, a, aq, \frac{\lambda b^2}{a^2}, \frac{\lambda c^2}{a^2}, \frac{\lambda d^2}{a^2}, \frac{\lambda^2 q^{2n+2}}{a^2}, q^{-2n} \\ \lambda^{\frac{1}{2}}, -\lambda^{\frac{1}{2}}, \frac{\lambda q^2}{a}, \frac{\lambda q}{a}, \frac{a^2q^2}{b^2}, \frac{a^2q^2}{c^2}, \frac{a^2q^2}{d^2}, \frac{a^2q^{-2n}}{\lambda}, \lambda q^{2n+2} \end{matrix} ; q^2, \frac{qa^2}{\lambda} \right]. \tag{3.10.15}$$

This is a q-analogue of Bailey [1935, 4.5(3)].

Similarly, setting $r = 3$ and choosing the parameters so that the Φ series on the right side of (3.10.2) can be summed by (3.10.7), we get the following q-analogue of Bailey [1935, 4.5(4)]

$$\Phi \left[\begin{matrix} a^2, aq^2, -aq^2, b^2, c^2, d^2 : -\lambda q^n/a, q^{-n} \\ a, -a, a^2q^2/b^2, a^2q^2/c^2, a^2q^2/d^2 : a^2q^{1-n}/\lambda, -aq^{n+1} \end{matrix} ; q^2, q; q \right]$$

$$= \frac{(\lambda/a, -aq; q)_n (\lambda/a^2; q^2)_n}{(\lambda/a^2, -q; q)_n (\lambda q^2; q^2)_n} q^n$$

$$\cdot {}_{10}\phi_9 \left[\begin{matrix} \lambda, q^2\lambda^{\frac{1}{2}}, -q^2\lambda^{\frac{1}{2}}, aq, aq^2, \frac{\lambda b^2}{a^2}, \frac{\lambda c^2}{a^2}, \frac{\lambda d^2}{a^2}, \frac{\lambda^2 q^{2n}}{a^2}, q^{-2n} \\ \lambda^{\frac{1}{2}}, -\lambda^{\frac{1}{2}}, \frac{\lambda q}{a}, \frac{\lambda}{a}, \frac{a^2q^2}{b^2}, \frac{a^2q^2}{c^2}, \frac{a^2q^2}{d^2}, \frac{a^2q^{2-2n}}{\lambda}, \lambda q^{2n+1} \end{matrix} ; q^2, \frac{qa^2}{\lambda} \right], \tag{3.10.16}$$

where $\lambda = (qa^2/bcd)^2$.

Exercises 3

3.1 Deduce from (3.10.13) that

$${}_3\phi_2 \left[\begin{matrix} a^2, b^2, z \\ abq^{\frac{1}{2}}, -abq^{\frac{1}{2}} \end{matrix} ; q, q \right] = {}_3\phi_2 \left[\begin{matrix} a^2, b^2, z^2 \\ a^2b^2q, 0 \end{matrix} ; q^2, q^2 \right],$$

provided that both series terminate. Show that this formula is a q-analogue of Gauss' quadratic transformation formula (3.1.7) when the series terminate.

3.2 Using the sum

$${}_2\phi_1 \left(q^{-n}, q^{1-n}; qb^2; q^2, q^2 \right) = \frac{(b^2; q^2)_n}{(b^2; q)_n} q^{-\binom{n}{2}},$$

prove that

(i) $${}_3\phi_2 \left[\begin{matrix} a, b, -b \\ b^2, az \end{matrix} ; q, -z \right] = \frac{(z; q)_\infty}{(az; q)_\infty} {}_2\phi_1 \left(a, aq; qb^2; q^2, z^2 \right), \quad |z| < 1.$$

Use this formula to prove that

(ii) $${}_3\phi_2 \left[\begin{matrix} a^2, & ab, & -ab \\ & a^2b^2, & -za^2 \end{matrix} ; q, z \right]$$

$$= \frac{(a^2z^2; q)_\infty}{(z, -a^2z; q)_\infty} {}_2\phi_2 \left[\begin{matrix} a^2, & b^2 \\ a^2b^2q, & a^2z^2 \end{matrix} ; q^2, a^2z^2q \right], \quad |z| < 1.$$

Formulas (i) and (ii), due to Jain [1981], are q-analogues of (3.1.5) and (3.1.6), respectively.

3.3 Show that

$$ {}_3\phi_2\left[\begin{matrix} a, q/a, z \\ c, -q \end{matrix}; q, q\right] = \frac{(-1, -qz/c; q)_\infty}{(-q/c, -z; q)_\infty}\, {}_3\phi_2\left[\begin{matrix} c/a, ac/q, z^2 \\ c^2, 0 \end{matrix}; q^2, q^2\right], $$

when the series terminate. This is a q-analogue of

$$ {}_2F_1\left(a, 1-a; c; z\right) = (1-z)^{c-1}\, {}_2F_1\left((c-a)/2, (c+a-1)/2; c; 4z(1-z)\right), $$

when the series terminate.

3.4 Show that

$$ \sum_{k=0}^{n} \frac{(q^{-n}, b, -b; q)_k}{(q, b^2; q)_k}\, q^{nk-\binom{k}{2}} $$

vanishes if n is an odd integer. Evaluate the sum when n is even. Hence, or otherwise, show that

$$ \sum_{m=0}^{\infty}\sum_{n=0}^{\infty} \frac{(a, c; q)_{m+n} (b^2; q^2)_m}{(q; q)_m (q; q)_n (d; q)_{m+n} (b^2; q)_m} (-z)^m z^n = {}_4\phi_3\left[\begin{matrix} a, aq, c, cq \\ d, dq, qb^2 \end{matrix}; q^2, z^2\right], \quad |z| < 1. $$

Deduce that

$$ {}_4\phi_3\left[\begin{matrix} q^{-n}, q^{1-n}, a, aq \\ qb^2, d, dq \end{matrix}; q^2, q^2\right] = \frac{(d/a; q)_n}{(d; q)_n} a^n\, {}_4\phi_3\left[\begin{matrix} q^{-n}, a, b, -b \\ b^2, aq^{1-n}/d, 0 \end{matrix}; q, -\frac{q}{d}\right]. $$

(Jain [1981])

3.5 By using Sears' summation formula (2.10.12) show that

$$ \begin{aligned} &\sum_{r=0}^{\infty} \frac{(a, aq/e; q)_r}{(q, abq/e, acq/e; q)_r} A_r \\ &= \frac{(aq/e, bq/e, cq/e, abcq/e; q)_\infty}{(q/e, abq/e, acq/e, bcq/e; q)_\infty} \\ &\quad \cdot \sum_{k=0}^{\infty} \frac{(a, b, c; q)_k q^k}{(q, e, abcq/e; q)_k} \sum_{r=0}^{\infty} \frac{(aq^k; q)_r}{(q, abcq^{k+1}/e; q)_r} A_r \\ &\quad + \frac{(a, b, c, abcq^2/e^2; q)_\infty}{(e/q, abq/e, acq/e, bcq/e; q)_\infty} \\ &\quad \cdot \sum_{k=0}^{\infty} \frac{(aq/e, bq/e, cq/e; q)_k}{(q, q^2/e, abcq^2/e^2; q)_k} q^k \sum_{r=0}^{\infty} \frac{(aq^{k+1}/e; q)_r}{(q, abcq^{k+2}/e^2; q)_r} A_r, \end{aligned} $$

where a, b, c, e are arbitrary parameters such that $e \neq q$, and $\{A_r\}$ is an arbitrary sequence such that the infinite series on both sides converge absolutely.

3.6 Prove that

$$
{}_3\phi_2\left[\begin{matrix} a, & b, & c \\ & d, & e \end{matrix}; q, q\right] + \frac{(q/e, a, b, c, dq/e; q)_\infty}{(e/q, aq/e, bq/e, cq/e, d; q)_\infty}
$$
$$
\cdot\, {}_3\phi_2\left[\begin{matrix} aq/e, & bq/e, & cq/e \\ & q^2/e, & dq/e \end{matrix}; q, q\right]
$$
$$
= \frac{(q/e, abq/e, acq/e, d/a; q)_\infty}{(d, aq/e, bq/e, cq/e; q)_\infty}\, {}_3\phi_2\left[\begin{matrix} a, aq/e, abcq/de \\ abq/e, acq/e \end{matrix}; q, \frac{d}{a}\right],
$$

where $e \neq q$ and $|d/a| < 1$.

3.7 Prove that

$$
{}_3\phi_2\left[\begin{matrix} a, aq/e, e/bc \\ abq/e, acq/e \end{matrix}; q, -\frac{bcq}{e}\right]
$$
$$
= \frac{(aq, aq/e, bq/e, cq/e, -q; q)_\infty}{(q/e, abq/e, acq/e, bcq/e, -bcq/e; q)_\infty}
$$
$$
\cdot \sum_{k=0}^{\infty} \frac{(a, b, c; q)_k\, (ab^2c^2q^{k+2}/e^2; q^2)_\infty}{(q, aq, e; q)_k (aq^{k+2}; q^2)_\infty}\, q^k,
$$

provided $|bcq/e| < 1$.

3.8 Assuming that $|x| < 1$ and $a/b \neq q^j$, $j = 0, \pm 1, \pm 2, \ldots$, prove that

$$
{}_2\phi_1(a, b; c; q, x)
$$
$$
= \frac{(b, c/a, ax; q)_\infty}{(b/a, c, x; q)_\infty}\, {}_3\phi_2\left[\begin{matrix} a, c/b, 0 \\ aq/b, ax \end{matrix}; q, q\right]
$$
$$
+ \frac{(a, c/b, bx; q)_\infty}{(a/b, c, x; q)_\infty}\, {}_3\phi_2\left[\begin{matrix} b, c/a, 0 \\ bq/a, bx \end{matrix}; q, q\right].
$$

Show that this is a q-analogue of the formula

$$
{}_2F_1(a, b; c; x)
$$
$$
= \frac{\Gamma(c)\Gamma(b-a)}{\Gamma(b)\Gamma(c-a)}(1-x)^{-a}\, {}_2F_1(a, c-b; a-b+1; (1-x)^{-1})
$$
$$
+ \frac{\Gamma(c)\Gamma(a-b)}{\Gamma(a)\Gamma(c-b)}(1-x)^{-b}\, {}_2F_1(b, c-a; b-a+1; (1-x)^{-1}).
$$

3.9 Show that

$$
{}_3\phi_2\left[\begin{matrix} a, \lambda q, b \\ \lambda, q\lambda^2/b \end{matrix}; q, \frac{\lambda^2}{ab^2}\right]
$$
$$
= \frac{1 - \lambda + \lambda/b(1-\lambda/a)}{(1-\lambda)(1+\lambda/b)}\, \frac{(\lambda^2/b^2, q\lambda^2/ab; q)_\infty}{(q\lambda^2/b, \lambda^2/ab^2; q)_\infty}, \quad |\lambda^2/ab^2| < 1.
$$

3.10 Show that

$$
{}_8W_7\left(-\lambda; q\lambda^{\frac{1}{2}}, -q\lambda^{\frac{1}{2}}, a, b, -b; q, \lambda/ab^2\right)
= \frac{(\lambda q, -\lambda q, \lambda/b^2, \lambda q/ab, -\lambda q/ab; q)_\infty}{(\lambda q/a, -\lambda q/a, \lambda q/b, -\lambda q/b, \lambda/ab^2; q)_\infty}, \quad |\lambda/ab^2| < 1.
$$

Show that this is a q-analogue of the formula

$$
{}_3F_2\left[\begin{matrix} a, & 1+\lambda/2, & b \\ & \lambda/2, & 1+\lambda-b \end{matrix}; 1\right]
= \frac{\Gamma(\lambda/2)\Gamma\left(1+\frac{\lambda-a}{2}\right)\Gamma(1+\lambda-b)\Gamma(\lambda-a-2b)}{\Gamma(1+\lambda/2)\Gamma\left(\frac{\lambda-a}{2}\right)\Gamma(1+\lambda-a-b)\Gamma(\lambda-2b)}, \quad \mathrm{Re}\,(\lambda-a-2b) > 0.
$$

3.11 Derive Jackson's [1941] product formula

$$
{}_2\phi_1(a^2, b^2; qa^2b^2; q^2, z)\ {}_2\phi_1(a^2, b^2; qa^2b^2; q^2, qz)
= {}_4\phi_3\left[\begin{matrix} a^2, b^2, ab, -ab \\ a^2b^2, abq^{\frac{1}{2}}, -abq^{\frac{1}{2}} \end{matrix}; q, z\right], \quad |z| < 1,\ |q| < 1,
$$

and show that it has Clausen's [1828] formula

$$
\left[{}_2F_1\left(a, b; a+b+\frac{1}{2}; z\right)\right]^2 = {}_3F_2\left[\begin{matrix} 2a, 2b, a+b \\ 2a+2b, a+b+\frac{1}{2} \end{matrix}; z\right], \quad |z| < 1,
$$

as a limit case. Additional q-analogues of Clausen's formula are given in §8.8.

3.12 Prove that

$$
\Phi\left[\begin{matrix} q^{-2n}, -q^{2-n}, b^2, c^2 : d, -q^{1-n}/w \\ -q^{-n}, q^{2-2n}/b^2, q^{2-2n}/c^2 : -q^{1-n}/d, w \end{matrix}; q^2, q; \frac{wq^{2-n}}{b^2c^2d}\right]
= \frac{(-1, w/d; q)_n}{(-d, w; q)_n} d^n
\cdot {}_5\phi_4\left[\begin{matrix} d^2, q^{2-2n}/b^2c^2, q^{2-2n}/w^2, q^{-n}, q^{1-n} \\ q^{2-2n}/b^2, q^{2-2n}/c^2, dq^{1-n}/w, dq^{2-n}/w \end{matrix}; q^2, q^2\right].
$$

3.13 If $\lambda = a^4q^2/b^2c^2d^2$, show that

$$
\Phi\left[\begin{matrix} a^2, -q^2a, b^2, c^2, d^2 : -\lambda q^n/a, q^{-n} \\ -a, a^2q^2/b^2, a^2q^2/c^2, a^2q^2/d^2 : a^2q^{1-n}/\lambda, -aq^{n+1} \end{matrix}; q^2, q; q^2\right]
= \frac{(-aq, \lambda/a; q)_n(\lambda/a^2, \lambda q^2/a; q^2)_n}{(-q, \lambda/a^2; q)_n(\lambda q^2, \lambda/a; q^2)_n}
\cdot {}_{10}W_9(\lambda; a, aq, \lambda b^2/a^2, \lambda c^2/a^2, \lambda d^2/a^2, \lambda^2q^{2n}/a^2, q^{-2n}; q^2, a^2q^2/\lambda)
$$

and

$$
\Phi\left[\begin{matrix} a^2, q^2a, -q^2a, b^2, c^2, d^2 : -\lambda q^{n-1}/a, q^{-n} \\ a, -a, a^2q^2/b^2, a^2q^2/c^2, a^2q^2/d^2 : a^2q^{2-n}/\lambda, -aq^{n+1} \end{matrix}; q^2, q; q^2\right]
= \frac{(-aq, \lambda/aq; q)_n(\lambda/a^2q^2; q^2)_n(1-\lambda q^{2n-1}/a)}{(-q, \lambda/a^2; q)_n(\lambda q^2; q^2)_n(1-\lambda q^{-1}/a)} q^n
\cdot {}_{10}W_9(\lambda; aq, aq^2, \lambda b^2/a^2, \lambda c^2/a^2, \lambda d^2/a^2, \lambda^2q^{2n-2}/a^2, q^{-2n}; q^2, a^2q^3/\lambda).
$$

(Nassrallah and Rahman [1981])

3.14 Using (3.4.7) show that the q-Bessel function defined in Ex. 1.24 can be expressed as

$$J_\nu^{(2)}(x;q) = \frac{(q^{\nu+1};q)_\infty}{(q,-x^2q^{\nu+1}/4;q)_\infty}\left(\frac{2}{x}\right)^\nu$$
$$\cdot \sum_{n=0}^\infty \frac{(-x^2q^\nu/4;q)_n(1+x^2q^{2n+\nu}/4)(-x^2/4;q)_n}{(q;q)_n(1+x^2q^\nu/4)(q^{\nu+1};q)_n}\left(-x^2q^{2\nu}/4\right)^n q^{2n^2}.$$

3.15 Using (3.5.4) show that

$$J_\nu^{(2)}(x;q) = \frac{1}{\Gamma_q(\nu+1)}\left(\frac{ix}{2};q^{\frac{1}{2}}\right)_\infty \left(\frac{x}{2(1-q)}\right)^\nu$$
$$\cdot\ {}_2\phi_1\left[\begin{matrix} q^{\nu/2+1/4}, & -q^{\nu/2+1/4} \\ & q^{\nu+\frac{1}{2}} \end{matrix}; q^{\frac{1}{2}}, \frac{ix}{2}\right],\quad |x|<2.$$

3.16 Show that

$${}_4\phi_3\left[\begin{matrix} a, -qa^{\frac{1}{2}}, b, c \\ -a^{\frac{1}{2}}, aq/b, aq/c \end{matrix}; q, \frac{axq}{bc}\right]$$
$$= \frac{(1-xa^{\frac{1}{2}})(axq;q)_\infty}{(x;q)_\infty}\ {}_5\phi_4\left[\begin{matrix} a^{\frac{1}{2}}, -qa^{\frac{1}{2}}, (aq)^{\frac{1}{2}}, -(aq)^{\frac{1}{2}}, aq/bc \\ aq/b, aq/c, q/x, axq \end{matrix}; q, q\right]$$
$$+ \frac{(1-a^{\frac{1}{2}})(aq, aq/bc, axq/b, axq/c;q)_\infty}{(aq/b, aq/c, axq/bc, 1/x;q)_\infty}$$
$$\cdot\ {}_5\phi_4\left[\begin{matrix} xa^{\frac{1}{2}}, -xqa^{\frac{1}{2}}, x(aq)^{\frac{1}{2}}, -x(aq)^{\frac{1}{2}}, axq/bc \\ axq/b, axq/c, qx, aqx^2 \end{matrix}; q, q\right].$$

3.17 If $\displaystyle \frac{(cz/ab;q^2)_\infty}{(z;q^2)_\infty}\ {}_2\phi_1(a,b;c;q,cz/abq) = \sum_{n=0}^\infty a_n z^n,$

show that

$${}_2\phi_1(c/a, c/b; cq; q^2, z)\ {}_2\phi_1(a,b;cq;q^2,cz/abq)$$
$$= \sum_{n=0}^\infty \frac{(c;q^2)_n}{(cq;q^2)_n} a_n z^n.$$

(Singh [1959], Nassrallah [1982])

3.18 If $\displaystyle \frac{(cqz/ab;q^2)_\infty}{(z;q^2)_\infty}\ {}_2\phi_1(a,b;c;q,cz/ab) = \sum_{n=0}^\infty a_n z^n,$

show that

$${}_2\phi_1(cq/a, cq/b; cq^2; q^2, z)\ {}_2\phi_1(a,b;c;q^2,cz/ab)$$
$$= \sum_{n=0}^\infty \frac{(cq;q^2)_n}{(cq^2;q^2)_n} a_n z^n.$$

(Singh [1959], Nassrallah [1982])

3.19 If $\dfrac{(cqz/ab;q^2)_\infty}{(z;q^2)_\infty}\ {}_2\phi_1(a/q,b;c/q;q,cz/ab)=\sum_{n=0}^{\infty}a_nz^n$, show that

$$
\begin{aligned}
&{}_2\phi_1(cq/a,c/bq;c;q^2,z)\ {}_2\phi_1(a,b;c;q^2,cz/ab)\\
&=\sum_{n=0}^{\infty}\frac{(c/q;q^2)_n}{(c;q^2)_n}a_nz^n.
\end{aligned}
$$

(Singh [1959], Nassrallah [1982])

3.20 Prove that

$$
\sum_{k=0}^{\infty}\frac{1-ap^kq^k}{1-a}\frac{(a;p)_k(b^{-1};q)_k}{(q;q)_k(abp;p)_k}b^k=0
$$

when $\max(|p|,|q|,|b|)<1$, and extend this to the bibasic transformation formulas

$$
\begin{aligned}
&\sum_{k=0}^{\infty}\frac{1-ap^kq^k}{1-a}\frac{(a;p)_k(c/b;q)_k}{(q;q)_k(abp;p)_k}b^k\\
&=\frac{1-c}{1-b}\sum_{k=0}^{\infty}\frac{(ap;p)_k(c/b;q)_k}{(q;q)_k(abp;p)_k}(bq)^k\\
&=\frac{1-c}{1-abp}\sum_{k=0}^{\infty}\frac{(ap;p)_k(cq/b;q)_k}{(q;q)_k(abp^2;p)_k}b^k\\
&=\frac{(1-c)(ap;p)_\infty}{(1-b)(abp;p)_\infty}\sum_{k=0}^{\infty}\frac{(b;p)_k(cqp^k;q)_\infty}{(p;p)_k(bqp^k;q)_\infty}(ap)^k
\end{aligned}
$$

when $\max(|p|,|q|,|ap|,|b|)<1$.
(Gasper [1989a])

3.21 Derive the quadbasic transformation formula

$$
\begin{aligned}
&\sum_{k=0}^{n}\frac{(1-ap^kq^k)(1-bp^kq^{-k})}{(1-a)(1-b)}\frac{(a,b;p)_k(c,a/bc;q)_k}{(q,aq/b;q)_k(ap/c,bcp;p)_k}\\
&\quad\cdot\frac{(CP^{-n}/A,P^{-n}/BC;P)_k(Q^{-n},BQ^{-n}/A;Q)_k}{(Q^{-n}/C,BCQ^{-n}/A;Q)_k(P^{-n}/A,P^{-n}/B;P)_k}q^k\\
&=\frac{(ap,bp;p)_n(cq,aq/bc;q)_n}{(q,aq/b;q)_n(ap/c,bcp;p)_n}\frac{(A,AQ/B;Q)_n(AP/C,BCP;P)_n}{(AP,BP;P)_n(CQ,AQ/BC;Q)_n}\\
&\quad\cdot\sum_{k=0}^{\infty}\frac{(1-AP^kQ^k)(1-BP^kQ^{-k})}{(1-A)(1-B)}\frac{(A,B;P)_k(C,A/BC;Q)_k}{(Q,AQ/B;Q)_k(AP/C,BCP;P)_k}\\
&\quad\cdot\frac{(cp^{-n}/a,p^{-n}/bc;p)_k(q^{-n},bq^{-n}/a;q)_k}{(q^{-n}/c,bcq^{-n}/a;q)_k(p^{-n}/a,p^{-n}/b;p)_k}Q^k
\end{aligned}
$$

for $n=0,1,\ldots$. Use it to derive the mixed bibasic and hypergeometric transformation formula

$$
\sum_{k=0}^{n}\frac{(1-ap^kq^k)(1-bp^kq^{-k})}{(1-a)(1-b)}\frac{(a,b;p)_k(c,a/bc;q)_k}{(q,aq/b;q)_k(ap/c,bcp;p)_k}
$$

$$\cdot \frac{(C-A-n)_k(-B-C-n)_k(\mu B-\mu A-n)_k(-n)_k}{(-\mu C-n)_k(\mu B+\mu C-\mu A-n)_k(-A-n)_k(-B-n)_k} q^k$$

$$= \frac{(ap,bp;p)_n(cq,aq/bc;q)_n}{(q,aq/b;q)_n(ap/c,bcp;p)_n} \frac{n!(\mu A+1-\mu B)_n(A+1-C)_n}{(A+1)_n(B+1)_n(\mu C+1)_n}$$

$$\cdot \frac{(B+C+1)_n}{(\mu A+1-\mu B-\mu C)_n}$$

$$\cdot \sum_{k=0}^{n} \frac{(A+k+k/\mu)(B+k-k\mu)}{AB} \frac{(A)_k(B)_k(\mu C)_k(\mu A-\mu B-\mu C)_k}{k!(\mu A+1-\mu B)_k(A+1-C)_k(B+C+1)_k}$$

$$\cdot \frac{(c/ap^n,1/bcp^n;p)_k(q^{-n},b/aq^n;q)_k}{(1/cq^n,bc/aq^n;q)_k(1/ap^n,1/bp^n;p)_k},$$

and the following transformation formula for a "*split-poised*" ${}_{10}\phi_9$ series

$${}_{10}\phi_9\left[\begin{matrix} a,q\sqrt{a},-q\sqrt{a},b,c,a/bc,C/Aq^n,q/BCq^n,B/Aq^n,q^{-n} \\ \sqrt{a},-\sqrt{a},aq/b,aq/c,bcq,1/Cq^n,BC/Aq^n,1/Bq^n,1/Aq^n \end{matrix};q,q\right]$$

$$= \frac{(aq,bq,cq,aq/bc,Aq/B,Aq/C,BCq;q)_n}{(Aq,Bq,Cq,Aq/BC,aq/b,aq/c,bcq;q)_n}$$

$$\cdot {}_{10}\phi_9\left[\begin{matrix} A,q\sqrt{A},-q\sqrt{A},B,C,A/BC,c/aq^n,1/bcq^n,b/aq^n,q^{-n} \\ \sqrt{A},-\sqrt{A},Aq/B,Aq/C,BCq,1/cq^n,bc/aq^n,1/bq^n,1/aq^n \end{matrix};q,q\right].$$

(Gasper [1989a])

3.22 Using the observation that, for arbitrary (fixed) positive integers $m_1,\ldots,m_r$,

$$\sum_{k_1,\ldots,k_r=0}^{\infty} \Lambda(k_1,\ldots,k_r) z_1^{k_1}\cdots z_r^{k_r}$$

$$= \sum_{M=0}^{\infty} \sum_{\substack{k_1m_1+\cdots+k_rm_r=M \\ k_1,\ldots,k_r\geq 0}} \Lambda(k_1,\ldots,k_r) z_1^{k_1}\cdots z_r^{k_r},$$

show that (3.7.14) implies the multivariable bibasic expansion formula

$$\sum_{k_1,\ldots,k_r=0}^{\infty} \Lambda(k_1,\ldots,k_r)\Omega_{k_1m_1+\cdots+k_rm_r}(x^{m_1}w^{m_1}z_1)^{k_1}\cdots(x^{m_r}w^{m_r}z_r)^{k_r}$$

$$= \sum_{n=0}^{\infty} \frac{(apq^n,bpq^{-n};p)_{n-1}}{(q,aq^n/b;q)_n}(-x)^n q^{n+\binom{n}{2}}$$

$$\cdot \sum_{j=0}^{\infty} \frac{(ap^nq^n,bp^nq^{-n};p)_j}{(q,aq^{2n+1}/b;q)_j}\Omega_{j+n}x^jq^j$$

$$\cdot \sum_{\substack{k_1m_1+\cdots+k_rm_r=M\leq n \\ k_1,\ldots,k_r\geq 0}} \frac{(1-ap^Mq^M)(1-bp^Mq^{-M})(q^{-n},aq^n/b;q)_M}{(apq^n,bpq^{-n};p)_M}$$

$$\cdot \Lambda(k_1,\ldots,k_r)(w^{m_1}z_1)^{k_1}\cdots(w^{m_r}z_r)^{k_r},$$

which is equivalent to (3.7.14) and extends Srivastava [1984, (10)]. (Gasper [1989a])

3.23 Prove the following q-Lagrange inversion theorem: If

$$G_n(x) = \sum_{j=n}^{\infty} b_{jn} x^j,$$

where b_{jn} is as defined in (3.6.20), and if

$$f(x) = \sum_{j=0}^{\infty} f_j x^j = \sum_{n=0}^{\infty} c_n G_n(x),$$

then

$$f_j = \sum_{n=0}^{j} b_{jn} c_n$$

and, vice versa,

$$c_n = \sum_{j=0}^{n} a_{nj} f_j,$$

where a_{nj} is as defined in (3.6.19). (Gasper [1989a])

3.24 Derive (3.8.19)–(3.8.22).

3.25 Prove Gosper's formula

$$\sum_{n=0}^{\infty} \frac{(a^2b^2c^2q^{-1};q^2)_n(1-a^2b^2c^2q^{3n-1})(abcd, abcd^{-1};q^2)_n(a^2,b^2,c^2;q)_n}{(q;q)_n(1-a^3b^2c^2q^{-1})(abcd^{-1},abcd;q)_n(b^2c^2q,c^2a^2q,a^2b^2q;q^2)_n} q^n$$
$$= \frac{(a^2b^2c^2q, a^2q, b^2q, c^2q, dq/abc, bcdq/a, cdaq/b, dabq/c;q^2)_\infty}{(q, b^2c^2q, c^2a^2q, a^2b^2q, abcdq, adq/bc, bdq/ac, cdq/ab;q^2)_\infty}$$
$$- \frac{(a^2,b^2,c^2;q)_\infty(a^2b^2c^2q;q^2)_\infty}{(q;q)_\infty(b^2c^2q, c^2a^2q, a^2b^2q, abcdq, abcd^{-1}q;q^2)_\infty}$$
$$\cdot \sum_{n=0}^{\infty}(1-d^2q^{4n+2})\frac{(bcdq/a, cdaq/b, dabq/c;q^2)_n}{(adq/bc, dbq/ac, cdq/ab;q^2)_{n+1}}\left(-\frac{d}{abc}\right)^n q^{(n+1)^2}.$$

(Rahman [1989d])

3.26 Show that

$$\sum_{k=0}^{\infty} \frac{(abcq;q)_k(1-abcq^{3k+1})(d,q/d;q)_k(abq,bcq,caq;q^2)_k}{(q^2;q^2)_k(1-abcq)(abcq^3/d, abcdq^2;q^2)_k(cq,aq,bq;q)_k} q^k$$
$$= {}_8\phi_7\left[\begin{matrix} abcq, q^2\sqrt{abcq}, -q^2\sqrt{abcq}, d, q/d, abq, bcq, caq \\ \sqrt{abcq}, -\sqrt{abcq}, abcq^3/d, abcdq^2, cq^2, aq^2, bq^2 \end{matrix}; q^2, q^2\right]$$
$$+ \frac{(abcq^3, abq, bcq, caq, d, q/d;q^2)_\infty}{(q^2, aq^2, bq^2, cq^2, abcq^3/d, abcdq^2;q^2)_\infty} \frac{q}{(1-aq)(1-bq)(1-cq)}$$
$$\cdot {}_4\phi_3\left[\begin{matrix} q^2, abcq^2, dq, q^2/d \\ aq^3, bq^3, cq^3 \end{matrix}; q^2, q^2\right].$$

3.27 Prove

$$\sum_{n=0}^{\infty} \frac{(bcdq^{-2};q^3)_n(1-bcdq^{4n-2})(b,c,d;q)_n}{(q;q)_n(1-bcdq^{-2})(cdq,bdq,bcq;q^3)_n} q^{n^2}$$

$$\cdot {}_4\phi_3\left[\begin{matrix} q^{-n},q^{1-n},q^{2-n},bcdq^{3n} \\ bq^2,cq^2,dq^2 \end{matrix};q^3,q^3\right]$$

$$= \frac{(bcdq,bq,cq,dq;q^3)_\infty}{(q,cdq,bdq,bcq;q^3)_\infty}.$$

(Rahman [1989d])

3.28 Show that

$$\sum_{n=0}^{\infty} \frac{(bcdq^{-1};q^3)_n(1-bcdq^{4n-1})(b,c,d;q)_n}{(q;q)_n(1-bcdq^{-1})(cdq^2,bdq^2,bcq^2;q^3)_n} q^{n^2+n}$$

$$\cdot {}_4\phi_3\left[\begin{matrix} q^{-n-1},q^{-n},q^{1-n},bcdq^{3n} \\ bq,cq,dq \end{matrix};q^3,q^3\right]$$

$$= \frac{(bcdq^2,bq^2,cq^2,dq^2;q^3)_\infty}{(q^2,cdq^2,bdq^2,bcq^2;q^3)_\infty}.$$

(Rahman [1989d])

3.29 Derive the summation formulas:

(i)

$$\sum_{k=0}^{\infty} \frac{(1-aq^{5k})(a,b;q^2)_k(ab^2/q^3;q^3)_k(q^2/b;q)_k(aq^3/b;q^6)_k}{(1-a)(q^3,aq^3/b;q^3)_k(q^5/b^2;q^2)_k(ab,abq^2;q^4)_k}\left(-\frac{q^2}{b}\right)^k q^{\binom{k+1}{2}}$$

$$= \frac{(aq^2,qa^3/b;q^2)_\infty(ab^2,q^9/b^3;q^6)_\infty}{(ab,q^5/b^2;q^2)_\infty(q^3,aq^6/b;q^6)_\infty},$$

(ii)

$$\sum_{k=0}^{\infty} \frac{(1-acq^{5k})(a,q^4/a;q^2)_k(q^5/ac;q^3)_k(ac/q^2;q)_k(ac/q^2;q)_k}{(1-ac)(cq^3,a^2c/q;q^3)_k(a^2c^2/q^3;q^2)_k}$$

$$\cdot \frac{(a^2c^2/q;q^6)_k}{(q^4,q^6;q^4)_k}\left(-\frac{ac}{q^2}\right)^k q^{\binom{k+1}{2}}$$

$$= \frac{(acq^2,ac/q;q^2)_\infty(q^6,a^3c^2/q^3,ac^2q,a^2c^2/q,aq^4,q^8/a;q^6)_\infty}{(q^4,a^2c^2/q^3;q^2)_\infty(cq^3,cq^6,a^2cq^2,a^2c/q,acq,acq^4;q^6)_\infty},$$

(iii)

$$\sum_{k=0}^{\infty} \frac{(1-acq^{5k})(a,q^2/a;q^2)_k(q/ac;q^3)_k(ac;q)_k(a^2c^2q;q^6)_k}{(1-ac)(cq^3,a^2cq;q^3)_k(a^2c^2q;q^2)_k(q^2,q^4;q^4)_k}(-ac)^k q^{\binom{k+1}{2}}$$

$$= \frac{(acq^2,acq;q^2)_\infty(q^6,a^3c^2q^3,ac^2q^5,a^2c^2q,aq^2,q^4/a;q^6)_\infty}{(q^2,a^2c^2q;q^2)_\infty(cq^3,cq^6,a^2cq^4,a^2cq,acq^2,acq^5;q^6)_\infty}.$$

(Rahman [1989e])

3.30 Derive the quartic summation formula

$$\sum_{k=0}^{\infty} \frac{1-aq^{5k}}{1-a} \frac{(a,b;q)_k(q/b,q^2/b,q^3/b;q^3)_k(a^2b^2/q^2;q^4)_k}{(q^4,aq^4/b;q^4)_k(abq,ab,ab/q;q^2)_k(q^3/ab^2;q)_k} q^k$$
$$+\frac{ab^3}{q^2}\frac{(aq,bq,1/b;q)_\infty(a^2b^2q^2;q^4)_\infty}{(ab,q^3/ab^2;q)_\infty(ab/q;q^2)_\infty(q^4,ab^3/q^2,aq^4/b;q^4)_\infty}$$
$$\cdot {}_1\phi_1\left[\begin{matrix} a^2b^2/q^2 \\ a^2b^2q^2 \end{matrix};q^4,ab^3q^2\right]$$
$$=\frac{(aq,ab^2/q^2;q)_\infty}{(ab;q)_\infty(ab/q;q^2)_\infty(aq^4/b,ab^3/q^2;q^4)_\infty}.$$

(Gasper [1989a])

3.31 Derive the cubic transformation formulas

$$\text{(i)}\sum_{k=0}^{n} \frac{1-acq^{4k}}{1-ac} \frac{(a,b;q)_k(cq/b;q)_{2k}(a^2bcq^{3n},q^{-3n};q^3)_k}{(cq^3,acq^3/b;q^3)_k(ab;q)_{2k}(q^{1-3n}/ab,acq^{3n+1};q)_k} q^k$$
$$=\frac{(1-acq^2)(1-ab/q^2)(1-abq^{3n})(1-acq^{3n})}{(1-acq^{3n+2})(1-abq^{3n-2})(1-ab)(1-ac)}$$
$$\cdot\sum_{k=0}^{n} \frac{1-acq^{6k+2}}{1-acq^2} \frac{(aq^2,bq^2,cq^2/b,cq^3/b,a^2bcq^{3n},q^{-3n};q^3)_k}{(cq^3,acq^3/b,abq^3,abq^2,q^{5-3n}/ab,acq^{3n+5};q^3)_k} q^{3k},$$

$$\text{(ii)}\sum_{k=0}^{n} \frac{(1-aq^{4k})(1-bq^{-2k})}{(1-a)(1-b)} \frac{(a,1/ab;q)_k(abq;q)_{2k}(c,a/bc;q^3)_k}{(q^3,a^2bq^3;q^3)_k(q/b;q)_{2k}(aq/c,bcq;q)_k} q^{3k}$$
$$=\frac{(aq,bq;q)_n(cq^3,aq^3/bc;q^3)_n}{(aq/c,bcq;q)_n(q^3,aq^3/b;q^3)_n}$$
$$\cdot {}_{10}W_9(a/b;q/ab^2,c,a/bc,aq^{n+1},aq^{n+2},aq^{n+3},q^{-3n};q^3,q^3),$$

where $n=0,1,\ldots$. (Gasper and Rahman [1989a])

3.32 Derive the cubic summation formula

$$\sum_{k=0}^{\infty} \frac{1-a^2q^{4k}}{1-a^2} \frac{(b,q^2/b;q)_k(a^2/q;q)_{2k}(c^3,a^2q^2/c^3;q^3)_k}{(a^2q^3/b,a^2bq;q^3)_k(q^2;q)_{2k}(a^2q/c^3,c^3/q;q)_k} q^k$$
$$=\frac{(bq^2,q^4/b,bc^3/q,c^3q/b,c^3/a^2,c^3q^2/a^2,a^2q,a^2q^3;q^3)_\infty}{(q^2,q^4,c^3q,bc^3/a^2,a^2q^3/b,a^2bq,c^3q^2/a^2b;q^3)_\infty}$$
$$-\frac{(b,bq,bq^2,q^2/b,q^3/b,q^4/b,a^2/q,a^2q,a^2q^3,c^3/a^2;q^3)_\infty}{(q^2,q^4,c^3/q,c^3q,a^2/c^3,a^2q/c^3,c^3q^3/a^2,c^3q^3/a^2b,a^3q^3/b;q^3)_\infty}$$
$$\cdot\frac{(c^6q/a^2;q^3)_\infty}{(a^2bq,bc^3/a^2;q^3)_\infty}\,{}_2\phi_1\left[\begin{matrix} bc^3/a^2,c^3q^2/a^2b \\ c^6q/a^2 \end{matrix};q^3,q^3\right].$$

and show that it has the $q \to 1^-$ limit case

$$
{}_7F_6\left[\begin{matrix} a-1/2, a, b, 2-b, c, (2a+2-3c)/3, a/2+1 \\ 3/2, (2a-b+3)/3, (2a+b+1)/3, 3c-1, 2a+1-3c, a/2 \end{matrix}; 1\right]
$$

$$
= \frac{\Gamma\left(\frac{2}{3}\right)\Gamma\left(\frac{4}{3}\right)\Gamma\left(c-\frac{1}{3}\right)\Gamma\left(c+\frac{1}{3}\right)\Gamma\left(\frac{2a-b+3}{3}\right)}{\Gamma\left(\frac{b+2}{3}\right)\Gamma\left(\frac{4-b}{3}\right)\Gamma\left(\frac{b+3c-1}{3}\right)\Gamma\left(\frac{3c-b+1}{3}\right)\Gamma\left(\frac{2a+1}{3}\right)}
\cdot \frac{\Gamma\left(\frac{2a+b+1}{3}\right)\Gamma\left(\frac{2a-3c+3}{3}\right)\Gamma\left(\frac{2a-3c+1}{3}\right)}{\Gamma\left(\frac{2a+3}{3}\right)\Gamma\left(\frac{2a-3c-b+3}{3}\right)\Gamma\left(\frac{2a-3c+b+1}{3}\right)}.
$$

(Gasper and Rahman [1989a])

3.33 Derive the quartic transformation formula

$$
\sum_{k=0}^{\infty} \frac{1-a^2b^2q^{5k-2}}{1-a^2b^2/q^2} \frac{(a,b;q)_k (ab/q, ab, abq; q^3)_k (a^2b^2/q^2; q^4)_k}{(ab^2q^2, a^2bq^2; q^4)_k (abq, ab, ab/q; q^2)_k (q;q)_k} q^k
$$

$$
= \frac{(aq,b;q)_\infty (a^2b^2q^2; q^4)_\infty}{(q;q)_\infty (abq;q^2)_\infty (b, ab^2q^2, a^2bq^2; q^4)_\infty} {}_1\phi_1\left[\begin{matrix} a \\ aq^4 \end{matrix}; q^4, bq^4\right].
$$

(Gasper and Rahman [1989a])

Notes 3

§3.4 Bressoud [1987] contains some transformation formulas for terminating ${}_{r+1}\phi_r(a_1, a_2, \ldots, a_{r+1}; b_1, \ldots, b_r; q; z)$ series that are *almost poised* in the sense that $b_k a_{k+1} = a_1 q^{\delta_k}$ with $\delta_k = 0, 1$ or 2 for $1 \le k \le r$. Transformations for *level basic series*, that is ${}_{r+1}\phi_r$ series in which $a_1 b_k = q a_{k+1}$ for $1 \le k \le r$, are considered in Gasper [1985].

§3.6 Agarwal and Verma [1967a,b] derived transformation formulas for certain sums of bibasic series by applying the theorem of residues to contour integrals of the form (4.9.2) considered in Chapter 4. Inversion formulas are also considered in Carlitz [1973] and, in connection with the Bailey lattice, in A.K. Agarwal, Andrews and Bressoud [1987].

§3.7 Jackson [1928] applied his q-analogue of the Euler's transformation formula (the $p = q$ case of (3.7.11)) to the derivation of transformation formulas and theta function series. Jackson [1942, 1944] and Jain [1980a] also derived q-analogues of some of the double hypergeometric function expansions in Burchnall and Chaundy [1940, 1941].

§3.8 Gosper [1988a] stated a strange q-series transformation formula containing bases q^2, q^3, and q^6. Krattenthaler [1989c] independently derived the terminating case of (3.8.18) and terminating special cases of some of the other summation formulas in this section.

§3.9 For multibasic series containing bibasic shifted factorials of the form $(a;p,q)_{r,s} = \prod_{j=0}^{r-1} \prod_{k=0}^{s-1} (1 - ap^j q^k)$ and connections with Schur functions and permutation statistics, see Désarménien and Foata [1985–1988].

§3.10 Jain and Verma [1986] contains nonterminating versions of some of Nassrallah and Rahman's [1981] transformation formulas.

Ex. 3.16 q-Differential equations for certain products of basic hypergeometric series are considered in Jackson [1911].

Exercises 3.17–3.19 These exercises are q-analogues of the Cayley [1858] and Orr [1899] theorems (also see Bailey [1935, Chapter X], Burchnall and Chaundy [1949], Edwards [1923], Watson [1924], and Whipple [1927, 1929]). Other q-analogues are given in N. Agarwal [1959], and Jain and Verma [1987].

Ex. 3.20 The formula obtained by writing the last series as a multiple of the series with argument bp is equivalent to the bibasic identity (21.9) in Fine [1988], and it is a special case of the Fundamental Lemma in Andrews [1966a, p. 65]. Applications of the Fundamental Lemma to mock theta functions and partitions are contained in Andrews [1966a,b]. Agarwal [1969a] extended Andrews' Fundamental Lemma and pointed out some expansion formulas that follow from his extension.

Ex. 3.23 For additional material on Lagrange inversion, see Andrews [1975b, 1979a], Bressoud [1983b], Cigler [1980], Fürlinger and Hofbauer [1985], Garsia [1981], Garsia and Remmel [1986], Gessel [1980], Gessel and Stanton [1983, 1986], Hofbauer [1982, 1984], Krattenthaler [1984, 1988, 1989a], Paule and Rother [1985], and Stanton [1988].

4

BASIC CONTOUR INTEGRALS

4.1 Introduction

Our first objective in this chapter is to give q-analogues of Barnes' [1908] contour integral representation for the hypergeometric function

$$ {}_2F_1(a,b;c;z) = \frac{\Gamma(c)}{\Gamma(a)\Gamma(b)} \frac{1}{2\pi i} \int_{-i\infty}^{i\infty} \frac{\Gamma(a+s)\Gamma(b+s)\Gamma(-s)}{\Gamma(c+s)} (-z)^s \, ds, \tag{4.1.1} $$

where $|\arg(-z)| < \pi$, Barnes' [1908] first lemma

$$ \begin{aligned} &\frac{1}{2\pi i} \int_{-i\infty}^{i\infty} \Gamma(a+s)\Gamma(b+s)\Gamma(c-s)\Gamma(d-s) \, ds \\ &= \frac{\Gamma(a+c)\Gamma(a+d)\Gamma(b+c)\Gamma(b+d)}{\Gamma(a+b+c+d)}, \end{aligned} \tag{4.1.2} $$

and Barnes' [1910] second lemma

$$ \begin{aligned} &\frac{1}{2\pi i} \int_{-i\infty}^{i\infty} \frac{\Gamma(a+s)\Gamma(b+s)\Gamma(c+s)\Gamma(1-d-s)\Gamma(-s)}{\Gamma(e+s)} \, ds \\ &= \frac{\Gamma(a)\Gamma(b)\Gamma(c)\Gamma(1-d+a)\Gamma(1-d+b)\Gamma(1-d+c)}{\Gamma(e-a)\Gamma(e-b)\Gamma(e-c)}, \end{aligned} \tag{4.1.3} $$

where $e = a+b+c-d+1$.

In (4.1.1) the contour of integration is the imaginary axis directed upward with indentations, if necessary, to ensure that the poles of $\Gamma(-s)$, i.e. $s = 0, 1, 2, \ldots$, lie to the right of the contour and the poles of $\Gamma(a+s)\Gamma(b+s)$, i.e. $s = -a-n, -b-n$, with $n = 0, 1, 2, \ldots$, lie to the left of the contour (as shown in Fig. 4.1 at the end of this section). The assumption that there exists such a contour excludes the possibility that a or b is zero or a negative integer. Similarly, in (4.1.2), (4.1.3) and the other contour integrals in this book it is assumed that the parameters are such that the contour of integration can be drawn separating the increasing and decreasing sequences of poles.

Barnes' first and second lemmas are integral analogues of Gauss' ${}_2F_1$ summation formula (1.2.11) and Saalschütz's formula (1.7.1), respectively. In Askey and Roy [1986] it was pointed out that Barnes' first lemma is an extension of the beta integral (1.11.8). To see this, replace b by $b - i\omega$, d by $d + i\omega$ and then set $s = \omega x$ in (4.1.2). Then let $\omega \to \infty$ and use Stirling's formula to obtain the beta integral in the form

$$ \int_{-\infty}^{\infty} x_+^{a+c-1} (1-x)_+^{b+d-1} dx = B(a+c, b+d), \tag{4.1.4} $$

where $\mathrm{Re}(a+c)>0$, $\mathrm{Re}(b+d)>0$ and

$$x_+ = \begin{cases} x & \text{if } x \geq 0, \\ 0 & \text{if } x < 0. \end{cases} \tag{4.1.5}$$

It is for this reason that Askey and Roy call (4.1.2) Barnes' beta integral. Following Watson [1910], we shall give a q-analogue of (4.1.1) in §4.2, that is, a Barnes-type integral representation for ${}_2\phi_1(a,b;c;q,z)$. It will be used in §4.3 to derive an analytic continuation formula for the ${}_2\phi_1$ series. We shall give q-analogues of (4.1.2) and (4.1.3) in §4.4. The rest of the chapter will be devoted to generalizations of these integral representations, other types of basic contour integrals, and to the use of these integrals to derive general transformation formulas for basic hypergeometric series.

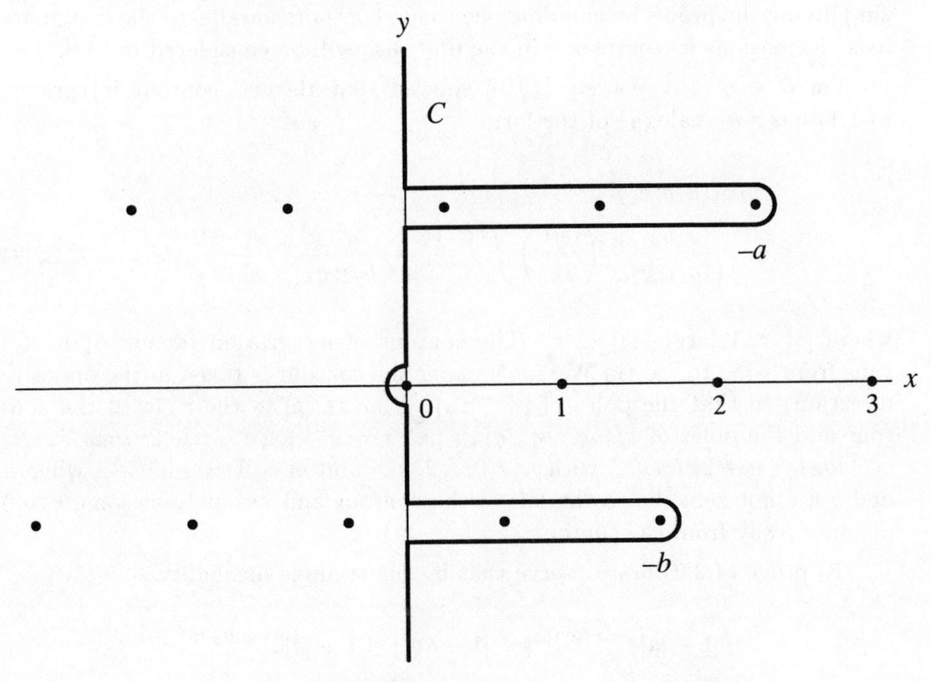

Fig. 4.1

4.2 Watson's contour integral representation for ${}_2\phi_1(a,b;c;q,z)$ series

For the sake of simplicity we shall assume in this and the following five sections that $0<q<1$ and write

$$q=e^{-\omega}, \quad \omega>0. \tag{4.2.1}$$

This is not a severe restriction for most applications because the results derived for $0<q<1$ can usually be extended to complex q in the unit disc by using anlaytic continuation. The restriction $0<q<1$ has the advantage of simplifying the proofs by enabling one to use contours parallel to the imaginary axis. Extensions to complex q in the unit disc will be considered in §4.8.

For $0<q<1$ Watson [1910] showed that Barnes' contour integral in (4.1.1) has a q-analogue of the form

$$\begin{aligned}&{}_2\phi_1(a,b;c;q,z)\\&=\frac{(a,b;q)_\infty}{(q,c;q)_\infty}\left(\frac{-1}{2\pi i}\right)\int_{-i\infty}^{i\infty}\frac{(q^{1+s},cq^s;q)_\infty}{(aq^s,bq^s;q)_\infty}\frac{\pi(-z)^s}{\sin\pi s}\,ds,\end{aligned} \tag{4.2.2}$$

where $|z|<1, |\arg(-z)|<\pi$. The contour of integration (denote it by C) runs from $-i\infty$ to $i\infty$ (in Watson's paper the contour is taken in the opposite direction) so that the poles of $(q^{1+s};q)_\infty/\sin\pi s$ lie to the right of the contour and the poles of $1/(aq^s,bq^s;q)_\infty$, i.e. $s=\omega^{-1}\log a-n+2\pi im\omega^{-1}, s=\omega^{-1}\log b-n+2\pi im\omega^{-1}$ with $n=0,1,2,\ldots$, and $m=0,\pm1,\pm2,\ldots$, when a and b are not zero, lie to the left of the contour and are at least some $\epsilon>0$ distance away from the contour.

To prove (4.2.2) first observe that by the triangle inequality,

$$|1-|a|e^{-\omega\,\mathrm{Re}\,(s)}|\le|1-aq^s|\le 1+|a|e^{-\omega\,\mathrm{Re}\,(s)}$$

and so

$$\begin{aligned}&\left|\frac{(q^{1+s},cq^s;q)_\infty}{(aq^s,bq^s;q)_\infty}\right|\\&\le\prod_{n=0}^{\infty}\frac{\left(1+e^{-(n+1+\,\mathrm{Re}\,(s))\omega}\right)\left(1+|c|e^{-(n+\,\mathrm{Re}\,(s))\omega}\right)}{\left(1-|a|e^{-(n+\,\mathrm{Re}\,(s))\omega}\right)\left(1-|b|e^{-(n+\,\mathrm{Re}\,(s))\omega}\right)},\end{aligned} \tag{4.2.3}$$

which is bounded on C. Hence the integral in (4.2.2) converges if $\mathrm{Re}[s\log(-z)-\log(\sin\pi s)]<0$ on C for large $|s|$, i.e. if $|\arg(-z)|<\pi$.

Now consider the integral in (4.2.2) with C replaced by a contour C_R consisting of a large clockwise-oriented semicircle of radius R with center at the origin that lies to the right of C, is terminated by C and is bounded away from the poles (as shown in Fig. 4.2).

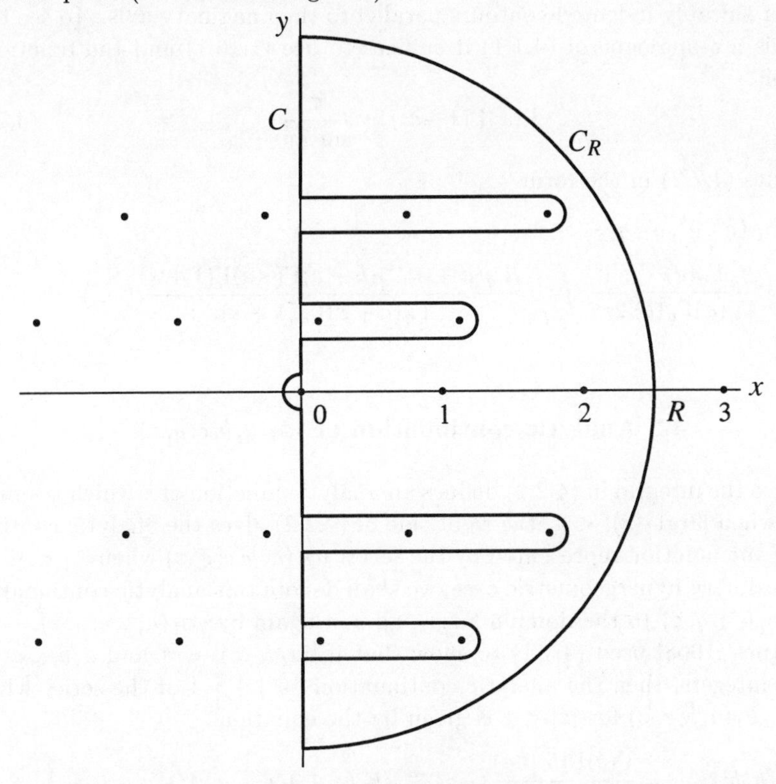

Fig. 4.2.

Setting $s = Re^{i\theta}$, we have for $|z| < 1$ that

$$\begin{aligned}
&\text{Re}\left[\log \frac{(-z)^s}{\sin \pi s}\right] \\
&= R\left[\cos\theta \, \log|z| - \sin\theta \, \arg(-z) - \pi|\sin\theta|\right] + O(1) \\
&\leq -R\left[\sin\theta \, \arg(-z) + \pi|\sin\theta|\right] + O(1).
\end{aligned}$$

Hence, when $|z| < 1$ and $|\arg(-z)| < \pi - \delta$, $0 < \delta < \pi$, we have

$$\frac{(-z)^s}{\sin \pi s} = O\left[\exp(-\delta R|\sin\theta|)\right] \tag{4.2.4}$$

on C_R as $R \to \infty$, and it follows that the integral in (4.2.2) with C replaced by C_R tends to zero as $R \to \infty$, under the above restrictions. Therefore, by applying Cauchy's theorem to the closed contour consisting of C_R and that part of C terminated above and below by C_R and letting $R \to \infty$, we obtain that $-\frac{1}{2\pi i}\int_{-i\infty}^{i\infty} \ldots ds$ equals the sum of the residues of the integrand at $n = 0, 1, 2, \ldots$. Since

$$\lim_{s \to n} (s - n) \frac{(q^{1+s}, cq^s; q)_\infty}{(aq^s, bq^s; q)_\infty} \frac{\pi(-z)^s}{\sin \pi s} = \frac{(q^{1+n}, cq^n; q)_\infty}{(aq^n, bq^n, q)_\infty} z^n,$$

this completes the proof of Watson's formula (4.2.2).

It should be noted that the contour of integration in (4.2.2) can be replaced by other suitably indented contours parallel to the imaginary axis. To see that (4.2.2) is a q-analogue of (4.1.1) it suffices to use (1.10.1) and the functional equation

$$\Gamma(x)\Gamma(1-x) = \frac{\pi}{\sin \pi x} \tag{4.2.5}$$

to rewrite (4.2.2) in the form

$$\begin{aligned} &{}_2\phi_1\left(q^a, q^b; q^c; q, z\right) \\ &= \frac{\Gamma_q(c)}{\Gamma_q(a)\Gamma_q(b)} \frac{1}{2\pi i} \int_{-i\infty}^{i\infty} \frac{\Gamma_q(a+s)\Gamma_q(b+s)\Gamma(-s)\Gamma(1+s)}{\Gamma_q(c+s)\Gamma_q(1+s)} (-z)^s \, ds. \end{aligned} \tag{4.2.6}$$

4.3 Analytic continuation of ${}_2\phi_1(a, b; c; q, z)$

Since the integral in (4.2.2) defines an analytic function of z which is single-valued when $|\arg(-z)| < \pi$, the right side of (4.2.2) gives the analytic continuation of the function represented by the series ${}_2\phi_1(a, b; c; q, z)$ when $|z| < 1$. As in the ordinary hypergeometric case, we shall denote this analytic continuation of ${}_2\phi_1(a, b; c; q, z)$ to the domain $|\arg(-z)| < \pi$ again by ${}_2\phi_1(a, b; c; q, z)$.

Barnes [1908] used (4.1.1) to show that if $|\arg(-z)| < \pi$ and $a, b, c, a-b$ are not integers, then the analytic continuation for $|z| > 1$ of the series which defines ${}_2F_1(a, b; c; z)$ for $|z| < 1$ is given by the equation

$$\begin{aligned} {}_2F_1(a, b; c; z) = &\frac{\Gamma(c)\Gamma(b-a)}{\Gamma(b)\Gamma(c-a)} (-z)^{-a} \, {}_2F_1(a, 1+a-c;\ 1+a-b; z^{-1}) \\ &+ \frac{\Gamma(c)\Gamma(a-b)}{\Gamma(a)\Gamma(c-b)} (-z)^{-b} \, {}_2F_1(b, 1+b-c;\ 1+b-a; z^{-1}), \end{aligned} \tag{4.3.1}$$

where, as elsewhere in this section, the symbol "=" is used in the sense "is the analytic continuation of". To illustrate the extension of Barnes' method to ${}_2\phi_1$ series we shall now give Watson's [1910] derivation of the following q-analogue of (4.3.1):

$$\begin{aligned} &{}_2\phi_1(a, b; c; q, z) \\ &= \frac{(b, c/a; q)_\infty (az, q/az; q)_\infty}{(c, b/a; q)_\infty (z, q/z; q)_\infty} \, {}_2\phi_1(a, aq/c; aq/b; q, cq/abz) \\ &\quad + \frac{(a, c/b; q)_\infty}{(c, a/b; q)_\infty} \frac{(bz, q/bz; q)_\infty}{(z, q/z; q)_\infty} \, {}_2\phi_1(b, bq/c; bq/a; q, cq/abz), \end{aligned} \tag{4.3.2}$$

provided that $|\arg(-z)| < \pi$, c and a/b are not integer powers of q, and $a, b, z \neq 0$.

First consider the integral

$$I_1 = \frac{1}{2\pi i} \int \frac{\left(q^{1+s}, cq^s; q\right)_\infty}{\left(aq^s, bq^s; q\right)_\infty} \frac{\pi(-z)^s}{\sin \pi s} \, ds \tag{4.3.3}$$

along three line-segments A_1, A_2, B, whose equations are:

$$A_1 : \operatorname{Im}(s) = m_1, \quad A_2 : \operatorname{Im}(s) = -m_2, \quad B : \operatorname{Re}(s) = -M, \tag{4.3.4}$$

where m_1, m_2, M are large positive constants chosen so that A_1, A_2, B are at least a distance $\epsilon > 0$ away from each pole and zero of

$$g(s) = \frac{(q^{1+s}, cq^s; q)_\infty}{(aq^s, bq^s; q)_\infty} \tag{4.3.5}$$

and it is assumed that A_1, A_2, B are terminated by each other and by the contour of the integral in (4.2.2), i.e. $\operatorname{Re}(s) = 0$ with suitable indentations.

From an asymptotic formula for $(a;q)_\infty$ with $q = e^{-\omega}, \omega > 0$, due to Littlewood [1907, §12], it follows that if $\operatorname{Re}(s) \to -\infty$ with $|s - s_0| > \epsilon$ for some fixed $\epsilon > 0$ and any zero s_0 of $(q^s;q)_\infty$, then

$$\operatorname{Re}[\log(q^s;q)_\infty] = \frac{\omega}{2}(\operatorname{Re}(s))^2 + \frac{\omega}{2}\operatorname{Re}(s) + O(1). \tag{4.3.6}$$

This implies that

$$g(s) = O\left(\left|\frac{ab}{cq}\right|^{\operatorname{Re}(s)}\right), \tag{4.3.7}$$

when $\operatorname{Re}(s) \to -\infty$ with s bounded away from the zeros and poles of $g(s)$. By using this asymptotic expansion and the method of §4.2 it can be shown that if $|z| > |cq/ab|$, then the value of the integral I_1 in (4.3.3) taken along the contours A_1, A_2, B tends to zero as $m_1, m_2, M \to \infty$.

Hence, by Cauchy's theorem, the value of I_1, taken along the contour C of §4.2, equals the sum of the residues of the integrand at its poles to the left of C when $|z| > |cq/ab|$. Set $\alpha = -\omega^{-1}\log a$, $\beta = -\omega^{-1}\log b$ so that $a = q^\alpha, b = q^\beta$. Since the residue of the integrand at $-\alpha - n + 2\pi i m\omega^{-1}$ is

$$\frac{(a^{-1}q^{1-n}, ca^{-1}q^{-n}, q^{n+1}; q)_\infty}{(q, q, ba^{-1}q^{-n}; q)_\infty}\pi\omega^{-1}(-z)^{-\alpha-n}q^{n(n+1)/2}$$
$$\cdot \exp\left\{2m\pi i\omega^{-1}\log(-z)\right\}\csc\left(2m\pi^2 i\omega^{-1} - \alpha\pi\right),$$

we have

$$\begin{aligned} I_1 = &\sum_{m=-\infty}^{\infty}\csc\left(2m\pi^2 i\omega^{-1} - \alpha\pi\right)\exp\left\{2m\pi i\omega^{-1}\log(-z)\right\} \\ &\cdot\frac{\pi\omega^{-1}(c/a, q/a; q)_\infty}{(b/a, q; q)_\infty}(-z)^{-\alpha}\,{}_2\phi_1(a, aq/c; aq/b; q, cq/abz) \\ &+ \text{idem }(a;b). \end{aligned} \tag{4.3.8}$$

Thus it remains to evaluate the above sums over m when $|\arg(-z)| < \pi$. Letting $c = b$ in (4.3.8) and using (4.2.2), we find that the analytic continuation of ${}_2\phi_1(a,b;b;q,z)$ is

$$\begin{aligned} &\sum_{m=-\infty}^{\infty}\csc\left(\alpha\pi - 2m\pi^2 i\omega^{-1}\right)\exp\left\{2m\pi i\omega^{-1}\log(-z)\right\} \\ &\cdot\frac{\pi\omega^{-1}(a, q/a; q)_\infty}{(q, q; q)_\infty}(-z)^{-\alpha}\,{}_2\phi_1(a, aq/b; aq/b; q, q/az). \end{aligned}$$

Since, by the q-binomial theorem,

$$ {}_2\phi_1(a, b; b; q, z) = \frac{(az;q)_\infty}{(z;q)_\infty} $$

and the products converge for all values of z, it follows that

$$ \sum_{m=-\infty}^{\infty} \csc\left(\alpha\pi - 2m\pi^2 i\omega^{-1}\right) \exp\left\{2m\pi i\omega^{-1}\log(-z)\right\} (-z)^{-\alpha} \\ = \frac{\omega(q, q, az, q/az; q)_\infty}{\pi(a, q/a, z, q/z; q)_\infty}. \tag{4.3.9} $$

Using (4.3.9) in (4.3.8) we finally obtain (4.3.2).

4.4 q-Analogues of Barnes' first and second lemmas

Assume, as before, that $0 < q < 1$, and consider the integral

$$ I_2 = \frac{1}{2\pi i}\int_{-i\infty}^{i\infty} \frac{\left(q^{1-c+s}, q^{1-d+s}; q\right)_\infty}{\left(q^{a+s}, q^{b+s}; q\right)_\infty} \frac{\pi q^s\, ds}{\sin\pi(c-s)\sin\pi(d-s)}, \tag{4.4.1} $$

where, as usual, the contour of integration runs from $-i\infty$ to $i\infty$ so that the increasing sequences of poles of the integrand (i.e. $c+n, d+n$ with $n = 0, 1, 2, \ldots$) lie to the right and the decreasing sequences of poles (i.e. the zeros of $\left(q^{a+s}, q^{b+s}; q\right)_\infty$) lie to the left of the contour. By using Cauchy's theorem as in §4.2 to evaluate this integral as the sum of the residues at the poles $c+n, d+n$ with $n = 0, 1, 2, \ldots$, we find that

$$ I_2 = \frac{\pi q^c}{\sin\pi(c-d)} \frac{\left(q, q^{1+c-d}; q\right)_\infty}{\left(q^{a+c}, q^{b+c}; q\right)_\infty} {}_2\phi_1\left(q^{a+c}, q^{b+c}; q^{1+c-d}; q, q\right) \\ + \text{idem } (c; d). \tag{4.4.2} $$

Applying the formula (2.10.13) to (4.4.2), we get

$$ \frac{1}{2\pi i}\int_{-i\infty}^{i\infty} \frac{\left(q^{1-c+s}, q^{1-d+s}; q\right)_\infty}{\left(q^{a+s}, q^{b+s}; q\right)_\infty} \frac{\pi q^s\, ds}{\sin\pi(c-s)\sin\pi(d-s)} \\ = \frac{q^c}{\sin\pi(c-d)} \frac{\left(q, q^{1+c-d}, q^{d-c}, q^{a+b+c+d}; q\right)_\infty}{\left(q^{a+c}, q^{a+d}, q^{b+c}, q^{b+d}; q\right)_\infty}, \tag{4.4.3} $$

which is Watson's [1910] q-analogue of Barnes' first lemma (4.1.2), as can be seen by rewriting it in terms of q-gamma functions.

A q-analogue of Barnes' second lemma (4.1.3) can be derived by proceeding as in Agarwal [1953b]. Set $c = n$ and $d = c - a - b$ in (4.4.3) to obtain

$$ \frac{1}{2\pi i}\int_{-i\infty}^{i\infty} \frac{\left(q^{1-n+s}, q^{1-c+a+b+s}; q\right)_\infty}{\left(q^{a+s}, q^{b+s}; q\right)_\infty} \frac{\pi q^s\, ds}{\sin\pi s\sin\pi(c-a-b-s)} \\ = \csc\pi(c-a-b)\frac{\left(q^{1+a+b-c}, q^{c-a-b}, q^c, q; q\right)_\infty}{\left(q^a, q^b, q^{c-a}, q^{c-b}; q\right)_\infty} \\ \cdot (-1)^n \frac{\left(q^a, q^b; q\right)_n}{(q^c; q)_n} q^{n(c-a-b)-\binom{n}{2}} \tag{4.4.4} $$

for $n = 0, 1, 2, \ldots$. Next, replace c by d in (4.4.4), multiply both sides by $(-1)^n q^{n(e-c)+\binom{n}{2}} (q^c;q)_n / (q,q^e;q)_n$, sum over n and change the order of integration and summation (which is easily justified if $|q^{e-c+s}| < 1$) to obtain

$$\csc \pi(d-a-b) \frac{(q^{1+a+b-d}, q^{d-a-b}, q^d, q; q)_\infty}{(q^a, q^b, q^{d-a}, q^{d-b}; q)_\infty}$$

$$\cdot {}_3\phi_2 (q^a, q^b, q^c; q^d, q^e; q, q^{d+e-a-b-c})$$

$$= \frac{1}{2\pi i} \int_{-i\infty}^{i\infty} \frac{(q^{1+s}, q^{1-d+a+b+s}; q)_\infty}{(q^{a+s}, q^{b+s}; q)_\infty} \frac{\pi q^s}{\sin \pi s \sin \pi(d-a-b-s)}$$

$$\cdot {}_2\phi_1 (q^{-s}, q^c; q^e; q, q^{e-c+s})\, ds$$

$$= \frac{(q^{e-c};q)_\infty}{(q^e;q)_\infty} \frac{1}{2\pi i} \int_{-i\infty}^{i\infty} \frac{(q^{1+s}, q^{e+s}, q^{1-d+a+b+s}; q)_\infty}{(q^{a+s}, q^{b+s}, q^{e-c+s}; q)_\infty}$$

$$\cdot \frac{\pi q^s\, ds}{\sin \pi s \sin \pi(d-a-b+s)}, \tag{4.4.5}$$

by the q-Gauss formula (1.5.1). Now take $c = d$. Then the series on the left of (4.4.5) can be summed by the q-Gauss formula to give, after an obvious change in parameters,

$$\frac{1}{2\pi i} \int_{-i\infty}^{i\infty} \frac{(q^{1+s}, q^{d+s}, q^{e+s}; q)_\infty}{(q^{a+s}, q^{b+s}, q^{c+s}; q)_\infty} \frac{\pi q^s\, ds}{\sin \pi s \sin \pi(d+s)}$$

$$= \csc \pi d\, \frac{(q, q^d, q^{1-d}, q^{e-a}, q^{e-b}, q^{e-c}; q)_\infty}{(q^a, q^b, q^c, q^{1+a-d}, q^{1+b-d}, q^{1+c-d}; q)_\infty}, \tag{4.4.6}$$

where $d + e = 1 + a + b + c$, which is Agarwal's q-analogue of Barnes' second lemma. This integral converges if q is so small that

$$\mathrm{Re}\, [s \log q - \log(\sin \pi s \sin \pi(d+s))] < 0 \tag{4.4.7}$$

on the contour for large $|s|$.

4.5 Analytic continuation of ${}_{r+1}\phi_r$ series

By employing Cauchy's theorem as in §4.2, we find that if $|z| < 1$ and $|\arg(-z)| < \pi$, then

$${}_{r+1}\phi_r \left[\begin{matrix} a_1, a_2, \ldots, a_{r+1} \\ b_1, \ldots, b_r \end{matrix} ; q, z \right]$$

$$= \frac{(a_1, a_2, \ldots, a_{r+1}; q)_\infty}{(q, b_1, \ldots, b_r; q)_\infty}$$

$$\cdot \left(\frac{-1}{2\pi i} \right) \int_{-i\infty}^{i\infty} \frac{(q^{1+s}, b_1 q^s, \ldots, b_r q^s; q)_\infty}{(a_1 q^s, a_2 q^s, \ldots, a_{r+1} q^s; w)_\infty} \frac{\pi (-z)^s\, ds}{\sin \pi s}, \tag{4.5.1}$$

where, as before, only the poles of the integrand at $0, 1, 2, \ldots,$ lie to the right of the contour. As in the $r = 1$ case, the right side of (4.5.1) gives the analytic continuation of the ${}_{r+1}\phi_r$ series on the left side to the domain $|\arg(-z)| < \pi$.

Also, as in §4.3, it can be shown that if $|z| > |b_1 \dots b_r q/a_1 \dots a_{r+1}|$, then the integral $I_3 = \frac{1}{2\pi i}\int_{-i\infty}^{i\infty} \dots ds$ is equal to the sum of the residues of the integrand at those poles of the integrand which lie on the left of the contour. Set $\alpha_1 = -\omega^{-1}\log a_1$. Since the residue of the integrand at $-\alpha_1 - n + 2\pi i m\omega^{-1}$ is

$$\frac{(q^{n+1}, a_1^{-1}q^{1-n}, b_1 a_1^{-1}q^{-n}, \dots, b_r a_1^{-1}q^{-n}; q)_\infty}{(q, q, a_2 a_1^{-1}q^{-n}, \dots, a_{r+1}a_1^{-1}q^{-n}; q)_\infty} \pi\omega^{-1}(-z)^{-\alpha_1 - n} q^{n(n+1)/2}$$
$$\cdot \csc(2m\pi^2 i\omega^{-1} - \alpha_1\pi)\exp\{2m\pi i\omega^{-1}\log(-z)\},$$

by proceeding as in the proof of (4.3.2) and using (4.3.9) and (4.5.1) we obtain the expansion

$$\begin{aligned} {}_{r+1}\phi_r &\left[\begin{matrix} a_1, a_2, \dots, a_{r+1} \\ b_1, \dots, b_r \end{matrix}; q, z\right] \\ &= \frac{(a_2, \dots, a_{r+1}, b_1/a_1, \dots, b_r/a_1, a_1 z, q/a_1 z; q)_\infty}{(b_1, \dots, b_r, a_2/a_1, \dots, a_{r+1}/a_1, z, q/z; q)_\infty} \\ &\cdot {}_{r+1}\phi_r\left[\begin{matrix} a_1, a_1 q/b_1, \dots, a_1 q/b_r \\ a_1 q/a_2, \dots, a_1 q/a_{r+1} \end{matrix}; q, \frac{q b_1 \cdots b_r}{z a_1 \cdots a_{r+1}}\right] \\ &+ \text{idem}\ (a_1; a_2, \dots, a_{r+1}), \end{aligned} \tag{4.5.2}$$

where the equality holds in the "is the analytic continuation of" sense. The symbol "idem $(a_1; a_2, \dots, a_{r+1})$" after an expression stands for the sum of the r expressions obtained from the preceeding expression by interchanging a_1 with each a_k, $k = 2, 3, \dots, r+1$.

4.6 Contour integrals representing well-poised series

Let us replace a, b, c, d and e in (4.4.6) by $a+n, b+n, c+n, d+n$ and $e+2n$, respectively, where

$$e = 1 + a + b + c - d, \tag{4.6.1}$$

and transform the integration variable s to $s - n$, where n is a non-negative integer. Then we get

$$\begin{aligned} \frac{1}{2\pi i}\int_{n-i\infty}^{n+i\infty} &\frac{(q^{1+s-n}, q^{d+s}, q^{e+s+n}; q)_\infty}{(q^{a+s}, q^{b+s}, q^{c+s}; q)_\infty} \frac{\pi q^{s-n}\, ds}{\sin \pi s \sin \pi(d+s)} \\ &= \csc \pi d\, \frac{(q, q^{d+n}, q^{1-d-n}, q^{e-a+n}, q^{e-b+n}, q^{e-c+n}; q)_\infty}{(q^{a+n}, q^{b+n}, q^{c+n}, q^{e-a-b}, q^{e-a-c}, q^{e-b-c}; q)_\infty}. \end{aligned} \tag{4.6.2}$$

The limits of integration $n \pm i\infty$ can be replaced by $\pm i\infty$ because we always indent the contour of integration to separate the increasing and decreasing sequences of poles. Thus, it follows from (4.6.2) that

$$\begin{aligned} &\csc(\pi d)\, \frac{(q, q^d, q^{1-d}, e^{e-a}, q^{e-b}, q^{e-c}; q)_\infty}{(q^a, q^b, q^c, q^{e-b-c}, q^{e-c-a}, q^{e-a-b}; q)_\infty} \frac{(q^a, q^b, q^c; q)_n}{(q^{e-a}, q^{e-b}, q^{e-c}; q)_n} q^{(1-d)n} \\ &= \frac{1}{2\pi i}\int_{-i\infty}^{i\infty} \frac{(q^{1+s}, q^{d+s}, q^{e+s}; q)_\infty}{(q^{a+s}, q^{b+s}, q^{c+s}; q)_\infty} \cdot \frac{\pi q^{(n+1)s}(q^{-s}; q)_n\, ds}{\sin \pi s \sin \pi(d+s)(q^{e+s}; q)_n}, \end{aligned} \tag{4.6.3}$$

and hence, by termwise integration, we obtain Agarwal's [1953b] formula

$$\csc(\pi d)\,\frac{(q,q^d,q^{1-d},q^{e-a},q^{e-b},q^{e-c};q)_\infty}{(q^a,q^b,q^c,q^{e-b-c},q^{e-c-a},q^{e-a-b};q)_\infty}$$

$$\cdot\,{}_{r+4}\phi_{r+3}\left[\begin{matrix}q^A,q^a,q^b,q^c,a_1,\ldots,a_r\\ q^{e-a},q^{e-b},q^{e-c},b_1,\ldots,b_r\end{matrix};q,zq^{-d}\right]$$

$$=\frac{1}{2\pi i}\int_{-i\infty}^{i\infty}\frac{(q^{1+s},q^{d+s},q^{e+s};q)_\infty}{(q^{a+s},q^{b+s},q^{c+s};q)_\infty}\frac{\pi q^s}{\sin\pi s\sin\pi(d+s)}$$

$$\cdot\,{}_{r+2}\phi_{r+1}\left[\begin{matrix}q^A,a_1,\ldots,a_r,q^{-s}\\ b_1,\ldots,b_r,q^{e+s}\end{matrix};q,zq^{s-1}\right]ds, \tag{4.6.4}$$

where $e=1+a+b+c-d$, $|z|<|q^d|$ and it is assumed that (4.4.7) holds.

Therefore, if we can sum the series on the right of (4.6.4), then we can find a simpler contour integral representing the series on the left. In particular, if we let

$$r=4,\ e=1+A,\ a_1=q^{1+A/2}=-a_2,\ a_3=q^d,\ a_4=q^e,$$
$$b_1=q^{A/2}=-b_2,\ b_3=q^{1-d+A},\ b_4=q^{1-e+A},\ z=q^{2-d-e+A},$$

then we get a very-well-poised ${}_6\phi_5$ series which can be summed by (2.7.1). This yields Agarwal's [1953b] contour integral representation for a very-well-poised ${}_8\phi_7$ series:

$${}_8\phi_7\left[\begin{matrix}q^A,q^{1+A/2},-q^{1+A/2},q^a,q^b,q^c,q^d,q^e\\ q^{A/2},-q^{A/2},q^{1+A-a},q^{1+A-b},q^{1+A-c},q^{1+A-d},q^{1+A-e}\end{matrix};q,q^B\right]$$

$$=\sin\pi(a+b+c-A)$$

$$\cdot\,\frac{(q^{1+A},q^a,q^b,q^c,q^{1+A-a-b},q^{1+A-b-c},q^{1+A-c-a},q^{1+A-d-e};q)_\infty}{(q,q^{a+b+c-A},q^{1+A-a-b-c},q^{1+A-a},q^{1+A-b},q^{1+A-c},q^{1+A-d},q^{1+A-e};q)_\infty}$$

$$\cdot\,\frac{1}{2\pi i}\int_{-i\infty}^{i\infty}\frac{(q^{1+s},q^{1+a-d+s},q^{1+A-e+s},q^{a+b+c-A+s};q)_\infty}{(q^{a+s},q^{b+s},q^{c+s},q^{1+A-d-e+s};q)_\infty}$$

$$\cdot\,\frac{\pi q^s\,ds}{\sin\pi s\sin\pi(a+b+c-A+s)}, \tag{4.6.5}$$

where $B=2+2A-a-b-c-d-e$, provided Re $B>0$ and

$$\mathrm{Re}\ [s\log q-\log(\sin\pi s\sin\pi(a+b+c-A+s))]<0.$$

If we evaluate the integral in (4.6.5) by considering the residues at the poles of $1/[\sin\pi s\sin\pi(a+b+c-A+s)]$ lying to the right of the contour, then we obtain the transformation (2.10.10) of a very-well-poised ${}_8\phi_7$ series in terms of the sum of two balanced ${}_4\phi_3$ series. In addition, if we replace A,d,e and a by $\lambda,\lambda+d-A,\lambda+e-A$ and $\lambda+a-A$, respectively, and take $\lambda+a+d+e=1+2A$ in (4.6.5), then the integral in (4.6.5) remains unchanged. This gives Bailey's transformation formula (2.10.1) between two very-well-poised ${}_8\phi_7$ series.

4.7 A contour integral analogue of Bailey's summation formula

By replacing A, a, b, c, d, e in (4.6.5) by a, d, e, f, b, c, respectively, we obtain the formula

$$
\begin{aligned}
{}_8W_7\left(q^a; q^b, q^c, q^d, q^e, q^f; q, q\right) &= \sin \pi(d+e+f-a) \\
&\cdot \frac{\left(q^{1+a}, q^d, q^e, q^f, q^{1+a-d-e}, q^{1+a-d-f}, q^{1+a-e-f}; q\right)_\infty}{\left(q, q^{1+a-b}, q^{1+a-c}, q^{1+a-d}, q^{1+a-e}, q^{1+a-f}, q^{1+a-d-e-f}; q\right)_\infty} \\
&\cdot \frac{1}{2\pi i}\int_{-i\infty}^{i\infty} \frac{\left(q^{1+s}, q^{1+a-b+s}, q^{1+a-c+s}; q\right)_\infty}{\left(q^{d+s}, q^{e+s}, q^{f+s}; q\right)_\infty} \frac{\pi q^s\, ds}{\sin \pi s \sin \pi(d+e+f-a+s)},
\end{aligned} \tag{4.7.1}
$$

provided the series is balanced, i.e.,

$$
1+2a = b+c+d+e+f. \tag{4.7.2}
$$

Since $1+2(2b-a) = b+(b+c-a)+(b+d-a)+(b+e-a)+(b+f-a)$ by (4.7.2), it follows that (4.7.2) remains unchanged if we replace a, c, d, e, f by $2b-a, b+c-a, b+d-a, b+e-a, b+f-a$, respectively, and keep b unaltered. Then (4.7.1) gives

$$
\begin{aligned}
{}_8W_7&\left(q^{2b-a}; q^b, q^{b+c-a}, q^{b+d-a}, q^{b+e-a}, q^{b+f-a}; q, q\right) = \sin \pi c \\
&\cdot \frac{\left(q^{1+2b-a}, q^{b+d-a}, q^{b+e-a}, q^{b+f-a}, q^{1+a-d-e}, q^{1+a-d-f}, q^{1+a-e-f}; q\right)_\infty}{\left(q, q^{1+b-a}, q^{1+b-c}, q^{1+b-d}, q^{1+b-e}, q^{1+b-f}, q^c; q\right)_\infty} \\
&\cdot \frac{1}{2\pi i}\int_{-i\infty}^{i\infty} \frac{\left(q^{1+b-a+s}, q^{1+b-c+s}, q^{1+s}; q\right)_\infty}{\left(q^{b+d-a+s}, q^{b+e-a+s}, q^{b+f-a+s}; q\right)_\infty} \frac{\pi q^s\, ds}{\sin \pi s \sin \pi(c-s)} \\
&= \sin \pi c\, \frac{\left(q^{1+2b-a}, q^{b+d-a}, q^{b+e-a}, q^{b+f-a}; q\right)_\infty}{\left(q, q^{1+b-a}, q^{1+b-c}, q^c; q\right)_\infty} \\
&\cdot \frac{\left(q^{1+a-d-e}, q^{1+a-d-f}, q^{1+a-e-f}; q\right)_\infty}{\left(q^{1+b-d}, q^{1+b-e}, q^{1+b-f}; q\right)_\infty} \frac{1}{2\pi i}\int_{-i\infty}^{i\infty} \frac{\left(q^{1+s}; q\right)_\infty}{\left(q^{d+s}; q\right)_\infty} \\
&\cdot \frac{\left(q^{1+a-c+s}, q^{1+a-b+s}; q\right)_\infty}{\left(q^{e+s}, q^{f+s}; q\right)_\infty} \frac{\pi q^s\, ds}{\sin \pi(a-b+s) \sin \pi(c+b-a-s)},
\end{aligned} \tag{4.7.3}
$$

where the second integral in (4.7.3) follows from the first by a change of the integration variable $s \to a-b+s$. Combining (4.7.1) and (4.7.3) and simplifying, we obtain

$$
\begin{aligned}
&\frac{\left(q^{1+a-b}, q^{1+a-c}, q^{1+a-d}, q^{1+a-e}, q^{1+a-f}; q\right)_\infty}{\left(q^{1+a}, q^c, q^d, q^e, q^f; q\right)_\infty} \\
&\quad \cdot {}_8W_7\left(q^a; q^b, q^c, q^d, q^e, q^f; q, q\right) \\
&\quad - q^{b-a} \frac{\left(q^{1+b-a}, q^{1+b-c}, q^{1+b-d}, q^{1+b-e}, q^{1+b-f}; q\right)_\infty}{\left(q^{1+2b-a}, q^{b+c-a}, q^{b+d-a}, q^{b+e-a}, q^{b+f-a}; q\right)_\infty}
\end{aligned}
$$

$$
\begin{aligned}
&\cdot {}_8W_7\left(q^{2b-a}; q^b, q^{b+c-a}, q^{b+d-a}, q^{b+e-a}, q^{b+f-a}; q, q\right) \\
&= \frac{\left(q^{1+a-d-e}, q^{1+a-e-f}, q^{1+a-d-f}; q\right)_\infty}{\left(q, q^c, q^{b+c-a}; q\right)_\infty} \\
&\quad \cdot \frac{1}{2\pi i}\int_{-i\infty}^{i\infty} \frac{\left(q^{1+s}, q^{1+a-b+s}, q^{1+a-c+s}; q\right)_\infty}{\left(q^{d+s}, q^{e+s}, q^{f+s}; q\right)_\infty} \frac{\pi q^s \sin\pi(a-b)\, ds}{\sin\pi s \sin\pi(a-b+s)},
\end{aligned} \tag{4.7.4}
$$

when (4.7.2) holds.

Evaluating the above integral via (4.4.6), we obtain Bailey's summation formula (2.11.7).

Agarwal's [1953b] formula

$$
\begin{aligned}
&\frac{1}{2\pi i}\int_{-i\infty}^{i\infty} \frac{\left(q^{1+s}, q^{\frac{1}{2}a+s}, -q^{\frac{1}{2}a+s}, q^{1+a-b+s}, q^{1+a-c+s}; q\right)_\infty}{\left(q^{a+s}, q^{1+\frac{1}{2}a+s}, -q^{1+\frac{1}{2}a+s}, q^{b+s}, q^{c+s}; q\right)_\infty} \\
&\quad \cdot \frac{\left(q^{1+a-d+s}, q^{1+a-e+s}, q^{1+a-f+s}; q\right)_\infty}{\left(q^{d+s}, q^{e+s}, q^{f+s}; q\right)_\infty} \frac{\pi q^s\, ds}{\sin\pi s \sin\pi(a-b+s)} \\
&= \csc\pi(a-b) \frac{\left(q, q^{1+a-b}, q^{b-a}, q^{1+a-d-e}, q^{1+a-e-f}; q\right)_\infty}{\left(q^b, q^c, q^d, q^e, q^f; q\right)_\infty} \\
&\quad \cdot \frac{\left(q^{1+a-d-f}, q^{1+a-c-d}, q^{1+a-c-e}, q^{1+a-c-f}; q\right)_\infty}{\left(q^{b+c-a}, q^{b+d-a}, q^{b+e-a}, q^{b+f-a}; q\right)_\infty},
\end{aligned} \tag{4.7.5}
$$

where $1 + 2a = b + c + d + e + f$, follows directly from (2.11.7) by considering the residues of the integrand of the above integral at the poles to the right of the contour, i.e. at $s = n, b - a + n$ with $n = 0, 1, 2, \ldots$. Thus (4.7.5) gives an integral analogue of Bailey's summation formula (2.11.7). The integral in (4.7.5) converges if q is so small that

$$
\mathrm{Re}\left[s \log q - \log(\sin\pi s \sin\pi(a-b+s))\right] < 0 \tag{4.7.6}
$$

on the contour for large $|s|$.

4.8 Extensions to complex q inside the unit disc

The previous basic contour integrals can be extended to complex q inside the unit disc by using suitable contours. For $0 < |q| < 1$, let

$$
\log q = -\left(\omega_1 + i\omega_2\right), \tag{4.8.1}
$$

where $\omega_1 = -\log|q| > 0$ and $\omega_2 = -\mathrm{Arg}\, q$.

Thus $q = e^{-(\omega_1 + i\omega_2)}$. Then a modification of the proof in §4.2 (see Watson [1910]) shows that if $0 < |q| < 1$ and $|z| < 1$, then formula (4.2.2) extends to

$$
\begin{aligned}
&{}_2\phi_1(a, b; c; q, z) \\
&= \frac{(a, b; q)_\infty}{(q, c; q)_\infty}\left(\frac{-1}{2\pi i}\right)\int_C \frac{\left(q^{1+s}, cq^s; q\right)_\infty}{\left(aq^s, bq^s; q\right)_\infty} \frac{\pi(-z)^s}{\sin\pi s}\, ds,
\end{aligned} \tag{4.8.2}
$$

where C is an upward directed contour parallel to the line $\mathrm{Re}(s(\omega_1 + i\omega_2)) = 0$ with indentations, to ensure that the increasing sequence of poles $0, 1, 2, \ldots,$ of the integrand lie to the right, and the decreasing sequences of poles lie to the left of C.

Since the above integral converges if $\mathrm{Re}[s \log(-z) - \log(\sin \pi s)] < 0$ on C for large $|s|$, i.e., if

$$|\arg(-z) - \omega_2\omega_1^{-1} \log|z|| < \pi, \tag{4.8.3}$$

it is required that z satisfies (4.8.3) in order for (4.8.2) to hold. This restriction means that the z-plane has a cut in the form of the spiral whose equation in polar coordinates is $r = e^{\omega_1\theta/\omega_2}$.

Analogously, when $0 < |q| < 1$, the contours in the q-analogues of Barnes' first and second lemmas given in §4.4 and the contours in the other integrals in §§4.4–4.7 must be replaced by upward directed contours parallel to the line $\mathrm{Re}(s(\omega_1 + i\omega_2)) = 0$ with indentations to separate the increasing and decreasing sequences of poles.

4.9 Other types of basic contour integrals

Let $q = e^{-\omega}$ with $\omega > 0$ and suppose that

$$P(z) = \frac{(a_1 z, \ldots, a_A z,\ b_1/z, \ldots, b_B/z; q)_\infty}{(c_1 z, \ldots, c_C z,\ d_1/z, \ldots, d_D/z; q)_\infty} \tag{4.9.1}$$

has only simple poles. During the 1950's Slater [1952c,d, 1955] considered contour integrals of the form

$$I_m \equiv I_m(A, B; C, D) = \frac{\omega}{2\pi i} \int_{-i\pi/\omega}^{i\pi/\omega} P(q^s) q^{ms} ds \tag{4.9.2}$$

with $m = 0$ or 1. However, here we shall let m be an arbitrary integer. It is assumed that none of the poles of $P(q^s)$ lie on the lines $\mathrm{Im} s = \pm\pi/\omega$ and that the contour of integration runs from $-i\pi/\omega$ to $i\pi/\omega$ and separates the increasing sequences of poles in $|\mathrm{Im}\ s| < \pi/\omega$ from those that are decreasing.

By setting $i\theta = -s\omega$ the integral I_m can also be written in the "exponential" form

$$I_m = \frac{1}{2\pi} \int_{-\pi}^{\pi} P\left(e^{i\theta}\right) e^{im\theta} d\theta \tag{4.9.3}$$

with suitable indentations, if necessary, in the contour of integration. Similarly, setting $z = q^s$ we obtain that

$$I_m = \frac{1}{2\pi i} \int_K P(z) z^{m-1} dz, \tag{4.9.4}$$

where the contour K is a deformation of the (positively oriented) unit circle so that the poles of $1/(c_1 z, \ldots, c_C z; q)_\infty$ lie outside the contour and the origin and poles of $1/(d_1/z, \ldots, d_D/z; q)_\infty$ lie inside the contour. Special cases of (4.9.3) and (4.9.4) have been considered by Askey and Roy [1986].

Although each of the above three types of integrals can be used to derive transformation formulas for basic hypergeometric series, we shall prefer to mainly use the contour integrals of the type in (4.9.4) since they are easier to work with, especially when the assumption in Slater [1952c,d, 1966] that $0 < q < 1$ is replaced by only assuming that $|q| < 1$, which is the case we wish to consider in the remainder of this chapter.

4.10 General basic contour integral formulas

Our main objective in this section is to see what formulas can be derived by applying Cauchy's theorem to the integrals I_m in (4.9.4).

Let $|q| < 1$ and let δ be a positive number such that $\delta \neq |d_j q^n|$ for $j = 1, 2, \ldots, D$, and $\delta \neq |c_j^{-1} q^{-n}|$ for $j = 1, 2, \ldots, C$ when $n = 0, 1, 2, \ldots$. Also let C_N be the circle $|z| = \delta |q|^N$, where N is a positive integer. Then C_N does not pass through any of the poles of $P(z)$ and we have that

$$\begin{aligned} |P(\delta q^N)(\delta q^N)^{m-1}| &= \left| \frac{(a_1\delta, \ldots, a_A\delta, b_1/\delta, \ldots, b_B/\delta; q)_\infty}{(c_1\delta, \ldots, c_C\delta, d_1/\delta, \ldots, d_D/\delta; q)_\infty} \right| \\ &\cdot \left| \frac{(c_1\delta, \ldots, c_C\delta, q\delta/b_1, \ldots, q\delta/b_B; q)_N}{(a_1\delta, \ldots, a_A\delta, q\delta/d_1, \ldots, q\delta/d_D; q)_N} \left(\frac{b_1 \cdots b_B q^{m-1}}{d_1 \cdots d_D} \right)^N \right| \\ &\cdot \delta^{m-1} \left| \delta^N q^{\binom{N+1}{2}} \right|^{D-B} = O\left(\left| \frac{b_1 \cdots b_B q^{m-1}}{d_1 \cdots d_D} \right|^N \left| \delta^N q^{\binom{N+1}{2}} \right|^{D-B} \right). \end{aligned} \tag{4.10.1}$$

Since C_N is of length $O(|q|^N)$ it follows from (4.10.1) that if $D > B$ or if $D = B$ and

$$\left| \frac{b_1 \cdots b_B q^m}{d_1 \cdots d_D} \right| < 1, \tag{4.10.2}$$

then

$$\lim_{N \to \infty} \int_{C_N} P(z) z^{m-1} dz = 0. \tag{4.10.3}$$

Hence, by applying Cauchy's residue theorem to the region between K and C_N for sufficiently large N and letting $N \to \infty$, we find that if $D > B$ or if $D = B$ and (4.10.2) holds, then I_m equals the sum of the residues of $P(z)z^{m-1}$ at the poles of $1/(d_1/z, \ldots, d_D/z; q)_\infty$. Therefore, since

$$\underset{z = dq^n}{\text{Residue}} \left(\frac{1}{(d/z; q)_\infty} \right) = \frac{(-1)^n d q^{2n + \binom{n}{2}}}{(q; q)_n (q; q)_\infty}, \qquad n = 0, 1, 2, \ldots, \tag{4.10.4}$$

it follows that

$$\begin{aligned} I_m = &\frac{(a_1 d_1, \ldots, a_A d_1, b_1/d_1, \ldots, b_B/d_1; q)_\infty}{(q, c_1 d_1, \ldots, c_C d_1, d_2/d_1, \ldots, d_D/d_1; q)_\infty} d_1^m \\ &\cdot \sum_{n=0}^{\infty} \frac{(c_1 d_1, \ldots, c_C d_1, q d_1/b_1, \ldots, q d_1/b_B; q)_n}{(q, a_1 d_1, \ldots, a_A d_1, q d_1/d_2, \ldots, q d_1/d_D; q)_n} \\ &\cdot \left(-d_1 q^{(n+1)/2} \right)^{n(D-B)} \left(\frac{b_1 \cdots b_B q^m}{d_1 \cdots d_D} \right)^n \\ &+ \text{idem } (d_1; d_2, \ldots, d_D) \end{aligned} \tag{4.10.5}$$

if $D > B$, or if $D = B$ and (4.10.2) holds.

In addition, by considering the residues of $P(z)z^{m-1}$ outside of K or by just using the inversion $z \to z^{-1}$ and renaming the parameters, we obtain

$$I_m = \frac{(b_1c_1, \ldots, b_Bc_1, a_1/c_1, \ldots, a_A/c_1; q)_\infty}{(q, d_1c_1, \ldots, d_Dc_1, c_2/c_1, \ldots, c_C/c_1; q)_\infty} c_1^{-m}$$

$$\cdot \sum_{n=0}^{\infty} \frac{(d_1c_1, \ldots, d_Dc_1, qc_1/a_1, \ldots, qc_1/a_A; q)_n}{(q, b_1c_1, \ldots, b_Bc_1, qc_1/c_2, \ldots, qc_1/c_C; q)_n}$$

$$\cdot \left(-c_1q^{(n+1)/2}\right)^{n(C-A)} \left(\frac{a_1 \cdots a_A q^{-m}}{c_1 \cdots c_C}\right)^n$$

$$+ \text{idem } (c_1; c_2, \ldots, c_C) \tag{4.10.6}$$

if $C > A$, or if $C = A$ and

$$\left|\frac{a_1 \cdots a_A q^{-m}}{c_1 \cdots c_C}\right| < 1. \tag{4.10.7}$$

In the special case when $C = A$ we can use the ${}_r\phi_s$ notation to write (4.10.5) in the form

$$I_m(A, B; A, D) = \frac{(a_1d_1, \ldots, a_Ad_1, b_1/d_1, \ldots, b_B/d_1; q)_\infty}{(q, c_1d_1, \ldots, c_Ad_1, d_2/d_1, \ldots, d_D/d_1; q)_\infty} d_1^m$$

$$\cdot {}_{A+B}\phi_{A+D-1}\left[\begin{matrix} c_1d_1, \ldots, c_Ad_1, qd_1/b_1, \ldots, qd_1/b_B \\ a_1d_1, \ldots, a_Ad_1, qd_1/d_2, \ldots, qd_1/d_D \end{matrix}; q, t(qd_1)^{D-B}\right]$$

$$+ \text{idem } (d_1; d_2, \ldots, d_D) \tag{4.10.8}$$

where $t = b_1b_2 \cdots b_Bq^m/d_1 \cdots d_D$, if $D > B$, or if $D = B$ and (4.10.2) holds.

Similarly, from the $D = B$ case of (4.10.6) we have

$$I_m(A, B; C, B) = \frac{(b_1c_1, \ldots, {}_Bc_1, a_1/c_1, \ldots, a_A/c_1; q)_\infty}{(q, d_1c_1, \ldots, d_Bc_1, c_2/c_1, \ldots, c_C/c_1; q)_\infty} c_1^{-m}$$

$$\cdot {}_{A+B}\phi_{B+C-1}\left[\begin{matrix} d_1c_1, \ldots, d_Bc_1, qc_1/a_1, \ldots, qc_1/a_A \\ b_1c_1, \ldots, b_Bc_1, qc_1/c_2, \ldots, qc_1/c_C \end{matrix}; q, u(qc_1)^{C-A}\right]$$

$$+ \text{idem } (c_1; c_2, \ldots, c_C) \tag{4.10.9}$$

if $C > A$, or if $C = A$ and (4.10.7) holds, where $u = a_1 \cdots a_A q^{-m}/c_1 \cdots c_C$.

Evaluations of I_m which follow from these formulas will be considered in §4.11.

From (4.10.8) and (4.10.9) it follows that if $C = A$ and $D = B$, then we have the transformation formula

$$\frac{(a_1d_1, \ldots, a_Ad_1, b_1/d_1, \ldots, b_B/d_1; q)_\infty}{(c_1d_1, \ldots, c_Ad_1, d_2/d_1, \ldots, d_B/d_1; q)_\infty} d_1^m$$

$$\cdot {}_{A+B}\phi_{A+B-1}\left[\begin{matrix} c_1d_1, \ldots, c_Ad_1, qd_1/b_1, \ldots, qd_1/b_B \\ a_1d_1, \ldots a_Ad_1, qd_1/d_2, \ldots, qd_1/d_B \end{matrix}; q, \frac{b_1 \cdots b_Bq^m}{d_1 \cdots d_B}\right]$$

$$+ \text{idem } (d_1; d_2, \ldots, d_B)$$

$$= \frac{(b_1c_1, \ldots, b_Bc_1, a_1/c_1, \ldots, a_A/c_1; q)_\infty}{(d_1c_1, \ldots, d_Bc_1, c_2/c_1, \ldots, c_A/c_1; q)_\infty} c_1^{-m}$$

$$\cdot {}_{A+B}\phi_{A+B-1}\left[\begin{matrix} d_1c_1,\ldots,d_Bc_1,qc_1/a_1,\ldots,qc_1/a_A \\ b_1c_1,\ldots,b_Bc_1,qc_1/c_2,\ldots,qc_1/c_A \end{matrix};q,\frac{a_1\cdots a_Aq^{-m}}{c_1\cdots c_A}\right]$$

$$+ \text{ idem } (c_1;c_2,\ldots,c_A) \tag{4.10.10}$$

provided that $|b_1\cdots b_Bq^m| < |d_1\cdots d_B|, |a_1\cdots a_Aq^{-m}| < |c_1\cdots c_A|$ and $m = 0,\pm1,\pm2,\ldots$.

In some applications it is useful to have a variable z in the argument of the series which is independent of the parameters in the series. This can be accomplished by replacing A by $A+1, B$ by $B+1$ and setting $b_{B+1} = z$ and $a_{A+1} = q/z$ in (4.10.10). More generally, doing this to the $m = 0$ case of (4.10.5) and of (4.10.6) gives the rather general transformation formula

$$\frac{(a_1d_1,\ldots,a_Ad_1,b_1/d_1,\ldots,b_B/d_1,z/d_1,qd_1/z;q)_\infty}{(c_1d_1,\ldots,c_Cd_1,d_2/d_1,\ldots,d_D/d_1;q)_\infty}$$

$$\cdot\sum_{n=0}^{\infty}\frac{(c_1d_1,\ldots,c_Cd_1,qd_1/b_1,\ldots,qd_1/b_B;q)_n}{(q,a_1d_1,\ldots,a_Ad_1,qd_1/d_2,\ldots,qd_1/d_D;q)_n}$$

$$\cdot\left(-d_1q^{(n+1)/2}\right)^{n(D-B-1)}\left(\frac{b_1\cdots b_Bz}{d_1\cdots d_D}\right)^n$$

$$+ \text{ idem } (d_1;d_2,\ldots,d_D)$$

$$= \frac{(b_1c_1,\ldots,b_Bc_1,a_1/c_1,\ldots,a_A/c_1,c_1z,q/c_1z;q)_\infty}{(d_1c_1,\ldots,d_Dc_1,c_2/c_1,\ldots,c_C/c_1;q)_\infty}$$

$$\cdot\sum_{n=0}^{\infty}\frac{(d_1c_1,\ldots,d_Dc_1,qc_1/a_1,\ldots,qc_1/a_A;q)_n}{(q,b_1c_1,\ldots,b_Bc_1,qc_1/c_2,\ldots,qc_1/c_C;q)_n}$$

$$\cdot\left(-c_1q^{(n+1)/2}\right)^{n(C-A-1)}\left(\frac{a_1\cdots a_Aq}{c_1\cdots c_Cz}\right)^n$$

$$+ \text{ idem } (c_1;c_2,\ldots,c_C), \tag{4.10.11}$$

where, for convergence,

(i) $\quad D > B+1$, or $D = B+1$ and $\left|\dfrac{b_1\cdots b_Bz}{d_1\cdots d_D}\right| < 1$

and

(ii) $\quad C > A+1$, or $C = A+1$ and $\left|\dfrac{a_1\cdots a_Aq}{c_1\cdots c_Cz}\right| < 1.$

This is formula (5.2.20) in Slater [1966]. Observe that by replacing z in (4.10.11) by zq^m and using the identity

$$(qd/zq^m, zq^m/d;q)_\infty = (-1)^m(d/z)^mq^{-\binom{m}{2}}(qd/z,zd;q)_\infty \tag{4.10.12}$$

we obtain from (4.10.11) the formula that would have been derived by using (4.10.5) and (4.10.6) with m an arbitrary integer.

4.11 Some additional extensions of the beta integral

Askey and Roy [1986] used Ramanujan's summation formula (2.10.17) to show that

$$\frac{1}{2\pi}\int_{-\pi}^{\pi}\frac{(ce^{i\theta}/\beta, qe^{i\theta}/c\alpha, c\alpha e^{-i\theta}, q\beta e^{-i\theta}/c;q)_\infty}{(ae^{i\theta}, be^{i\theta}, \alpha e^{-i\theta}, \beta e^{-i\theta};q)_\infty}d\theta$$
$$= \frac{(ab\alpha\beta, c, q/c, c\alpha/\beta, q\beta/c\alpha;q)_\infty}{(a\alpha, a\beta, b\alpha, b\beta, q;q)_\infty}, \tag{4.11.1}$$

where $\max(|q|, |a|, |b|, |\alpha|, |\beta|) < 1$ and $c\alpha\beta \neq 0$; and they extended it to the contour integral form

$$\frac{1}{2\pi i}\int_K \frac{(cz/\beta, qz/c\alpha, c\alpha/z, q\beta/cz;q)_\infty}{(az, bz, \alpha/z, \beta/z, ;q)_\infty}\frac{dz}{z}$$
$$= \frac{ab\alpha\beta, c, q/c, c\alpha/\beta, q\beta/c\alpha;q)_\infty}{(a\alpha, a\beta, b\alpha, b\beta, q;q)_\infty} \tag{4.11.2}$$

where $a\alpha, a\beta, b\alpha, b\beta \neq q^{-n}, n = 0, 1, 2, \ldots, c\alpha\beta \neq 0$, and K is a deformation of the unit circle as described in §4.9. These formulas can also be derived from the $A = B = D = 2, m = 0$ case of (4.10.8) by setting $a_1 = c/\beta, a_2 = q/c\alpha, b_1 = c\alpha, b_2 = q\beta/c, c_1 = a, c_2 = b, d_1 = \alpha$ and $d_2 = \beta$ and then using the summation formula (2.10.13) for the sum of the two ${}_2\phi_1$ series resulting on the right side. In Askey and Roy [1986] it is also shown how Barnes' beta integral (4.1.2) can be obtained as a limit case of (4.11.1).

Analogously, application of the summation formula (2.10.11) to the $A = 3, B = D = 2, m = 0$ case of (4.10.8) gives

$$\frac{1}{2\pi i}\int_K \frac{(\delta z, qz/\gamma, \gamma z/\alpha\beta, \gamma/z, q\alpha\beta/\gamma z;q)_\infty}{(az, bz, cz, \alpha/z, \beta/z;q)_\infty}\frac{dz}{z}$$
$$= \frac{(\gamma/\alpha, q\alpha/\gamma, \gamma/\beta, q\beta/\gamma, \delta/a, \delta/b, \delta/c;q)_\infty}{(a\alpha, a\beta, b\alpha, b\beta, c\alpha, c\beta, q;q)_\infty}, \tag{4.11.3}$$

where $\delta = abc\alpha\beta, abc\alpha\beta\gamma \neq 0$, and

$$a\alpha, a\beta, b\alpha, b\beta, c\alpha, c\beta \neq q^{-n}, \qquad n = 0, 1, 2, \ldots$$

Note that (4.11.2) follows from the $c \to 0$ case of (4.11.3).

In addition, application of Bailey's summation formula (2.11.7) gives the more general formula

$$\frac{1}{2\pi i}\int_K \frac{\left(za^{\frac{1}{2}}, -za^{\frac{1}{2}}, qaz/b, qaz/c, qaz/d, qaz/f;q\right)_\infty}{\left(qza^{\frac{1}{2}}, -qza^{\frac{1}{2}}, bz, cz, dz, fz;q\right)_\infty}$$
$$\cdot\frac{(qz/\gamma, \gamma z/\alpha\beta, \gamma/z, q\alpha\beta/\gamma z;q)_\infty}{(a\alpha z, a\beta z, \alpha/z, \beta/z;q)_\infty}\frac{dz}{z}$$
$$= \frac{(\gamma/\alpha, q\alpha/\gamma, \gamma/\beta, q\beta/\gamma, aq/cd, aq/bd, aq/bc, aq/bf, aq/cf, aq/df;q)_\infty}{(a\alpha\beta, b\alpha, c\alpha, d\alpha, f\alpha, b\beta, c\beta, d\beta, f\beta, q;q)_\infty} \tag{4.11.4}$$

where $aq = bcdf\alpha\beta, bcdf\alpha\beta\gamma \neq 0$,

$$a\alpha\beta, b\alpha, c\alpha, d\alpha, f\alpha, b\beta, c\beta, d\beta, f\beta \neq q^{-n}, \qquad n = 0, 1, 2, \ldots,$$

and K is as described in §4.9; see Gasper [1989c].

4.12 Sears' transformations of well-poised series

Sears [1951d, (7.2)] used series manipulations of well-poised series to derive the transformation formula

$$\begin{aligned}
&\frac{\left(qa_1/a_{M+2}, \ldots, qa_1/a_{2M}, q/a_{M+2}, \ldots, q/a_{2M}, a_1^{\frac{1}{2}}, -a_1^{\frac{1}{2}}, q/a_1^{\frac{1}{2}}, -q/a_1^{\frac{1}{2}}; q\right)_\infty}{(a_1, \ldots, a_{M+1}, a_2/a_1, \ldots, a_{M+1}/a_1; q)_\infty} \\
&\cdot {}_{2M}\phi_{2M-1}\left[\begin{matrix} a_1, a_2, \ldots, a_{2M} \\ qa_1/a_2, \ldots, qa_1/a_{2M} \end{matrix}; q, -x\right] \\
&= a_2 \frac{(qa_2/a_{M+2}, \ldots, qa_2/a_{2M}, qa_1/a_2a_{M+2}, \ldots, qa_1/a_2a_{2M}; q)_\infty}{(a_1/a_2, a_2, a_3/a_2, \ldots, a_{M+1}/a_2, a_2^2/a_1, a_2a_3/a_1, \ldots, a_2a_{M+1}/a_1; q)_\infty} \\
&\cdot \left(a_1^{\frac{1}{2}}/a_2, -a_1^{\frac{1}{2}}/a_2, qa_2/a_1^{\frac{1}{2}}, -qa_2/a_1^{\frac{1}{2}}; q\right)_\infty \\
&\cdot {}_{2M}\phi_{2M-1}\left[\begin{matrix} a_2^2/a_1, a_2, a_2a_3/a_1, \ldots, a_2a_{2M}/a_1 \\ qa_2/a_1, qa_2/a_3, \ldots, qa_2/a_{2M} \end{matrix}; q, -x\right] \\
&+ \text{idem } (a_2; a_3, \ldots, a_{2M}),
\end{aligned} \tag{4.12.1}$$

where $x = (qa_1)^M / a_1a_2 \cdots a_{2M}$. Slater [1952c] observed that this formula could also be derived from (4.10.10) by taking $A = B = M + 1, m = 1$, choosing the parameters such that $P(z)$ in (4.9.1) becomes

$$\begin{aligned}
&\frac{\left(qa_1z/a_{M+2}, \ldots, qa_1/za_{2M}, qza_1^{\frac{1}{2}}, -qza_1^{\frac{1}{2}}; q\right)_\infty}{(a_1z, \ldots, a_{M+1}z; q)_\infty} \\
&\cdot \frac{\left(q/za_{M+2}, \ldots, q/za_{2M}, 1/za_1^{\frac{1}{2}}, -1/za_1^{\frac{1}{2}}; q\right)_\infty}{(1/z, a_2/za_1, \ldots, a_{M+1}/za_1; q)_\infty},
\end{aligned} \tag{4.12.2}$$

and then using the fact that

$$\begin{aligned}
&(a, -a, q/a, -q/a; q)_\infty - a^2(qa, -qa, 1/a, -1/a; q)_\infty \\
&= 2(a, -a, q/a, -q/a; q)_\infty
\end{aligned} \tag{4.12.3}$$

to combine the terms with the same ${}_{2M}\phi_{2M-1}$ series.

Similarly, taking $A = B = M + 2$ and $m = 1$ in (4.10.10) and choosing the parameters such that $P(z)$ in (4.9.1) becomes

$$\begin{aligned}
&\frac{\left(qa_1z/a_{M+3}, \ldots, qa_1z/a_{2M}, qza_1^{\frac{1}{2}}, -qza_1^{\frac{1}{2}}, z(qa_1)^{\frac{1}{2}}, -z(qa_1)^{\frac{1}{2}}, q\right)_\infty}{(a_1z, \ldots, a_{M+2}z; q)_\infty} \\
&\cdot \frac{\left(q/za_{M+3}, \ldots, q/za_{2M}, 1/za_1^{\frac{1}{2}}, -1/za_1^{\frac{1}{2}}, q^{\frac{1}{2}}/za_1^{\frac{1}{2}}, -q^{\frac{1}{2}}/za_1^{\frac{1}{2}}; q\right)_\infty}{(1/z, a_2/za_1, \ldots, a_{M+2}/za_1; q)_\infty},
\end{aligned} \tag{4.12.4}$$

we obtain

$$\frac{(qa_1/a_{M+3},\ldots,qa_1/a_{2M},q/a_{M+3},\ldots,q/a_{2M},q/a_1;q)_\infty}{(a_2,\ldots,a_{M+2},a_2/a_1,\ldots,a_{M+2}/a_1;q)_\infty}$$
$$\cdot {}_{2M}\phi_{2M-1}\left[\begin{matrix}a_1,a_2,\ldots,a_{2M}\\ qa_1/a_2,\ldots,qa_1/a_{2M}\end{matrix};q,x\right]$$
$$= a_2\frac{(qa_2/a_{M+3},\ldots,qa_2/a_{2M},qa_1/a_2a_{M+3},\ldots,qa_1/a_2a_{2M};q)_\infty}{(a_1/a_2,a_3/a_2,\ldots,a_{M+2}/a_2,a_2a_3/a_1,\ldots,a_2a_{M+2}/a_1;q)_\infty}$$
$$\cdot\frac{(a_1/a_2^2,qa_2^2/a_1;q)_\infty}{(a_2,a_2^2/a_1;q)_\infty}\,{}_{2M}\phi_{2M-1}\left[\begin{matrix}a_2^2/a_1,a_2,a_2a_3/a_1,\ldots,a_2a_{2M}/a_1\\ qa_2/a_1,qa_2/a_3,\ldots,qa_2/a_{2M}\end{matrix};q,x\right]$$
$$+\text{ idem }(a_2;a_3,\ldots,a_{M+2}), \tag{4.12.5}$$

where $x=(qa_1)^M/a_1\cdots a_{2M}$, which is formula (7.3) in Sears [1951d].

Finally, if we take $A=B=M+2$ and $m=1$ in (4.10.10) and choose the parameters such that $P(z)$ in (4.9.1) becomes

$$\frac{\left(qa_1z/a_{M+3},\ldots,qa_1z/a_{2M+1},qza_1^{\frac12},-qza_1^{\frac12},\pm q^{\frac12}za_1^{\frac12};q\right)_\infty}{(a_1z,\ldots,a_{M+2}z;q)_\infty}$$
$$\cdot\frac{\left(q/za_{M+3},\ldots,q/za_{2M+1},1/za_1^{\frac12},-1/za_1^{\frac12},\pm q^{\frac12}/za_1^{\frac12};q\right)_\infty}{(1/z,a_2/za_1,\ldots,a_{M+2}/za_1;q)_\infty}, \tag{4.12.6}$$

we obtain

$$\frac{(qa_1/a_{M+3},\ldots,qa_1/a_{2M+1},q/a_{M+3},\ldots,q/a_{2M+1};q)_\infty}{(a_1,\ldots,a_{M+2},a_2/a_1,\ldots,a_{M+2}/a_1;q)_\infty}$$
$$\cdot\left(a_1^{\frac12},-a_1^{\frac12},q/a_1^{\frac12},-q/a_1^{\frac12},\pm(a_1q)^{\frac12},\pm(q/a_1)^{\frac12};q\right)_\infty$$
$$\cdot {}_{2M+1}\phi_{2M}\left[\begin{matrix}a_1,a_2,\ldots,a_{2M+1}\\ qa_1/a_2,\ldots,qa_1/a_{2M+1}\end{matrix};q,\mp y\right]$$
$$= a_2\frac{(qa_2/a_{M+3},\ldots,qa_2/a_{2M+1},qa_1/a_2a_{M+3},\ldots,qa_1/a_2a_{2M+1};q)_\infty}{(a_1/a_2,a_3/a_2,\ldots,a_{M+2}/a_2,a_2,a_2^2/a_1,a_2a_3/a_1,\ldots,a_2a_{M+2}/a_1;q)_\infty}$$
$$\cdot\left(a_1^{\frac12}/a_2,-a_1^{\frac12}/a_2,qa_2/a_1^{\frac12},-qa_2/a_1^{\frac12},\pm a_2(q/a_1)^{\frac12},\pm(qa_1)^{\frac12}/a_2;q\right)_\infty$$
$$\cdot {}_{2M+1}\phi_{2M}\left[\begin{matrix}a_2^2/a_1,a_2,a_2a_3/a_1,\ldots,a_2a_{2M+1}/a_1\\ qa_2/a_1,qa_2/a_3,\ldots,qa_2/a_{2M+1}\end{matrix};q,\mp y\right]$$
$$+\text{ idem }(a_2;a_3,\ldots,a_{M+2}), \tag{4.12.7}$$

where $y=(qa_1)^{M+\frac12}/a_1\cdots a_{2M+1}$, which are formulas (7.4) and (7.5) in Sears [1951d].

Exercises 4

4.1 Let Re $c > 0$, Re $d > 0$, and $\mathrm{Re}(x+y) > 1$. Show that Cauchy's [1825] beta integral

$$\frac{1}{2\pi i}\int_{-i\infty}^{i\infty}\frac{ds}{(1+cs)^x(1-ds)^y} = \frac{\Gamma(x+y-1)(1+d/c)^{1-y}(1+c/d)^{1-x}}{(c+d)\Gamma(x)\Gamma(y)}$$

has a q-analogue of the form

$$\begin{aligned}&\frac{1}{2\pi i}\int_{-i\infty}^{i\infty}\frac{(-csq^x, dsq^y;q)_\infty}{(-cs, ds;q)_\infty}\,ds\\ &= \frac{\Gamma_q(x+y-1)}{\Gamma_q(x)\Gamma_q(y)}\,\frac{(-cq^x/d, -dq^y/c;q)_\infty}{(c+d)(-cq/d, -dq/c;q)_\infty},\end{aligned}$$

where $0 < q < 1$.
(Wilson [1985])

4.2 Prove that

$$\begin{aligned}&\frac{1}{2\pi i}\int_{-i\infty}^{i\infty}\frac{(q^{1+s}, -q^{a+s}, q^{a-b+1+s}, -q^{a-b+1+s}, q^{c+d+e-a+s};q)_\infty}{(q^{c+s}, q^{d+s}, q^{e+s}, -q^{a+1+s}, -q^{a-2b+s};q)_\infty}\\ &\quad\cdot\frac{\pi q^s\,ds}{\sin\pi s\sin\pi(c+d+e-a+s)}\\ &= \csc\pi(c+d+e-a)\frac{(q, q^{c+d+e-a}, q^{1+a-c-d-e}, q^{1+a-b};q)_\infty}{(q^{1+a}, -q^{1+a}, -q^{a-2b}, q^c;q)_\infty}\\ &\quad\cdot\frac{(-q^{1+a-b}, -q^a, q^{1+a-c}, q^{1+a-d}, q^{1+a-e};q)_\infty}{(q^d, q^e, q^{1+a-c-d}, q^{1+a-c-e}, q^{1+a-d-e};q)_\infty}\\ &\quad\cdot{}_{10}W_9\left(q^a; iq^{1+a/2}, -iq^{1+a/2}, q^b, -q^b, q^c, q^d, q^e; q, -q^{1+2a-2b-c-d-e}\right),\end{aligned}$$

where $1+2a-2b > c+d+e$.

4.3 Show that

$$\begin{aligned}&{}_3\phi_2\left[\begin{matrix}ac, bc, ad\\ abcg, acdh\end{matrix};q, gh\right]\\ &= \frac{(q, ac, bc, ag, bg, ch;q)_\infty}{(f, q/f, cf/g, qg/cf, abcg, acdh;q)_\infty}\\ &\quad\cdot\frac{1}{2\pi}\int_{-\pi}^{\pi}\frac{(fe^{i\theta}/g, qe^{i\theta}/cf, dhe^{i\theta}, qge^{-i\theta}/f, cfe^{-i\theta};q)_\infty}{(ae^{i\theta}, be^{i\theta}, he^{i\theta}, ce^{-i\theta}, ge^{-i\theta};q)_\infty}\,d\theta.\end{aligned}$$

4.4 Prove that

$$\frac{1}{2\pi}\int_{-\pi}^{\pi}\frac{\left(fe^{i\theta},ke^{i\theta}/d,qde^{-i\theta}/k,cke^{-i\theta},qe^{i\theta}/ck,abcdghe^{i\theta}/f;q\right)_\infty}{(ae^{i\theta},be^{i\theta},ce^{-i\theta},de^{-i\theta},ge^{i\theta},he^{i\theta};q)_\infty}\,d\theta$$
$$=\frac{(k,q/k,ck/d,qd/ck,cf,df,acdg,bcdg,cdgh,abcdh/f;q)_\infty}{(q,ac,ad,bc,bd,cg,dg,ch,dh,cdfg;q)_\infty}$$
$$\cdot\,{}_8W_7(cdfg/q;cg,dg,f/a,f/b,f/h;q,abcdh/f).$$

4.5 Prove that

$${}_4\phi_2\left[\begin{matrix}q^{-n},abcdq^{n-1},ae^{i\theta},ae^{-i\theta}\\ ab,ac,ad\end{matrix};q,q\right]$$
$$=\frac{(q;q)_\infty}{(q^{\frac12},q^{\frac12};q)_\infty}\left|\frac{\left(ae^{i\theta},be^{i\theta},ce^{i\theta};q\right)_\infty}{\left(q^{\frac12}e^{2i\theta};q\right)_\infty}\right|^2$$
$$\cdot\frac{1}{2\pi}\int_{-\pi}^{\pi}\frac{\left(q^{\frac12}e^{i\theta+i\phi}/\sigma,q^{\frac12}\sigma e^{-i\theta-i\phi},\sigma q^{\frac12}e^{i\theta-i\phi},q^{\frac12}e^{i\phi-i\theta}/\sigma,abce^{i\phi}/\sigma;q\right)_\infty}{(ae^{i\phi}/\sigma,be^{i\phi}/\sigma,ce^{i\phi}/\sigma,\sigma e^{i\theta-i\phi},\sigma e^{-i\theta-i\phi};q)_\infty}$$
$$\cdot\frac{(d\sigma e^{-i\phi},bc;q)_n}{(abce^{i\phi}/\sigma,ad;q)_n}\left(\frac{a}{\sigma}e^{i\phi}\right)^n\,d\phi$$

for $n=0,1,2,\ldots$.

4.6 Prove that

$$a^{-1}\frac{(aq/e,aq/f,aq/g,aq/h,q/ae,q/af,q/ag,q/ah;q)_\infty}{(qa^2,ab,ac,ad,b/a,c/a,d/a;q)_\infty}$$
$$\cdot\,{}_{10}W_9\left(a^2;ab,ac,ad,ae,af,ag,ah;q,q^3/abcdefgh\right)$$
$$+\text{ idem }(a;b,c,d)=0,$$

where $|q^3|<|abcdefgh|$.

4.7 Prove that

$$a_1^{-1}\frac{(a_1q/b_1,a_1q/b_2,\ldots,a_1q/b_r,q/a_1b_1,\ldots,q/a_1b_r;q)_\infty}{(qa_1^2,a_1a_2,\ldots,a_1a_r,a_2/a_1,\ldots,a_r/a_1;q)_\infty}$$
$$\cdot\,{}_{2r+2}W_{2r+1}\left(a_1^2;a_1a_2,\ldots,a_1a_r,a_1b_1,\ldots,a_1b_r;q,q^{r-1}/a_1\cdots a_rb_1\cdots b_r\right)$$
$$+\text{ idem }(a_1;a_2,\ldots,a_r)=0,$$

where $r=1,2,\ldots$, and $|q^{r-1}|<|a_1\cdots a_rb_1\cdots b_r|$.

4.8 Show that

$$\frac{1}{2\pi i}\int_{-i\infty}^{i\infty}\frac{(-csq^{n+1},bds,\alpha s;q)_\infty}{(-cs,ds,b\alpha sq^{n-1};q)_\infty}\,ds$$
$$=\frac{(-\alpha/c,-bd/c;q)_\infty}{(c+d)(-dq/c,-b\alpha/cq;q)_\infty}\frac{(b,\alpha/d;q)_n}{(q,-cq/d;q)_n},$$

where $\mathrm{Re}(c, d, b\alpha) > 0$ and $n = 0, 1, \ldots$. Show that the q-Cauchy beta integral in Ex. 4.1 follows from this formula by letting $n \to \infty$ and then setting $b = q^y, \alpha = -cq^x$.

4.9 Extend the integral in Ex. 4.8 to

$$\frac{1}{2\pi i}\int_{-i\infty}^{i\infty} \frac{(-acs, ac^2s/\alpha, ac^2s/\beta, ac^2s/\gamma, ac^2s/\delta, ac^2s/\lambda; q)_\infty}{(-cs, \alpha s, \beta s, \gamma s, \delta s, \lambda s; q)_\infty} \cdot \left(1 - \frac{a^2c^2s^2}{q}\right) ds$$
$$= \frac{(a/q, -ac/\alpha, -ac/\beta, -ac/\gamma, -ac/\delta, -ac/\lambda; q)_\infty}{c(q, -\alpha/c, -\beta/c, -\gamma/c, -\delta/c, -\lambda/c; q)_\infty} \cdot \frac{(q^2/a, ac^2/\alpha\beta, ac^2/\alpha\gamma, ac^2/\beta\gamma; q)_n}{(-cq/\alpha, -cq/\beta, -cq/\gamma, -a^2c^3/\alpha\beta\gamma q; q)_n},$$

where $\mathrm{Re}(c, \alpha, \beta, \gamma, \delta, \lambda) > 0$, $a^3c^5 = -\alpha\beta\gamma\delta\lambda q^2$, $ac = -\lambda q^{n+1}$, and $n = 1, 2 \ldots$.

4.10 Show that

$$\frac{1}{2\pi i}\int_K \frac{(q^2z/a\alpha\gamma, q^2z/b\alpha\gamma, q^2z/c\alpha\gamma, \gamma/z; q)_\infty}{(az, bz, cz, \alpha/z; q)_\infty} \frac{1 - qz^2/\alpha\gamma}{z} dz$$
$$= \frac{(a\gamma, b\gamma, c\gamma, \alpha q/\gamma; q)_\infty}{(a\alpha, b\alpha, c\alpha, q; q)_\infty},$$

where $q^2 = abc\alpha q^2, |\gamma/\alpha| < 1$, and the contour K is as defined in §4.9.

4.11 Show that

$$\frac{1}{2\pi i}\int_K \frac{(bqz, qz/\gamma, \gamma/z; q)_\infty}{(az, bz, \alpha/z; q)_\infty} \cdot (qz/\gamma q^m; q)_m (a_1z; q)_{m_1} \cdots (a_rz; q)_{m_r} \frac{dz}{z}$$
$$= \frac{(\gamma/\alpha, \alpha q/\gamma, bq/z; q)_\infty}{(a\alpha, q/a\alpha, b\alpha; q)_\infty} \cdot (q/b\gamma q^m; q)_m (a_1/b; q)_{m_1} \cdots (a_r/b; q)_{m_r} (b\alpha)^{m+m_1+\cdots+m_r},$$

provided $|\gamma/\alpha| < 1$, where $m, m_1, \ldots, m_r$ are nonnegative integers, $q = a\gamma q^{m+m_1+\cdots+m_r}$ and K is as defined in §4.9.

4.12 Show that

$$\frac{1}{2\pi i}\int_{-i\infty}^{i\infty} \frac{(-acs, dqs; q)_\infty}{(-cs, ds; q)_\infty}(a_1s; q)_{m_1} \cdots (a_rs; q)_{m_r}\, ds$$
$$= \frac{(-ac/d; q)_\infty}{(c+d)(-cq/d; q)_\infty}(a_1/d; q)_{m_1} \cdots (a_r/d; q)_{m_r}$$

provided $|aq^{-(m_1+\cdots+m_r)}| < 1$, Re $(c,d,a_1,\cdots,a_r) > 0$, and $m_1,\ldots,m_r$ are nonnegative integers.

4.13 Show that

$$\frac{1}{2\pi i}\int_{-i\infty}^{i\infty} \frac{(-acs, -ac^2s/f, ac^2s/\alpha, ac^2s/\beta;q)_\infty}{(-cs,-fs,\alpha s,\beta s;q)_\infty} \cdot \frac{(ac^2s/\gamma, ac^2s/\delta;q)_\infty}{(\gamma s,\delta s;q)_\infty}(1-ac^2s^2/q)\,ds$$
$$= \frac{(a/q, ac/f, -ac/\alpha, -ac/\beta, -ac/\gamma, -ac/\delta;q)_\infty}{c(q,f/c,-\alpha/c,-\beta/c,-\gamma/c,-\delta/c;q)_\infty} \cdot \frac{(q^2/a, ac^2/\alpha\beta, ac^2/\alpha\gamma, ac^2/\beta\gamma;q)_n}{(-cq/\alpha,-cq/\beta,-cq/\gamma,-a^2c^3/\alpha\beta\gamma q;q)_n}$$
$$+ \frac{(ac^2/f^2q, ac/f, -ac^2/f\alpha, -ac^2/f\beta, -ac^2/f\gamma, -ac^2/f\delta;q)_\infty}{f(q,c/f,-\alpha/f,-\beta/f,-\gamma/f,-\delta/f;q)_\infty} \cdot \frac{(f^2q^2/ac^2, ac^2/\alpha\beta, ac^2/\alpha\gamma, ac^2/\beta\gamma;q)_n}{(-fq/\alpha,-fq/\beta,-fq/\gamma,-a^2c^4/f\alpha\beta\gamma q;q)_n}$$

provided $ac = fq^{n+1}, a^3c^5 = f\alpha\beta\gamma\delta q^2$, Re $(c,f,\alpha,\beta,\gamma,\delta) > 0$, the integrand has only simple poles, and $n = 0,1,\ldots$.

(For the formulas in Exercises 4.8 – 4.13, and related formulas, see Gasper [1989c])

Notes 4

§4.4 Kalnins and Miller [1989b] exploited symmetry (recurrence relation) techniques similar to those used by Nikiforov and Suslov [1986], Nikiforov, Suslov and Uvarov [1985], and Nikiforov and Uvarov [1988] to give another proof of (4.4.3) and of (4.11.1).

§4.6 Contour integrals of the types considered in this section were used by Agarwal [1953c] to give simple proofs of the two-term and three-term transformation formulas for ${}_8\phi_7$ series.

§4.12 Sears [1951b] also derived the hypergeometric limit cases of the transformation formulas in this section. Applications of (4.12.1) to some formulas in partition theory are given in M. Jackson [1949].

5

BILATERAL BASIC HYPERGEOMETRIC SERIES

5.1 Notations and definitions

The general bilateral basic hypergeometric series in base q with r numerator and s denominator parameters is defined by

$$
\begin{aligned}
{}_r\psi_s(z) \equiv {}_r\psi_s\left[\begin{matrix} a_1, a_2, \ldots, a_r \\ b_1, b_2, \ldots, b_s \end{matrix}; q, z\right] \\
= \sum_{n=-\infty}^{\infty} \frac{(a_1, a_2, \ldots, a_r; q)_n}{(b_1, b_2, \ldots, b_s; q)_n} (-1)^{(s-r)n} q^{(s-r)\binom{n}{2}} z^n.
\end{aligned}
\tag{5.1.1}
$$

In (5.1.1) it is assumed that q, z and the parameters are such that each term of the series is well-defined (i.e., the denominator factors are never zero, $q \neq 0$ if $s < r$, and $z \neq 0$ if negative powers of z occur). Note that a bilateral basic hypergeometric series is a series $\sum_{n=-\infty}^{\infty} v_n$ such that $v_0 = 1$ and v_{n+1}/v_n is a rational function of q^n. By applying (1.2.28) to the terms with negative n, we obtain that

$$
\begin{aligned}
{}_r\psi_s(z) = \sum_{n=0}^{\infty} \frac{(a_1, a_2, \ldots, a_r; q)_n}{(b_1, b_2, \ldots, b_s; q)_n} (-1)^{(s-r)n} q^{(s-r)\binom{n}{2}} z^n \\
+ \sum_{n=1}^{\infty} \frac{(q/b_1, q/b_2, \ldots, q/b_s; q)_n}{(q/a_1, q/a_2, \ldots, q/a_r; q)_n} \left(\frac{b_1 \cdots b_s}{a_1 \cdots a_r z}\right)^n.
\end{aligned}
\tag{5.1.2}
$$

Let $R = |b_1 \cdots b_s / a_1 \cdots a_r|$. If $s < r$ and $|q| < 1$, then the first series on the right side of (5.1.2) diverges for $z \neq 0$; if $s < r$ and $|q| > 1$, then the first series converges for $|z| < R$ and the second series converges for all $z \neq 0$. When $r < s$ and $|q| < 1$ the first series converges for all z, but the second series converges only when $|z| > R$. If $r < s$ and $|q| > 1$, the second series diverges for all $z \neq 0$. If $r = s$, which is the most important case, and $|q| < 1$, the first series converges when $|z| < 1$ and the second when $|z| > R$; on the other hand, if $|q| > 1$ the second series converges when $|z| > 1$ and the first when $|z| < R$.

We shall assume throughout this chapter that $|q| < 1$, so that the region of convergence of the bilateral series

$$
\begin{aligned}
{}_r\psi_r(z) &= \sum_{n=-\infty}^{\infty} \frac{(a_1, \ldots, a_r; q)_n}{(b_1, \ldots, b_r; q)_n} z^n \\
&= \sum_{n=0}^{\infty} \frac{(a_1, \ldots, a_r; q)_n}{(b_1, \ldots, b_r; q)_n} z^n \\
&\quad + \sum_{n=1}^{\infty} \frac{(q/b_1, \ldots, q/b_r; q)_n}{(q/a_1, \ldots, q/a_r; q)_n} \left(\frac{b_1 \cdots b_r}{a_1 \cdots a_r z}\right)^n
\end{aligned}
\tag{5.1.3}
$$

is the annulus

$$\left|\frac{b_1\cdots b_r}{a_1\cdots a_r}\right| < |z| < 1. \tag{5.1.4}$$

When $b_j = q$ for some j, the second series on the right sides of (5.1.2) and (5.1.3) vanish and the first series become basic hypergeometric series. If we replace the index of summation n in (5.1.1) by $k+n$, where k is an integer, then it follows that

$$\begin{aligned}
&{}_r\psi_s\left[\begin{matrix} a_1,\ldots,a_r \\ b_1,\ldots,b_s \end{matrix};q,z\right] \\
&= \frac{(a_1,\ldots,a_r;q)_k}{(b_1,\ldots,b_s;q)_k} z^k \left[(-1)^k q^{\binom{k}{2}}\right]^{s-r} \\
&\quad \cdot {}_r\psi_s\left[\begin{matrix} a_1q^k,\ldots,a_rq^k \\ b_1q^k,\ldots,b_sq^k \end{matrix};q,zq^{k(s-r)}\right].
\end{aligned} \tag{5.1.5}$$

When r and s are small we shall frequently use the single-line notation

$${}_r\psi_s(z) \equiv {}_r\psi_s(a_1,\ldots,a_r;b_1,\ldots,b_s;q,z).$$

An ${}_r\psi_r$ series will be called *well-poised* if $a_1b_1 = a_2b_2 = \cdots = a_rb_r$, and *very-well-poised* if it is well-poised and $a_1 = -a_2 = qb_1 = -qb_2$.

5.2 Ramanujan's sum for ${}_1\psi_1(a;b;q,z)$

The bilateral summation formula

$${}_1\psi_1(a;b;q,z) = \frac{(q,b/a,az,q/az;q)_\infty}{(b,q/a,z,b/az;q)_\infty}, \qquad |b/a| < |z| < 1, \tag{5.2.1}$$

which is an extension of the q-binomial formula (1.3.2), was first given by Ramanujan (see Hardy [1940]). In Chapter 2 we saw that this formula follows as a special case of Sears' ${}_3\phi_2$ summation formula (2.10.12). Andrews [1969, 1970a], Hahn [1949b], M. Jackson [1950b], Ismail [1977] and Andrews and Askey [1978] published different proofs of (5.2.1). The proof given here is due to Andrews and Askey [1978].

The first step is to regard ${}_1\psi_1(a;b;q,z)$ as a function of b, say, $f(b)$. Then

$$\begin{aligned}
f(b) &= {}_1\psi_1(a;b;q,z) \\
&= \sum_{n=0}^{\infty} \frac{(a;q)_n}{(b;q)_n} z^n + \sum_{n=1}^{\infty} \frac{(q/b;q)_n}{(q/a;q)_n} (b/az)^n
\end{aligned} \tag{5.2.2}$$

so that, by (5.1.4), the two series are convergent when $|b/a| < |z| < 1$. As a function of b, $f(b)$ is clearly analytic for $|b| < \min(1,|az|)$ when $|z| < 1$. Since

$$\begin{aligned}
&{}_1\psi_1(a;b;q,z) - a\,{}_1\psi_1(a;b;q,qz)\\
&= \sum_{n=-\infty}^{\infty} \left\{ \frac{(a;q)_n}{(b;q)_n} - \frac{aq^n(a;q)_n}{(b;q)_n} \right\} z^n\\
&= \sum_{n=-\infty}^{\infty} \frac{(a;q)_{n+1}}{(b;q)_n} z^n\\
&= z^{-1}(1-b/q) \sum_{n=-\infty}^{\infty} \frac{(a;q)_{n+1}}{(b/q;q)_{n+1}} z^{n+1}\\
&= z^{-1}(1-b/q)\,{}_1\psi_1(a;b/q;q,z),
\end{aligned}$$

we get

$$f(bq) - z^{-1}(1-b)f(b) = a\,{}_1\psi_1(a;bq;q,qz). \tag{5.2.3}$$

However,

$$\begin{aligned}
&a\,{}_1\psi_1(a;bq;q,qz)\\
&= a \sum_{n=-\infty}^{\infty} \frac{(a;q)_n}{(bq;q)_n} (qz)^n\\
&= -ab^{-1} \sum_{n=-\infty}^{\infty} \frac{(a;q)_n(1-bq^n-1)}{(bq;q)_n} z^n\\
&= -ab^{-1}(1-b)f(b) + ab^{-1}f(bq).
\end{aligned} \tag{5.2.4}$$

Combining (5.2.3) and (5.2.4) gives the functional equation

$$(1-ab^{-1})f(bq) = (1-b)(z^{-1}-ab^{-1})f(b),$$

that is,

$$f(b) = \frac{1-b/a}{(1-b)(1-b/az)} f(bq). \tag{5.2.5}$$

Iterating (5.2.5) $n-1$ times we get

$$f(b) = \frac{(b/a;q)_n}{(b,b/az;q)_n} f(bq^n). \tag{5.2.6}$$

Since $f(b)$ is analytic for $|b| < \min(1,|az|)$, by letting $n \to \infty$ we obtain

$$f(b) = \frac{(b/a;q)_\infty}{(b,b/az;q)_\infty} f(0). \tag{5.2.7}$$

However, since

$$f(q) = \sum_{n=0}^{\infty} \frac{(a;q)_n}{(q;q)_n} z^n = \frac{(az;q)_\infty}{(z;q)_\infty}$$

by (1.3.2), on setting $b = q$ in (5.2.7) we find that

$$f(0) = \frac{(q,q/az;q)_\infty}{(q/a;q)_\infty} f(q) = \frac{(q,q/az,az;q)_\infty}{(q/a,z;q)_\infty}.$$

Substituting this in (5.2.7) we obtain formula (5.2.1).

Jacobi's triple-product identity (1.6.1) is a limit case of Ramanujan's sum. First replace a and z in (5.2.1) by a^{-1} and az, respectively, to obtain

$$\sum_{n=-\infty}^{\infty} \frac{(a^{-1};q)_n}{(b;q)_n}(az)^n = \frac{(q, ab, z, q/z;q)_\infty}{(b, aq, az, b/z;q)_\infty}, \tag{5.2.8}$$

when $|b| < |z| < |a^{-1}|$. Now set $b = 0$, replace q by q^2, z by zq, and then take $a \to 0$ to get (1.6.1).

5.3 Bailey's sum of a very-well-poised $_6\psi_6$ series

Bailey [1936] proved that

$$\begin{aligned} &{}_6\psi_6\left[\begin{matrix} qa^{\frac{1}{2}}, -qa^{\frac{1}{2}}, b, c, d, e \\ a^{\frac{1}{2}}, -a^{\frac{1}{2}}, aq/b, aq/c, aq/d, aq/e \end{matrix}; q, \frac{qa^2}{bcde}\right] \\ &= \frac{(aq, aq/bc, aq/bd, aq/be, aq/cd, aq/ce, aq/de, q, q/a;q)_\infty}{(aq/b, aq/c, aq/d, aq/e, q/b, q/c, q/d, q/e, qa^2/bcde;q)_\infty}; \end{aligned} \tag{5.3.1}$$

provided $|qa^2/bcde| < 1$. Since this $_6\psi_6$ reduces to a very-well-poised $_6\phi_5$ series when one of the parameters b, c, d, e equals a, (5.3.1) can be regarded as an extension of the $_6\phi_5$ summation formula (2.7.1).

There are several known proofs of (5.3.1). Bailey's proof depends crucially on the identity

$$\begin{aligned} &\frac{(aq/d, aq/e, aq/f, q/ad, q/ae, q/af;q)_\infty}{a(qa^2, ab, ac, b/a, c/a;q)_\infty} \\ &\cdot {}_8W_7(a^2; ab, ac, ad, ae, af; q, q^2/abcdef) + \text{ idem } (a; b, c) = 0, \end{aligned} \tag{5.3.2}$$

when $|q^2/abcdef| < 1$, which is easily proved by using the q-integral representation (2.10.19) of an $_8\phi_7$ series (see Exercise 2.15). If we set $c = q/a$, the first and third series in (5.3.2) combine to give, via (5.1.3),

$$\begin{aligned} &\frac{(aq/d, aq/e, aq/f, q/ad, q/ae, q/af;q)_\infty}{a(qa^2, q/a^2, q, ab, b/a;q)_\infty} \\ &\cdot {}_6\psi_6\left[\begin{matrix} qa, -qa, ab, ad, ae, af \\ a, -a, aq/b, aq/d, aq/e, aq/f \end{matrix}; q, \frac{q}{bdef}\right], \end{aligned}$$

while, by (2.7.1), the second series reduces to

$$\begin{aligned} &{}_6\phi_5\left[\begin{matrix} b^2, bq, -bq, bd, be, bf \\ b, -b, bq/d, bq/e, bq/f \end{matrix}; q, \frac{q}{bdef}\right] \\ &= \frac{(qb^2, q/de, q/df, q/ef;q)_\infty}{(bq/d, bq/e, bq/f, q/bdef;q)_\infty}. \end{aligned}$$

This gives (5.3.1) after we replace a^2, ab, ad, ae, af by a, b, c, d, e, respectively, and use the same square root of a everywhere.

Slater and Lakin [1956] gave a proof of (5.3.1) via a Barnes type integral and a second proof via a q-difference operator. Andrews [1974a] gave a simpler proof and Askey [1984c] showed that it can be obtained from a simple difference equation. The simplest proof was given by Askey and Ismail [1979] who only used the ${}_6\phi_5$ sum (2.7.1) and an argument based on the properties of analytic functions.

Setting $e = a^{\frac{1}{2}}$ in (5.3.1), we obtain

$$
\begin{aligned}
&{}_4\psi_4\left[\begin{matrix}-qa^{\frac{1}{2}},\ b,\ c,\ d\\ -a^{\frac{1}{2}},\ aq/b,\ aq/c,\ aq/d\end{matrix};q,\frac{qa^{\frac{3}{2}}}{bcd}\right]\\
&\quad=\frac{(aq,aq/bc,aq/bd,aq/cd,qa^{\frac{1}{2}}/b,qa^{\frac{1}{2}}/c,qa^{\frac{1}{2}}/d,q,q/a;q)_\infty}{(aq/b,aq/c,aq/d,q/b,q/c,q/d,qa^{\frac{1}{2}},qa^{-\frac{1}{2}},qa^{\frac{3}{2}}/bcd;q)_\infty}
\end{aligned}
\tag{5.3.3}
$$

provided $|qa^{\frac{3}{2}}/bcd| < 1$. This is an extension of the q-Dixon formula (2.7.2).

If we set $d = a^{\frac{1}{2}}$, $e = -a^{\frac{1}{2}}$ in (5.3.1) and simplify, we get the sum of a very-well-poised ${}_2\psi_2$ series

$$
\begin{aligned}
&{}_2\psi_2(b,c;aq/b,aq/c;q,-aq/bc)\\
&\quad=\frac{(aq/bc;q)_\infty(aq^2/b^2,aq^2/c^2,q^2,aq,q/a;q^2)_\infty}{(aq/b,aq/c,q/b,q/c,-aq/bc;q)_\infty},\quad \left|\frac{aq}{bc}\right|<1.
\end{aligned}
\tag{5.3.4}
$$

5.4 A general transformation formula for an ${}_r\psi_r$ series

In this section we shall derive a transformation formula for an ${}_r\psi_r$ series from those for ${}_{r+1}\phi_r$ series in Chapter 4. First observe that (5.1.3) gives

$$
\begin{aligned}
&{}_r\psi_r(a_1,a_2,\ldots,a_r;b_1,b_2,\ldots,b_r;q,z)\\
&\quad={}_{r+1}\phi_r\left[\begin{matrix}q,\ a_1,\ldots,\ a_r\\ b_1,\ldots,b_r\end{matrix};q,z\right]+z^{-1}\prod_{k=1}^{r}\frac{b_k-q}{a_k-q}\\
&\qquad\cdot{}_{r+1}\phi_r\left[\begin{matrix}q,\ q^2/b_1,\ldots,q^2/b_r\\ q^2/a_1,\ldots,q^2/a_r\end{matrix};q,\frac{b_1\cdots b_r}{a_1\cdots a_r z}\right].
\end{aligned}
\tag{5.4.1}
$$

In (4.10.11) let us now make the following specialization of the parameters

$$
\begin{aligned}
&C=A+1,\qquad D=B+1,\qquad A=B=r,\\
&c_1d_1=c_2d_2=\cdots=c_{A+1}d_{A+1}=q,\\
&qd_1/b_1=\alpha_1,qd_1/b_2=\alpha_2,\ldots,qd_1/b_A=\alpha_A,\\
&a_1d_1=\beta_1,a_2d_1=\beta_2,\ldots,a_Ad_1=\beta_A,\\
&\frac{b_1\cdots b_A z}{d_1\cdots d_{A+1}}=x.
\end{aligned}
\tag{5.4.2}
$$

Then, combining the pairs of the resulting ${}_{r+1}\phi_r$ series in (4.10.11) via (5.4.1), simplifying the coefficients and relabelling the parameters, we obtain Slater's [1952b, (4)] transformation formula

$$
\begin{aligned}
&\frac{(b_1,b_2,\ldots,b_r,q/a_1,q/a_2,\ldots,q/a_r,dz,q/dz;q)_\infty}{(c_1,c_2,\ldots,c_r,q/c_1,q/c_2,\ldots,q/c_r;q)_\infty}\,{}_r\psi_r\left[\begin{matrix}a_1,\ a_2,\ldots,a_r\\ b_1,\ b_2,\ldots,b_r\end{matrix};q,z\right]\\
&=\frac{q}{c_1}\,\frac{(c_1/a_1,c_1/a_2,\ldots,c_1/a_r,qb_1/c_1,qb_2/c_1,\ldots,qb_r/c_1,dc_1z/q,q^2/dc_1z;q)_\infty}{(c_1,q/c_1,c_1/c_2,\ldots,c_1/c_r,qc_2/c_1,\ldots,qc_r/c_1;q)_\infty}
\end{aligned}
$$

$$\cdot {}_r\psi_r \left[\begin{matrix} qa_1/c_1, qa_2/c_1, \ldots, qa_r/c_1 \\ qb_1/c_1, qb_2/c_1, \ldots, qb_r/c_1 \end{matrix} ; q, z \right]$$

$$+ \text{ idem } (c_1; c_2, \ldots, c_r), \tag{5.4.3}$$

where $d = a_1a_2 \cdots a_r/c_1c_2 \cdots c_r$, $\left| \frac{b_1 \cdots b_r}{a_1 \cdots a_r} \right| < |z| < 1$.

Note that the c's are absent in the ${}_r\psi_r$ series on the left side of (5.4.3). This gives us the freedom to choose the c's in any convenient way. For example, if we set $c_j = qa_j$, where j is an integer between 1 and r, then the jth series on the right becomes an ${}_r\phi_{r-1}$ series. So if we set $c_j = qa_j$, $j = 1, 2, \ldots, r$, in (5.4.3), then we get an expansion of an ${}_r\psi_r$ series in terms of r ${}_r\phi_{r-1}$ series:

$$\frac{(b_1, b_2, \ldots, b_r, q/a_1, q/a_2, \ldots, q/a_r, z, q/z; q)_\infty}{(qa_1, qa_2, \ldots, qa_r, 1/a_1, 1/a_2, \ldots, 1/a_r; q)_\infty} {}_r\psi_r \left[\begin{matrix} a_1, a_2, \ldots, a_r \\ b_1, b_2, \ldots, b_r \end{matrix} ; q, z \right]$$

$$= \frac{a_1^{r-1}(q, qa_1/a_2, \ldots, qa_1/a_r, b_1/a_1, b_2/a_1, \ldots, b_r/a_1, a_1 z, q/a_1 z; q)_\infty}{(qa_1, 1/a_1, a_1/a_2, \ldots, a_1/a_r, qa_2/a_1, \ldots, qa_r/a_1; q)_\infty}$$

$$\cdot {}_r\phi_{r-1} \left[\begin{matrix} qa_1/b_1, qa_1/b_2, \ldots, qa_1/b_r \\ qa_1/a_2, \ldots, qa_1/a_r \end{matrix} ; q, \frac{b_1 \cdots b_r}{a_1 \cdots a_r z} \right]$$

$$+ \text{ idem } (a_1; a_2, \ldots, a_r), \tag{5.4.4}$$

provided $\left| \frac{b_1 \cdots b_r}{a_1 \cdots a_r} \right| < |z| < 1$.

On the other hand, if we set $c_j = b_j$, $j = 1, 2, \ldots, r$ in (5.4.3), then we obtain the expansion formula

$$\frac{(q/a_1, q/a_2, \ldots, q/a_r, dz, q/dz; q)_\infty}{(q/b_1, q/b_2, \ldots, q/b_r; q)_\infty} {}_r\psi_r \left[\begin{matrix} a_1, a_2, \ldots, a_r \\ b_1, b_2, \ldots, b_r \end{matrix} ; q, z \right]$$

$$= \frac{q}{b_1} \frac{(q, b_1/a_1, b_1/a_2, \ldots, b_1/a_r, db_1 z/q, q^2/db_1 z; q)_\infty}{(b_1, q/b_1, b_1/b_2, \ldots, b_1/b_r; q)_\infty} \tag{5.4.5}$$

$$\cdot {}_r\phi_{r-1} \left[\begin{matrix} qa_1/b_1, \ldots, qa_r/b_1 \\ qb_2/b_1, \ldots, qb_r/b_1 \end{matrix} ; q, z \right] + \text{ idem } (b_1; b_2, \ldots, b_r),$$

with $d = a_1a_2 \cdots a_r/b_1b_2 \cdots b_r$.

5.5 A general transformation formula for a very-well-poised ${}_{2r}\psi_{2r}$ series

Using (4.12.1) and (5.4.1) as in §5.4, we obtain Slater's [1952b] expansion of a well-poised ${}_{2r}\psi_{2r}$ series in terms of r other well-poised ${}_{2r}\psi_{2r}$ series:

$$\frac{(q/b_1, \ldots, q/b_{2r}, aq/b_1, \ldots, aq/b_{2r}, a^{\frac{1}{2}}, -a^{\frac{1}{2}}, qa^{-\frac{1}{2}}, -qa^{-\frac{1}{2}}; q)_\infty}{(a, a_1, \ldots, a_r, aq/a_1, \ldots, aq/a_r, q/a, q/a_1, \ldots, q/a_r, a_1/a, \ldots, a_r/a; q)_\infty}$$

$$\cdot {}_{2r}\psi_{2r} \left[\begin{matrix} b_1, b_2, \ldots, b_{2r} \\ aq/b_1, aq/b_2, \ldots, aq/b_{2r} \end{matrix} ; q, -\frac{a^r q^r}{b_1 \cdots b_{2r}} \right]$$

$$= \frac{a_1(a_1q/b_1, \ldots, a_1q/b_{2r}, aq/a_1b_1, \ldots, aq/a_1b_{2r}; q)_\infty}{(a_1, q/a_1, a/a_1, qa_1/a, a_2/a_1, \ldots, a_r/a_1, qa_1/a_2, \ldots, qa_1/a_r; q)_\infty}$$

$$\cdot \frac{(a^{\frac{1}{2}}/a_1, -a^{\frac{1}{2}}/a_1, qa_1a^{-\frac{1}{2}}, -qa_1a^{-\frac{1}{2}}; q)_\infty}{(a_1^2/a, qa/a_1^2, a_1a_2/a, \ldots, a_1a_r/a, qa/a_1a_2, \ldots, qa/a_1a_r; q)_\infty}$$

$$\cdot {}_{2r}\psi_{2r}\left[\begin{matrix} a_1b_1/a,\ a_1b_2/a, \ldots, a_1b_{2r}/a \\ a_1q/b_1,\ a_1q/b_2, \ldots, a_1q/b_{2r} \end{matrix}; q, -\frac{a^rq^r}{b_1\cdots b_{2r}}\right]$$

$$+ \operatorname{idem}(a_1; a_2, \ldots, a_r). \tag{5.5.1}$$

For the very-well-poised case when $a_1 = b_1 = qa^{\frac{1}{2}}$, $a_2 = b_2 = -qa^{\frac{1}{2}}$, the first two terms on the right side vanish and we get

$$\frac{(q/b_3, \ldots, q/b_{2r}, aq/b_3, \ldots, aq/b_{2r}; q)_\infty}{(aq, q/a, a_3, \ldots, a_r, q/a_3, \ldots, q/a_r; q)_\infty}$$

$$\cdot (a_3/a, \ldots, a_r/a, aq/a_3, \ldots, aq/a_r; q)_\infty^{-1}$$

$$\cdot {}_{2r}\psi_{2r}\left[\begin{matrix} qa^{\frac{1}{2}}, -qa^{\frac{1}{2}}, b_3, \ldots, b_{2r} \\ a^{\frac{1}{2}}, -a^{\frac{1}{2}}, aq/b_3, \ldots, aq/b_{2r} \end{matrix}; q, \frac{a^{r-1}q^{r-2}}{b_3\cdots b_{2r}}\right]$$

$$= \frac{(a_3q/b_3, \ldots, a_3q/b_{2r}, aq/a_3b_3, \ldots, aq/a_3b_{2r}; q)_\infty}{(a_3, q/a_3, a_3/a, aq/a_3, qa_3^2/a, aq/a_3^2; q)_\infty}$$

$$\cdot (a_4/a_3, \ldots, a_r/a_3, qa_3/a_4, \ldots, qa_3/a_r; q)_\infty^{-1}$$

$$\cdot (a_3a_4/a, \ldots, a_3a_r/a, aq/a_3a_4, \ldots, aq/a_3a_r; q)_\infty^{-1}$$

$$\cdot {}_{2r}\psi_{2r}\left[\begin{matrix} qa_3a^{-\frac{1}{2}}, -qa_3a^{-\frac{1}{2}}, a_3b_3/a, \ldots, a_3b_{2r}/a \\ a_3a^{-\frac{1}{2}}, -a_3a^{-\frac{1}{2}}, qa_3/b_3, \ldots, qa_3/b_{2r} \end{matrix}; q, \frac{a^{r-1}q^{r-2}}{b_3\cdots b_{2r}}\right]$$

$$+ \operatorname{idem}\,(a_3; a_4, \ldots, a_r). \tag{5.5.2}$$

In particular, for $r = 3$ we have

$$\frac{(q/b_3, \ldots, q/b_6, aq/b_3, \ldots, aq/b_6; q)_\infty}{(aq, q/a, a_3, q/a_3, a_3/a, aq/a_3; q)_\infty}$$

$$\cdot {}_6\psi_6\left[\begin{matrix} qa^{\frac{1}{2}},\ -qa^{\frac{1}{2}},\ b_3,\ b_4,\ b_5,\ b_6 \\ a^{\frac{1}{2}},\ -a^{\frac{1}{2}}, aq/b_3,\ aq/b_4,\ aq/b_5,\ aq/b_6 \end{matrix}; q, \frac{a^2q}{b_3b_4b_5b_6}\right]$$

$$= \frac{(a_3q/b_3, \ldots, a_3q/b_6, aq/a_3b_3, \ldots, aq/a_3b_6; q)_\infty}{(a_3, q/a_3, a_3/a, aq/a_3, qa_3^2/a, aq/a_3^2; q)_\infty}$$

$$\cdot {}_6\psi_6\left[\begin{matrix} qa_3a^{-\frac{1}{2}}, -qa_3a^{-\frac{1}{2}}, a_3b_3/a, a_3b_4/a, a_3b_5/a, a_3b_6/a \\ a_3a^{-\frac{1}{2}}, -a_3a^{-\frac{1}{2}}, qa_3/b_3, qa_3/b_4, qa_3/b_5, qa_3/b_6 \end{matrix}; q, \frac{a^2q}{b_3b_4b_5b_6}\right]. \tag{5.5.3}$$

If we now set $a_3 = b_6$, then the ${}_6\psi_6$ series on the right side becomes a ${}_6\phi_5$ with sum

$$\frac{(qb_6^2/a, aq/b_3b_4, aq/b_3b_5, aq/b_4b_5; q)_\infty}{(qb_6/b_3, qb_6/b_4, qb_6/b_5, qa^2/b_3b_4b_5b_6; q)_\infty}.$$

This provides another derivation of the ${}_6\psi_6$ sum (5.3.1); see M. Jackson [1950a]. As in Slater [1952b], Sears' formulas (4.12.5) and (4.12.7) can be used to obtain the transformation formulas:

$$\frac{(q/b_1, \ldots, q/b_{2r}, aq/b_1, \ldots, aq/b_{2r}; q)_\infty}{(a_1, \ldots, a_{r+1}, q/a_1, \ldots, q/a_{r+1}, a_1/a, \ldots, a_{r+1}/a, aq/a_1, \ldots, aq/a_{r+1}; q)_\infty}$$

$$\cdot {}_{2r}\psi_{2r}\left[\begin{matrix} b_1,\ldots,b_{2r} \\ aq/b_1,\ldots,aq/b_{2r} \end{matrix};q,\frac{a^r q^r}{b_1 b_2\cdots b_{2r}}\right]$$

$$= \frac{(a_1q/b_1,\ldots,a_1q/b_{2r},aq/a_1b_1,\ldots,aq/a_1b_{2r};q)_\infty}{(a_2/a_1,\ldots,a_{r+1}/a_1,qa_1/a_2,\ldots,qa_1/a_{r+1},aq/a_1a_2,\ldots,aq/a_1a_{r+1};q)_\infty}$$

$$\cdot (a_1,q/a_1,aq/a_1,a_1/a,a_1a_2/a,\ldots,a_1a_{r+1}/a;q)_\infty^{-1}$$

$$\cdot {}_{2r}\psi_{2r}\left[\begin{matrix} a_1b_1/a,a_1b_2/a,\ldots,a_1b_{2r}/a \\ a_1q/b_1,a_1q/b_2,\ldots,a_1q/b_{2r} \end{matrix};q,\frac{q^r a^r}{b_1b_2\cdots b_{2r}}\right]$$

$$+\ \text{idem}\ (a_1;a_2,\ldots,a_{r+1}), \tag{5.5.4}$$

and

$$\frac{(q/b_1,\ldots,q/b_{2r-1},aq/b_1,\ldots,aq/b_{2r-1};q)_\infty}{(a,a_1,\ldots,a_r,q/a,q/a_1,\ldots,q/a_r;q)_\infty}$$

$$\cdot\frac{(a^{\frac{1}{2}},-a^{\frac{1}{2}},q/a^{\frac{1}{2}},-q/a^{\frac{1}{2}},\pm(aq)^{\frac{1}{2}},\pm(q/a)^{\frac{1}{2}};q)_\infty}{(a_1/a,\ldots,a_r/a,aq/a_1,\ldots,aq/a_r;q)_\infty}$$

$$\cdot {}_{2r-1}\psi_{2r-1}\left[\begin{matrix} b_1,\ldots,b_{2r-1} \\ aq/b_1,\ldots,aq/b_{2r-1} \end{matrix};q,\frac{\mp q^{r-\frac{1}{2}}a^{r-\frac{1}{2}}}{b_1b_2\cdots b_{2r-1}}\right]$$

$$= \frac{a_1(a_1q/b_1,\ldots,a_1q/b_{2r-1},aq/a_1b_1,\ldots,aq/a_1b_{2r-1};q)_\infty}{(aq/a_1^2,aq/a_1a_2,\ldots,aq/a_1a_r,a_2/a_1,\ldots,a_r/a_1;q)_\infty}$$

$$\cdot\frac{(a^{\frac{1}{2}}/a_1,-a^{\frac{1}{2}}/a_1,a_1q/a^{\frac{1}{2}},-a_1q/a^{\frac{1}{2}};q)_\infty}{(a/a_1,q/a_1,a_1^2/a,a_1a_2/a,\ldots,a_1a_r/a;q)_\infty}$$

$$\cdot\frac{(\pm(aq)^{\frac{1}{2}}/a_1,\pm a_1(q/a)^{\frac{1}{2}};q)_\infty}{(qa_1/a,a_1,qa_1/a_2,\ldots,qa_1/a_r;q)_\infty}$$

$$\cdot {}_{2r-1}\psi_{2r-1}\left[\begin{matrix} a_1b_1/a,\ldots,a_1b_{2r-1}/a \\ qa_1/b_1,\ldots,qa_1/b_{2r-1} \end{matrix};q,\frac{\mp q^{r-\frac{1}{2}}a^{r-\frac{1}{2}}}{b_1b_2\cdots b_{2r-1}}\right]$$

$$+\ \text{idem}\ (a_1;a_2,\ldots,a_r). \tag{5.5.5}$$

5.6 Transformation formulas for very-well-poised ${}_8\psi_8$ and ${}_{10}\psi_{10}$ series

In this section we consider two special cases of (5.5.2) that may be regarded as extensions of the transformation formulas for very-well-poised ${}_8\phi_7$ and ${}_{10}\phi_9$ series derived in Chapter 2. First, set $r=4$ in (5.5.2) and replace b_3, b_4, b_5, b_6, b_7, b_8 by b, c, d, e, f, g, respectively, choose $a_3=f$, $a_4=g$ and simplify to get

$$\frac{(aq/b,aq/c,aq/d,aq/e,q/b,q/c,q/d,q/e;q)_\infty}{(f,g,f/a,g/a,aq,q/a;q)_\infty}$$

$$\cdot {}_8\psi_8\left[\begin{matrix} qa^{\frac{1}{2}},-qa^{\frac{1}{2}},b,c,d,e,f,g \\ a^{\frac{1}{2}},-a^{\frac{1}{2}},aq/b,aq/c,aq/d,aq/e,aq/f,aq/g \end{matrix};q,\frac{a^3q^2}{bcdefg}\right]$$

$$= \frac{(q, aq/bf, aq/cf, aq/df, aq/ef, fq/b, fq/c, fq/d, fq/e; q)_\infty}{(f, q/f, aq/f, f/a, g/f, fg/a, qf^2/a; q)_\infty}$$
$$\cdot {}_8\phi_7 \left[\begin{matrix} f^2/a, qfa^{-\frac{1}{2}}, -qfa^{-\frac{1}{2}}, fb/a, fc/a, fd/a, fe/a, fg/a \\ fa^{-\frac{1}{2}}, -fa^{-\frac{1}{2}}, fq/b, fq/c, fq/d, fq/e, fq/g \end{matrix} ; q, \frac{a^3q^2}{bcdefg} \right]$$
$$+ \text{idem } (f; g), \tag{5.6.1}$$

where $\left| \frac{a^3q^2}{bcdefg} \right| < 1.$

Replacing a, b, c, d, e, f, g by a^2, ba, ca, da, ea, fa, ga, respectively, we may rewrite (5.6.1) as

$$\frac{(aq/b, aq/c, aq/d, aq/e, q/ab, q/ac, q/ad, q/ae; q)_\infty}{(fa, ga, f/a, g/a, qa^2, q/a^2; q)_\infty}$$
$$\cdot {}_8\psi_8 \left[\begin{matrix} qa, -qa, ba, ca, da, ea, fa, ga \\ a, -a, aq/b, aq/c, aq/d, aq/e, aq/f, aq/g \end{matrix} ; q, \frac{q^2}{bcdefg} \right]$$
$$= \frac{(q, q/bf, q/cf, q/df, q/ef, qf/b, qf/c, qf/d, qf/e; q)_\infty}{(fa, q/fa, aq/f, f/a, g/f, fg, qf^2; q)_\infty}$$
$$\cdot {}_8\phi_7 \left[\begin{matrix} f^2, qf, -qf, fb, fc, fd, fe, fg \\ f, -f, fq/b, fq/c, fq/d, fq/e, fq/g \end{matrix} ; q, \frac{q^2}{bcdefg} \right]$$
$$+ \text{idem } (f; g), \tag{5.6.2}$$

provided $|q^2/bcdefg| < 1$. Note that no a's appear in the ${}_8\phi_7$ series on the right side of (5.6.2). This is essentially the same as eq. (2.2) in M. Jackson [1950a].

For the next special case of (5.5.2) we take $r = 5$ and replace $b_3, \ldots, b_{10}$ by b, c, d, e, f, g, h, k, respectively, choose $a_3 = g$, $a_4 = h$, $a_5 = k$, and finally, replace $a, b, \ldots, k$ by $a^2, ba, \ldots, ka$ and simplify. This gives

$$\frac{(aq/b, aq/c, aq/d, aq/e, aq/f, q/ab, q/ac, q/ad, q/ae, q/af; q)_\infty}{(ag, ah, ak, g/a, h/a, k/a, qa^2, q/a^2; q)_\infty}$$
$$\cdot {}_{10}\psi_{10} \left[\begin{matrix} qa, -qa, ba, ca, da, ea, fa, ga, ha, ka \\ a, -a, aq/b, aq/c, aq/d, aq/e, aq/f, aq/g, aq/h, aq/k \end{matrix} ; q, \frac{q^3}{bcdefghk} \right]$$
$$= \frac{(q, q/bg, q/cg, q/dg, q/eg, q/fg, qg/b, qg/c, qg/d, qg/e, qg/f; q)_\infty}{(gh, gk, k/g, h/g, ag, q/ag, g/a, aq/g, qg^2; q)_\infty}$$
$$\cdot {}_{10}\phi_9 \left[\begin{matrix} g^2, qg, -qg, gb, gc, gd, ge, gf, gh, gk \\ g, -g, qg/b, qg/c, qg/d, qg/e, qg/f, qg/h, qg/k \end{matrix} ; q, \frac{q^3}{bcdefghk} \right]$$
$$+ \text{idem } (g; h, k), \tag{5.6.3}$$

where $|q^3/bcdefghk| < 1.$

Exercises 5

5.1 Show that

$$\sum_{n=-\infty}^{\infty} (-1)^n q^{n^2} = \frac{(q;q)_\infty}{(-q;q)_\infty}.$$

5.2 Letting $c \to \infty$ in (5.3.4) and setting $a = 1$, $b = -1$, show that

$$1 + 4\sum_{n=1}^{\infty} \frac{(-1)^n q^{n(n+1)/2}}{1+q^n} = \left[\frac{(q;q)_\infty}{(-q;q)_\infty}\right]^2.$$

5.3 In (5.4.1) set $b = a$, $d = c$, $e = -1$ and then let $a \to 1$ to show that

$$1 + 8\sum_{n=1}^{\infty} \frac{(-q)^n}{(1+q^n)^2} = \left[\frac{(q;q)_\infty}{(-q;q)_\infty}\right]^4.$$

See section 8.11 for applications of Exercises 5.1-5.3 to Number Theory.

5.4 Set $b = c = d = e = -1$ and then let $a \to 1$ in (5.3.1) to obtain

$$1 + 16\sum_{n=1}^{\infty} \frac{q^{2n}(4 - q^n - q^{-n})}{(1+q^n)^4}$$
$$= \left[\frac{(q;q)_\infty}{(-q;q)_\infty}\right]^8.$$

5.5 Show that

$$\sum_{n=-\infty}^{\infty} q^{4n^2} z^{2n}(1 + zq^{4n+1}) = (q^2, -zq, -q/z; q^2)_\infty, \qquad z \neq 0.$$

5.6 Prove the quintuple product identity

$$\sum_{n=-\infty}^{\infty} (-1)^n q^{n(3n-1)/2} z^{3n}(1 + zq^n)$$
$$= (q, -z, -q/z; q)_\infty (qz^2, q/z^2; q^2)_\infty, \qquad z \neq 0.$$

(See the Notes for this exercise)

5.7 Show that

$$\sum_{n=-\infty}^{\infty} \frac{(1-q^{10n+4})}{(1-q^{5n+1})(1-q^{5n+3})^2} q^{5n+1}$$
$$= \frac{q(1-q^4)}{(1-q)^2(1-q^3)^2} {}_6\psi_6\left[\begin{matrix} q^7, & -q^7, & q, & q, & q^3, & q^3 \\ q^2, & -q^2, & q^8, & q^8, & q^6, & q^6 \end{matrix}; q^5, q^5\right].$$

Deduce that

$$\sum_{n=0}^{\infty} \left\{ \frac{q^{5n+1}}{(1-q^{5n+1})^2} - \frac{q^{5n+2}}{(1-q^{5n+2})^2} - \frac{q^{5n+3}}{(1-q^{5n+3})^2} + \frac{q^{5n+4}}{(1-q^{5n+4})^2} \right\}$$
$$= q\frac{(q^5;q^5)_\infty^5}{(q;q)_\infty}, \qquad |q| < 1.$$

See Andrews [1974a] for the above formulas.

5.8 Deduce (5.4.4) directly from (4.5.2).

5.9 Deduce (5.3.1) from (5.4.5) by using (2.7.1).

5.10 Show that

$$\frac{q}{e}\frac{(e/a,e/b,e/ab,qc/e,q^2/e,q^2f/e;q)_\infty}{(e,q/e,e/f,qf/e;q)_\infty}\,{}_2\psi_2\left[\begin{matrix}e/c,e/q\\e/a,e/b\end{matrix};q,q\right]+\text{idem }(e;f)$$
$$=\frac{(q,q/a,q/b,c/a,c/b,c/ef,qef/c;q)_\infty}{(c,f,q/e,q/f,c/ab;q)_\infty}$$

5.11 Show that

$${}_8\psi_8\left[\begin{matrix}qa^{\frac12},-qa^{\frac12},c,d,e,f,aq^{-n},q^{-n}\\a^{\frac12},-a^{\frac12},aq/c,aq/d,aq/e,aq/f,q^{n+1},aq^{n+1}\end{matrix};q,\frac{a^2q^{2n+2}}{cdef}\right]$$
$$=\frac{(aq,q/a,aq/cd,aq/ef;q)_n}{(q/c,q/d,aq/e,aq/f;q)_n}$$
$$\cdot\,{}_4\psi_4\left[\begin{matrix}e,f,aq^{n+1}/cd,q^{-n}\\aq/c,aq/d,q^{n+1},ef/aq^n\end{matrix};q,q\right],\qquad n=0,1,\ldots,$$

and deduce the limit cases

$${}_2\psi_2\left[\begin{matrix}e,f\\aq/c,aq/d\end{matrix};q,\frac{aq}{ef}\right]$$
$$=\frac{(q/c,q/d,aq/e,aq/f;q)_\infty}{(aq,q/a,aq/cd,aq/ef;q)_\infty}$$
$$\cdot\sum_{n=-\infty}^{\infty}\frac{(1-aq^{2n})(c,d,e,f;q)_n}{(1-a)(aq/c,aq/d,aq/e,aq/f;q)_n}\left(\frac{qa^3}{cdef}\right)^n q^{n^2}$$

and

$$\sum_{n=0}^{\infty}\frac{a^nq^{n^2}}{(q;q)_n}=\frac{1}{(aq;q)_\infty}\sum_{n=0}^{\infty}\frac{(a;q)_n}{(q;q)_n}\frac{1-aq^{2n}}{1-a}(-1)^na^{2n}q^{n(5n-1)/2}.$$

(Bailey [1950a])

5.12 Using (5.6.2) and (2.11.7), show that

$${}_8\psi_8\left[\begin{matrix}qa,-qa,ab,ac,ad,ae,af,ag\\a,-a,aq/b,aq/c,aq/d,aq/e,aq/f,aq/g\end{matrix};q,q\right]$$
$$=\frac{(q,qa^2,q/a^2,ag,g/a,q/bc,q/bd,q/be,q/bf,q/cd,q/ce,q/cf;q)_\infty}{(bg,cg,dg,eg,fg,aq/b,aq/c,aq/d,aq/e,aq/f,q/ab,q/ac;q)_\infty}$$
$$\cdot\frac{(q/de,q/df,q/ef;q)_\infty}{(q/ad,q/ae,q/af;q)_\infty},$$

provided $bcdefg = q$ and

$$\frac{(bf, q/bf, cf, q/cf, df, q/df, ef, q/ef, ag, q/ag, g/a, aq/g; q)_\infty}{(bg, q/bg, cg, q/cg, dg, q/dg, eg, q/eg, af, q/af, f/a, aq/f; q)_\infty} = 1.$$

Following Gosper [1988b], we may call this the bilateral Jackson formula.

5.13 Deduce from Ex. 5.12 the bilateral q-Saalschütz formula

$$\sum_{n=-\infty}^{\infty} \frac{(a, b, c; q)_n}{(d, e, f; q)_n} q^n = \frac{(q, d/a, d/b, d/c, e/a, e/b, e/c, q/f; q)_\infty}{(d, e, aq/f, bq/f, cq/f, q/a, q/b, q/c; q)_\infty},$$

provided $def = abcq^2$ and

$$(e/a, aq/e, e/b, bq/e, e/c, cq/e, f, q/f; q)_\infty = (f/a, aq/f, f/b, bq/f, f/c, cq/f, e, q/e; q)_\infty.$$

5.14 Show that

$$\int_{-c}^{d} \frac{(-qt/c, qt/d; q)_\infty}{(-at/c, bt/d; q)_\infty} d_qt = \frac{d(1-q)}{1-b} \, {}_1\psi_1(q/a; bq; q, -ad/c) = \frac{cd(1-q)(q, ab, -c/d, -d/c; q)_\infty}{(c+d)(a, b, -bc/d, -ad/c; q)_\infty}$$

when $|ab| < |ad/c| < 1$.
(Andrews and Askey [1981])

5.15 Show that

$$\int_{-\infty}^{\infty} \frac{(ct, -dt; q)_\infty}{(at, -bt; q)_\infty} d_qt = \frac{2(1-q)(c/a, d/b, -c/b, -d/a, ab, q/ab; q)_\infty (q^2; q^2)_\infty^2}{(cd/abq, q; q)_\infty (a^2, q^2/a^2, b^2, q^2/b^2; q^2)_\infty}$$

(Askey [1981])

5.16 Show that

$$\int_0^{\infty} \frac{(\alpha at, a/t, \alpha bt, b/t, \alpha ct, c/t, \alpha dt, d/t; q)_\infty}{(\alpha qt^2, q/\alpha t^2; q)_\infty} \frac{d_qt}{t}$$
$$= \frac{(1-q)(\alpha a, a, \alpha b, b, \alpha c, c, \alpha d, d; q)_\infty}{(\alpha q, q/\alpha; q)_\infty} \cdot {}_6\psi_6 \left[\begin{matrix} q\sqrt{\alpha}, -q\sqrt{\alpha}, q/a, q/b, q/c, q/d \\ \sqrt{\alpha}, -\sqrt{\alpha}, \alpha a, \alpha b, \alpha c, \alpha d \end{matrix} ; q, \frac{\alpha^2 abcd}{q^3} \right]$$
$$= \frac{(1-q)(q, \alpha ab/q, \alpha ac/q, \alpha ad/q, \alpha bc/q, \alpha bd/q, \alpha cd/q; q)_\infty}{(\alpha^2 abcd/q^3; q)_\infty}$$

when $|\alpha^2 abcd/q^3| < 1$.

5.17 Show that

$$\int_0^\infty \frac{(a_1t,\ldots,a_rt,b_1/t,\ldots,b_s/t;q)_\infty}{(c_1t,\ldots,c_rt,d_1/t,\ldots,d_s/t;q)_\infty} t^{\gamma-1} d_qt$$
$$= \frac{(1-q)(a_1,\ldots,a_r,b_1,\ldots,b_s;q)_\infty}{(c_1,\ldots,c_r,d_1,\ldots,d_s;q)_\infty}$$
$$\cdot {}_{r+s}\psi_{r+s}\left[\begin{matrix} c_1,\ldots,c_r,q/b_1,\ldots,q/b_s \\ a_1,\ldots,a_r,q/d_1,\ldots,q/d_s \end{matrix};q,\frac{b_1\cdots b_s}{d_1\cdots d_s}q^\gamma\right]$$

when

$$\left|\frac{b_1\cdots b_s}{d_1\cdots d_s}q^\gamma\right| < 1 < \left|\frac{c_1\cdots c_r}{a_1\cdots a_r}q^\gamma\right|.$$

5.18 Derive Bailey's [1950b] summation formulas:

(i) $${}_3\psi_3\left[\begin{matrix} b,c,d \\ q/b,q/c,q/d \end{matrix};q,\frac{q}{bcd}\right] = \frac{(q,q/bc,q/bd,q/cd;q)_\infty}{(q/b,q/c,q/d,q/bcd;q)_\infty},$$

(ii) $${}_3\psi_3\left[\begin{matrix} b,c,d \\ q^2/b,q^2/c,q^2/d \end{matrix};q,\frac{q^2}{bcd}\right] = \frac{(q,q^2/bc,q^2/bd,q^2/cd;q)_\infty}{(q^2/b,q^2/c,q^2/d,q^2/bcd;q)_\infty},$$

(iii) $${}_5\psi_5\left[\begin{matrix} b,c,d,e,q^{-n} \\ q/b,q/c,q/d,q/e,q^{n+1} \end{matrix};q,q\right]$$
$$= \frac{(q,q/bc,q/bd,q/cd;q)_n}{(q/b,q/c,q/d,q/bcd;q)_n},$$

(iv) $${}_5\psi_5\left[\begin{matrix} b,c,d,e,q^{-n} \\ q^2/b,q^2/c,q^2/d,q^2/e,q^{n+2} \end{matrix};q,q\right]$$
$$= \frac{(1-q)(q^2,q^2/bc,q^2/bd,q^2/cd;q)_n}{(q^2/b,q^2/c,q^2/d,q^2/bcd;q)_n},$$

where $n = 0,1,\ldots$.

5.19 Show that

(i) $$\sum_{k=-n}^{n} (-1)^k \begin{bmatrix} 2n \\ n+k \end{bmatrix}_q^3 q^{k(3k+1)/2} = \frac{(q;q)_{3n}}{(q,q,q;q)_n}$$

and

(ii) $$\sum_{k=-n-1}^{n} (-1)^k \begin{bmatrix} 2n+1 \\ n+k+1 \end{bmatrix}_q^3 q^{k(3k+1)/2} = \frac{(q;q)_{3n+1}}{(q,q,q,;q)_n}$$

for $n = 0,1,\ldots$. (Bailey [1950b])

5.20 Derive the ${}_2\psi_2$ transformation formulas

(i) $${}_2\psi_2\left[\begin{matrix} a,b \\ c,d \end{matrix};q,z\right] = \frac{(az,d/a,c/b,dq/abz;q)_\infty}{(z,d,q/b,cd/abz;q)_\infty}$$
$$\cdot {}_2\psi_2\left[\begin{matrix} a,abz/d \\ az,c \end{matrix};q,\frac{d}{a}\right],$$

(ii) $${}_2\psi_2\left[\begin{matrix} a,b \\ c,d \end{matrix};q,z\right] = \frac{(az,bz,cq/abz,dq/abz;q)_\infty}{(q/a,q/b,c,d;q)_\infty}$$

$$\cdot {}_2\psi_2 \left[\begin{matrix} abz/c, abz/d \\ az, bz \end{matrix} ; q, \frac{cd}{abz} \right].$$

(Bailey [1950a])

5.21 Verify that

$$\begin{aligned}
&(b/a, aq/b, df/a, aq/df, ef/a, aq/ef, bde/a, aq/bde; q)_\infty \\
&= (f/a, aq/f, bd/a, aq/bd, be/a, aq/be, def/a, aq/def; q)_\infty \\
&\quad - \frac{b}{a}(d, q/d, e, q/e, f/b, qb/f, bdef/a^2, a^2q/bdef; q)_\infty.
\end{aligned}$$

(Bailey [1936])

5.22 Extend the above identity to

$$\begin{aligned}
&ab\left(\frac{bc}{a}, \frac{aq}{bc}, \frac{bd}{a}, \frac{aq}{bd}, \frac{be}{a}, \frac{aq}{be}, \frac{bf}{a}, \frac{aq}{bf}, \frac{g}{a}, \frac{aq}{g}, \frac{h}{a}, \frac{aq}{h}, \frac{g}{h}, \frac{hq}{g}; q\right)_\infty \\
&- ab\left(\frac{ch}{a}, \frac{aq}{ch}, \frac{dh}{a}, \frac{aq}{dh}, \frac{eh}{a}, \frac{aq}{eh}, \frac{fh}{a}, \frac{aq}{fh}, \frac{b}{a}, \frac{aq}{b}, \frac{g}{a}, \frac{aq}{g}, \frac{g}{b}, \frac{bq}{g}; q\right)_\infty \\
&= ag\left(\frac{cg}{a}, \frac{aq}{cg}, \frac{dg}{a}, \frac{aq}{dg}, \frac{eg}{a}, \frac{aq}{eg}, \frac{fg}{a}, \frac{aq}{fg}, \frac{b}{a}, \frac{aq}{b}, \frac{h}{a}, \frac{aq}{h}, \frac{b}{h}, \frac{hq}{b}; q\right)_\infty \\
&- bh\left(c, \frac{q}{c}, d, \frac{q}{d}, e, \frac{q}{e}, f, \frac{q}{f}, \frac{b}{h}, \frac{hq}{b}, \frac{g}{h}, \frac{hq}{g}, \frac{g}{b}, \frac{bq}{g}; q\right)_\infty,
\end{aligned}$$

where $a^3q^2 = bcdefgh$. (Slater [1954a])

5.23 More generally, show that it follows from the general formula for sigma functions in Whittaker and Watson [1965, p. 451, Example 3] that

$$\sum_{k=1}^{n} \frac{(a_k/b_1, a_k/b_2, \ldots, a_k/b_n; q)_\infty}{(a_k/a_1, a_k/a_2, \ldots, a_k/a_{k-1}, a_k/a_{k+1}, \ldots, a_k/a_n; q)_\infty} \cdot \frac{(qb_1/a_k, qb_2/a_k, \ldots, qb_n/a_k; q)_\infty}{(qa_1/a_k, qa_2/a_k, \ldots, qa_{k-1}/a_k, qa_{k+1}/a_k, \ldots, qa_n/a_k; q)_\infty} = 0,$$

where $a_1a_2\cdots a_n = b_1b_2\cdots b_n$. (Slater 1954a])

Notes 5

§5.2 Andrews [1979c] used Ramanujan's sum (5.2.1) to prove a continued fraction identity that appeared in Ramanujan's [1988] "lost" notebook. Formal Laurent series and Ramanujan's sum are considered in Askey [1987]. A probabilistic proof of (5.2.1) can be found in Kadell [1987b]. Milne [1986, 1988a, 1989c] derived multidimensional $U(n)$ generalizations of (5.2.1).

§5.3 Gustafson [1987b, 1989a,b] derived a multilateral generalization of (5.2.1), (5.3.1) and related formulas by employing contour integration and Milne's [1985d,e, 1989a,b] work on $U(n)$ generalizations of the q-Gauss, q-Saalschütz, and very-well-poised ${}_6\phi_5$ summation formulas.

§5.4 M. Jackson [1954] employed (5.4.3) to derive transformation formulas for ${}_3\psi_3$ series.

§5.6 A transformation formula between certain ${}_4\phi_3$ and ${}_8\psi_8$ series was found by Jain [1980b], along with transformation formulas for particular ${}_7\psi_7$ series, and then used to deduce identities of Rogers-Ramanujan type with moduli 5, 6, 8, 12, 16, 20 and 24.

Ex. 5.6 Watson [1929b] derived this identity in an equivalent form. For various proofs of the quintuple product identity (and of its equivalent forms) and applications to number theory, Lie algebras, etc., see Adiga, Berndt, Bhargava and Watson [1985], Andrews [1974a], Atkin and Swinnerton-Dyer [1954], Bailey [1951], Carlitz and Subbarao [1972], Gordon [1961], Hirschhorn [1988], Kac [1978, 1985], Sears [1952], and Subbarao and Vidyasagar [1970].

Ex. 5.18 Using (i) and (ii) one can show that the formula in Ex. 2.31 holds even when a_1, a_2, a_3 are not nonnegative integers, provided that $|q^{a_1+a_2+a_3+1}| < 1$ and $|q| < 1$.

Exercises 5.21–5.23 Additional identities connecting sums of infinite products are given in Slater [1951, 1954b, 1966] and Watson [1929b].

6

THE ASKEY-WILSON q-BETA INTEGRAL AND SOME ASSOCIATED FORMULAS

6.1 The Askey-Wilson q-extension of the beta integral

It should be clear by now that the beta integral and extensions of it that can be evaluated compactly are important. A significant extension of the beta integral was found by Askey and Wilson [1985]. Since it has five degrees of freedom, four free parameters and the parameter q from basic hypergeometric functions, it has enough flexibility to be useful in many situations. This integral is

$$\int_{-1}^{1} \frac{h(x; 1, -1, q^{\frac{1}{2}}, -q^{\frac{1}{2}})}{h(x; a, b, c, d)} \frac{dx}{\sqrt{1-x^2}} \\ = \frac{2\pi (abcd; q)_\infty}{(q, ab, ac, ad, bc, bd, cd; q)_\infty}, \tag{6.1.1}$$

where

$$\begin{aligned} h(x; a_1, a_2, \ldots, a_m) &= h(x; a_1, a_2, \ldots, a_m; q) \\ &= h(x; a_1) h(x; a_2) \cdots h(x; a_m), \\ h(x; a) = h(x; a; q) &= \prod_{n=0}^{\infty} (1 - 2axq^n + a^2 q^{2n}) \\ &= \left(ae^{i\theta}, ae^{-i\theta}; q\right)_\infty, \quad x = \cos\theta, \end{aligned} \tag{6.1.2}$$

and

$$\max\left(|a|, |b|, |c|, |d|, |q|\right) < 1. \tag{6.1.3}$$

As in (6.1.1), we shall use the h notation without the base q displayed when the base is q.

Askey and Wilson deduced (6.1.1) from the contour integral

$$\frac{1}{2\pi i} \int_K \frac{(z^2, z^{-2}; q)_\infty}{(az, az^{-1}, bz, bz^{-1}, cz, cz^{-1}, dz, dz^{-1}; q)_\infty} \frac{dz}{z} \\ = \frac{2(abcd; q)_\infty}{(q, ab, ac, ad, bc, bd, cd; q)_\infty}, \tag{6.1.4}$$

where the contour K is as defined in §4.9 and the parameters a, b, c, d are no longer restricted by (6.1.3), but by the milder restriction that their pairwise products are not of the form q^{-j}, $j = 0, 1, 2, \ldots$. Askey and Wilson's original proof of (6.1.4) required a number of interim assumptions that had to be removed by continuity and analytic continuation arguments. In their paper they also provided a direct evaluation of the reduced integral

$$\int_{-1}^{1} \frac{h(x;1,-1)}{h(x;a,b)} \frac{dx}{\sqrt{1-x^2}}$$
$$= \frac{2\pi(-abq;q)_\infty}{(q,-q,aq^{\frac{1}{2}},-aq^{\frac{1}{2}},bq^{\frac{1}{2}},-bq^{\frac{1}{2}},ab;q)_\infty} \tag{6.1.5}$$

by using summation formulas for ${}_1\psi_1$ and ${}_4\psi_4$ series. Simpler proofs of (6.1.1) were subsequently found by Rahman [1984] and Ismail and Stanton [1988]. In the following section we shall give Rahman's proof since it only uses formulas that we have already proved, whereas the Ismail and Stanton proof uses some results for certain orthogonal polynomials which will not be covered until Chapter 7.

We shall conclude this section by showing that the beta integral

$$\int_{-1}^{1} (1-x)^\alpha (1+x)^\beta dx = 2^{\alpha+\beta+1} \frac{\Gamma(\alpha+1)\Gamma(\beta+1)}{\Gamma(\alpha+\beta+2)} \tag{6.1.6}$$

is a limit case of (6.1.5).

Let $0 < q < 1,\ a = q^{\alpha+\frac{1}{2}},\ b = -q^{\beta+\frac{1}{2}}$ and use the notation

$$(z;q)_\alpha = \frac{(z;q)_\infty}{(zq^\alpha;q)_\infty} \tag{6.1.7}$$

and the definition (1.10.1) of the q-gamma function to express the right side of (6.1.5) in the form

$$2^{2\alpha+2\beta+2} \frac{\Gamma_q(\alpha+1)\Gamma_q(\beta+1)}{\Gamma_q(\alpha+\beta+2)} \frac{\pi}{\Gamma_q^2\left(\frac{1}{2}\right)} \frac{(-q;q)_{\alpha+\beta}(-q^{\frac{1}{2}};q)_{\alpha+\frac{1}{2}}(-q^{\frac{1}{2}};q)_{\beta+\frac{1}{2}}}{2^{2\alpha+2\beta+1}}.$$

By (1.10.3) this tends to $2^{2\alpha+2\beta+1}\ \Gamma(\alpha+1)\ \Gamma(\beta+1)/\Gamma(\alpha+\beta+2)$ as $q \to 1^-$, since $\Gamma\left(\frac{1}{2}\right) = \sqrt{\pi}$.

For the integrand in (6.1.5) we have

$$\frac{h(x;1,-1)}{h(x;a,b)} = \left(e^{i\theta};q\right)_{\alpha+\frac{1}{2}} \left(e^{-i\theta};q\right)_{\alpha+\frac{1}{2}} \left(-e^{i\theta};q\right)_{\beta+\frac{1}{2}} \left(-e^{-i\theta};q\right)_{\beta+\frac{1}{2}}$$

and hence

$$\lim_{q\to 1^-} \frac{h(x;1,-1)}{h(x;q^{\alpha+\frac{1}{2}},-q^{\beta+\frac{1}{2}})}$$
$$= \left[\left(1-e^{i\theta}\right)\left(1-e^{-i\theta}\right)\right]^{\alpha+\frac{1}{2}} \left[\left(1+e^{i\theta}\right)\left(1+e^{-i\theta}\right)\right]^{\beta+\frac{1}{2}}$$
$$= 2^{\alpha+\beta+1}(1-\cos\theta)^{\alpha+\frac{1}{2}}(1+\cos\theta)^{\beta+\frac{1}{2}},$$

which shows that (6.1.6) is a limit of (6.1.5).

Formula (6.1.1) is substantially more general than (6.1.5) since it contains two more parameters. It is the freedom provided by these extra parameters which will enable us to prove a number of important results in this and the subsequent chapters.

6.2 Proof of formula (6.1.1)

Denote the integral in (6.1.1) by $I(a,b,c,d)$. Since $x = \cos\theta$ is an even function of θ, one can write

$$I(a,b,c,d) = \frac{1}{2}\int_{-\pi}^{\pi} \frac{h(x;1,-1,\sqrt{q},-\sqrt{q})}{h(x;a,b,c,d)}\,d\theta. \tag{6.2.1}$$

Let us assume, for the moment, that a,b,c,d and their pairwise products and quotients are not of the form q^{-j}, $j = 0,1,2,\ldots$. It is easy to check that, by (2.10.18),

$$\begin{aligned} &h(x;1)/h(x;a,b) \\ &= \frac{(a^{-1},b^{-1};q)_\infty}{b(1-q)(q,a/b,bq/a,ab;q)_\infty}\int_a^b \frac{(qu/a,qu/b,u;q)_\infty}{(u/ab;q)_\infty}\,\frac{d_qu}{h(x;u)}, \end{aligned} \tag{6.2.2}$$

$$\begin{aligned} &h(x;-1)/h(x;c,d) \\ &= \frac{(-c^{-1},-d^{-1};q)_\infty}{d(1-q)(q,c/d,dq/c,cd;q)_\infty}\int_c^d \frac{(qv/c,qv/d,-v;q)_\infty}{(-v/cd;q)_\infty}\,\frac{d_qv}{h(x;v)}, \end{aligned} \tag{6.2.3}$$

and

$$\begin{aligned} h(x;-q^{\frac{1}{2}})/h(x;u,v) &= \frac{q^{\frac{1}{2}}(-q^{\frac{1}{2}}u^{-1},-q^{\frac{1}{2}}v^{-1};q)_\infty}{v(1-q)(q,u/v,vq/u,uv;q)_\infty} \\ &\cdot \int_{uq^{-\frac{1}{2}}}^{vq^{-\frac{1}{2}}} \frac{(tq^{\frac{3}{2}}/u,tq^{\frac{3}{2}}/v,-qt;q)_\infty}{(-qt/uv;q)_\infty}\,\frac{d_qt}{h(x;tq^{\frac{1}{2}})}. \end{aligned} \tag{6.2.4}$$

Also,

$$\begin{aligned} &\frac{1}{2}\int_{-\pi}^{\pi} \frac{h(x;q^{\frac{1}{2}})}{h(x;tq^{\frac{1}{2}})}\,d\theta \\ &= \frac{1}{2}\int_{-\pi}^{\pi} \frac{(q^{\frac{1}{2}}e^{i\theta},q^{\frac{1}{2}}e^{-i\theta};q)_\infty}{(tq^{\frac{1}{2}}e^{i\theta},tq^{\frac{1}{2}}e^{-i\theta};q)_\infty}\,d\theta \\ &= \frac{1}{2}\int_{-\pi}^{\pi}\sum_{k=0}^{\infty}\sum_{\ell=0}^{\infty} \frac{(t^{-1};q)_k(t^{-1};q)_\ell}{(q;q)_k(q;q)_\ell}\left(tq^{\frac{1}{2}}\right)^{k+\ell} e^{i(k-\ell)\theta}\,d\theta \\ &= \frac{1}{2}\sum_{k=0}^{\infty}\sum_{\ell=0}^{\infty} \frac{(t^{-1};q)_k(t^{-1};q)_\ell}{(q;q)_k(q;q)_\ell}\left(tq^{\frac{1}{2}}\right)^{k+\ell}\int_{-\pi}^{\pi} e^{i(k-\ell)\theta}\,d\theta \\ &= \pi\sum_{k=0}^{\infty} \frac{(t^{-1},t^{-1};q)_k}{(q,q;q)_k}(qt^2)^k = \pi\frac{(qt,qt;q)_\infty}{(q,qt^2;q)_\infty} \end{aligned} \tag{6.2.5}$$

for $|tq^{\frac{1}{2}}| < 1$, by (1.5.1). Since

$$(qt^2;q)_\infty = (qt^2,q^2t^2;q^2)_\infty = (tq^{\frac{1}{2}},-tq^{\frac{1}{2}},qt,-qt;q)_\infty, \tag{6.2.6}$$

we have

$$\frac{1}{2}\int_{-\pi}^{\pi} \frac{h(x;q^{\frac{1}{2}})}{h(x;tq^{\frac{1}{2}})}\,d\theta = \frac{\pi(qt;q)_\infty}{(q,tq^{\frac{1}{2}},-tq^{\frac{1}{2}},-tq;q)_\infty}. \tag{6.2.7}$$

Thus

$$I(a,b,c,d)$$

$$= \frac{\pi q^{\frac{1}{2}}(a^{-1},b^{-1},-c^{-1},-d^{-1};q)_\infty}{bd(1-q)^3(q;q)_\infty^4(a/b,bq/a,ab,c/d,dq/c,cd;q)_\infty}$$

$$\cdot \int_a^b d_qu \frac{(qu/a,qu/b,u;q)_\infty}{(u/ab;q)_\infty} \int_c^d d_qv \frac{(qv/c,qv/d,-v,-q^{\frac{1}{2}}/u,-q^{\frac{1}{2}}/v;q)_\infty}{v(-v/cd,vq/u,u/v,uv;q)_\infty}$$

$$\cdot \int_{uq^{-\frac{1}{2}}}^{vq^{-\frac{1}{2}}} d_qt \frac{(tq^{\frac{3}{2}}/u,tq^{\frac{3}{2}}/v,qt;q)_\infty}{(tq^{\frac{1}{2}},-tq^{\frac{1}{2}},-qt/uv;q)_\infty}$$

$$= \frac{\pi(a^{-1},b^{-1},-c^{-1},-d^{-1};q)_\infty}{bd(1-q)^2(q;q)_\infty^3(a/b,bq/a,ab,c/d,dq/c,cd;q)_\infty}$$

$$\cdot \int_a^b d_qu \frac{(qu/a,qu/b;q)_\infty}{(-u,u/ab;q)_\infty} \int_c^d d_qv \frac{(qv/c,qv/d,-uv;q)_\infty}{(v,uv,-v/cd;q)_\infty}$$

$$= \frac{\pi(a^{-1},b^{-1},q^{\frac{1}{2}},-q^{\frac{1}{2}},-1;q)_\infty}{b(1-q)(q;q)_\infty^2(a/b,bq/a,ab,c,d,cd;q)_\infty} \int_a^b d_qu \frac{(qu/a,qu/b,cdu;q)_\infty}{(cu,du,u/ab;q)_\infty}$$

$$= \frac{\pi(-1,q^{\frac{1}{2}},-q^{\frac{1}{2}},abcd;q)_\infty}{(q,ab,ac,ad,bc,bd,cd;q)_\infty}, \tag{6.2.8}$$

by repeated applications of (2.10.18).

Since $(-1,q^{\frac{1}{2}},-q^{\frac{1}{2}};q)_\infty = 2(q^{\frac{1}{2}},-q^{\frac{1}{2}},-q;q)_\infty = 2$, which follows from (6.2.6) by setting $t=1$, we get (6.1.1). By analytic continuation, the restrictions on a,b,c,d mentioned above may be removed.

6.3 Integral representations for very-well-poised ${}_8\phi_7$ series

Formulas (2.10.18) and (2.10.19) enable us to use the Askey-Wilson q-beta integral (6.1.1) to derive Riemann integral representations for very-well-poised ${}_8\phi_7$ series.

Let us first set

$$w(x;a,b,c,d) = (1-x^2)^{-\frac{1}{2}} \frac{h(x;1,-1,q^{\frac{1}{2}},-q^{\frac{1}{2}})}{h(x;a,b,c,d)} \tag{6.3.1}$$

and

$$J(a,b,c,d,f,g) = \int_{-1}^1 w(x;a,b,c,d) \frac{h(x;g)}{h(x;f)}\, dx, \tag{6.3.2}$$

where $\max(|a|,|b|,|c|,|d|,|f|,|q|) < 1$ and g is arbitrary. Since, by (2.10.18),

$$\frac{h(x;g)}{h(x;d,f)} = \frac{(g/d,g/f;q)_\infty}{f(1-q)(q,d/f,qf/d,fd;q)_\infty} \cdot \int_d^f d_qu \frac{(qu/d,qu/f,gu;q)_\infty}{(gu/df;q)_\infty h(x;u)}, \tag{6.3.3}$$

we have

$$J(a,b,c,d,f,g) = \frac{(g/d,g/f;q)_\infty}{f(1-q)(q,d/f,qf/d,df;q)_\infty}$$

$$\cdot \int_d^f d_q u \frac{(qu/d, qu/f, gu;q)_\infty}{(gu/df;q)_\infty} \int_{-1}^1 w(x;a,b,c,u)\,dx. \tag{6.3.4}$$

By (6.1.1),

$$\int_{-1}^1 w(x;a,b,c,u)\,dx = \frac{2\pi(abcu;q)_\infty}{(q,ab,ac,bc;q)_\infty(au,bu,cu;q)_\infty}. \tag{6.3.5}$$

Substituting this into (6.3.4), we obtain

$$J(a,b,c,d,f,g) = \frac{2\pi(g/d, g/f;q)_\infty}{f(1-q)(q;q)_\infty^2(d/f, qf/d, df, ab, ac, bc;q)_\infty}$$
$$\cdot \int_d^f d_q u \frac{(qu/d, qu/f, gu, abcu;q)_\infty}{(au,bu,cu,gu/df;q)_\infty}. \tag{6.3.6}$$

The parameters in this q-integral are such that (2.10.19) can be applied to obtain

$$J(a,b,c,d,f,g) = \frac{2\pi(g/f, fg, abcf, bcdf, cdaf, dabf;q)_\infty}{(q,ad,bd,cd,af,bf,cf,df,ab,ac,bc,abcdf^2;q)_\infty}$$
$$\cdot {}_8W_7(abcdf^2q^{-1}; af, bf, cf, df, abcdfg^{-1}; q, g/f), \tag{6.3.7}$$

provided $|g/f| < 1$, if the series does not terminate. By virtue of the transformation formula (2.10.1) many different forms of (6.3.7) can be written down. Two particularly useful ones are

$$J(a,b,c,d,f,g) = \frac{2\pi(ag,bg,cg,abcd,abcf;q)_\infty}{(q,ab,ac,ad,af,bc,bd,bf,cd,cf,abcg;q)_\infty}$$
$$\cdot {}_8W_7(abcgq^{-1}; ab, ac, bc, g/d, g/f; q, df), \tag{6.3.8}$$

which was derived in Nassrallah and Rahman [1985], and

$$J(a,b,c,d,f,g) = \frac{2\pi(ag,bg,cg,dg,fg,abcdf/g;q)_\infty}{(q,ab,ac,ad,af,bc,bd,bf,cd,cf,df,g^2;q)_\infty}$$
$$\cdot {}_8W_7(g^2q^{-1}; g/a, g/b, g/c, g/d, g/f; q, abcdfg^{-1}). \tag{6.3.9}$$

If the series in (6.3.9) does not terminate, then we must impose the condition $|abcdfg^{-1}| < 1$ so that it converges.

Note that, if in (6.3.8) we let $0 < q < 1$, $a = -b = q^{\frac{1}{2}}$, $c = q^{\alpha+\frac{1}{2}}$, $d = z$, $f = -q^{\beta+\frac{1}{2}}$, $g = zq^\gamma$ with $\mathrm{Re}(\alpha,\beta) > -\frac{1}{2}$ and then take the limit $q \to 1^-$, we obtain, after some simplification,

$${}_2F_1(\gamma, \alpha+1; \alpha+\beta+2; z) = \frac{\Gamma(\alpha+\beta+2)}{\Gamma(\alpha+1)\Gamma(\beta+1)} \int_0^1 x^\alpha(1-x)^\beta(1-xz)^{-\gamma}dx. \tag{6.3.10}$$

This shows that (6.3.8) is a q-analogue of Euler's integral representation (1.11.10).

Another limiting case of (6.3.8) was pointed out in Rahman [1985]. To derive it, replace a, b, c, d, f, g in (6.3.8) by $q^a, q^b, q^c, q^d, q^f, q^g$, respectively. Also, replace x in the integral (6.3.2) by $\cos(t \log q)$, which corresponds to replacing $e^{i\theta}$ by q^{it}. Now let $q \to 1^-$ and use (1.10.1) and (1.10.3) to get the formula

$$\begin{aligned}
&\frac{1}{4\pi}\int_{-\infty}^{\infty} \frac{\Gamma(a+it)\Gamma(a-it)\Gamma(b+it)\Gamma(b-it)\Gamma(c+it)\Gamma(c-it)}{\Gamma(2it)\Gamma(-2it)} \\
&\quad\cdot \frac{\Gamma(d+it)\Gamma(d-it)\Gamma(f+it)\Gamma(f-it)}{\Gamma(g+it)\Gamma(g-it)}\,dt \\
&= \frac{\Gamma(a+b)\Gamma(a+c)\Gamma(a+d)\Gamma(a+f)\Gamma(b+c)\Gamma(b+d)}{\Gamma(a+g)\Gamma(b+g)\Gamma(c+g)} \\
&\quad\cdot \frac{\Gamma(b+f)\Gamma(c+d)\Gamma(c+f)\Gamma(a+b+c+g)}{\Gamma(a+b+c+d)\Gamma(a+b+c+f)} \\
&\quad\cdot {}_7F_6\left[\begin{matrix} a+b+c+g-1,\ \frac{1}{2}(a+b+c+g+1),\ a+b,\ a+c,\ b+c \\ \frac{1}{2}(a+b+c+g-1),\ c+g,\ b+g,\ a+g, \end{matrix}\right. \\
&\qquad\qquad \left.\begin{matrix} g-d, & g-f, \\ a+b+c+d, & a+b+c+f \end{matrix};1\right],
\end{aligned} \tag{6.3.11}$$

where Re $(a,b,c,d,f) > 0$.

6.4 Integral representations for very-well-poised ${}_{10}\phi_9$ series

If we set $g = abcdf$ in (6.3.7), then the ${}_8W_7$ series collapses to one term with value 1 and so we have the formula

$$\begin{aligned}
&\int_{-1}^{1} \frac{h(x;1,-1,q^{\frac{1}{2}},-q^{\frac{1}{2}},abcdf)}{h(x;a,b,c,d,f)} \frac{dx}{\sqrt{1-x^2}} \\
&= \frac{2\pi(abcd, abcf, bcdf, abdf, acdf;q)_\infty}{(q,ab,ac,ad,af,bc,bd,bf,cd,cf,df;q)_\infty} \\
&= g_0(a,b,c,d,f), \quad \text{say},
\end{aligned} \tag{6.4.1}$$

where $(\max |a|,|b|,|c|,|d|,|f|,|q|) < 1$. This is a q-analogue of the formula

$$\int_0^1 x^{a-1}(1-x)^{b-1}(1-tx)^{-a-b}\,dx = \frac{\Gamma(a)\Gamma(b)}{\Gamma(a+b)}(1-t)^{-a}, \ \text{Re}(a,b) > 0.$$

Replace f by fq^n in (6.4.1), where n is a nonnegative integer, to get

$$\begin{aligned}
&\int_{-1}^{1} v(x;a,b,c,d,f)\frac{(fe^{i\theta}, fe^{-i\theta};q)_n}{(abcdfe^{-i\theta}, abcdfe^{i\theta};q)_n}\,dx \\
&= g_0(a,b,c,d,f)\frac{(af,bf,cf,df;q)_n}{(bcdf,acdf,abdf,abcf;q)_n},
\end{aligned} \tag{6.4.2}$$

where

$$v(x;a,b,c,d,f) = \left(1-x^2\right)^{-\frac{1}{2}} \frac{h(x;1,-1,q^{\frac{1}{2}},-q^{\frac{1}{2}},abcdf)}{h(x;a,b,c,d,f)}. \tag{6.4.3}$$

Let $\sigma = abcdf$. If $|z| < 1$, then (6.4.2) gives the formula

$$\begin{aligned}
&\int_{-1}^{1} v(x;a,b,c,d,f)\ {}_{r+5}W_{r+4}(\sigma f q^{-1}; a_1,\ldots,a_r, fe^{i\theta}, fe^{-i\theta}; q,z)\,dx \\
&= g_0(a,b,c,d,f)\ {}_{r+7}W_{r+6}(\sigma f q^{-1}; a_1,\ldots,a_r, af,bf,cf,df;q,z).
\end{aligned} \tag{6.4.4}$$

In particular, for $r=3$ and $z=\sigma^2/a_1a_2a_3$, we have the formula

$$\int_{-1}^{1} v(x;a,b,c,d,f)\,{}_8W_7(\sigma fq^{-1};a_1,a_2,a_3,fe^{i\theta},fe^{-i\theta};q,\sigma^2/a_1a_2a_3)\,dx$$
$$= g_0(a,b,c,d,f)\,{}_{10}W_9(\sigma fq^{-1};a_1,a_2,a_3,af,bf,cf,df;q,\sigma^2/a_1a_2a_3), \tag{6.4.5}$$

where $|\sigma^2/a_1a_2a_3|<1$, if the series do not terminate.

Let us assume that

$$a_1a_2a_3q=\sigma^2 \tag{6.4.6}$$

which ensures that the very-well-poised series on either side of (6.4.5) are balanced. Then, by (2.11.7)

$$\begin{aligned}&{}_8W_7(\sigma fq^{-1};a_1,a_2,a_3,fe^{i\theta},fe^{-i\theta};q,q)\\&+\frac{(\sigma f,qa_1/\sigma f,a_2,qa_1/a_2,a_3,qa_1/a_3;q)_\infty h(x;f,qa_1/f)}{(\sigma f/qa_1,\sigma f/a_2,\sigma f/a_3,qa_1a_2/\sigma f,qa_1a_3/\sigma f,q^2a_1^2/\sigma f;q)_\infty h(x;\sigma,qa_1/\sigma)}\\&\cdot{}_8W_7(qa_1^2/\sigma f;a_1,qa_1a_2/\sigma f,qa_1a_3/\sigma f,qa_1e^{i\theta}/\sigma,qa_1e^{-i\theta}/\sigma;q,q)\\&=\frac{(\sigma f,\sigma f/a_2a_3,qa_1/\sigma f,\sigma/f;q)_\infty h(x;\sigma/a_2,\sigma/a_3)}{(\sigma f/a_2,\sigma f/a_3,qa_1a_2/\sigma f,qa_1a_3/\sigma f;q)_\infty h(x;\sigma,\sigma/a_2a_3)},\end{aligned} \tag{6.4.7}$$

and hence

$$\begin{aligned}&\int_{-1}^{1}\frac{h(x;1,-1,q^{\frac12},-q^{\frac12},\sigma/a_2,\sigma/a_3)}{h(x;a,b,c,d,f,\sigma/a_2a_3)}\frac{dx}{\sqrt{1-x^2}}\\&=g_0(a,b,c,d,f)\left\{\frac{(\sigma f/a_2,\sigma f/a_3,\sigma/fa_2,\sigma/fa_3;q)_\infty}{(\sigma f,\sigma f/a_2a_3,\sigma/f,\sigma/fa_2a_3;q)_\infty}\right.\\&\quad\cdot{}_{10}W_9(\sigma fq^{-1};a_1,a_2,a_3,af,bf,cf,df;q,q)\\&\quad+\frac{(af,bf,cf,df,a_2,a_3,qa_1/a_2,qa_1/a_3;q)_\infty}{(\sigma/a,\sigma/b,\sigma/c,\sigma/d,qaa_1/\sigma,qba_1/\sigma,qca_1/\sigma,qda_1/\sigma;q)_\infty}\\&\quad\cdot\frac{(qa_1/af,qa_1/bf,qa_1/cf,qa_1/df,qa_1/\sigma f;q)_\infty}{(\sigma/f,\sigma f/qa_1,\sigma f/a_2a_3,\sigma/fa_2a_3,q^2a_1^2/\sigma f;q)_\infty}\\&\quad\left.\cdot{}_{10}W_9(qa_1^2/\sigma f;a_1,\sigma/fa_2,\sigma/fa_3,qaa_1/\sigma,qba_1/\sigma,qca_1/\sigma,qda_1/\sigma;q,q)\right\},\end{aligned} \tag{6.4.8}$$

where $\sigma=abcdf$ and $a_1a_2a_3q=\sigma^2$.

Since the integrand on the left side of (6.4.8) is symmetric in a,b,c,d and f, the expression on the right side must have the same property. This provides an alternate proof of Bailey's four-term transformation formula (2.12.9) for very-well-poised ${}_{10}\phi_9$ series which are balanced and nonterminating.

If we set $a_3=q^{-n}$, $n=0,1,2,\ldots$, then the coefficient of the second ${}_{10}W_9$ on the right side of (6.4.8) vanishes and we obtain

$$\begin{aligned}&\int_{-1}^{1}\frac{h(x;1,-1,q^{\frac12},-q^{\frac12},\sigma)}{h(x;a,b,c,d,f)}\frac{(\sigma e^{i\theta}/g,\sigma e^{-i\theta}/g;q)_n}{(\sigma e^{i\theta},\sigma e^{-i\theta};q)_n}\frac{dx}{\sqrt{1-x^2}}\\&=g_0(a,b,c,d,f)\frac{(\sigma f/g,\sigma/fg;q)_n}{(\sigma f,\sigma/f;q)_n}\end{aligned}$$

$$\cdot {}_{10}\phi_9\left[\begin{matrix} \nu, q\sqrt{\nu}, -q\sqrt{\nu}, & g, & af, & bf, & cf, & df, & \sigma^2 q^{n-1}/g, & q^{-n} \\ \sqrt{\nu}, & -\sqrt{\nu}, & \sigma f/g, & \sigma/a, & \sigma/b, & \sigma/c, & \sigma/d, & fgq^{1-n}/\sigma, \sigma f q^n \end{matrix}; q, q\right], \tag{6.4.9}$$

where $\nu = \sigma f q^{-1}$. By applying the iteration of the transformation formula (2.9.1) given in Exercise 2.19, this can be written in the more symmetric form

$$\int_{-1}^{1} \frac{h(x; 1, -1, q^{\frac{1}{2}}, -q^{\frac{1}{2}}, \sigma q^n, \tau)}{h(x; a, b, c, d, f, \tau q^n)} \frac{dx}{\sqrt{1-x^2}}$$
$$= g_0(a, b, c, d, f) \frac{(\tau a, \tau b, \tau c, \tau d, \sigma/\tau; q)_n}{(\sigma/a, \sigma/b, \sigma/c, \sigma/d, \tau^2; q)_n}$$
$$\cdot {}_{10}\phi_9\left[\begin{matrix} \tau^2 q^{-1}, \tau q^{\frac{1}{2}}, -\tau q^{\frac{1}{2}}, \tau/a, \tau/b, \tau/c, \tau/d, \tau/f, \sigma\tau q^{n-1}, q^{-n} \\ \tau q^{-\frac{1}{2}}, -\tau q^{-\frac{1}{2}}, \tau a, \tau b, \tau c, \tau d, \tau f, \tau q^{1-n}/\sigma, \tau^2 q^n \end{matrix}; q, q\right], \tag{6.4.10}$$

where $\sigma = abcdf$ and τ is arbitrary. Similarly, by applying (2.12.9) twice one can rewrite (6.4.8) in the form

$$\int_{-1}^{1} \frac{h(x; 1, -1, q^{\frac{1}{2}}, -q^{\frac{1}{2}}, \lambda, \mu)}{h(x; a, b, c, d, f, g)} \frac{dx}{\sqrt{1-x^2}}$$
$$= \frac{2\pi(\lambda\mu/af, \lambda\mu/bf, \lambda\mu/cf, \lambda\mu/df, \lambda g, \mu g, \lambda/g, \mu/g; q)_\infty}{(q, ab, ac, ad, ag, bc, bd, bg, cd, cg, dg, fg, f/g, \lambda\mu g/f; q)_\infty}$$
$$\cdot {}_{10}\phi_9\left[\begin{matrix} \nu_1, \sqrt{\nu_1}, -q\sqrt{\nu_1}, & ag, & bg, & cg, & dg, & \lambda/f, \mu/f, \lambda\mu/q \\ \sqrt{\nu_1}, -\sqrt{\nu_1}, & \lambda\mu/af, & \lambda\mu/bf, & \lambda\mu/cf, & \lambda\mu/df, & \mu g, \lambda g, gq/f \end{matrix}; q, q\right]$$
$$+ \frac{2\pi(\lambda\mu/ag, \lambda\mu/bg, \lambda\mu/cg, \lambda\mu/dg, \lambda f, \mu f, \lambda/f, \mu/f; q)_\infty}{(q, ab, ac, ad, af, bc, bd, bf, cd, cf, df, gf, g/f, \lambda\mu f/g; q)_\infty}$$
$$\cdot {}_{10}\phi_9\left[\begin{matrix} \nu_2, q\sqrt{\nu_2}, -q\sqrt{\nu_2}, & af, & bf, & cf, & df, & \lambda/g, \mu/g, \lambda\mu/q \\ \sqrt{\nu_2}, -\sqrt{\nu_2}, & \lambda\mu/ag, & \lambda\mu/bg, & \lambda\mu/cg, & \lambda\mu/dg, & \mu f, \lambda f, fq/g \end{matrix}; q, q\right] \tag{6.4.11}$$

where $\nu_1 = \lambda\mu g/fq$, $\nu_2 = \lambda\mu f/gq$, $\lambda\mu = abcdfg$ and $\max(|a|, |b|, |c|, |d|, |f|, |g|, |q|) < 1$. For these and other results see Rahman [1986b].

If λ or μ equals a, b, c or d, then the right side of (6.4.11) is summable by (2.11.7), while the integral on the left side gives the sum by (6.4.1). For this reason the integral in (6.4.1) may be considered as an integral analogue of Bailey's formula (2.11.7) for the sum of two balanced very-well-poised ${}_8\phi_7$ series. Likewise, the integral in (6.4.2) is an integral analogue of Jackson's sum (2.6.2). Indeed, a basic integral may be called well-poised if it can be written in the form

$$\int_{-\pi}^{\pi} \frac{\prod_{j=1}^{n} (a_j e^{i\theta}, a_j e^{-i\theta}; q)_\infty}{\prod_{j=1}^{m} (b_j e^{i\theta}, b_j e^{-i\theta}; q)_\infty} d\theta. \tag{6.4.12}$$

The integrals in (6.1.1), (6.3.2), (6.4.1) and (6.4.2) are all well-poised.

6.5 A quadratic transformation formula for very-well-poised balanced ${}_{10}\phi_9$ series

In (6.4.11) let us set $\mu = -\lambda$, $g = -f$ and $b = -a$ so that $\lambda^2 = -a^2cdf^2$. Then the expression on the right side of (6.4.11) becomes

$$\frac{2\pi(\lambda^2/cf, \lambda^2/df; q)_\infty (\lambda^4/a^2f^2, \lambda^2f^2, \lambda^2/f^2; q^2)_\infty}{(q, -1, -a^2, -f^2, \lambda^2; q)_\infty (a^2c^2, a^2d^2, a^2f^2; q^2)_\infty (cd, cf, df; q)_\infty}$$
$$\cdot\, {}_{10}W_9(\lambda^2q^{-1}; -\lambda^2q^{-1}, af, -af, \lambda/f, -\lambda/f, cf, df; q, q)$$
$$+ \frac{2\pi(-\lambda^2/cf, -\lambda^2/df; q)_\infty (\lambda^4/a^2f^2, \lambda^2f^2, \lambda^2/f^2; q^2)_\infty}{(q, -1, -a^2, -f^2, \lambda^2; q)_\infty (a^2c^2, a^2d^2, a^2f^2; q^2)_\infty (cd, -cf, -df; q)_\infty}$$
$$\cdot\, {}_{10}W_9(\lambda^2q^{-1}; -\lambda^2q^{-1}, af, -af, \lambda/f, -\lambda/f, -cf, -df; q, q). \tag{6.5.1}$$

We now turn to the integral on the left side of (6.4.11). Observing that

$$h(x; a, -a) = (a^2e^{2i\theta}, a^2e^{-2i\theta}; q^2)_\infty = h(\xi; a^2; q^2), \tag{6.5.2}$$

where $x = \cos\theta$ and $\xi = \cos 2\theta = 2x^2 - 1$, it follows from (2.10.18) that

$$\frac{h(x; \lambda - \lambda)}{h(x; a, -a, f, -f)} = \frac{h(\xi; \lambda^2; q^2)}{h(\xi; a^2, f^2; q^2)}$$
$$= \frac{(\lambda^2/a^2, \lambda^2/f^2; q^2)_\infty}{f^2(1-q^2)(q^2, a^2/f^2, q^2f^2/a^2, a^2f^2; q^2)_\infty}$$
$$\cdot \int_{a^2}^{f^2} \frac{(q^2u/a^2, q^2u/f^2, \lambda^2u; q^2)_\infty}{(\lambda^2u/a^2f^2; q^2)_\infty h(\xi; u; q^2)}\, d_{q^2}u. \tag{6.5.3}$$

Hence the integral on the left side of (6.4.11) can be expressed as

$$\frac{(\lambda^2/a^2, \lambda^2/f^2; q^2)_\infty}{f^2(1-q^2)(q^2, a^2/f^2, q^2f^2/a^2, a^2f^2; q^2)_\infty}$$
$$\cdot \int_{a^2}^{f^2} \frac{(q^2u/a^2, q^2u/f^2, \lambda^2u; q^2)_\infty}{(\lambda^2u/a^2f^2; q^2)_\infty}\, d_{q^2}u \int_{-1}^{1} w(x; c, d, u^{\frac{1}{2}}, -u^{\frac{1}{2}})\, dx$$
$$= \frac{2\pi(\lambda^2/a^2, \lambda^2/f^2; q^2)_\infty}{f^2(1-q^2)(q^2, a^2/f^2, q^2f^2/a^2, a^2f^2; q^2)_\infty (q, cd; q)_\infty}$$
$$\cdot \int_{a^2}^{f^2} \frac{(q^2u/a^2, q^2u/f^2, \lambda^2u, -cdu, -cdqu; q^2)_\infty}{(\lambda^2u/a^2f^2, c^2u, d^2u, -u, -uq; q^2)_\infty}\, d_{q^2}u \tag{6.5.4}$$

by (6.1.1). Since $\lambda^2 = -a^2cdf^2$, the q-integral on the right side of (6.5.4) reduces to

$$\int_{a^2}^{f^2} \frac{(q^2u/a^2, q^2u/f^2, -cdqu, -a^2cdf^2u; q^2)_\infty}{(c^2u, d^2u, -u, -uq; q^2)_\infty}\, d_{q^2}u$$
$$= \frac{f^2(1-q^2)(q^2, a^2/f^2, q^2f^2/a^2; q^2)_\infty}{(a^2c^2, a^2d^2, c^2f^2, d^2f^2; q^2)_\infty}$$
$$\cdot \frac{(-a^2c^2f^2, -a^2d^2f^2, a^2c^2d^2f^2, -cdqf^2, -a^2cdf^4; q^2)_\infty}{(-a^2, -f^2, -qf^2, -a^2c^2d^2f^4; q^2)_\infty}$$
$$\cdot\, {}_8W_7(-a^2c^2d^2f^4q^{-2}; c^2f^2, d^2f^2, -f^2, cd, cda^2f^2q^{-1}; q^2, -qa^2), \tag{6.5.5}$$

by (2.10.19).

Using this in (6.5.4) and equating with (6.5.1) we obtain the desired quadratic transformation formula

$$\begin{aligned}
&\frac{(-a^2cf,-a^2df;q)_\infty}{(df,cf;q)_\infty}\,{}_{10}W_9(\lambda^2q^{-1};-\lambda^2q^{-1},af,-af,\lambda/f,-\lambda/f,cf,df;q,q)\\
&\quad+\frac{(a^2cf,a^2df;q)_\infty}{(-df,-cf;q)_\infty}\,{}_{10}W_9(\lambda^2q^{-1};-\lambda^2q^{-1},af,-af,\lambda/f,-\lambda/f,-cf,-df;q,q)\\
&=(-1,-cdf^2,-cda^2f^2;q)_\infty\,\frac{(-qa^2,-a^2c^2f^2,-a^2d^2f^2;q^2)_\infty}{(c^2f^2,d^2f^2,-a^2c^2d^2f^4;q^2)_\infty}\\
&\quad\cdot{}_8W_7(-a^2c^2d^2f^4q^{-2};c^2f^2,d^2f^2,-f^2,cd,cda^2f^2q^{-1};q^2,-qa^2),
\end{aligned}\tag{6.5.6}$$

with $\lambda^2=-cda^2f^2$. This gives a nonterminating extension of the transformation formula (3.10.3) when the bibasic Φ series is a balanced ${}_{10}\phi_9$. For an extension of (3.10.3) when the Φ series is not balanced, see Nassrallah and Rahman [1986].

6.6 The Askey-Wilson integral when max $(|a|,|b|,|c|,|d|)\geq 1$

Our aim in this section is to extend the Askey-Wilson formula (6.1.1) to cases in which $|q|<1$ and the absolute value of at least one of the parameters is greater than or equal to 1. Since the integral in (6.1.1), which we have already denoted by $I(a,b,c,d)$, is symmetric in a,b,c,d, without loss of generality we may assume that $|a|\geq 1$.

Let us first consider

$$|a|\geq 1>\max(|b|,|c|,|d|).\tag{6.6.1}$$

If $a=\pm1$, then the functions $h(x;\pm1)$ and $h(x;a)$ in the integrand in (6.1.1) cancel and by continuity it follows that

$$I(\pm1,b,c,d)=\frac{2\pi(\pm bcd;q)_\infty}{(q,\pm b,\pm c,\pm d,bc,bd,cd;q)_\infty}.\tag{6.6.2}$$

However, if $|a|=1$ and $a\neq\pm1$, then $h(x;a)=0$ for some x in the interval $(-1,1)$ and so the integral in (6.1.1) does not converge. Similarly, this integral does not converge if $|aq^n|=1$ and $aq^n\neq\pm1$ for some positive integer n.

If there is a nonnegative interger m such that

$$|aq^{m+1}|<1<|aq^m|\tag{6.6.3}$$

and if ab,ac and ad are not of the form q^{-n} for any nonnegative integer n, then the integral in (6.1.1) converges and we can evaluate it by the following technique.

Observe that, since

$$\begin{aligned}
h(x;a)&=(ae^{i\theta},ae^{-i\theta};q)_{m+1}\,h(x;aq^{m+1})\\
&=a^{2m+2}q^{m+m^2}\,h(x;aq^{m+1},q^{-m}/a)/h(x;q/a),
\end{aligned}\tag{6.6.4}$$

we have

$$I(a,b,c,d) = a^{-2m-2}q^{-m-m^2} \cdot \int_{-1}^{1} \frac{h(x;1,-1,q^{\frac{1}{2}}-q^{\frac{1}{2}},q/a)}{h(x;b,c,d,aq^{m+1},q^{-m}/a)} \frac{dx}{\sqrt{1-x^2}}, \tag{6.6.5}$$

where the parameters $b, c, d, aq^{m+1}, q^{-m}/a$ in the denominator of the integrand are now all less than 1 in absolute value. By (6.3.8),

$$\begin{aligned} I(a,b,c,d) = {} & \frac{2\pi(qb/a, qc/a, qd/a, abcdq^{m+1}, bcdq^{-m}/a;q)_\infty}{(q,bc,bd,cd,abq^{m+1},acq^{m+1},adq^{m+1},bcdq/a;q)_\infty} \\ & \cdot \frac{a^{-2m-2}q^{-m-m^2}}{(bq^{-m}/a, cq^{-m}/a, dq^{-m}/a;q)_\infty} \\ & \cdot {}_8W_7(bcda^{-1}; bc, bd, cd, q^{-m}a^{-2}, q^{m+1}; q, q). \end{aligned} \tag{6.6.6}$$

The series in (6.6.6) is balanced and so we can apply Bailey's summation formula (2.11.7). After some simplification we find that

$$I(a,b,c,d) = \frac{2\pi(abcd;q)_\infty}{(q,ab,ac,ad,bc,bd,cd;q)_\infty} + L_m(a;b,c,d), \tag{6.6.7}$$

where

$$\begin{aligned} L_m(a;b,c,d) = {} & \frac{2\pi(aq/d, bq/d, cq/d, q^{m+1}, a^2bcq^{m+1}, q^{-m}/a^2, bcq^{-m};q)_\infty}{(q,ab,ac,bc,abq^{m+1},acq^{m+1},adq^{m+1},aq^{m+1}/d,abcq/d;q)_\infty} \\ & \cdot \frac{a^{-2m-1}d^{-1}q^{-m-m^2}}{(bq^{-m}/a, cq^{-m}/a, dq^{-m}/a, q^{-m}/ad;q)_\infty} \\ & \cdot {}_8W_7(abcd^{-1}; bc, q^{-m}/ad, aq^{m+1}/d, ab, ac; q, q). \end{aligned} \tag{6.6.8}$$

By (2.10.1),

$$\begin{aligned} & {}_8W_7(abcd^{-1}; bc, q^{-m}/ad, aq^{m+1}/d, ab, ac; q, q) \\ & = \frac{(abcq/d, q/ad, abq, acq;q)_\infty}{(bq/d, cq/d, q, a^2bcq;q)_\infty} \\ & \quad \cdot {}_8W_7(a^2bc; abcd, ab, ac, a^2q^{m+1}, q^{-m}; q, q/ad). \end{aligned} \tag{6.6.9}$$

Since m is a nonnegative integer, the series on the right side of (6.6.9) terminates and hence, by Watson's formula (2.5.1),

$$\begin{aligned} & {}_8W_7(a^2bc; abcd, ab, ac, a^2q^{m+1}, q^{-m}; q, q/ad) = \frac{(a^2bcq, q;q)_m}{(abq, acq;q)_m} \\ & \quad \cdot {}_4\phi_3\left[\begin{matrix} q^{-m},\ ab,\ ac,\ q^{-m}/ad \\ bcq^{-m},\ aq/d,\ q^{-m} \end{matrix}; q, q\right] = \frac{(a^2bcq, q, aq/b, aq/c;q)_m}{(qa^2, abq, acq, q/bc;q)_m} \\ & \quad \cdot {}_8\phi_7\left[\begin{matrix} a^2, qa, -qa, ab, ac, ad, a^2q^{m+1}, q^{-m} \\ a, -a, aq/b, aq/c, aq/d, q^{-m}, a^2q^{m+1} \end{matrix}; q, \frac{q}{abcd}\right] \\ & = \frac{(a^2bcq, q, aq/b, aq/c;q)_m}{(qa^2, abq, acq, q/bc;q)_m} \\ & \quad \cdot \sum_{k=0}^{m} \frac{(a^2;q)_k(1-a^2q^{2k})(ab,ac,ad;q)_k}{(q;q)_k(1-a^2)(aq/b,aq/c,aq/d;q)_k} \left(\frac{q}{abcd}\right)^k. \end{aligned} \tag{6.6.10}$$

Using (6.6.10) and (6.6.9) in (6.6.8), we obtain

$$L_m(a;b,c,d) = -\frac{2\pi(a^{-2};q)_\infty}{(q,ab,ac,ad,b/a,c/a,d/a;q)_\infty} \cdot \sum_{k=0}^{m} \frac{(a^2;q)_k(1-a^2q^{2k})(ab,ac,ad;q)_k}{(q;q)_k(1-a^2)(aq/b,aq/c,aq/d;q)_k}\left(\frac{q}{abcd}\right)^k. \tag{6.6.11}$$

Hence

$$\int_{-1}^{1} \frac{h(x;1,-1,q^{\frac{1}{2}},-q^{\frac{1}{2}})}{h(x;a,b,c,d)} \frac{dx}{\sqrt{1-x^2}} + \frac{2\pi(a^{-2};q)_\infty}{(q,ab,ac,ad,b/a,c/a,d/a;q)_\infty} \cdot \sum_{k=0}^{m} \frac{(a^2;q)_k(1-a^2q^{2k})(ab,ac,ad;q)_k}{(q;q)_k(1-a^2)(aq/b,aq/c,aq/d;q)_k}\left(\frac{q}{abcd}\right)^k$$
$$= \frac{2\pi(abcd;q)_\infty}{(q,ab,ac,ad,bc,bd,cd;q)_\infty}, \tag{6.6.12}$$

where $\max(|b|,|c|,|d|,|q|) < 1$, $|aq^{m+1}| < 1 < |aq^m|$ for some nonnegative integer m, and the products ab, ac, ad are not of the form $q^{-n}, n = 1, 2, \ldots$. Askey and Wilson [1985] proved this formula by using contour integration. By continuity, formula (6.6.12) also holds if the restriction (6.6.3) is replaced by $aq^m = \pm 1$.

Note that, if one of the products ab, ac or ad is of the form q^{-n} for some nonnegative integer n, the integral in (6.6.12) converges even though the denominator on the right side of (6.6.12) equals zero as does the denominator in the coefficient of the sum in (6.6.12). If we let ab tend to q^{-n} then, since $|b| < 1$ and $|aq^{m+1}| < 1 < |aq^m|$, we must have $n \leq m$. We may then multiply (6.6.12) by $1 - abq^n$ and take the limit $ab \to q^{-n}$. The result is a terminating ${}_6\phi_5$ series on the left side and its sum on the right, giving the summation formula (2.4.2).

If $\max(|c|,|d|,|q|) < 1$ and there are nonnegative integers m and r such that

$$|aq^{m+1}| < 1 < |aq^m|, \ |bq^{r+1}| < 1 < |bq^r|, \tag{6.6.13}$$

then the above technique can be extended to evaluate $I(a,b,c,d)$ provided the products ab, ac, ad, bc, bd are not of the form q^{-n} and $a/b \neq q^{\pm n}$ for any nonnegative integer n. Splitting $h(x;b)$ in the same way as in (6.6.4) and using (6.3.4) we get

$$\begin{aligned} &I(a,b,c,d) \\ &= b^{-2r-2}q^{-r-r^2}\, J(a, bq^{r+1}, q^{-r}/b, c, d, q/b) \\ &= b^{-2r-2}q^{-r-r^2} \frac{(q/bc, q/bd; q)_\infty}{d(1-q)(q, c/d, dq/c, cd; q)_\infty} \\ &\cdot \int_c^d d_q u \frac{(qu/c, qu/d, qu/b; q)_\infty}{(qu/bcd; q)_\infty} \int_{-1}^{1} w(x; a, bq^{r+1}, q^{-r}/b, u)\, dx. \end{aligned} \tag{6.6.14}$$

However, by (6.6.12),

$$\int_{-1}^{1} w(x;q,bq^{r+1},q^{-r}/b,u)\,dx$$
$$= \frac{2\pi(aqu;q)_\infty}{(q,q,abq^{r+1},aq^{-r}/b;q)_\infty(au,buq^{r+1},uq^{-r}/b;q)_\infty}$$
$$- \frac{2\pi(a^{-2};q)_\infty}{(q,abq^{r+1},q^{-r}/ab,bq^{r+1}/a;q)_\infty(au,u/a;q)_\infty}$$
$$\cdot \sum_{k=0}^{m} \frac{(a^2;q)_k(1-a^2q^{2k})(au;q)_k}{(q;q)_k(1-a^2)(aq/u;q)_k}(au)^{-k}. \tag{6.6.15}$$

Since by (2.10.18),

$$\int_c^d d_qu \frac{(qu/c,qu/d,qu/b;q)_\infty}{(qu/bcd,au,u/a;q)_\infty}\frac{(au;q)_k}{(aq/u;q)_k}(qu)^{-k}$$
$$= (-1)^k a^{-2k} q^{-k(k+1)/2}\int_c^d d_qu \frac{(qu/c,qu/d,qu/b;q)_\infty}{(qu/bcd,auq^k,uq^{-k}/a;q)_\infty}$$
$$= \frac{d(1-q)(q,c/d,dq/c,cd,q/ab,aq/b;q)_\infty}{(q/bc,q/bd,ac,ad,c/a,d/a;q)_\infty}$$
$$\cdot \frac{(ab,ac,ad;q)_k}{(aq/b,aq/c,aq/d;q)_k}\left(\frac{q}{abcd}\right)^k, \tag{6.6.16}$$

we find that

$$I(a,b,c,d) = L_m(a;b,c,d)$$
$$+ \frac{2\pi(adq,cdq,dq/b,q^{r+2},q^{1-r}/b^2;q)_\infty b^{-2r-2}q^{-r-r^2}}{(q,q,ad,cd,dq^2/b,abq^{r+1},bcq^{r+1},bdq^{r+1},aq^{-r}/b,cq^{-r}/b,dq^{-r}/b;q)_\infty}$$
$$\cdot {}_8W_7(dq/b;bdq^{r+1},q,dq^{-r}/b,q/bc,q/ab;q,ac), \tag{6.6.17}$$

where $L_m(a;b,c,d)$ is as defined in (6.6.11). The reduction of this ${}_8W_7$ will be done in two stages. First we use (2.10.1) twice to reduce it to a balanced ${}_8W_7$ and then apply (2.11.7) to obtain

$${}_8W_7(dq/b;bdq^{r+1},q,dq^{-r}/b,q/bc,q/ab;q,ac)$$
$$= \frac{(dq^2/b,q/bd,abcdq^{r+1},acdq^{-r}/b;q)_\infty}{(q^{1-r}/b^2,q^{r+2},ac,qacd^2;q)_\infty}$$
$$\cdot {}_8W_7(acd^2;abcd,bdq^{r+1},dq^{-r}/b,ad,cd;q,q/bd)$$
$$= \frac{(dq^2/b,aq/b,cq/b,q,abcdq^{r+1},acdq^{-r}/b;q)_\infty}{(ac,adq,cdq,acdq/b,q^{r+2},q^{1-r}/b^2;q)_\infty}$$
$$\cdot {}_8W_7(acdb^{-1};cd,ad,ac,q^{r+1},q^{-r}/b^2;q,q)$$
$$= \frac{(dq^2/b,aq/b,b/a,q,q/ab,abcd,bcq^{r+1},bdq^{r+1},cq^{-r}/b,dq^{-r}/b;q)_\infty}{(ac,adq,cdq,dq/b,bd,bc,q^{r+2},bq^{r+1}/a,q^{-r}/ab,q^{1-r}/b^2;q)_\infty}$$
$$+ \frac{b}{a}\frac{(dq^2/b,q,bq/a,cq/a,dq/a,ad,q^{r+1},b^2cdq^{r+1},cdq^{-r},q^{-r}/b^2;q)_\infty}{(bcdq/a,dq/b,bc,bd,adq,cdq,q^{r+2},bq^{r+1}/a,q^{-r}/ab,q^{1-r}/b^2;q)_\infty}$$
$$\cdot {}_8W_7(bcda^{-1};bc,bd,cd,bq^{r+1}/a,q^{-r}/ab;q,q). \tag{6.6.18}$$

Substituting (6.6.18) into (6.6.17) and simplifying the coefficients we get

$$I(a,b,c,d) - L_m(a;b,c,d) - \frac{2\pi(abcd;q)_\infty}{(q,ab,ac,ad,bc,bd,cd;q)_\infty}$$
$$= \frac{2\pi(bq/a,cq/a,dq/a,q^{r+1},b^2cdq^{r+1},cdq^{-r},q^{-r}/b^2;q)_\infty}{(q,bc,bd,cd,abq^{r+1},bcq^{r+1},bq^{r+1}/a,aq^{-r}/b,cq^{-r}/b,dq^{-r};q)_\infty}$$
$$\cdot \frac{b^{-2r-1}a^{-1}q^{-r-r^2}}{(q^{-r}/ab,bcdq/a;q)_\infty}\ {}_8W_7(bcda^{-1};bc,bd,cd,bq^{r+1}/a,q^{1-r}/ab;q,q). \tag{6.6.19}$$

The expression on the right side of (6.6.19) is the same as that in (6.6.8) with a,b,c,d,m replaced by b,d,c,a and r, respectively, and so has the value

$$-\frac{2\pi(b^{-2};q)_\infty}{(q,ba,bc,bd,a/b,c/b,d/b;q)_\infty}$$
$$\cdot \sum_{k=0}^{r} \frac{(b^2;q)_k(1-b^2q^{2k})(ba,bc,bd;q)_k}{(q;q)_k(1-b^2)(bq/a,bq/c,bq/d;q)_k}\left(\frac{q}{abcd}\right)^k$$
$$= L_r(b;c,d,a), \tag{6.6.20}$$

by (6.6.11). So we find that

$$I(a,b,c,d) - L_m(a;b,c,d) - L_r(b;c,d,a)$$
$$= \frac{2\pi(abcd;q)_\infty}{(q,ab,ac,ad,bc,bd,cd;q)_\infty}, \tag{6.6.21}$$

where the parameters satisfy the conditions stated earlier.

It is now clear that we can handle the cases of three or all four of the parameters a,b,c,d exceeding 1 in absolute value in exactly the same way. For example, in the extreme case when $\min(|a|,|b|,|c|,|d|) > 1 > |q|$ with

$$|aq^{m+1}| < 1 < |aq^m|, \quad |bq^{r+1}| < 1 < |bq^r|,$$
$$|cq^{s+1}| < 1 < |cq^s|, \quad |dq^{t+1}| < 1 < |dq^t|, \tag{6.6.22}$$

for some nonnegative integers m,r,s,t such that the pairwise products of a,b,c,d are not of the form q^{-n} and the pairwise ratios of a,b,c,d are not of the form $q^{\pm n}$ for $n = 0,1,2,\ldots$, we have the formula

$$I(a,b,c,d) - L_m(a;b,c,d) - L_r(b;c,d,a) - L_s(c;d,a,b)$$
$$- L_t(d;a,b,c) = \frac{2\pi(abcd;q)_\infty}{(q,ab,ac,ad,bc,bd,cd;q)_\infty}, \tag{6.6.23}$$

where $L_s(c;d,a,b)$ and $L_t(d;a,b,c)$ are the same type of finite series as those in (6.6.11) and (6.6.20), and can be written down by obvious replacement of the parameters.

Exercises 6

6.1 Prove that

$$\int_0^\pi \frac{\sin^2\theta\, d\theta}{\prod_{j=1}^{4}(1-2a_j\cos\theta+a_j^2)} = \frac{\pi(1-a_1a_2a_3a_4)}{2\prod_{1\le j<k\le 4}(1-a_ja_k)}$$

when $\max(|a_1|,|a_2|,|a_3|,|a_4|) < 1$.

6.2 Use the q-binomial theorem and an appropriate transformation formula for the ${}_2\phi_1$ series to show that

$$\int_{-1}^{1} \frac{h(x;1,-1)}{h(x;a,-a)} \frac{dx}{\sqrt{1-x^2}} = \frac{2\pi(qa^2;q)_\infty}{(q,-q,aq^{\frac{1}{2}},-aq^{\frac{1}{2}},aq^{\frac{1}{2}},-aq^{\frac{1}{2}},-a^2;q)_\infty}, \quad |a|<1.$$

6.3 Prove that

$$\int_0^\pi \frac{(e^{2i\theta},e^{-2i\theta};q^2)_\infty}{(qe^{2i\theta},qe^{-2i\theta};q^2)_\infty} \frac{d\theta}{1-2z\cos\theta+z^2} = \frac{2\pi(q,qz^2,q^3z^2;q^2)_\infty}{(q^2,q^2,q^2z^2;q^2)_\infty} \sum_{n=0}^\infty \frac{1-q}{1-q^{2n+1}}(qz^2)^n,$$

where $0<q<1$ and $|zq^{\frac{1}{2}}|<1$.

6.4 If $0<q<1$ and $\max(|a|,|b|,|c|,|d|)<1$, show that

$$\int_{-1}^{1} \frac{h(x;1,-1,q^{\frac{1}{2}},f)}{h(x;a,b,c,d)} \frac{dx}{\sqrt{1-x^2}} = \frac{2\pi(af,bf,cf,-abcq^{\frac{1}{2}},abcd;q)_\infty}{(q,ab,ac,ad,bc,bd,cd,-aq^{\frac{1}{2}},-bq^{\frac{1}{2}},-cq^{\frac{1}{2}},abcf;q)_\infty} \cdot {}_8W_7(abcfq^{-1};ab,ac,bc,-fq^{-\frac{1}{2}},f/d;q,-dq^{\frac{1}{2}}).$$

6.5 If $\max(|a|,|b|,|c|,|d|,|q|)<1$, show that

$$\int_{-1}^{1} \frac{h(x;1,-1,q^{\frac{1}{2}},-q^{\frac{1}{2}},-q/c)}{h(x;a,-a,b,-b,c)} \frac{dx}{\sqrt{1-x^2}} = \frac{2\pi(-q;q)_\infty(-a^2b^2c^2,-q^2/c^2;q^2)_\infty}{(q;q)_\infty(-a^2,-b^2,a^2b^2,a^2c^2,b^2c^2;q^2)_\infty}.$$

6.6 If $b=a^{-1}, |c|<1, |d|<1$ and $|aq^{m+1}|<1<|aq^m|$ for some nonnegative integer m such that ac, ad are not of the form $q^{-n}, n=0,1,2,\ldots$, show that

$$I(a,a^{-1},c,d) + \frac{2\pi(qa^2)^{-1}}{(q,q,acq,adq,c/aq,d/aq;q)_\infty}$$
$$\cdot \sum_{k=0}^{m-1} \frac{(1-a^2q^{2k+2})(acq,adq;q)_k(q/cd)^k}{(1-q^{k+1})(1-a^2q^{k+1})(aq^2/c,aq^2/d;q)_k}$$
$$= \frac{2\pi}{(q,q,ac,ad,c/a,d/a;q)_\infty}$$
$$\cdot \sum_{r=0}^{\infty} \left\{ \frac{1}{a^2-q^r} + \frac{1}{(cd)^{-1}-q^r} - \frac{1}{ac^{-1}-q^r} - \frac{1}{ad^{-1}-q^r} \right\} q^r,$$

where $I(a,b,c,d)$ is as defined in (6.2.1). If $m=0$, the series on the left side is to be interpreted as zero.

6.7 Applying (2.12.9) twice deduce from(6.4.8) that

$$\int_0^\pi \frac{h(\cos\theta;1,-1,q^{\frac{1}{2}},-q^{\frac{1}{2}},\lambda,\mu)}{h(\cos\theta;a_1,a_2,a_3,a_4,a_5,a_6)}\,d\theta$$
$$= A\frac{(\mu^2;q)_\infty}{(\mu/\lambda;q)_\infty} \prod_{j=1}^{6} \frac{(\lambda a_j;q)_\infty}{(\lambda/a_j;q)_\infty}$$
$$\cdot {}_{10}W_9(\lambda^2q^{-1};\lambda\mu/q,\lambda/a_1,\lambda/a_2,\lambda/a_3,\lambda/a_4,\lambda/a_5,\lambda/a_6;q,q)$$
$$+ A\frac{(\lambda^2;q)_\infty}{(\lambda/\mu;q)_\infty} \prod_{j=1}^{6} \frac{(\mu a_j;q)_\infty}{(\mu/a_j;q)_\infty}$$
$$\cdot {}_{10}W_9(\mu^2q^{-1};\lambda\mu/q,\mu/a_1,\mu/a_2,\mu/a_3,\mu/a_4,\mu/a_5,\mu/a_6;q,q),$$

where $\lambda\mu = \prod_{j=1}^{6} a_j$, $\max(|q|,|a_1|,\ldots,|a_6|) < 1$, and

$$A = \frac{2\pi \prod_{j=1}^{6} (\lambda/a_j,\mu/a_j;q)_\infty}{(q,\lambda^2,\mu^2;q)_\infty \prod_{1\le j<k\le 6} (a_ja_k;q)_\infty}.$$

(Rahman [1986b, 1988b])

6.8 Prove that

$${}_8W_7\left(abczq^{-1};az,bz,cz,abcdq^{n-1},q^{-n};q,\frac{q}{dz}\right)$$
$$= \frac{(q,az,a/z,bz,b/z,cz,abcz,abc/\lambda z,\lambda\mu^2q/abdz,\mu^2,ab/\mu^2;q)_\infty}{2\pi(ab,ac,bc,q/dz,zq^{1-n}/d,abczq^n,q\mu^2/dz,aq^{1-n}/d\mu^2;q)_\infty}$$

$$\cdot\left(\frac{\lambda q}{bd}, \frac{a^2bc}{\lambda\mu^2}, bcq^n; q\right)_\infty \int_{-1}^{1} \frac{h(x;1,-1,q^{\frac{1}{2}},-q^{\frac{1}{2}},\frac{\mu q}{d},\frac{q^{1-n}}{d\mu})}{h(x;\frac{a}{\mu},\frac{b}{\mu},\mu z,\frac{\mu}{z},\frac{abc}{\lambda\mu},\frac{\lambda\mu q}{abd})}$$

$$\cdot {}_8W_7\left(\frac{aq^{-n}}{d\mu^2}; \frac{ae^{i\theta}}{\mu}, \frac{ae^{-i\theta}}{\mu}, \frac{q^{1-n}}{bd}, \frac{abq^{-n}}{\lambda\mu^2}, \frac{\lambda q^{1-n}}{abcd}; q, bcq^n\right)$$

$$\cdot {}_8W_7\left(\frac{\mu^2}{dz}; \frac{\mu e^{i\theta}}{z}, \frac{\mu e^{-i\theta}}{z}, \frac{q}{dz}, \frac{ab}{\lambda}, \frac{\lambda\mu^2 q}{abcd}; q, cz\right)\frac{dx}{\sqrt{1-x^2}},$$

where $n = 0,1,2,\ldots$, $|z| = 1$, and $\max(|a|,|b|,|c|,|d|, |\mu|,|a/\mu|,|b/\mu|, |abc/\lambda\mu|,|\lambda\mu q/abd|,|q|) < 1$.

6.9 Choosing $\lambda = abcdq^{-1}$, deduce from Ex. 6.8 that

$$ {}_4\phi_3\left[\begin{matrix} q^{-n}, & abcdq^{n-1}, & a\mu e^{i\psi}, & a\mu e^{-i\psi} \\ & ab\mu^2, & ac, & ad \end{matrix}; q, q\right]$$

$$= \frac{(q,ab,\mu^2;q)_\infty}{2\pi(ab\mu^2;q)_\infty} h(z;a\mu,b\mu) \int_{-1}^{1} \frac{h(x;1,-1,q^{\frac{1}{2}},-q^{\frac{1}{2}})}{h(x;a,b,\mu e^{i\psi},\mu e^{-i\psi})}$$

$$\cdot {}_4\phi_3\left[\begin{matrix} q^{-n}, & abcdq^{n-1}, & ae^{i\theta}, & ae^{-i\theta} \\ & ab, & ac, & ad \end{matrix}; q, q\right]\frac{dx}{\sqrt{1-x^2}},$$

where $z = \cos\psi$ and $\max(|a|,|b|,|\mu|,|q|) < 1$.

6.10 Show that

$$\int_{-\infty}^{\infty} \frac{(iae^u, -iae^{-u}, ibe^u, -ibe^{-u}, ice^u, -ice^{-u}; q)_\infty}{(iq^{\frac{1}{2}}e^u, -iq^{\frac{1}{2}}e^{-u}, iqe^u, -iqe^{-u}, -iq^{\frac{1}{2}}e^u, iq^{\frac{1}{2}}e^{-u}; q)_\infty}$$

$$\cdot \frac{(ide^u, -ide^{-u}; q)_\infty}{(-iqe^u, iqe^{-u}; q)_\infty} du$$

$$= \frac{(\log q^{-1})(ab/q, ac/q, ad/q, bc/q, bd/q, cd/q, q; q)_\infty}{(abcd/q^3; q)_\infty}$$

when $|abcd/q^3| < 1$ and $|q| < 1$.
(Askey [1988b])

6.11 Show that

$$\int_0^\infty \int_0^\infty t_1^{x-1} t_2^{x-1} \frac{(-t_1q^{x+y+2k}, -t_2q^{x+y+2k}; q)_\infty}{(-t_1, -t_2; q)_\infty} t_1^{2k}(t_2q^{1-k}/t_1; q)_{2k} dt_1 dt_2$$

$$= \frac{\Gamma_q(x)\Gamma_q(x+k)\Gamma_q(y)\Gamma_q(y+k)\Gamma_q(k+1)\Gamma_q(2k+1)}{\Gamma_q(x+y+k)\Gamma_q(x+y+2k)\Gamma_q(k+1)\Gamma_q(k+1)}$$

$$\cdot \frac{[\Gamma(x)\Gamma(1-x)]^2}{\Gamma_q(x)\Gamma_q(1-x)\Gamma_q(x+2k)\Gamma_q(1-x-2k)}$$

when $\mathrm{Re}\, x > 0$, $\mathrm{Re}\, y > 0$, $|q| < 1$, and $k = 0,1,\ldots$.
(Askey [1980b])

6.12 Show that

$$\int_0^\infty \int_0^\infty t_1^{x-1} t_2^{x-1} \frac{(-at_1 q^{x+y+2k}, -at_2 q^{x+y+2k}; q)_\infty}{(-at_1, -at_2; q)_\infty}$$
$$\cdot\, t_1^{2k} (t_2 q^{1-k}/t_1; q)_{2k}\, d_q t_1 d_q t_2$$
$$= \frac{\Gamma_q(x)\Gamma_q(x+k)\Gamma_q(y)\Gamma_q(y+k)\Gamma_q(k+1)\Gamma_q(2k+1)}{\Gamma_q(x+y+k)\Gamma_q(x+y+2k)\Gamma_q(k+1)\Gamma_q(k+1)}$$
$$\cdot \frac{(-aq^x, -aq^{x+2k}, -q^{1-x}/a, -q^{1-x-2k}/a; q)_\infty}{(-a, -a, -q/a, -q/a; q)_\infty}$$

when $\mathrm{Re}\, x > 0$, $\mathrm{Re}\, y > 0$, $|q| < 1$, and $k = 0, 1, \ldots$.
(Askey [1980b])

6.13 With $w(x; a, b, c, d)$ defined as in (6.3.1), prove that

$$\int_{-1}^1 \int_{-1}^1 w(x; a, aq^{\frac{1}{2}}, b, bq^{\frac{1}{2}})\, w(y; a, aq^{\frac{1}{2}}, b, bq^{\frac{1}{2}})$$
$$\cdot \left|(q^{\frac{1}{2}} e^{i(\theta+\phi)}, q^{\frac{1}{2}} e^{i(\theta-\phi)}; q)_k\right|^2 dx dy$$
$$= \prod_{j=1}^2 \frac{2\pi}{[(1-ab)(q, abq^{(j-1)k+1/2}, abq^{(j-1)k+1}; q)_\infty]^2}$$
$$\cdot \prod_{j=1}^2 \frac{(a^2 b^2 q^{jk+1}, q^{k+1}, q; q)_\infty}{(a^2 q^{(j-1)k+1/2}, b^2 q^{(j-1)k+1/2}, q^{jk+1}; q)_\infty},$$

where $x = \cos\theta$, $y = \cos\phi$, $|q| < 1$, and $k = 0, 1, \ldots$.
(Rahman [1986a])

6.14 Let $C.T.\ f(x)$ denote the constant term in the Laurent expansion of a function $f(x)$. Prove that if j and k are nonnegative integers, then

$$C.T.\ (qx, x^{-1}; q)_j (qx^2, qx^{-2}; q^2)_k$$
$$= \frac{1}{2\pi} \int_0^\pi (qe^{2i\theta}, e^{-2i\theta}; q)_j (qe^{4i\theta}, qe^{-4i\theta}; q^2)_k\, d\theta$$
$$= \frac{(q;q)_{2j} (q^2;q^2)_{2k} (q^{2j+1}; q^2)_k}{(q;q)_j (q^2;q^2)_k (q;q)_{j+2k}}.$$

(Askey [1982b])

6.15 Verify Ramanujan's identities

(i)
$$\int_{-\infty}^\infty e^{-x^2+2mx} (-aqe^{2kx}, -bqe^{-2kx}; q)_\infty\, dx$$
$$= \frac{\sqrt{\pi}(abq; q)_\infty e^{m^2}}{(ae^{2mk}\sqrt{q}, be^{-2mk}\sqrt{q}; q)_\infty},$$

(ii)
$$\int_{-\infty}^\infty \frac{e^{-x^2+2mx}\, dx}{(ae^{2ikx}\sqrt{q}, be^{-2ikx}\sqrt{q}; q)_\infty}$$
$$= \frac{\sqrt{\pi}\, e^{m^2} (-aqe^{2imk}, -bqe^{-2imk}; q)_\infty}{(abq; q)_\infty},$$

where $q = e^{-2k^2}$ and $|q| < 1$.
(See Ramanujan [1988] and Askey [1982a])

6.16 Derive the q-beta integral formulas

(i)
$$\int_0^\infty \frac{(-tq^b, -q^{a+1}/t; q)_\infty}{(-t, -q/t; q)_\infty} \frac{d_q t}{t} = \frac{\Gamma_q(a)\Gamma_q(b)}{\Gamma_q(a+b)}$$

and

(ii)
$$\int_0^\infty \frac{(-tq^b, -q^{a+1}/t; q)_\infty}{(-t, -q/t; q)_\infty} \frac{dt}{t} = \frac{-\log q}{1-q} \frac{\Gamma_q(a)\Gamma_q(b)}{\Gamma_q(a+b)},$$

where $0 < q < 1$, Re $a > 0$ and Re $b > 0$.
(Askey and Roy [1986], Gasper [1987])

6.17 Extend the above q-beta integral formulas to

(i)
$$\int_0^\infty t^{c-1} \frac{(-tq^b, -q^{a+1}/t; q)_\infty}{(-t, -q/t; q)_\infty} d_q t$$
$$= \frac{(-q^c, -q^{1-c}; q)_\infty}{(-1, -q; q)_\infty} \frac{\Gamma_q(a+c)\Gamma_q(b-c)}{\Gamma_q(a+b)}$$

and

(ii)
$$\int_0^\infty t^{c-1} \frac{(-tq^b, -q^{a+1}/t; q)_\infty}{(-t, -q/t; q)_\infty} dt$$
$$= \frac{\Gamma(c)\Gamma(1-c)\Gamma_q(a+c)\Gamma_q(b-c)}{\Gamma_q(c)\Gamma_q(1-c)\Gamma_q(a+b)},$$

where $0 < q < 1$, Re $(a+c) > 0$ and Re $(b-c) > 0$.
(Ramanujan [1915], Askey and Roy [1986], Gasper [1987])

Notes 6

Ex. 6.7 Setting $a_1 = -a_2 = a$, $a_3 = -a_4 = aq^{\frac{1}{2}}$, $a_5 = b$, $a_6 = c, \lambda = a^2bq$, $\mu = a^2c$ and using Ex. 2.25, Rahman [1988b] evaluated this integral in closed form and found the corresponding systems of biorthogonal rational functions.

Ex. 6.8 This may be regarded as a q-extension of the terminating case of Erdélyi's [1953, 2.4(3)] formula

$$F(a, b; c; x) = \frac{\Gamma(c)}{\Gamma(\mu)\Gamma(c-\mu)} \int_0^1 t^{\mu-1}(1-t)^{c-\mu-1}(1-tx)^{\lambda-a-b}$$
$$\cdot F(\lambda-a, \lambda-b; \mu; tx)\; F(a+b-\lambda, \lambda-\mu; c-\mu; (1-t)x/(1-tx))\, dt,$$

where $|x| < 1$, $0 <$ Re $\mu <$ Re c.

Ex. 6.10 Askey [1988b] found the ${}_4\phi_3$ polynomials which are orthogonal with respect to the weight function associated with this integral.

Exercises 6.11–6.14 The double integrals in these exercises are q-analogues of the $n = 2$ case of Selberg's [1944] important multivariable extension of the beta integral:

$$\int_0^1 \cdots \int_0^1 \prod_{i=1}^{n} t_i^{x-1}(1-t_i)^{y-1} \left| \prod_{1\le i<j\le n} (t_i - t_j) \right|^{2z} dt_1 \cdots dt_n$$
$$= \prod_{j=1}^{n} \frac{\Gamma(x+(j-1)z)\Gamma(y+(j-1)z)\Gamma(jz+1)}{\Gamma(x+y+(n+j-2)z)\Gamma(z+1)},$$

where Re $x > 0$, Re $y > 0$, Re $z > -\min(1/n,$ Re $x/(n-1)$, Re $y/(n-1))$.

Aomoto [1987] considered a generalization of Selberg's integral and utilized the extra freedom that he had in his integral to give a short elegant proof of it. Habsieger [1989] and Kadell [1988b] proved a q-analogue of Selberg's integral that was conjectured in Askey [1980b]. For conjectured multivariable extensions of the integrals in Exercises 6.11–6.13, other conjectured q-analogues of Selberg's integral, and related constant term identities that come from root systems associated with Lie algebras, see Andrews [1986, 1988], Askey [1980b, 1982b, 1985, 1989f,g], Evans, Ismail and Stanton [1982], Garvan [1989], Garvan and Gonnet [1989], Habsieger [1989], Kadell [1988a–1989c], Macdonald [1972–1989], Milne [1985a, 1989c], Morris [1982], Rahman [1986a], Stanton [1986b, 1989], and Zeilberger [1987, 1988, 1989a].

7

APPLICATIONS TO ORTHOGONAL POLYNOMIALS

7.1 Orthogonality

Let $\alpha(x)$ be a non-constant, non-decreasing, real-valued bounded function defined on $(-\infty, \infty)$ such that its *moments*

$$\mu_n = \int_{-\infty}^{\infty} x^n \, d\alpha(x), \quad n = 0, 1, 2, \ldots, \tag{7.1.1}$$

are finite. A finite or infinite sequence $p_0(x), p_1(x), \ldots$ of polynomials, where $p_n(x)$ is of degree n in x, is said to be *orthogonal with respect to the measure* $d\alpha(x)$ and called an *orthogonal system of polynomials* if

$$\int_{-\infty}^{\infty} p_m(x) p_n(x) \, d\alpha(x) = 0, \quad m \neq n. \tag{7.1.2}$$

In view of the definition of $\alpha(x)$ the integrals in (7.1.1) and (7.1.2) exist in the Lebesgue-Stieltjes sense. If $\alpha(x)$ is absolutely continuous and $d\alpha(x) = w(x)dx$, then the orthogonality relation reduces to

$$\int_{-\infty}^{\infty} p_m(x) p_n(x) w(x) \, dx = 0, \quad m \neq n, \tag{7.1.3}$$

and the sequence $\{p_n(x)\}$ is said to be *orthogonal with respect to the weight function* $w(x)$.

If $\alpha(x)$ is a step function (usually taken to be right-continuous) with jumps w_j at $x = x_j, j = 0, 1, 2, \ldots$, then (7.1.2) reduces to

$$\sum_j p_m(x_j) p_n(x_j) w_j = 0, \quad m \neq n. \tag{7.1.4}$$

In this case the polynomials are said to be *orthogonal with respect to a jump function* and are usually referred to as *orthogonal polynomials of a discrete variable.*

Every orthogonal system of real valued polynomials $\{p_n(x)\}$ satisfies a *three-term recurrence relation* of the form

$$x p_n(x) = A_n p_{n+1}(x) + B_n p_n(x) + C_n p_{n-1}(x) \tag{7.1.5}$$

with $p_{-1}(x) \equiv 0, p_0(x) \equiv 1$, where A_n, B_n, C_n are real and $A_n C_{n+1} > 0$. Conversely, if (7.1.5) holds for a sequence of polynomials $\{p_n(x)\}$ such that $p_{-1}(x) \equiv 0, p_0(x) \equiv 1$ and A_n, B_n, C_n are real with $A_n C_{n+1} > 0$, then there exists a positive measure $d\alpha(x)$ such that

$$\int_{-\infty}^{\infty} p_m(x) p_n(x) d\alpha(x) = \begin{cases} 0, & m \neq n, \\ v_n^{-1} \int_{-\infty}^{\infty} d\alpha(x), & m = n, \end{cases} \tag{7.1.6}$$

where

$$v_n = \prod_{k=1}^{n} \frac{A_{k-1}}{C_k}, \quad v_0 = 1. \tag{7.1.7}$$

If $\{p_n(x)\} = \{p_n(x)\}_{n=0}^{\infty}$ and $A_n C_{n+1} > 0$ for $n = 0, 1, 2, \ldots$, then the measure has infinitely many points of support, (7.1.5) holds for $n = 0, 1, 2, \ldots$, and (7.1.6) holds for $m, n = 0, 1, 2, \ldots$. If $\{p_n(x)\} = \{p_n(x)\}_{n=0}^{N}$ and $A_n C_{n+1} > 0$ for $n = 0, 1, 2, \ldots, N-1$, where N is a fixed positive integer, then the measure can be taken to have support on $N+1$ points $x_0, x_1, \ldots, x_N$, (7.1.5) holds for $n = 0, 1, \ldots, N-1$, and (7.1.6) holds for $m, n = 0, 1, 2, \ldots, N$.

This characterization theorem of orthogonal polynomials is usually attributed to Favard [1935], but it appeared earlier in published works of Perron [1929], Wintner [1929] and Stone [1932]. For a detailed discussion of this theorem see, for example, Atkinson [1964], Chihara [1978], Freud [1971] and Szegő [1975].

In the finite discrete case the recurrence relation (7.1.5) is a discrete analogue of a Sturm-Liouville two-point boundary-value problem with boundary conditions $p_{-1}(x) = 0, p_{N+1}(x) = 0$. If $x_0, x_1, \ldots, x_N$ are the zeros of $p_{N+1}(x)$, which can be easily proved to be real and distinct (see e.g., Atkinson [1964] for a complete proof), then the orthogonality relation (7.1.6) can be written in the form

$$\sum_{j=0}^{N} p_m(x_j) p_n(x_j) w_j = v_n^{-1} \sum_{j=0}^{N} w_j \; \delta_{m,n}, \tag{7.1.8}$$

$m, n = 0, 1, \ldots, N$, where w_j is the positive jump at x_j and v_n is as defined in (7.1.7). The *dual orthogonality* relation

$$\sum_{n=0}^{N} p_n(x_j) p_n(x_k) v_n = w_j^{-1} \sum_{n=0}^{N} w_n \; \delta_{j,k}, \tag{7.1.9}$$

$j, k = 0, 1, \ldots, N$, follows from the fact that a matrix that is orthogonal by rows is also orthogonal by columns. It can be shown that

$$w_j = \left[A_N v_N p_N(x_j) p'_{N+1}(x_j)\right]^{-1} \sum_{n=0}^{N} w_n, \quad j = 0, 1, \ldots, N, \tag{7.1.10}$$

where the prime indicates the first derivative.

In general, the measure in (7.1.6) is not unique and, given a recurrence relation, it may not be possible to find an explicit formula for $\alpha(x)$. Even though the classical orthogonal polynomials, which include the Jacobi polynomials

$$P_n^{(\alpha,\beta)}(x) = \frac{(\alpha+1)_n}{n!} \, {}_2F_1\left(-n, n+\alpha+\beta+1; \alpha+1; \frac{1-x}{2}\right), \tag{7.1.11}$$

and the ultraspherical polynomials

$$\begin{aligned} C_n^{\lambda}(x) &= \frac{(2\lambda)_n}{\left(\lambda+\frac{1}{2}\right)_n} P_n^{(\lambda-\frac{1}{2},\lambda-\frac{1}{2})}(x) \\ &= \sum_{k=0}^{n} \frac{(\lambda)_k (\lambda)_{n-k}}{k!(n-k)!} e^{i(n-2k)\theta}, \quad x = \cos\theta, \end{aligned} \tag{7.1.12}$$

are orthogonal with respect to unique measures (see Szegő [1975]), it is not easy to discover these measures from the corresponding recurrence relations (see e.g., Askey and Ismail [1984]). However, for a wide class of discrete orthogonal polynomials it is possible to use the recurrence relation (7.1.5) and the formulas (7.1.8)–(7.1.10) to compute the jumps w_j and hence the measure. We shall illustrate this in the next section by considering the q-Racah polynomials (Askey and Wilson [1979]).

7.2 The finite discrete case: the q-Racah polynomials and some special cases

Suppose $\{p_n(x)\}$ is a finite discrete orthogonal polynomial sequence which satisfies a three-term recurrence relation of the form (7.1.5) and the orthogonality relations (7.1.8) and (7.1.9) with the weights w_j and the normalization constants v_n given by (7.1.10) and (7.1.7), respectively. We shall now assume, without any loss of generality, that $p_n(x_0) = 1$ for $n = 0, 1, \ldots, N$. This enables us to rewrite (7.1.5) in the form

$$(x - x_0)p_n(x) = A_n \left[p_{n+1}(x) - p_n(x)\right] - C_n \left[p_n(x) - p_{n-1}(x)\right], \tag{7.2.1}$$

where $n = 0, 1, \ldots, N$. Setting $j = k = 0$ in (7.1.9) we find that

$$\sum_{n=0}^{N} v_n = w_o^{-1} \sum_{n=0}^{N} w_n. \tag{7.2.2}$$

It is clear that in order to obtain solutions of (7.2.1) which are representable in terms of basic series it would be helpful if v_n and $\sum\limits_{n=0}^{N} v_n$ were equal to quotients of products of q-shifted factorials. Therefore, with the ${}_6\phi_5$ sum (2.4.2) in mind, let us take

$$\begin{aligned} v_n &= \frac{(abq;q)_n \left(1 - abq^{2n+1}\right)(aq, cq, bdq; q)_n}{(q;q)_n (1 - abq)(bq, abq/c, aq/d; q)_n} (cdq)^{-n} \\ &= \prod_{k=1}^{n} \frac{\left(1 - abq^k\right)\left(1 - abq^{2k+1}\right)(1 - aq^k)(1 - cq^k)(1 - bdq^k)}{(1 - q^k)\left(1 - abq^{2k-1}\right)(1 - bq^k)(1 - abq^k/c)(1 - aq^k/d)cdq}, \end{aligned} \tag{7.2.3}$$

where $bdq = q^{-N}, 0 < q < 1$, so that

$$\begin{aligned} \sum_{n=0}^{N} v_n &= {}_6\phi_5 \left[\begin{matrix} abq, q(abq)^{\frac{1}{2}}, -q(abq)^{\frac{1}{2}}, aq, cq, q^{-N} \\ (abq)^{\frac{1}{2}}, -(abq)^{\frac{1}{2}}, bq, abq/c, abq^{N+2} \end{matrix} ; q, \frac{bq^N}{c} \right] \\ &= \frac{(abq^2, b/c; q)_N}{(bq, abq/c; q)_N}, \end{aligned} \tag{7.2.4}$$

where it is assumed that a, b, c, d are such that $v_n > 0$ for $n = 0, 1, \ldots, N$. In view of (7.1.7) we can take

$$A_{k-1} = \frac{(1 - abq^k)(1 - aq^k)(1 - cq^k)(1 - bdq^k)}{(1 - abq^{2k-1})} r_k, \tag{7.2.5}$$

$$C_k = \frac{cdq(1-q^k)(1-bq^k)\left(1-abq^k/c\right)\left(1-aq^k/d\right)}{(1-abq^{2k+1})} r_k, \tag{7.2.6}$$

where $\{r_k\}_{k=1}^N$ is an arbitrary sequence with $r_k \neq 0, 1 \leq k \leq N$. Since $C_0 = 0$ and $A_0 = (1-aq)(1-cq)(1-bdq)r_1$, we have from the $n=0$ case of (7.2.1) that

$$p_1(x_j) = 1 - \frac{(1-q^{-1})(x_j - x_0)qr_1^{-1}}{(1-q)(1-aq)(1-cq)(1-bdq)}. \tag{7.2.7}$$

This suggests that we should look for a basic series representation of $p_n(x_j)$ whose $(k+1)$-th term has $(q, aq, cq, bdq; q)_k$ as its denominator, which in turn suggests considering a terminating ${}_4\phi_3$ series. In view of the product $(-n)_k(n+\alpha+\beta+1)_k$ in the numerator of the $(k+1)$-th term in the hypergeometric series representation of $P_n^{(\alpha,\beta)}(x)$ in (7.1.11) and the dual orthogonality relations (7.1.8) and (7.1.9) required for $p_n(x_j)$ it is natural to look for a ${}_4\phi_3$ series whose $(k+1)$-th term has the numerator

$$\left(q^{-n}, q^{n+\alpha+\beta+1}, q^{-j}, q^{j+\gamma+\delta+1}; q\right)_k q^k.$$

Replacing $q^\alpha, q^\beta, q^\gamma, q^\delta$ by a, b, c, d, respectively, we are then led to consider a ${}_4\phi_3$ series of the form

$${}_4\phi_3\left[\begin{matrix} q^{-n}, abq^{n+1}, q^{-j}, cdq^{j+1} \\ aq, cq, bdq \end{matrix}; q, q\right]. \tag{7.2.8}$$

Observing that $\left(q^{-j}, cdq^{j+1}; q\right)_k$ is a polynomial of degree k in the variable $q^{-j} + cdq^{j+1}$, we find that if we take

$$x_j = q^{-j} + cdq^{j+1} \tag{7.2.9}$$

then $x_j - x_0 = -(1-q^{-j})(1-cdq^{j+1})$, and so (7.2.7) is satisfied with

$$r_k = \left(1-abq^{2k}\right)^{-1}. \tag{7.2.10}$$

Then $A_kC_{k+1} > 0$ for $0 \leq k \leq N-1$ if, for example, a, b, c are real, $d = b^{-1}q^{-N-1}, bc < 0, \max(|aq|, |bq|, |cq|, |ab/c|) < 1$ and $0 < q < 1$.

We shall now verify that

$$p_n(x_j) = {}_4\phi_3\left[\begin{matrix} q^{-n}, abq^{n+1}, q^{-j}, cdq^{j+1} \\ aq, cq, bdq \end{matrix}; q, q\right] \tag{7.2.11}$$

satisfies (7.2.1) with $x = x_j$. A straightforward calculation gives

$$\begin{aligned} &p_n(x_j) - p_{n-1}(x_j) \\ &= \frac{-q^{1-n}\left(1-abq^{2n}\right)\left(1-q^{-j}\right)\left(1-cdq^{j+1}\right)}{(1-aq)(1-cq)(1-bdq)} \\ &\quad \cdot {}_4\phi_3\left[\begin{matrix} q^{1-n}, abq^{n+1}, q^{1-j}, cdq^{j+2} \\ aq^2, cq^2, bdq^2 \end{matrix}; q, q\right], \end{aligned} \tag{7.2.12}$$

$0 \leq n \leq N$. So we need to verify that

$$
\begin{aligned}
&{}_4\phi_3\left[\begin{matrix} q^{-n}, abq^{n+1}, q^{-j}, cdq^{j+1} \\ aq, cq, bdq \end{matrix}; q, q\right] \\
&= \frac{A_n q^{-n}\left(1-abq^{2n+2}\right)}{(1-aq)(1-cq)(1-bdq)}\, {}_4\phi_3\left[\begin{matrix} q^{-n}, abq^{n+2}, q^{1-j}, cdq^{j+2} \\ aq^2, cq^2, bdq^2 \end{matrix}; q, q\right] \\
&\quad - \frac{C_n q^{1-n}\left(1-abq^{2n}\right)}{(1-aq)(1-cq)(1-bdq)}\, {}_4\phi_3\left[\begin{matrix} q^{1-n}, abq^{n+1}, q^{1-j}, cdq^{j+2} \\ aq^2, cq^2, bdq^2 \end{matrix}; q, q\right],
\end{aligned}
\tag{7.2.13}
$$

where A_n and C_n are given by (7.2.5) and (7.2.6) with r_k as defined in (7.2.10). Use of (2.10.4) on both sides of (7.2.13) reduces the problem to verifying that

$$
\begin{aligned}
&{}_4\phi_3\left[\begin{matrix} q^{-n}, abq^{n+1}, bq^{-j}/c, bdq^{j+1} \\ bq, abq/c, bdq \end{matrix}; q, q\right] \\
&= \frac{\left(1-abq^{n+1}\right)\left(1-bdq^{n+1}\right)}{\left(1-abq^{2n+1}\right)(1-bdq)} \\
&\quad \cdot {}_4\phi_3\left[\begin{matrix} q^{-n}, abq^{n+2}, bq^{-j}/c, bdq^{j+1} \\ bq, abq/c, bdq^2 \end{matrix}; q, q\right] \\
&\quad - \frac{\left(1-q^{n}\right)\left(1-aq^{n}/d\right)}{\left(1-abq^{2n+1}\right)(1-bdq)} bdq \\
&\quad \cdot {}_4\phi_3\left[\begin{matrix} q^{1-n}, abq^{n+1}, bq^{-j}/c, bdq^{j+1} \\ bq, abq/c, bdq^2 \end{matrix}; q, q\right],
\end{aligned}
\tag{7.2.14}
$$

which follows immediately from the fact that

$$
\begin{aligned}
&\frac{\left(q^{-n};q\right)_k \left(abq^{n+1};q\right)_{k+1}\left(1-bdq^{n+1}\right)}{\left(1-abq^{2n+1}\right)\left(1-bdq^{k+1}\right)} \\
&+ \frac{\left(q^{-n};q\right)_{k+1}\left(abq^{n+1};q\right)_k \left(1-\frac{aq^n}{d}\right) bdq^{n+1}}{\left(1-abq^{2n+1}\right)\left(1-bdq^{k+1}\right)} = \left(q^{-n}, abq^{n+1};q\right)_k .
\end{aligned}
$$

The verification that (7.2.11) satisfies the boundary condition $p_{N+1}(x_j) = 0$ for $0 \leq j \leq N$ is left as an exercise (Ex. 7.3).

Note that the ${}_4\phi_3$ series in (7.2.11) remains unchanged if we switch n, a, b, respectively, with j, c, d. This implies that the polynomials $p_n(x_j)$ are self-dual in the sense that they are of degree n in $x_j = q^{-j} + cdq^{j+1}$ and of degree j in $y_n = q^{-n} + abq^{n+1}$ and that the weights w_j are obtained from the v_n in (7.2.3) by replacing n, a, b, c, d, by j, c, d, a, b, respectively, i.e.,

$$
w_j = \frac{(cdq;q)_j\left(1-cdq^{2j+1}\right)(cq, aq, bdq;q)_j}{(q;q)_j(1-cdq)\left(dq, cdq/a, cq/b;q\right)_j}(abq)^{-j},
\tag{7.2.15}
$$

$0 \leq j \leq N$. Thus $\{p_n(x_j)\}_{n=0}^{N}$ is orthogonal with respect to the weights w_j, while $\{p_j(y_n)\}_{j=0}^{N}$ is orthogonal with respect to the weights v_n (see Ex. 7.4). The calculations have so far been done with the assumption that $bdq = q^{-N}$, but the same results will hold if we assume that one of aq, cq and bdq is q^{-N}.

The polynomials $p_n(x_j)$ were discovered by Askey and Wilson [1979] as q-analogues of the *Racah polynomials*

$$p_n(\lambda_j) = {}_4F_3\left[\begin{matrix} -n, n+\alpha+\beta+1, -j, j+\gamma+\delta+1 \\ \alpha+1, \gamma+1, \beta+\delta+1 \end{matrix}; 1\right], \tag{7.2.16}$$

where $\lambda_j = j(j+\gamma+\delta+1)$, which were named after the physicist Racah, who worked out their orthogonality without apparently being aware of the polynomial orthogonality. In the physicist's language the $p_n(\lambda_j)$'s are known as Racah coefficients or 6-j symbols. The connection between the 6-j symbols and the ${}_4F_3$ polynomials (7.2.16) was first made in Wilson's [1978] thesis. The q-analogues in (7.2.11) are called the *q-Racah polynomials.*

We shall now point out some important special cases of the q-Racah polynomials. First, let us set $d = b^{-1}q^{-N-1}$ in (7.2.11) and replace c by bc to rewrite the q-Racah polynomials in the more standard notation

$$W_n(x; a, b, c, N; q) = {}_4\phi_3\left[\begin{matrix} q^{-n}, abq^{n+1}, q^{-x}, cq^{x-N} \\ aq, q^{-N}, bcq \end{matrix}; q, q\right]. \tag{7.2.17}$$

Then, by (7.1.8), (7.2.3) and (7.2.15),

$$\sum_{x=0}^{N} \rho(x;q) W_m(x;q) W_n(x;q) = \frac{\delta_{m,n}}{h_n(q)}, \tag{7.2.18}$$

where

$$\begin{aligned} W_n(x;q) &\equiv W_n(x; a, b, c, N; q), \\ \rho(x;q) &\equiv \rho(x; a, b, c, N; q) \\ &= \frac{(cq^{-N};q)_x \left(1-cq^{2x-N}\right)(aq, bcq, q^{-N};q)_x}{(q;q)_x(1-cq^{-N})(ca^{-1}q^{-N}, b^{-1}q^{-N}, cq;q)_x}(abq)^{-x} \end{aligned} \tag{7.2.19}$$

and

$$\begin{aligned} h_n(q) &\equiv h_n(a, b, c, N; q) \\ &= \frac{(bq, aq/c;q)_N}{(abq^2, 1/c;q)_N} \frac{(abq;q)_n\left(1-abq^{2n+1}\right)(aq, bcq, q^{-N};q)_n}{(q;q)_n(1-abq)(bq, aq/c, abq^{N+2};q)_n}\left(q^N/c\right)^n. \end{aligned} \tag{7.2.20}$$

Setting $c = 0$ in (7.2.17) gives the *q-Hahn polynomials*

$$Q_n(x) \equiv Q_n(x; a, b, N; q) = {}_3\phi_2\left[\begin{matrix} q^{-n}, abq^{n+1}, q^{-x} \\ aq, q^{-N} \end{matrix}; q, q\right], \tag{7.2.21}$$

which were introduced by Hahn [1949a] and, by (7.2.18), satisfy the orthogonality relation

$$\begin{aligned} &\sum_{x=0}^{N} Q_m(x)Q_n(x)\frac{(aq;q)_x(bq;q)_{N-x}}{(q;q)_x(q;q)_{N-x}}(aq)^{-x} \\ &= \frac{(abq^2;q)_N(aq)^{-N}}{(q;q)_N}\frac{(q;q)_n(1-abq)(bq, abq^{N+2};q)_n}{(abq;q)_n(1-abq^{2n+1})(aq, q^{-N};q)_n}(-aq)^n q^{\binom{n}{2}-Nn}\delta_{m,n}, \end{aligned} \tag{7.2.22}$$

$m, n = 0, 1, \ldots, N$.

Setting $a = 0$ in (7.2.17) we obtain the *dual q-Hahn polynomials* (Hahn [1949b])

$$\begin{aligned} R_n(\mu(x)) &\equiv R_n(\mu(x); b, c, N; q) \\ &= {}_3\phi_2 \left[\begin{matrix} q^{-n}, q^{-x}, cq^{x-N} \\ q^{-N}, bcq \end{matrix} ; q, q \right], \end{aligned} \tag{7.2.23}$$

where $\mu(x) = q^{-x} + cq^{x-N}$, which satisfy the orthogonality relation

$$\sum_{x=0}^{N} R_m(\mu(x)) R_n(\mu(x)) \frac{(cq^{-N}; q)_x \left(1 - cq^{2x-N}\right) (bcq, q^{-N}; q)_x}{(q; q)_x (1 - cq^{-N})(b^{-1}q^{-N}, cq; q)_x} q^{Nx - \binom{x}{2}} (-bcq)^{-x}$$

$$= \frac{(1/c; q)_N}{(bq; q)_N} \frac{(q, bq; q)_n}{(bcq, q^{-N}; q)_n} \left(cq^{-N}\right)^n \delta_{m,n}, \tag{7.2.24}$$

$m, n = 0, 1, \ldots, N$. For some applications of q-Hahn, dual q-Hahn polynomials, and their limit cases, see Delsarte [1976a,b, 1978], Delsarte and Goethals [1975], Dunkl [1977–1980] and Stanton [1977–1986c].

7.3 The infinite discrete case: the little and big q-Jacobi polynomials

As a q-analogue of the Jacobi polynomials (7.1.11), Hahn [1949a] (also see Andrews and Askey [1977]) introduced the polynomials

$$p_n(x; a, b; q) = {}_2\phi_1(q^{-n}, abq^{n+1}; aq; q, xq). \tag{7.3.1}$$

It can be easily verified that

$$\lim_{q \to 1} p_n \left(\frac{1-x}{2}; q^{\alpha}, q^{\beta}; q \right) = \frac{P_n^{(\alpha,\beta)}(x)}{P_n^{(\alpha,\beta)}(1)}. \tag{7.3.2}$$

He proved that

$$\sum_{x=0}^{\infty} p_m(q^x; a, b; q) p_n(q^x; a, b; q) \frac{(bq; q)_x}{(q; q)_x} (aq)^x = \frac{\delta_{m,n}}{h_n(a, b; q)}, \tag{7.3.3}$$

where $0 < q, aq < 1$ and

$$h_n(a, b; q) = \frac{(abq; q)_n (1 - abq^{2n+1})(aq; q)_n (aq; q)_\infty}{(q; q)_n (1 - abq)(bq; q)_n (abq^2; q)_\infty} (aq)^{-n}. \tag{7.3.4}$$

Observe that (7.3.3) and (7.3.4) also follow from (7.2.22) when we replace x by $N - x$ and then let $N \to \infty$. To prove (7.3.3), assume, as we may, that $0 \le m \le n$, and observe that

$$\sum_{x=0}^{\infty} \frac{(bq; q)_x}{(q; q)_x} (aq)^x q^{xk} p_m(q^x; a, b; q)$$

$$= \sum_{j=0}^{m} \frac{\left(q^{-m}, abq^{m+1}; q\right)_j}{(q, aq; q)_j} q^j \, {}_1\phi_0 \left(bq; -; q, aq^{1+k+j}\right)$$

$$= \frac{(abq^{k+2};q)_\infty}{(aq^{k+1};q)_\infty} \, {}_3\phi_2 \left[\begin{matrix} q^{-m}, & abq^{m+1}, & aq^{k+1} \\ & aq, & abq^{k+2} \end{matrix} ; q, q \right],$$

$$= \frac{(q;bq;q)_m (abq^2;q)_\infty}{(abq^2;q)_{2m}(aq;q)_\infty} (-aq)^m q^{\binom{m}{2}} \delta_{k,m}, \quad 0 \le k \le m, \tag{7.3.5}$$

by the q-binomial and q-Saalschütz formulas. Then the orthogonality relation (7.3.3) follows immediately by using (1.5.2) and (7.3.5).

It is easy to verify that $p_n(x;a,b;q)$ satisfies the three-term recurrence relation

$$xp_n(x) = A_n[p_{n+1}(x) - p_n(x)] - C_n[p_n(x) - p_{n-1}(x)], \tag{7.3.6}$$

for $n \ge 0$, where

$$A_n = \frac{(1-abq^{n+1})(1-aq^{n+1})}{(1-abq^{2n+1})(1-abq^{2n+2})}(-q^n), \tag{7.3.7}$$

$$C_n = \frac{(1-q^n)(1-bq^n)}{(1-abq^{2n})(1-abq^{2n+1})}(-aq^n), \tag{7.3.8}$$

so the condition that $A_n C_{n+1} > 0$ for $n = 0, 1, \ldots,$ is satisfied if $0 < q, aq < 1$ and $bq < 1$. When $b < 0$ the polynomials $p_n(x;a,b;q)$ give a q-analogue of the Laguerre polynomials $L_n^{(\alpha)}(x)$ since

$$\lim_{q\to 1} p_n\left((1-q)x; q^\alpha, -q^\beta; q\right) = L_n^{(\alpha)}(x)/L_n^{(\alpha)}(0). \tag{7.3.9}$$

Andrews and Askey [1985] introduced a second q-analogue of the Jacobi polynomials,

$$P_n(x;a,b,c;q) = {}_3\phi_2 \left[\begin{matrix} q^{-n}, abq^{n+1}, x \\ aq, cq \end{matrix} ; q, q \right], \tag{7.3.10}$$

which has the property that

$$\lim_{q\to 1} P_n(x;q^\alpha, q^\beta, -q^\gamma; q) = \frac{P_n^{(\alpha,\beta)}(x)}{P_n^{(\alpha,\beta)}(1)}, \tag{7.3.11}$$

where γ is real. In view of the third free parameter in (7.3.10) they called the $P_n(x;a,b,c;q)$ the *big q-Jacobi* and the $p_n(x;a,b;q)$ the *little q-Jacobi polynomials*.

We shall now prove that the big q-Jacobi polynomials satisfy the orthogonality relation

$$\int_{cq}^{aq} P_m(x;a,b,c;q) P_n(x;a,b,c;q) \frac{(x/a, x/c;q)_\infty}{(x, bx/c;q)_\infty} d_q x$$

$$= \frac{\delta_{m,n}}{h_n(a,b,c;q)}, \tag{7.3.12}$$

where

$$h_n(a,b,c;q) = M^{-1} \frac{(abq;q)_n \left(1-abq^{2n+1}\right)(aq,cq;q)_n}{(q;q)_n(1-abq)(bq,abq/c;q)_n} (-ac)^n q^{-\binom{n}{2}} \tag{7.3.13}$$

and

$$M = \int_{cq}^{aq} \frac{(x/a, x/c; q)_\infty}{(x, bx/c; q)_\infty} d_q x$$
$$= \frac{aq(1-q)(q, c/a, aq/c, abq^2; q)_\infty}{(aq, bq, cq, abq/c; q)_\infty} \tag{7.3.14}$$

by (2.10.20). Since

$$\int_{cq}^{aq} (bx/c; q)_j (x; q)_k \frac{(x/a, x/c; q)_\infty}{(x, bx/c; q)_\infty} d_q x$$
$$= \int_{cq}^{aq} \frac{(x/a, x/c; q)_\infty}{(xq^k, bxq^j/c; q)_\infty} d_q x$$
$$= M \frac{(bq, abq/c; q)_j (aq, cq; q)_k}{(abq^2; q)_{j+k}},$$

the left side of (7.3.12) becomes

$$M \frac{(bq, abq/c; q)_m}{(aq, cq; q)_m} (c/b)^m$$
$$\cdot \sum_{j=0}^{m} \sum_{k=0}^{n} \frac{(q^{-m}, abq^{m+1}; q)_j \, (q^{-n}, abq^{n+1}; q)_k}{(q; q)_j (q; q)_k (abq^2; q)_{j+k}} q^{j+k}, \tag{7.3.15}$$

where we used (7.3.10) and the observation that, by (3.2.2) and (3.2.5),

$$P_m(x; a, b, c; q)$$
$$= \frac{(bq, abq/c; q)_m}{(aq, cq; q)_m} (c/b)^m \; {}_3\phi_2 \left[\begin{matrix} q^{-m}, abq^{m+1}, bx/c \\ bq, abq/c \end{matrix} ; q, q \right]. \tag{7.3.16}$$

Assume that $0 \le m \le n$. Since, by (1.5.3)

$$\sum_{j=0}^{m} \frac{(q^{-m}, abq^{m+1}; q)_j}{(q, abq^{k+2}; q)_j} q^j = \frac{\left(q^{1+k-m}; q\right)_m}{\left(abq^{k+2}; q\right)_m} \left(abq^{m+1}\right)^m,$$

the double sum in (7.3.15) equals

$$\frac{(q^{-n}, abq^{n+1}; q)_m}{(abq^2; q)_{2m}} \left(abq^{m+2}\right)^m \sum_{k=0}^{n-m} \frac{(q^{m-n}, abq^{m+n+1}; q)_k}{(q, abq^{2m+2}; q)_k} q^k$$
$$= \frac{(q^{-n}, abq^{n+1}; q)_n}{(abq^2; q)_{2n}} \left(abq^{n+2}\right)^n \delta_{m,n}. \tag{7.3.17}$$

Substituting this into (7.3.15), we obtain (7.3.12).

7.4 An absolutely continuous measure: the continuous q-ultraspherical polynomials

In this and the following section we shall give two important examples of orthogonal polynomials which are orthogonal with respect to an absolutely continuous measure $d\alpha(x) = w(x)dx$.

In his work of the 1890's, in which he discovered the now-famous Rogers-Ramanujan identities, Rogers [1893b, 1894, 1895] introduced a set of orthogonal polynomials that are representable in terms of basic hypergeometric series and have the ultraspherical polynomials (7.1.12) as limits when $q \to 1$. Following Askey and Ismail [1983], we shall call these polynomials the *continuous q-ultraspherical polynomials* and define them by the generating function

$$\frac{\left(\beta te^{i\theta}, \beta te^{-i\theta}; q\right)_\infty}{\left(te^{i\theta}, te^{-i\theta}; q\right)_\infty} = \sum_{n=0}^{\infty} C_n(x; \beta|q) t^n, \tag{7.4.1}$$

where $x = \cos\theta$, $0 \le \theta \le \pi$ and $\max(|q|, |t|) < 1$. Using the q-binomial theorem, it follows from (7.4.1) that

$$\begin{aligned} C_n(x; \beta|q) &= \sum_{k=0}^{n} \frac{(\beta; q)_k (\beta; q)_{n-k}}{(q; q)_k (q; q)_{n-k}} e^{i(n-2k)\theta} \\ &= \frac{(\beta; q)_n}{(q; q)_n} e^{in\theta} {}_2\phi_1\left(q^{-n}, \beta; \beta^{-1} q^{1-n}; q, q\beta^{-1} e^{-2i\theta}\right). \end{aligned} \tag{7.4.2}$$

Note that

$$\begin{aligned} \lim_{q \to 1} C_n(x; q^\lambda|q) &= \sum_{k=0}^{n} \frac{(\lambda)_k (\lambda)_{n-k}}{k!(n-k)!} e^{i(n-2k)\theta} \\ &= C_n^\lambda(x). \end{aligned} \tag{7.4.3}$$

Before considering the orthogonality relation for $C_n(x; \beta|q)$, we shall first derive some important formulas for these polynomials. For $0 < \theta < \pi, |\beta| < 1$, set $a = q\beta^{-1}e^{2i\theta}$, $b = q\beta^{-1}$, $c = qe^{2i\theta}$ and $z = \beta^2 q^n$ in (3.3.5) to obtain

$$\begin{aligned} &{}_2\phi_1\left(q^{-n}, \beta; \beta^{-1} q^{1-n}; q, q\beta^{-1} e^{-2i\theta}\right) \\ &= \frac{\left(\beta q^n, \beta e^{-2i\theta}; q\right)_\infty}{\left(q^{n+1}, e^{-2i\theta}; q\right)_\infty} {}_2\phi_1\left(q\beta^{-1}, q\beta^{-1} e^{2i\theta}; qe^{2i\theta}; q, \beta^2 q^n\right) \\ &\quad + \frac{\left(\beta, q\beta^{-1}, \beta q^n e^{2i\theta}, \beta^{-1} q^{1-n} e^{-2i\theta}; q\right)_\infty}{\left(q^{n+1}, e^{2i\theta}, q\beta^{-1} e^{-2i\theta}, \beta^{-1} q^{1-n}; q\right)_\infty} \\ &\quad \cdot {}_2\phi_1\left(q\beta^{-1}, q\beta^{-1} e^{-2i\theta}; qe^{-2i\theta}; q, \beta^2 q^n\right). \end{aligned} \tag{7.4.4}$$

Then, application of the transformation formula (1.4.3) to the two ${}_2\phi_1$ series on the right side of (7.4.4) gives

$$\begin{aligned} &C_n(x; \beta|q) \\ &= \frac{(\beta; q)_\infty}{(\beta^2; q)_\infty} W_\beta^{-1}(x|q) \frac{(\beta^2; q)_n}{(q; q)_n} \\ &\quad \cdot \left\{ \frac{\left(e^{2i\theta}; q\right)_\infty}{\left(\beta e^{2i\theta}; q\right)_\infty} e^{in\theta} {}_2\phi_1\left(\beta, \beta e^{2i\theta}; qe^{2i\theta}; q, q^{n+1}\right) \right. \\ &\quad \left. + \frac{\left(e^{-2i\theta}; q\right)_\infty}{\left(\beta e^{-2i\theta}; q\right)_\infty} e^{-in\theta} {}_2\phi_1\left(\beta, \beta e^{-2i\theta}; qe^{-2i\theta}; q, q^{n+1}\right) \right\}, \end{aligned} \tag{7.4.5}$$

where

$$W_\beta(x|q) = \frac{(e^{2i\theta}, e^{-2i\theta}; q)_\infty}{(\beta e^{2i\theta}, \beta e^{-2i\theta}; q)_\infty}, \quad x = \cos\theta. \tag{7.4.6}$$

Rewriting the right side of (7.4.5) as a q-integral we obtain the formula

$$C_n(x;\beta|q) = \frac{2i\sin\theta}{(1-q)W_\beta(x|q)} \frac{(\beta,\beta;q)_\infty}{(q,\beta^2;q)_\infty} \frac{(\beta^2;q)_n}{(q;q)_n} \cdot \int_{e^{i\theta}}^{e^{-i\theta}} \frac{(que^{i\theta}, que^{-i\theta};q)_\infty}{(\beta ue^{i\theta}, \beta ue^{-i\theta};q)_\infty} u^n \, d_q u, \tag{7.4.7}$$

which was found by Rahman and Verma [1986a].

Now use (1.4.1) to obtain from (7.4.5) that

$$\begin{aligned} C_n(x;\beta|q) &= \frac{(\beta,\beta q;q)_\infty}{(q,\beta^2;q)_\infty} W_\beta^{-1}(x|q) \frac{(\beta^2;q)_n}{(\beta q;q)_n} \\ &\quad \cdot \left\{ \left(1-e^{2i\theta}\right) e^{in\theta} \, {}_2\phi_1\left(q\beta^{-1}, q^{n+1}; \beta q^{n+1}; q, \beta e^{2i\theta}\right) \right. \\ &\quad \left. + \left(1-e^{-2i\theta}\right) e^{-in\theta} \, {}_2\phi_1\left(q\beta^{-1}, q^{n+1}; \beta q^{n+1}; q, \beta e^{-2i\theta}\right) \right\} \\ &= 4\sin\theta \, W_\beta^{-1}(x|q) \sum_{k=0}^{\infty} b(k,n;\beta) \sin(n+2k+1)\theta, \end{aligned} \tag{7.4.8}$$

where $0 < \theta < \pi$, $|\beta| < 1$ and

$$b(k,n;\beta) = \frac{(\beta,\beta q;q)_\infty}{(q,\beta^2;q)_\infty} \frac{(\beta^2;q)_n (q\beta^{-1};q)_k (q;q)_{n+k}}{(q;q)_n (q;q)_k (\beta q;q)_{n+k}} \beta^k. \tag{7.4.9}$$

The series on the right side of (7.4.8) is absolutely convergent if $|\beta| < 1$. For $|x| < 1$, $|q| < 1$ and large n it is clear from (7.4.5) that the leading term in the asymptotic expansion of $C_n(\cos\theta;\beta|q)$ is given by

$$\begin{aligned} C_n(\cos\theta;\beta|q) &\sim \frac{(\beta;q)_\infty}{(q;q)_\infty} \left\{ \frac{(\beta e^{2i\theta};q)_\infty}{(e^{2i\theta};q)_\infty} e^{-in\theta} + \frac{(\beta e^{-2i\theta};q)_\infty}{(e^{-2i\theta};q)_\infty} e^{in\theta} \right\} \\ &= 2\frac{(\beta;q)_\infty}{(q;q)_\infty} \left|A(e^{i\theta})\right| \cos(n\theta - \alpha), \end{aligned} \tag{7.4.10}$$

where

$$A(z) = \frac{(\beta z^2;q)_\infty}{(z^2;q)_\infty} \quad \text{and} \quad \alpha = \arg A(e^{i\theta}). \tag{7.4.11}$$

For further results on the asymptotics of $C_n(x;\beta|q)$, see Askey and Ismail [1980] and Rahman and Verma [1986a,b, 1987].

If we use (3.5.4) to express ${}_2\phi_1\left(q^{-n}, \beta; \beta^{-1}q^{1-n}; q, q\beta^{-1}e^{-2i\theta}\right)$ as a terminating very-well-poised ${}_8\phi_7$ series in base $q^{\frac{1}{2}}$ and then apply (2.5.1), we obtain

$$\begin{aligned} {}_2\phi_1\left(q^{-n}, \beta; \beta^{-1}q^{1-n}; q, q\beta^{-1}e^{-2i\theta}\right) &= \frac{(q^{\frac{1}{2}-n}e^{-2i\theta};q)_n (-q^{(1-n)/2}\beta^{-1};q^{\frac{1}{2}})_n}{(\beta^{-1}q^{\frac{1}{2}-n};q^{\frac{1}{2}})_n} \\ &\quad \cdot {}_4\phi_3\left[\begin{matrix} q^{-n/2}, \beta^{\frac{1}{2}}e^{-i\theta}, -\beta^{\frac{1}{2}}e^{-i\theta}, -q^{-n/2} \\ -\beta, q^{\frac{1}{4}-n/2}e^{-i\theta}, -q^{\frac{1}{4}-n/2}e^{-i\theta} \end{matrix}; q^{\frac{1}{2}}, q^{\frac{1}{2}}\right] \end{aligned}$$

$$= \frac{(\beta^2;q)_n}{(\beta;q)_n}\beta^{-n/2}e^{-in\theta}\ {}_4\phi_3\left[\begin{matrix} q^{-n/2},\beta q^{n/2},\beta^{\frac{1}{2}}e^{i\theta},\beta^{\frac{1}{2}}e^{-i\theta} \\ -\beta,\beta^{\frac{1}{2}}q^{\frac{1}{4}},-\beta^{\frac{1}{2}}q^{\frac{1}{4}} \end{matrix};q^{\frac{1}{2}},q^{\frac{1}{2}}\right] \tag{7.4.12}$$

by (2.10.4). However, by (3.10.13),

$$\begin{aligned} &{}_4\phi_3\left[\begin{matrix} q^{-n/2},\beta q^{n/2},\beta^{\frac{1}{2}}e^{i\theta},\beta^{\frac{1}{2}}e^{-i\theta} \\ -\beta,\beta^{\frac{1}{2}}q^{\frac{1}{4}},-\beta^{\frac{1}{2}}q^{\frac{1}{4}} \end{matrix};q^{\frac{1}{2}},q^{\frac{1}{2}}\right] \\ &= {}_4\phi_3\left[\begin{matrix} q^{-n},\beta^2 q^{n},\beta^{\frac{1}{2}}e^{i\theta},\beta^{\frac{1}{2}}e^{-i\theta} \\ \beta q^{\frac{1}{2}},-\beta,-\beta q^{\frac{1}{2}} \end{matrix};q,q\right], \end{aligned} \tag{7.4.13}$$

and hence, from (7.4.2), (7.4.12) and (7.4.13), we have

$$\begin{aligned} &C_n(\cos\theta;\beta|q) \\ &= \frac{(\beta^2;q)_n}{(q;q)_n}\beta^{-n/2}\ {}_4\phi_3\left[\begin{matrix} q^{-n},\beta^2 q^{n},\beta^{\frac{1}{2}}e^{i\theta},\beta^{\frac{1}{2}}e^{-i\theta} \\ \beta q^{\frac{1}{2}},-\beta,-\beta q^{\frac{1}{2}} \end{matrix};q,q\right]. \end{aligned} \tag{7.4.14}$$

Since $W_\beta(\cos\theta|q) = |A(e^{i\theta})|^{-2}$ for real β, it follows from Theorem 40 in Nevai [1979] and the asymptotic formula (7.4.10) that the polynomials $C_n(\cos\theta;\beta|q)$ are orthogonal on $[0,\pi]$ with respect to the measure $W_\beta(\cos\theta|q)d\theta$, $-1 < \beta < 1$. One can also guess the weight function by setting $\beta = q^\lambda$ and comparing the generating function (7.4.1) and the expansion (7.4.8) with the $q \to 1$ limit cases and the weight function $(1-e^{2i\theta})^\lambda(1-e^{-2i\theta})^\lambda$ for the ultraspherical polynomials $C_n^\lambda(\cos\theta)$.

We shall now give a direct proof of the orthogonality relation

$$\int_0^\pi C_m(\cos\theta;\beta|q)C_n(\cos\theta;\beta|q)W_\beta(\cos\theta|q)\,d\theta = \frac{\delta_{m,n}}{h_n(\beta|q)}, \tag{7.4.15}$$

where $|q| < 1, |\beta| < 1$ and

$$h_n(\beta|q) = \frac{(q,\beta^2;q)_\infty}{2\pi(\beta,\beta q;q)_\infty}\frac{(q;q)_n(1-\beta q^n)}{(\beta^2;q)_n(1-\beta)}. \tag{7.4.16}$$

As we shall see in the next section, (7.4.15) can be proved by using (7.4.14) and the Askey-Wilson q-beta integral (6.1.1); but here we shall give a direct proof by using (7.4.2), (1.9.10) and (1.9.11), as in Gasper [1981b], to evaluate the integral. Since the integrand in (7.4.15) is even in θ, it suffices to prove that

$$\int_{-\pi}^\pi C_m(\cos\theta;\beta|q)C_n(\cos\theta;\beta|q)W_\beta(\cos\theta|q)\,d\theta = \frac{2\delta_{m,n}}{h_n(\beta|q)}, \tag{7.4.17}$$

when $0 \le m \le n$. We first show that, for any integer k,

$$\int_{-\pi}^\pi e^{ik\theta}W_\beta(\cos\theta|q)\,d\theta = \begin{cases} 0, & \text{if } k \text{ is odd}, \\ c_{k/2}(\beta|q), & \text{if } k \text{ is even}, \end{cases} \tag{7.4.18}$$

where

$$c_j(\beta|q) = \frac{2\pi(\beta,\beta q;q)_\infty}{(q,\beta^2;q)_\infty}\frac{(\beta^{-1};q)_j}{(\beta q;q)_j}(1+q^j)\beta^j. \tag{7.4.19}$$

By the q-binomial theorem,

$$\int_{-\pi}^{\pi} e^{ik\theta} W_\beta(\cos\theta|q)\, d\theta$$

$$= \sum_{r=0}^{\infty}\sum_{s=0}^{\infty} \frac{(\beta^{-1};q)_r(\beta^{-1};q)_s}{(q;q)_r(q;q)_s} \beta^{r+s} \int_{-\pi}^{\pi} e^{i(k+2r-2s)\theta} d\theta,$$

which equals zero when k is odd and equals

$$2\pi \sum_{s=0}^{\infty} \frac{(\beta^{-1};q)_s(\beta^{-1};q)_{s+j}}{(q;q)_s(q;q)_{s+j}} \beta^{j+2s}$$

$$= 2\pi \frac{(\beta^{-1};q)_j}{(q;q)_j} \beta^j \, {}_2\phi_1 \left(\beta^{-1}, \beta^{-1}q^j; q^{j+1}; q, \beta^2\right)$$

when $k = 2j$, $j = 0, 1, \ldots$. By (1.4.5), the above ${}_2\phi_1$ series equals

$$\frac{(\beta, \beta q^{j+1}; q)_\infty}{(\beta^2, q^{j+1}; q)_\infty} \, {}_2\phi_1 \left(q^{-1}, \beta^{-1}; \beta; q, \beta q^{j+1}\right)$$

$$= \frac{(\beta, \beta q; q)_\infty}{(q, \beta^2; q)_\infty} \frac{(q;q)_j}{(\beta q;q)_j} (1 + q^j).$$

From this and the fact that $W_\beta(\cos\theta|q)$ is symmetric in θ, so that we can handle negative k's, we get (7.4.18). Hence, from (7.4.2),

$$\int_{-\pi}^{\pi} e^{ik\theta} C_n(\cos\theta; \beta|q) W_\beta(\cos\theta|q)\, d\theta \tag{7.4.20}$$

equals zero when $n - k$ is odd and equals

$$\frac{2\pi(\beta, \beta q; q)_\infty}{(q, \beta^2; q)_\infty} \frac{(\beta;q)_n(\beta^{-1};q)_j}{(q;q)_n(\beta q;q)_j} (1 + q^j)\beta^j$$

$$\cdot \, {}_4\phi_3 \left[\begin{matrix} q^{-n}, \beta, \beta^{-1}q^{-j}, -q^{1-j} \\ \beta q^{1-j}, \beta^{-1}q^{1-n}, -q^{-j} \end{matrix} ; q, q \right] \tag{7.4.21}$$

when $n - k = 2j$ is even. From (1.9.11) it follows that this ${}_4\phi_3$ series and hence the integral (7.4.20) are equal to zero when $n > |k|$. Hence (7.4.17) holds when $m \neq n$. If $k = \pm n \neq 0$, then (1.9.10) gives

$$\int_{-\pi}^{\pi} e^{ik\theta} C_n(\cos\theta; \beta|q) W_\beta(\cos\theta|q)\, d\theta$$

$$= \frac{2\pi(\beta, \beta q; q)_\infty(\beta^2;q)_n}{(q, \beta^2; q)_\infty(\beta q;q)_n} \tag{7.4.22}$$

from which it follows that (7.4.17) also holds when $m = n$.

7.5 The Askey-Wilson polynomials

In view of the ${}_4\phi_3$ series representation (7.4.14) for the continuous q-ultraspherical polynomials it is natural to consider the more general polynomials

$$r_n(x) = {}_4\phi_3 \left[\begin{matrix} q^{-n}, \alpha q^n, \beta e^{i\theta}, \beta e^{-i\theta} \\ \gamma, \delta, \epsilon \end{matrix} ; q, q \right] \tag{7.5.1}$$

which are of degree n in $x = \cos\theta$, and to try to determine the values of $\alpha, \beta, \gamma, \delta, \epsilon$ for which these polynomials are orthogonal. Because terminating balanced ${}_4\phi_3$ series can be transformed to other balanced ${}_4\phi_3$ series and to very-well-poised ${}_8\phi_7$ series which satisfy three-term transformation formulas (see, e.g., (7.2.13), (2.11.1), Exercise 2.15 and the three-term recurrence relation for the q-Racah polynomials), one is led to consider balanced ${}_4\phi_3$ series. From Sears' transformation formula (2.10.4) it follows that if we set $\alpha = abcdq^{-1}, \beta = a, \gamma = ab, \delta = ac$ and $\epsilon = ad$, then the polynomials

$$\begin{aligned} p_n(x) &\equiv p_n(x; a, b, c, d|q) \\ &= (ab, ac, ad; q)_n a^{-n} \, {}_4\phi_3 \left[\begin{matrix} q^{-n}, abcdq^{n-1}, ae^{i\theta}, ae^{-i\theta} \\ ab, ac, ad \end{matrix} ; q, q \right] \end{aligned} \tag{7.5.2}$$

are symmetric in a, b, c, d. In addition, for real θ these polynomials are analytic functions of a, b, c, d and are, in view of the coefficient $(ab, ac, ad; q)_n a^{-n}$, real-valued when a, b, c, d are real or, if complex, occur in conjugate pairs.

Askey and Wilson [1985] introduced these polynomials as q-analogues of the ${}_4F_3$ polynomials of Wilson [1978, 1980]. Since they derived the orthogonality relation, three-term recurrence relation, difference equation and other properties of $p_n(x; a, b, c, d|q)$, these polynomials are now called the *Askey-Wilson polynomials.*

Since the three-term recurrence relation (7.2.1) for the q-Racah polynomials continues to hold without the restriction $bdq = q^{-N}$, by translating it into the notation for $p_n(x; a, b, c, d|q)$ we find, as in Askey and Wilson [1985], that the recurrence relation for these polynomials can be written in the form

$$2xp_n(x) = A_n p_{n+1}(x) + B_n p_n(x) + C_n p_{n-1}(x), \quad n \geq 0, \tag{7.5.3}$$

with $p_{-1}(x) = 0, p_0(x) = 1$, where

$$A_n = \frac{1 - abcdq^{n-1}}{(1 - abcdq^{2n-1})(1 - abcdq^{2n})}, \tag{7.5.4}$$

$$\begin{aligned} C_n = &\frac{(1 - q^n)(1 - abq^{n-1})(1 - acq^{n-1})(1 - adq^{n-1})}{(1 - abcdq^{2n-2})(1 - abcdq^{2n-1})} \\ &\cdot (1 - bcq^{n-1})(1 - bdq^{n-1})(1 - cdq^{n-1}), \end{aligned} \tag{7.5.5}$$

and

$$\begin{aligned} B_n = a + a^{-1} &- A_n a^{-1}(1 - abq^n)(1 - acq^n)(1 - adq^n) \\ &- C_n a/(1 - abq^{n-1})(1 - acq^{n-1})(1 - adq^{n-1}). \end{aligned} \tag{7.5.6}$$

It is clear that A_n, B_n, C_n are real if a, b, c, d are real or, if complex, occur in conjugate pairs. Also $A_n C_{n+1} > 0, n = 0, 1, \ldots,$ if the pairwise products of a, b, c, d are less than 1 in absolute value. So by Favard's theorem, there exists a measure $d\alpha(x)$ with respect to which $p_n(x; a, b, c, d|q)$ are orthogonal. In order to determine this measure let us assume that $\max(|a|, |b|, |c|, |d|, |q|) < 1$. Then, by (2.5.1),

$$p_n(\cos\theta; a, b, c, d|q)$$

$$= \frac{(ab, ac, bc, de^{-i\theta}, ; q)_n}{(abce^{i\theta}; q)_n} e^{in\theta}$$
$$\cdot {}_8W_7\left(abce^{i\theta}q^{-1}; ae^{i\theta}, be^{i\theta}, ce^{i\theta}, abcdq^{n-1}, q^{-n}; q, qd^{-1}e^{-i\theta}\right), \quad (7.5.7)$$

and, by (2.11.1),

$$ {}_8W_7\left(abce^{i\theta}q^{-1}; ae^{i\theta}, be^{i\theta}, ce^{i\theta}, abcdq^{n-1}, q^{-n}; qd^{-1}e^{-i\theta}\right)$$
$$= \frac{(abce^{i\theta}, bcde^{i\theta}q^n, be^{i\theta}q^{n+1}, ce^{i\theta}q^{n+1}, ae^{-i\theta}, be^{-i\theta}, ce^{-i\theta}, de^{-i\theta}q^n; q)_\infty}{(ab, ac, bc, q^{n+1}, bdq^n, cdq^n, bcq^{n+1}e^{2i\theta}, e^{-2i\theta}; q)_\infty}$$
$$\cdot {}_8W_7\left(bcq^n e^{2i\theta}; bcq^n, be^{i\theta}, ce^{i\theta}, qa^{-1}e^{i\theta}, qd^{-1}e^{i\theta}; q, adq^n\right)$$
$$+ \frac{(abce^{i\theta}, abce^{-i\theta}q^n, bcde^{-i\theta}q^n, be^{-i\theta}q^{n+1}, ce^{-i\theta}q^{n+1}; q)_\infty}{(q^{n+1}, bdq^n, cdq^n, abce^{i\theta}q^n, bcq^{n+1}e^{-2i\theta}; q)_\infty}$$
$$\cdot \frac{(ae^{i\theta}, be^{i\theta}, ce^{i\theta}, de^{i\theta}; q)_\infty}{(ab, ac, ad, e^{2i\theta}; q)_\infty} e^{-2in\theta}$$
$$\cdot {}_8W_7\left(bcq^n e^{-2i\theta}; bcq^n, be^{-i\theta}, ce^{-i\theta}, qa^{-1}e^{-i\theta}, qd^{-1}e^{-i\theta}; q, adq^n\right), \quad (7.5.8)$$

where $0 < \theta < \pi$. Hence

$$p_n(\cos\theta; a, b, c, d|q)$$
$$= (bc, bd, cd; q)_n \left\{Q_n\left(e^{i\theta}; a, b, c, d|q\right) + Q_n\left(e^{-i\theta}; a, b, c, d|q\right)\right\}, \quad (7.5.9)$$

where

$$Q_n(z; a, b, c, d|q)$$
$$= \frac{(abczq^n, bcdzq^n, bzq^{n+1}, czq^{n+1}, a/z, b/z, c/z, d/z; q)_\infty}{(bc, bd, cd, abq^n, acq^n, q^{n+1}, bcz^2q^{n+1}, z^{-2}; q)_\infty} z^n$$
$$\cdot {}_8W_7\left(bcq^n z^2; bcq^n, bz, cz, qa^{-1}z, qd^{-1}z; q, adq^n\right). \quad (7.5.10)$$

It is clear from (7.5.10) that

$$Q_n(z; a, b, c, d|q) \sim z^n B(z^{-1})/(bc, bd, cd; q)_\infty, \quad (7.5.11)$$

where

$$B(z) = (az, bz, cz, dz; q)_\infty/(z^2; q)_\infty \quad (7.5.12)$$

as $n \to \infty$, uniformly for z, a, b, c, d in compact sets avoiding the poles $z^2 = q^{-k}$, $k = 0, 1, \ldots$. Using (7.5.9) we find that

$$p_n(\cos\theta; a, b, c, d|q)$$
$$\sim e^{in\theta}B(e^{-i\theta}) + e^{-in\theta}B(e^{i\theta})$$
$$= 2|B(e^{i\theta})|\cos(n\theta - \beta), \quad (7.5.13)$$

where $\beta = \arg B(e^{i\theta})$ and $0 < \theta < \pi$ (see Rahman [1986c]). Then

$$\left|B(e^{i\theta})\right|^2 = [\sin\theta\ w(\cos\theta; a, b, c, d|q)]^{-1}, \quad (7.5.14)$$

where, in order to be consistent with the $p_n(x; a, b, c, d|q)$ notation, we have used $w(x; a, b, c, d|q)$ to denote the weight function $w(x; a, b, c, d)$ defined in (6.3.1). It follows from Theorem 40 in Nevai [1979] that the polynomials

$p_n(x;a,b,c,d|q)$ are orthogonal on $[-1,1]$ with respect to the measure $w(x;a,b,c,d|q)dx$ when $\max(|a|,|b|,|c|,|d|,|q|)<1$.

We shall now give a direct proof of the orthogonality relation

$$\int_{-1}^{1} p_m(x)p_n(x)w(x)\,dx=\frac{\delta_{m,n}}{h_n}, \tag{7.5.15}$$

where $w(x)\equiv w(x;a,b,c,d|q)$ and

$$\begin{aligned} h_n &\equiv h_n(a,b,c,d|q)\\ &=\kappa^{-1}(a,b,c,d|q)\frac{(abcdq^{-1};q)_n\left(1-abcdq^{2n-1}\right)}{(q;q)_n\left(1-abcdq^{-1}\right)(ab,ac,ad,bc,bd,cd;q)_n}, \end{aligned} \tag{7.5.16}$$

with

$$\begin{aligned} \kappa(a,b,c,d|q) &= \int_{-1}^{1} w(x;a,b,c,d|q)\,dx\\ &= \frac{2\pi(abcd;q)_\infty}{(q,ab,ac,ad,bc,bd,cd;q)_\infty}, \end{aligned} \tag{7.5.17}$$

by (6.1.1). First observe that, by (7.5.17),

$$\begin{aligned} &\int_{-1}^{1}\left(ae^{i\theta},ae^{-i\theta};q\right)_j\left(be^{i\theta},be^{-i\theta};q\right)_k w(x;a,b,c,d|q)\,dx\\ &=\int_{-1}^{1} w(x;aq^j,bq^k,c,d|q)\,dx\\ &=\kappa(aq^j,bq^k,c,d|q). \end{aligned} \tag{7.5.18}$$

By using (7.5.2) and the fact that $p_n(x;a,b,c,d|q)=p_n(x;b,a,c,d|q)$ we find that the left side of (7.5.15) equals

$$\begin{aligned} &\kappa(a,b,c,d|q)(ab,ac,ad;q)_m(ba,bc,bd;q)_n a^{-m}b^{-n}\\ &\cdot\sum_{j=0}^{m}\frac{(q^{-m},abcdq^{m-1};q)_j}{(q,abcd;q)_j}q^j\,{}_3\phi_2\left[\begin{matrix}q^{-n},abcdq^{n-1},abq^j\\ abcdq^j,ab\end{matrix};q,q\right]. \end{aligned} \tag{7.5.19}$$

Assuming that $0\le n\le m$ and using the q-Saalschütz formula to sum the ${}_3\phi_2$ series, the sum over j in (7.5.19) gives

$$\begin{aligned} &\frac{(cd,abcdq^{m-1},q^{-m};q)_n}{(q^{1-n}/ab;q)_n\,(abcd;q)_{2n}}q^n\,{}_2\phi_1\left(q^{n-m},abcdq^{n+m-1};abcdq^{2n};q,q\right)\\ &=\frac{(cd,abcdq^{m-1},q^{-m};q)_n}{(q^{1-n}/ab;q)_n\,(abcd;q)_{2n}}\frac{(q^{1+n-m};q)_{m-n}}{(abcdq^{2n};q)_{m-n}}\\ &=\frac{(q,cd;q)_n\left(1-abcdq^{-1}\right)}{(abcdq^{-1},ab;q)_n\left(1-abcdq^{2n-1}\right)}(ab)^n\delta_{m,n}. \end{aligned} \tag{7.5.20}$$

Combining (7.5.19) and (7.5.20) completes the proof of (7.5.15).

Askey and Wilson proved a more general orthogonality relation by using contour integration. They showed that if $|q| < 1$ and the pairwise products and quotients of a, b, c, d are not of the form q^{-k}, $k = 0, 1, \ldots$, then

$$\int_{-1}^{1} p_m(x)p_n(x)w(x)\,dx + 2\pi \sum_k p_m(x_k)p_n(x_k)w_k$$

$$= \frac{\delta_{m,n}}{h_n(a,b,c,d|q)}, \tag{7.5.21}$$

where x_k are the points $\frac{1}{2}\left(fq^k + f^{-1}q^{-k}\right)$ with f equal to any of the parameters a, b, c, d whose absolute value is greater than one, the sum is over the k with $|fq^k| > 1$, and

$$w_k \equiv w_k(a,b,c,d|q)$$

$$= \frac{(a^{-2};q)_\infty}{(q, ab, ac, ad, b/a, c/a, d/a; q)_\infty}$$

$$\cdot \frac{(a^2;q)_k(1-a^2q^{2k})(ab, ac, ad; q)_k}{(q;q)_k(1-a^2)(aq/b, aq/c, aq/d; q)_k}\left(\frac{q}{abcd}\right)^k \tag{7.5.22}$$

when $x_k = \frac{1}{2}\left(aq^k + a^{-1}q^{-k}\right)$. For a proof and complete discussion, see Askey and Wilson [1985]. Also, see Ex. 7.31.

In order to get a q-analogue of Jacobi polynomials, Askey and Wilson set

$$a = q^{(2\alpha+1)/4}, b = q^{(2\alpha+3)/4}, c = -q^{(2\beta+1)/4}, d = -q^{(2\beta+3)/4} \tag{7.5.23}$$

and defined the continuous q-Jacobi polynomials by

$$P_n^{(\alpha,\beta)}(x|q) = \frac{(q^{\alpha+1};q)_n}{(q;q)_n}$$

$$\cdot {}_4\phi_3\left[\begin{matrix} q^{-n}, q^{n+\alpha+\beta+1}, q^{(2\alpha+1)/4}e^{i\theta}, q^{(2\alpha+1)/4}e^{-i\theta} \\ q^{\alpha+1}, -q^{(\alpha+\beta+1)/2}, -q^{(\alpha+\beta+2)/2} \end{matrix}; q, q\right]. \tag{7.5.24}$$

On the other hand, Rahman [1981] found it convenient to work with an apparently different q-analogue, namely,

$$P_n^{(\alpha,\beta)}(x;q) = \frac{\left(q^{\alpha+1}, -q^{\beta+1}; q\right)_n}{(q, -q; q)_n}$$

$$\cdot {}_4\phi_3\left[\begin{matrix} q^{-n}, q^{n+\alpha+\beta+1}, q^{\frac{1}{2}}e^{i\theta}, q^{\frac{1}{2}}e^{-i\theta} \\ q^{\alpha+1}, -q^{\beta+1}, -q \end{matrix}; q, q\right]. \tag{7.5.25}$$

However, as Askey and Wilson pointed out, these two q-analogues are not really different since, by the quadratic transformation (3.10.13),

$$P_n^{(\alpha,\beta)}(x|q^2) = \frac{(-q;q)_n}{(-q^{\alpha+\beta+1};q)_n} q^{\alpha n} P_n^{(\alpha,\beta)}(x;q). \tag{7.5.26}$$

Note that

$$\lim_{q\to 1} P_n^{(\alpha,\beta)}(x|q) = \lim_{q\to 1} P_n^{(\alpha,\beta)}(x;q) = P_n^{(\alpha,\beta)}(x). \tag{7.5.27}$$

The orthogonality relations for these q-analogues are

$$\int_0^\pi P_m^{(\alpha,\beta)}(\cos\theta|q)P_n^{(\alpha,\beta)}(\cos\theta|q)w(\theta;q^{\frac{1}{2}})\,d\theta = \frac{\delta_{m,n}}{a_n(\alpha,\beta|q)}, \tag{7.5.28}$$

and

$$\int_0^\pi P_m^{(\alpha,\beta)}(\cos\theta;q)P_n^{(\alpha,\beta)}(\cos\theta;q)w(\theta;q)\,d\theta = \frac{\delta_{m,n}}{b_n(\alpha,\beta;q)}, \tag{7.5.29}$$

where $0 < q < 1$, $\alpha \geq -\frac{1}{2}, \beta \geq -\frac{1}{2}$,

$$w(\theta;q) = \left| \frac{\left(e^{i\theta}, -e^{i\theta};q\right)_\infty}{(q^{\alpha+\frac{1}{2}}e^{i\theta}, -q^{\beta+\frac{1}{2}}e^{i\theta};q)_\infty} \right|^2, \tag{7.5.30}$$

$$\begin{aligned} a_n(\alpha,\beta|q) &= \frac{\left(q, q^{\alpha+1}, q^{\beta+1}, -q^{(\alpha+\beta+1)/2}, -q^{(\alpha+\beta+2)/2};q\right)_\infty}{2\pi\left(q^{(\alpha+\beta+2)/2}, q^{(\alpha+\beta+3)/2};q\right)_\infty} \\ &\quad \cdot \frac{\left(1-q^{2n+\alpha+\beta+1}\right)\left(q, q^{\alpha+\beta+1}, -q^{(\alpha+\beta+1)/2};q\right)_n}{\left(1-q^{\alpha+\beta+1}\right)\left(q^{\alpha+1}, q^{\beta+1}, -q^{(\alpha+\beta+3)/2};q\right)_n} q^{-n(2\alpha+1)/4} \end{aligned} \tag{7.5.31}$$

and

$$\begin{aligned} b_n(\alpha,\beta;q) &= \frac{\left(q, q^{\alpha+1}, q^{\beta+1}, -q^{\alpha+1}, -q^{\beta+1} - q^{\alpha+\beta+1}, -q;q\right)_\infty}{2\pi\,(q^{\alpha+\beta+2};q)_\infty} \\ &\quad \cdot \frac{\left(1-q^{2n+\alpha+\beta+1}\right)(q^{\alpha+\beta+1}, q, -q, -q, -q;q)_n\, q^{-n}}{(1-q^{\alpha+\beta+1})(q^{\alpha+1}, q^{\beta+1}, -q^{\alpha+1}, -q^{\beta+1}, -q^{\alpha+\beta+1};q)_n}. \end{aligned} \tag{7.5.32}$$

From (7.4.14), (7.5.24) and (7.5.25) it is obvious that

$$\begin{aligned} &C_n(\cos\theta;q^\lambda|q) \\ &= \frac{\left(q^{2\lambda};q\right)_n}{\left(q^{\lambda+\frac{1}{2}};q\right)_n} q^{-n\lambda/2} P_n^{(\lambda-\frac{1}{2},\lambda-\frac{1}{2})}(\cos\theta|q) \end{aligned} \tag{7.5.33}$$

$$= \frac{\left(q^\lambda, -q^{\frac{1}{2}};q^{\frac{1}{2}}\right)_n}{\left(q^{(2\lambda+1)/4}, -q^{(2\lambda+1)/4};q^{\frac{1}{2}}\right)_n} q^{-n/4} P_n^{(\lambda-\frac{1}{2},\lambda-\frac{1}{2})}(\cos\theta;q^{\frac{1}{2}}). \tag{7.5.34}$$

It can also be shown that

$$C_{2n}(x;q^\lambda|q) = \frac{\left(q^\lambda, -q;q\right)_n}{\left(q^{\frac{1}{2}}, -q^{\frac{1}{2}};q\right)_n} q^{-n/2} P_n^{(\lambda-\frac{1}{2},-\frac{1}{2})}(2x^2-1;q), \tag{7.5.35}$$

$$C_{2n+1}(x;q^\lambda|q) = x\frac{\left(q^\lambda, -1;q\right)_{n+1}}{\left(q^{\frac{1}{2}}, -q^{\frac{1}{2}};q\right)_{n+1}} q^{-n/2} P_n^{(\lambda-\frac{1}{2},\frac{1}{2})}(2x^2-1;q), \tag{7.5.36}$$

which are q-analogues of the quadratic transformations

$$C_{2n}^\lambda(x) = \frac{(\lambda)_n}{(\frac{1}{2})_n} P_n^{(\lambda-\frac{1}{2},-\frac{1}{2})}(2x^2-1) \tag{7.5.37}$$

and

$$C_{2n+1}^{\lambda}(x) = x\frac{(\lambda)_{n+1}}{(\frac{1}{2})_{n+1}} P_n^{(\lambda-\frac{1}{2},\frac{1}{2})}(2x^2-1), \tag{7.5.38}$$

respectively. To prove (7.5.35), observe that from (7.4.2)

$$\begin{aligned} &C_{2n}(\cos\theta; q^{\lambda}|q) \\ &= \frac{(q^{\lambda};q)_{2n}}{(q;q)_{2n}} e^{2in\theta}\ {}_2\phi_1\left(q^{-2n}, q^{\lambda}; q^{1-\lambda-2n}; q, q^{1-\lambda}e^{-2i\theta}\right), \end{aligned}$$

and hence, by the Sears-Carlitz formula (Ex. 2.26),

$$\begin{aligned} &C_{2n}(\cos\theta; q^{\lambda}|q) \\ &= \frac{(q^{\lambda};q)_{2n}}{(q;q)_{2n}} \left(q^{\frac{1}{2}}e^{-2i\theta}, q^{\frac{1}{2}-n}e^{-2i\theta}; q\right)_n e^{2in\theta} \\ &\quad \cdot {}_4\phi_3\left[\begin{matrix} q^{-n}, -q^{-n}, -q^{\frac{1}{2}-n}, q^{\frac{1}{2}-\lambda-n} \\ q^{1-\lambda-2n}, q^{\frac{1}{2}-n}e^{2i\theta}, q^{\frac{1}{2}-n}e^{-2i\theta} \end{matrix}; q, q\right]. \end{aligned} \tag{7.5.39}$$

Reversing the ${}_4\phi_3$ series, we obtain

$$\begin{aligned} &C_{2n}(\cos\theta; q^{\lambda}|q) \\ &= \frac{\left(q^{\lambda}, q^{\lambda+\frac{1}{2}}; q\right)_n}{\left(q, q^{\frac{1}{2}}; q\right)_n} q^{-n/2}\ {}_4\phi_3\left[\begin{matrix} q^{-n}, q^{n+\lambda}, q^{\frac{1}{2}}e^{2i\theta}, q^{\frac{1}{2}}e^{-2i\theta} \\ q^{\lambda+\frac{1}{2}}, -q^{\frac{1}{2}}, -q \end{matrix}; q, q\right]. \end{aligned} \tag{7.5.40}$$

This, together with (7.5.26), yields (7.5.35). The proof of (7.5.36) is left as an exercise.

Following Askey and Wilson [1985] we shall now obtain another interesting special case of the Askey-Wilson polynomials. First note that the orthogonality relation (7.5.15) can be written in the form

$$\begin{aligned} &\int_{-\pi}^{\pi} p_m(\cos\theta; a,b,c,d|q) p_n(\cos\theta; a,b,c,d|q) w(\cos\theta; a,b,c,d|q) \sin\theta\, d\theta \\ &= \frac{2\delta_{m,n}}{h_n(a,b,c,d|q)}. \end{aligned} \tag{7.5.41}$$

Replace θ by $\theta+\phi$, a by $ae^{i\phi}$, b by $ae^{-i\phi}$ and then set $c = be^{i\phi}, d = be^{-i\phi}$ to find by periodicity that if $-1 < a, b < 1$ or if $b = \bar{a}$ and $|a| < 1$, then

$$\begin{aligned} &\int_{-\pi}^{\pi} p_m(\cos(\theta+\phi); a,b|q) p_n(\cos(\theta+\phi); a,b|q) W(\theta)\, d\theta \\ &= \frac{\delta_{m,n}}{\rho_n(a,b|q)}, \end{aligned} \tag{7.5.42}$$

where

$$\begin{aligned} p_n(\cos(\theta+\phi); a,b|q) &= (a^2, ab, abe^{2i\phi}; q)_n (ae^{i\phi})^{-n} \\ &\quad \cdot {}_4\phi_3\left[\begin{matrix} q^{-n}, a^2b^2q^{n-1}, ae^{2i\phi+i\theta}, ae^{-i\theta} \\ a^2, ab, abe^{2i\phi} \end{matrix}; q, q\right], \end{aligned} \tag{7.5.43}$$

$$W(\theta) = \left|\frac{\left(e^{2i(\theta+\phi)}; q\right)_\infty}{\left(ae^{i\theta}, be^{i\theta}, ae^{i(\theta+2\phi)}, be^{i(\theta+2\phi)}; q\right)_\infty}\right|^2, \tag{7.5.44}$$

and

$$\rho_n(a,b|q) = \frac{\left(q, a^2, b^2, ab, ab, abe^{2i\phi}, abe^{-2i\phi}; q\right)_\infty}{4\pi(a^2b^2;q)_\infty} \cdot \frac{\left(1-a^2b^2q^{2n-1}\right)\left(a^2b^2q^{-1};q\right)_n}{\left(1-a^2b^2q^{-1}\right)\left(q, a^2, b^2, ab, ab, abe^{2i\phi}, abe^{-2i\phi}; q\right)_n}. \tag{7.5.45}$$

The recurrence relation for these polynomials is

$$\begin{aligned} &2\cos(\theta+\phi)p_n(\cos(\theta+\phi);a,b|q) \\ &= A_n p_{n+1}(\cos(\theta+\phi);a,b|q) + B_n p_n(\cos(\theta+\phi);a,b|q) \\ &+ C_n p_{n-1}(\cos(\theta+\phi);a,b|q) \end{aligned} \tag{7.5.46}$$

for $n = 0, 1, \ldots$, where $p_{-1}(x;a,b|q) = 0$,

$$A_n = \frac{1-a^2b^2q^{n-1}}{\left(1-a^2b^2q^{2n-1}\right)\left(1-a^2b^2q^{2n}\right)}, \tag{7.5.47}$$

$$\begin{aligned} C_n &= \frac{(1-q^n)(1-a^2q^{n-1})(1-b^2q^{n-1})(1-abq^{n-1})(1-abq^{n-1})}{\left(1-a^2b^2q^{2n-2}\right)\left(1-a^2b^2q^{2n-1}\right)}, \\ &\cdot (1-abe^{2i\phi}q^{n-1})(1-abe^{-2i\phi}q^{n-1}) \end{aligned} \tag{7.5.48}$$

and

$$\begin{aligned} B_n &= ae^{i\phi} + a^{-1}e^{-i\phi} - A_n a^{-1}e^{-i\phi}(1-a^2q^n)(1-abq^n)(1-abe^{2i\phi}q^n) \\ &- C_n ae^{i\phi}/(1-a^2q^{n-1})(1-abe^{2i\phi}q^{n-1})(1-abq^{n-1}). \end{aligned} \tag{7.5.49}$$

If we set $a = q^\alpha, b = q^\beta, \theta = x\log q$ in (7.5.42) and then let $q \to 1$ we obtain

$$\int_{-\infty}^{\infty} p_m(x;\alpha,\beta)p_n(x;\alpha,\beta)|\Gamma(\alpha+ix)\Gamma(\beta+ix)|^2 dx = 0, \quad m \neq n, \tag{7.5.50}$$

where

$$p_n(x;\alpha,\beta) = i^n \, {}_3F_2\left[\begin{matrix} -n, n+2\alpha+2\beta-1, \alpha-ix \\ \alpha+\beta, 2\alpha \end{matrix}; 1\right]. \tag{7.5.51}$$

See Suslov [1982, 1987] and Askey and Wilson [1982, 1985] for further details.

7.6 Connection coefficients

Suppose $f_n(x)$ and $g_n(x), n = 0, 1, \ldots$, are polynomials of exact degree n in x. Sometimes it is of interest to express one of these sequences as a linear combination of the polynomials in the other sequence, say,

$$g_n(x) = \sum_{k=0}^{n} c_{k,n} f_k(x). \tag{7.6.1}$$

The numbers $c_{k,n}$ are called the *connection coefficients*. If the polynomials $f_n(x)$ happen to be orthogonal on an interval I with respect to a measure

$d\alpha(x)$, then $c_{k,n}$ is the k-th Fourier coefficient of $g_n(x)$ with respect to the orthogonal polynomials $f_k(x)$ and hence can be expressed as a multiple of the integral $\int_I f_k(x)g_n(x)d\alpha(x)$.

A particularly interesting problem is to determine the conditions under which the connection coefficients are nonnegative for particular systems of orthogonal polynomials. Formula (7.6.1) is sometimes called a *projection formula* when all of the coefficients are nonnegative. See the applications to positive definite functions, isometric embeddings of metric spaces, and inequalities in Askey [1970, 1975], Askey and Gasper [1971], Gangolli [1967] and Gasper [1975a]. As an illustration we shall consider the coefficients $c_{k,n}$ in the relation

$$p_n(x;\alpha,\beta,\gamma,d|q) = \sum_{k=0}^{n} c_{k,n} p_k(x;a,b,c,d|q). \tag{7.6.2}$$

Askey and Wilson [1985] showed that

$$\begin{aligned} c_{k,n} = {} & \frac{(\alpha d,\beta d,\gamma d,q;q)_n\,(\alpha\beta\gamma d q^{n-1};q)_k}{(\alpha d,\beta d,\gamma d,q,abcdq^{k-1};q)_k\,(q;q)_{n-k}} q^{k^2-nk} d^{k-n} \\ & \cdot {}_5\phi_4\left[\begin{matrix} q^{k-n},\alpha\beta\gamma d q^{n+k-1},adq^k,bdq^k,cdq^k \\ abcdq^{2k},\alpha dq^k,\beta dq^k,\gamma dq^k \end{matrix};q,q\right]. \end{aligned} \tag{7.6.3}$$

To prove (7.6.3), temporarily assume that max $(|a|,|b|,|c|,|d|,|q|) < 1$, and observe that, by orthogonality,

$$b_{k,j} = \int_{-1}^{1} w(x;a,b,c,d|q)p_k(x;a,b,c,d|q)(de^{i\theta},de^{-i\theta};q)_j\, dx \tag{7.6.4}$$

vanishes if $j < k$, and that

$$\begin{aligned} b_{k,j} &= \kappa(a,b,c,d|q)(ab,ac,ad;q)_k a^{-k} \\ &\quad \cdot \frac{(ad;q)_j}{(abcdq;q)_j}\, {}_3\phi_2\left[\begin{matrix} q^{-k},abcdq^{k-1},adq^j \\ abcdq^j,ad \end{matrix};q,q\right] \\ &= \kappa(a,b,c,d|q)(ab,ac,bc,q^{-j};q)_k(ad;q)_j(dq^j)^k/(abcd;q)_{j+k} \end{aligned} \tag{7.6.5}$$

if $j \ge k$. Since

$$\begin{aligned} & p_n(x;\alpha,\beta,\gamma,d|q) \\ &= (\alpha d,\beta d,\gamma d;q)_n d^{-n} \sum_{j=0}^{n} \frac{(q^{-n},\alpha\beta\gamma d q^{n-1},de^{i\theta},de^{-i\theta};q)_j}{(q,\alpha d,\beta d,\gamma d;q)_j} q^j, \end{aligned} \tag{7.6.6}$$

we find that

$$\begin{aligned} c_{k,n} &= h_k(a,b,c,d|q)\int_{-1}^{1} w(x;a,b,c,d|q)p_k(x;a,b,c,d|q)p_n(x;\alpha,\beta,\gamma,d|q)\, dx \\ &= h_k(a,b,c,d|q)(\alpha d,\beta d,\gamma d;q)_n d^{-n}\sum_{j=0}^{n}\frac{(q^{-n},\alpha\beta\gamma d q^{n-1};q)_j}{(q,\alpha d,\beta d,\gamma d;q)_j} q^j b_{k,j} \end{aligned} \tag{7.6.7}$$

and hence (7.6.3) follows from (7.5.16), (7.5.17) and (7.6.5).

The ${}_5\phi_4$ series in (7.6.3) is balanced but, in general, cannot be transformed in a simple way, so one cannot hope to say much about the nonnegativity of $c_{k,n}$ unless the parameters are related in some way. One of the simplest cases is when the ${}_5\phi_4$ series reduces to a ${}_3\phi_2$, which can be summed by the q-Saalschütz formula. Thus for $\beta = b$ and $\gamma = c$ we get

$$p_n(x;\alpha,b,c,d|q) = \sum_{k=0}^{n} c_{k,n} p_k(x;a,b,c,d|q) \tag{7.6.8}$$

with

$$c_{k,n} = \frac{(\alpha/a;q)_{n-k}(\alpha bcdq^{n-1};q)_k(bc,bd,cd;q)_n a^{n-k}}{(q,bc,bd,cd;q)_k(abcdq^{k-1};q)_k(q,abcdq^{2k};q)_{n-k}}. \tag{7.6.9}$$

It is clear that $c_{k,n} > 0$ when $0 < \alpha < a < 1$, $0 < q < 1$ and $\max(bc,bd,cd,abcd) < 1$.

Another simple case is when the ${}_5\phi_4$ series in (7.6.3) reduces to a summable ${}_4\phi_3$ series. For example, this happens when we set $d = q^{\frac{1}{2}}, c = -q^{\frac{1}{2}} = \gamma, b = -a$ and $\beta = -\alpha$. The ${}_5\phi_4$ series then reduces to

$${}_4\phi_3\left[\begin{matrix} q^{k-n}, \alpha^2 q^{n+k}, aq^{k+\frac{1}{2}}, -aq^{k+\frac{1}{2}} \\ \alpha q^{k+\frac{1}{2}}, -\alpha q^{k+\frac{1}{2}}, a^2 q^{2k+1} \end{matrix}; q, q\right] \tag{7.6.10}$$

which equals

$$\frac{\left(\alpha^2 q^{n+k+1}, q^{k-n+1}, a^2 q^{k+n+2}, a^2 q^{k-n+2}/\alpha^2; q^2\right)_\infty}{\left(q, \alpha^2 q^{2k+1}, a^2 q^{2k+2}, a^2 q^2/\alpha^2; q^2\right)_\infty}\left(\alpha^2 q^{n+k}\right)^{(n-k)/2} \tag{7.6.11}$$

by Andrews' formula (Ex. 2.8). Clearly, this vanishes when $n-k$ is an odd integer and equals

$$\frac{\left(q\alpha^2, q^2 a^2; q^2\right)_{n-2j}\left(q, \alpha^2/a^2; q^2\right)_j}{\left(q\alpha^2, q^2 a^2; q^2\right)_{n-j}}\left(a^2 q^{2n+1-4j}\right)^j \tag{7.6.12}$$

when $n-k = 2j, j = 0, 1, \ldots, [\frac{n}{2}]$. Since by (3.10.12) and (7.4.14)

$$\begin{aligned} &p_n(x;a,-a,-q^{\frac{1}{2}},q^{\frac{1}{2}}|q) \\ &\quad = \frac{\left(q^2, qa^2; q^2\right)_n}{(a^2;q)_n} C_n(x;a^2|q^2), \end{aligned} \tag{7.6.13}$$

we obtain, after some simplification, Rogers' formula

$$C_n(x;\gamma|q) = \sum_{k=0}^{[n/2]} \frac{\beta^k(\gamma/\beta;q)_k(\gamma;q)_{n-k}(1-\beta q^{n-2k})}{(q;q)_k(\beta q;q)_{n-k}(1-\beta)} C_{n-2k}(x;\beta|q) \tag{7.6.14}$$

after replacing α^2, a^2, q^2 by γ, β and q, respectively.

It is left as an exercise (Ex. 7.15) to show that formulas (7.6.2) and (7.6.14) are special cases of the q-analogue of the Fields-Wimp formula given in (3.7.9).

For other applications of the connection coefficient formula (7.6.2), see Askey and Wilson [1985].

7.7 A difference equation and a Rodrigues-type formula for the Askey-Wilson polynomials

The polynomials $p_n(x;a,b,c,d|q)$, unlike the Jacobi polynomials, do not satisfy a differential equation; but, as Askey and Wilson [1985] showed, they satisfy a second-order difference equation. Define

$$E_q^{\pm} f(e^{i\theta}) = f\left(q^{\pm\frac{1}{2}} e^{i\theta}\right), \tag{7.7.1}$$

$$\delta_q f\left(e^{i\theta}\right) = \left(E_q^+ - E_q^-\right) f\left(e^{i\theta}\right) \tag{7.7.2}$$

$$\text{and} \quad D_q f(x) = \frac{\delta_q f(x)}{\delta_q x}, \quad x = \cos\theta. \tag{7.7.3}$$

Clearly,

$$\delta_q(\cos\theta) = \frac{1}{2}\left(q^{\frac{1}{2}} - q^{-\frac{1}{2}}\right)\left(e^{i\theta} - e^{-i\theta}\right) = -iq^{-\frac{1}{2}}(1-q)\sin\theta, \tag{7.7.4}$$

$$\text{and } \delta_q\left(ae^{i\theta}, ae^{-i\theta}; q\right)_n = 2aiq^{-\frac{1}{2}}(1-q^n)\sin\theta\,(aq^{\frac{1}{2}}e^{i\theta}, aq^{\frac{1}{2}}e^{-i\theta}; q)_{n-1}, \tag{7.7.5}$$

so that

$$D_q\left(ae^{i\theta}, ae^{-i\theta}; q\right)_n = -\frac{2a(1-q^n)}{(1-q)}(aq^{\frac{1}{2}}e^{i\theta}, aq^{\frac{1}{2}}e^{-i\theta}; q)_{n-1}, \tag{7.7.6}$$

which implies that the divided difference operator D_q plays the same role for $\left(ae^{i\theta}, ae^{-i\theta}; q\right)_n$ as d/dx does for x^n. When $q \to 1$, formula (7.7.6) becomes

$$\frac{d}{dx}\left(1 - 2ax + a^2\right)^n = -2an(1 - 2ax + a^2)^{n-1}.$$

Generally, for a differentiable function $f(x)$ we have

$$\lim_{q\to 1} D_q f(x) = \frac{d}{dx} f(x).$$

Following Askey and Wilson [1985], we shall now use the operator D_q and the recurrence relation (7.5.3) to derive a Rodrigues-type formula for $p_n(x;a,b,c,d|q)$. First note that by (7.5.2) and (7.7.5)

$$\begin{aligned}&\delta_q p_n(x;a,b,c,d|q)\\ &= -2iq^{-n/2}\sin\theta\,(1-q^n)(1-q^{n-1}abcd)p_{n-1}(x;aq^{\frac{1}{2}}, bq^{\frac{1}{2}}, cq^{\frac{1}{2}}, dq^{\frac{1}{2}}|q).\end{aligned} \tag{7.7.7}$$

If we define

$$A(\theta) = \frac{\left(1-ae^{i\theta}\right)\left(1-be^{i\theta}\right)\left(1-ce^{i\theta}\right)\left(1-de^{i\theta}\right)}{\left(1-e^{2i\theta}\right)\left(1-qe^{2i\theta}\right)} \tag{7.7.8}$$

and

$$r_n(e^{i\theta}) = {}_4\phi_3\left[\begin{matrix} q^{-n}, abcdq^{n-1}, ae^{i\theta}, ae^{-i\theta} \\ ab, ac, ad \end{matrix}; q, q\right], \tag{7.7.9}$$

then the recurrence relation (7.5.3) can be written as

$$\begin{aligned}&q^{-n}(1-q^n)(1-abcdq^{n-1})r_n(e^{i\theta})\\ &= A(-\theta)\left[r_n(q^{-1}e^{i\theta}) - r_n(e^{i\theta})\right] + A(\theta)\left[r_n(qe^{i\theta}) - r_n(e^{i\theta})\right].\end{aligned} \tag{7.7.10}$$

Also, setting

$$V(e^{i\theta}) = \frac{(e^{2i\theta};q)_\infty}{(ae^{i\theta}, be^{i\theta}, ce^{i\theta}, de^{i\theta};q)_\infty} \tag{7.7.11}$$

$$\text{and} \quad W(e^{i\theta}) \equiv W(e^{i\theta};a,b,c,d|q) = V(e^{i\theta})V(e^{-i\theta}) \tag{7.7.12}$$

we find that (7.7.10) can be expressed in the form

$$\begin{aligned} &q^{-n}(1-q^n)(1-abcdq^{n-1})V(e^{i\theta})V(e^{-i\theta})p_n(x) \\ &= \delta_q\left[\left\{E_q^+V(e^{i\theta})\right\}\left\{E_q^-V(e^{-i\theta})\right\}\left\{\delta_q p_n(x)\right\}\right]. \end{aligned} \tag{7.7.13}$$

Combining (7.7.7) and (7.7.13) we have

$$\begin{aligned} &-q^{-n/2}V(e^{i\theta})V(e^{-i\theta})p_n(x;a,b,c,d|q) \\ &= \delta_q\left[\left\{E_q^+V(e^{i\theta})\right\}\left\{E_q^-V(e^{-i\theta})\right\}\left(e^{i\theta}-e^{-i\theta}\right)\right. \\ &\quad \left.\cdot\, p_{n-1}\left(x;aq^{\frac{1}{2}},bq^{\frac{1}{2}},cq^{\frac{1}{2}},dq^{\frac{1}{2}}|q\right)\right]. \end{aligned} \tag{7.7.14}$$

Since

$$\begin{aligned} &\left(e^{i\theta}-e^{-i\theta}\right)E_q^+V(e^{i\theta})E_q^-V(e^{-i\theta}) \\ &= \frac{\left(e^{i\theta}-e^{-i\theta}\right)\left(qe^{2i\theta},qe^{-2i\theta};q\right)_\infty}{h(\cos\theta;aq^{\frac{1}{2}})h(\cos\theta;bq^{\frac{1}{2}})h(\cos\theta;cq^{\frac{1}{2}})h(\cos\theta;dq^{\frac{1}{2}})} \\ &= -\frac{W\left(e^{i\theta};aq^{\frac{1}{2}},bq^{\frac{1}{2}},cq^{\frac{1}{2}},dq^{\frac{1}{2}}|q\right)}{e^{i\theta}-e^{-i\theta}}, \end{aligned}$$

(7.7.14) can be written in a slightly better form

$$\begin{aligned} &q^{-n/2}W(e^{i\theta};a,b,c,d|q)p_n(x;a,b,c,d|q) \\ &= \delta_q\left[(e^{i\theta}-e^{-i\theta})^{-1}W(e^{i\theta};aq^{\frac{1}{2}},bq^{\frac{1}{2}},cq^{\frac{1}{2}},dq^{\frac{1}{2}}|q)\right. \\ &\quad \left.\cdot\, p_{n-1}(x;aq^{\frac{1}{2}},bq^{\frac{1}{2}},cq^{\frac{1}{2}},dq^{\frac{1}{2}}|q)\right]. \end{aligned} \tag{7.7.15}$$

Observing that,

$$w(x;a,b,c,d|q) = W(e^{i\theta};a,b,c,d|q)$$

we find by iterating (7.7.15) that

$$\begin{aligned} &\sqrt{1-x^2}\,w(x;a,b,c,d|q)p_n(x;a,b,c,d|q) \\ &= (-1)^k\left(\frac{1-q}{2}\right)^k q^{nk/2-k(k+1)/4} \\ &\quad \cdot D_q^k\left[\sqrt{1-x^2}\,w(x;aq^{\frac{k}{2}},bq^{\frac{k}{2}},cq^{\frac{k}{2}},dq^{\frac{k}{2}}|q)p_{n-k}(x;aq^{\frac{k}{2}},bq^{\frac{k}{2}},cq^{\frac{k}{2}},dq^{\frac{k}{2}}|q)\right] \\ &= (-1)^n\left(\frac{1-q}{2}\right)^n q^{(n^2-n)/4}D_q^n\left[\sqrt{1-x^2}\,w(x;aq^{n/2},bq^{n/2},cq^{n/2},dq^{n/2}|q)\right]. \end{aligned} \tag{7.7.16}$$

This gives a Rodrigues-type formula for the Askey-Wilson polynomials.

By combining (7.7.7) and (7.7.15) it can be easily seen that the polynomials $p_n(x) = p_n(x; a, b, c, d|q)$ satisfy the second-order difference equation

$$D_q\left[\sqrt{1-x^2}w(x; aq^{\frac{1}{2}}, bq^{\frac{1}{2}}, cq^{\frac{1}{2}}, dq^{\frac{1}{2}}|q)D_q p_n(x)\right] + \lambda_n\sqrt{1-x^2}w(x; a, b, c, d|q)p_n(x) = 0, \tag{7.7.17}$$

where

$$\lambda_n = -4q(1-q^{-n})(1-abcdq^{n-1})(1-q)^{-2}. \tag{7.7.18}$$

Exercises 7

7.1 If $\{p_n(x)\}$ is an orthogonal system of polynomials on $(-\infty, \infty)$ with respect to a positive measure $d\alpha(x)$ that has infinitely many points of support, prove that they satisfy a three-term recurrence relation of the form

$$xp_n(x) = A_n p_{n+1}(x) + B_n p_n(x) + C_n p_{n-1}(x), \quad n \geq 0,$$

with $p_{-1}(x) = 0$, $p_0(x) = 1$, where A_n, B_n, C_n are real and $A_n C_{n+1} > 0$ for $n \geq 0$.

7.2 Let $p_0(x), p_1(x), \ldots, p_N(x)$ be a system of polynomials that satisfies a three-term recurrence relation

$$xp_n(x) = A_n p_{n+1}(x) + B_n p_n(x) + C_n p_{n-1}(x),$$

$n = 0, 1, \ldots, N$, where $p_{-1}(x) = 0, p_0(x) = 1$. Prove the Christoffel-Darboux formula

$$(x-y)\sum_{j=0}^{n} p_j(x)p_j(y)v_j = A_n v_n \left[p_{n+1}(x)p_n(y) - p_n(x)p_{n+1}(y)\right],$$

$0 \leq n \leq N$, where $v_0 = 1$ and $v_n C_n = v_{n-1}A_{n-1}, 1 \leq n \leq N$. Deduce that

$$\sum_{j=0}^{n} p_j^2(x)v_j = A_n v_n \left[p'_{n+1}(x)p_n(x) - p'_n(x)p_{n+1}(x)\right]$$

and hence

$$\sum_{j=0}^{N} p_j^2(x_k)v_j = A_N v_N p_N(x_k)p'_{N+1}(x_k)$$

if x_k is a zero of $p_{N+1}(x)$.

7.3 Show that when $n = N$, the recurrence relation (7.2.1) reduces to

$$\left(1-q^{-j}\right)\left(1-cdq^{j+1}\right)p_N(x_j) = C_N\left[p_N(x_j) - p_{N-1}(x_j)\right],$$

where $p_n(x_j)$ is given by (7.2.11), x_j by (7.2.9), and A_n and C_n by (7.2.5) and (7.2.6). Hence show that (7.2.1) holds with $x = x_j, j = 0, 1, \ldots, N$. (Askey and Wilson [1979])

7.4 If $p_n(x_j)$, v_n and w_j are defined by (7.2.11), (7.2.3) and (7.2.15), respectively, prove directly (i.e. without the use of Favard's theorem) that

$$\sum_{j=0}^{N} p_m(x_j)p_n(x_j)w_j = v_n^{-1}\sum_{j=0}^{N} w_j\ \delta_{m,n}$$

and

$$\sum_{n=0}^{N} p_n(x_j)p_n(x_k)v_n = w_j^{-1}\sum_{n=0}^{N} w_n\ \delta_{j,k}.$$

[Hint: First transform one of the polynomials, say $p_n(x)$, to be a multiple of the ${}_4\phi_3$ series on the left side of (7.2.14)].

7.5 Let one of a, b, c, d be a nonnegative integer power of q^{-1} and let

$$\phi(a,b) = {}_4\phi_3\left[\begin{matrix} a, b, c, d \\ e, f, g \end{matrix}; q, q\right],$$

where $efg = abcdq$. Prove the following contiguous relation

$$A\phi(aq^{-1}, bq) + B\phi(a,b) + C\phi(aq, bq^{-1}) = 0,$$

where

$$\begin{aligned} A &= b(1-b)(aq-b)(a-e)(a-f)(a-g), \\ B &= ab(a-bq)(a-b)(aq-b)(1-c)(1-d) \\ &\quad - b(1-b)(aq-b)(a-e)(a-f)(a-g) \\ &\quad + a(1-a)(a-bq)(e-b)(f-b)(g-b), \\ C &= -a(1-a)(a-bq)(e-b)(f-b)(g-b). \end{aligned}$$

(Askey and Wilson [1979])

7.6 Determine the conditions that a, b, c, d must satisfy so that $A_n C_{n+1} > 0$ for $0 \le n \le N-1$, where A_n and C_n are as defined in (7.2.5) and (7.2.6) and one of aq, cq, bdq is q^{-N}, N a nonnegative integer.

7.7 Prove (7.2.22) directly by using the appropriate transformation and summation formulas derived in Chapters 1-3.

7.8 (i) Prove that the *q-Krawtchouk polynomials*

$$K_n(x; a, N; q) = {}_3\phi_2\left[\begin{matrix} q^{-n}, x, -a^{-1}q^n \\ q^{-N}, 0 \end{matrix}; q, q\right]$$

satisfy the orthogonality relation

$$\begin{aligned} &\sum_{x=0}^{N} K_m(q^{-x}; a, N; q)K_n(q^{-x}; a, N; q)\frac{(q^{-N}; q)_x}{(q; q)_x}(-a)^x \\ &= \left(-qa^{-1}; q\right)_N a^N q^{-\binom{N+1}{2}} \frac{(q; q)_n(1+a^{-1})(-a^{-1}q^{N+1}; q)_n}{(-a^{-1}; q)_n(1+a^{-1}q^{2n})(q^{-N}; q)_n} \\ &\quad \cdot \left(-aq^{N+1}\right)^{-n} q^{n(n+1)}\delta_{m,n}, \end{aligned}$$

and find their three-term recurrence relation. (Stanton [1980b])

(ii) Let

$$K_n(x;a,N|q) = {}_2\phi_1(q^{-n}, x; q^{-N}; q, aq^{n+1})$$

be another family of *q-Krawtchouk polynomials.* Prove that they satisfy the orthogonality relation

$$\sum_{x=0}^{N} K_m(q^{-x};a,N|q)\ K_n(q^{-x};a,N|q)\frac{(aq;q)_{N-x}(-1)^{N-x}q^{\binom{x}{2}}}{(q;q)_x(q;q)_{N-x}}$$

$$= \frac{(q,aq;q)_n(q;q)_{N-n}}{(q,q;q)_N}(-1)^n a^N q^{\binom{N+n+1}{2}-n(n+1)}\delta_{m,n}.$$

7.9 Prove that

$$xp_n(x) = A_n\left[p_{n+1}(x) - p_n(x)\right] - C_n\left[p_n(x) - p_{n-1}(x)\right], \quad n \geq 0,$$

where $p_n(x) = p_n(x;a,b;q)$ are the little q-Jacobi polynomials and

$$A_n = \frac{-q^n\,(1-aq^{n+1)})\,(1-abq^{n+1})}{(1-abq^{2n+1})\,(1-abq^{2n+2})},\ C_n = \frac{(1-q^n)\,(1-bq^n)\,(-aq^n)}{(1-abq^{2n})\,(1-abq^{2n+1})}.$$

7.10 Prove that

$$(x-1)P_n(x) = A_n\left[P_{n+1}(x) - P_n(x)\right] + C_n\left[P_n(x) - P_{n-1}(x)\right], \quad n \geq 0,$$

where $P_n(x) = P_n(x;a,b,c;q)$ are the big q-Jacobi polynomials and

$$A_n = \frac{(1-aq^{n+1})\,(1-cq^{n+1})\,(1-abq^{n+1})}{(1-abq^{2n+1})\,(1-abq^{2n+2})},$$

$$C_n = \frac{(1-q^n)\,(1-bq^n)\,(1-abc^{-1}q^n)}{(1-abq^{2n})\,(1-abq^{2n+1})}acq^{n+1}.$$

7.11 The *affine q-Krawtchouk polynomials* are defined by

$$K_n^{Aff}\,(x;a,N;q) = {}_3\phi_2\left[\begin{matrix} q^{-n}, & x, & 0 \\ & aq, & q^{-N} \end{matrix}; q, q\right],\ 0 < aq < 1.$$

Prove that they satisfy the orthogonality relation

$$\sum_{x=0}^{N} K_m^{Aff}\,(q^{-x};a,N;q)\ K_n^{Aff}\,(q^{-x};a,N;q)\ \frac{(aq,q^{-N};q)_x}{(q;q)_x}\left(-\frac{q^{N-1}}{a}\right)^x q^{-\binom{x}{2}}$$

$$= \frac{\delta_{m,n}}{h_n}, \quad m,n = 0,1,\ldots,N,$$

where

$$h_n = \frac{(aq,q^{-N};q)_n}{(q;q)_n}(-1)^n(aq)^{N-n}q^{Nn-\binom{n}{2}}.$$

(Delsarte [1976a,b], Dunkl [1977])

7.12 The *q-Meixner polynomials* are defined by

$$M_n(x;a,c;q) = {}_2\phi_1\left(q^{-n}, x; aq; q, -q^{n+1}/c\right),$$

with $0 < aq < 1$ and $c > 0$. Show that they satisfy the orthogonality relation

$$\sum_{x=0}^{\infty} M_m(q^{-x}; a, c; q) M_n(q^{-x}; a, c; q) \frac{(aq; q)_x}{(q, -acq; q)_x} c^x q^{\binom{x}{2}} = \frac{\delta_{m,n}}{h_n},$$

where

$$h_n = \frac{(-acq; q)_\infty (aq; q)_n}{(-c; q)_\infty (q, -qc^{-1}; q)_n} q^n.$$

(When $a = q^{-r-1}$, the q-Meixner polynomials reduce to the q-Krawtchouk polynomials considered in Koornwinder [1989c].)

7.13 The *q-Charlier polynomials* are defined by

$$c_n(x; a; q) = {}_2\phi_1\left(q^{-n}, x; 0; q, -q^{n+1}/a\right).$$

Show that

$$\sum_{x=0}^{\infty} c_m(q^{-x}; a; q) c_n(q^{-x}; a; q) \frac{a^x}{(q; q)_x} q^{\binom{x}{2}}$$

$$= \frac{q^n \delta_{m,n}}{(-a; q)_\infty (q, -qa^{-1}; q)_n}.$$

7.14 Show that, for $x = \cos\theta$,

$$C_n(x; q|q) = \frac{\sin(n+1)\theta}{\sin\theta} = U_n(x), \quad n \geq 0$$

and

$$\lim_{\beta\to 1} \frac{1-q^n}{2(1-\beta)} C_n(x; \beta|q) = \cos n\theta = T_n(x), \quad n \geq 1,$$

where $T_n(x)$ and $U_n(x)$ are the Tchebichef polynomials of the first and second kind, respectively.

7.15 Verify that formulas (7.6.2) and (7.6.14) follow from the q-analogue of the Fields-Wimp formula (3.7.9).

7.16 Let $x = \cos\theta, |t| < 1, |q| < 1$. Show that

$$\sum_{n=0}^{\infty} C_n(x; \beta|q) \frac{(\lambda; q)_n}{(\beta^2; q)_n} t^n = \frac{2i\sin\theta}{(1-q)W_\beta(x|q)} \frac{(\beta, \beta; q)_\infty}{(q, \beta^2; q)_\infty}$$

$$\cdot \int_{e^{i\theta}}^{e^{-i\theta}} \frac{\left(que^{i\theta}, que^{-i\theta}, \lambda tu; q\right)_\infty}{\left(\beta ue^{i\theta}, \beta ue^{-i\theta}, tu; q\right)_\infty} d_q u$$

and deduce that

(*i*)
$$\sum_{n=0}^{\infty} C_n(x; \beta|q) \frac{t^n}{(\beta^2; q)_n}$$
$$= (te^{-i\theta}; q)_\infty^{-1} \, {}_2\phi_1\left(\beta, \beta e^{-2i\theta}; \beta^2; q, te^{i\theta}\right).$$

(*ii*)
$$\sum_{n=0}^{\infty} C_n(x; \beta|q) q^{\binom{n}{2}} \frac{(\beta t)^n}{(\beta^2; q)_n}$$
$$= (-te^{-i\theta}; q)_\infty \, {}_2\phi_1\left(\beta, \beta e^{2i\theta}; \beta^2; q, -te^{-i\theta}\right).$$

7.17 Using (1.8.1), or otherwise, prove that

$$C_n(0;\beta|q) = \begin{cases} 0, & \text{if } n \text{ is odd}, \\ (-1)^{n/2}\dfrac{(\beta^2;q^2)_{n/2}}{(q^2;q^2)_{n/2}}, & \text{if } n \text{ is even}. \end{cases}$$

7.18 If $-1 < q, \beta < 1$, show that

$$|C_n(x;\beta|q)| \leq C_n(1;\beta|q).$$

7.19 Derive the recurrence relation

$$2xC_n(x;\beta|q) = \frac{1-q^{n+1}}{1-\beta q^n}C_{n+1}(x;\beta|q) + \frac{1-\beta^2 q^{n-1}}{1-\beta q^n}C_{n-1}(x;\beta|q), n \geq 0,$$

with $C_{-1}(x;\beta|q) = 0$, $C_0(x;\beta|q) = 1$.

7.20 Prove that

$$\int_0^\pi C_n(\cos\theta;\beta|q)\cos(n+2k)\theta\, W_\beta(\cos\theta|q)\,d\theta = \frac{\pi(\beta,\beta q;q)_\infty}{(q,\beta^2;q)_\infty}\beta^k$$
$$\cdot \frac{(\beta^{-1};q)_k(\beta^2;q)_n(q;q)_{n+k}}{(q;q)_k(q;q)_n(\beta q;q)_{n+k}}\frac{1-q^{n+2k}}{1-q^{n+k}}, \quad n,k \geq 0,$$

where $W_\beta(x|q)$ is defined in (7.4.6).

7.21 Using (7.4.15) and (7.6.14) prove that

$$\frac{h(x;\gamma)}{h(x;\beta)}C_n(x;\beta|q) = \frac{(\gamma^2,\beta,\beta q;q)_\infty}{(\gamma,\gamma q,\beta^2;q)_\infty}\sum_{k=0}^\infty d_{k,n}C_{n+2k}(x;\gamma|q),$$

where $h(x;a)$ is as defined in (6.1.2) and

$$d_{k,n} = \frac{\beta^k(\gamma/\beta;q)_k(q;q)_{n+2k}(\beta^2;q)_n(\gamma;q)_{n+k}(1-\gamma q^{n+2k})}{(q;q)_k(\gamma^2;q)_{n+2k}(q;q)_n(\beta q;q)_{n+k}(1-\gamma)}.$$

(Askey and Ismail [1983])

7.22 Prove that the continuous q-Hermite polynomials defined in Ex. 1.28 satisfy the orthogonality relation

$$\int_0^\pi H_m(\cos\theta|q)H_n(\cos\theta|q)\left|(e^{2i\theta};q)_\infty\right|^2 d\theta = \frac{2\pi(q;q)_n}{(q;q)_\infty}\delta_{m,n}.$$

7.23 Setting

$$C_n(x;\beta|q) = \frac{(\beta^2;q)_n}{(q;q)_n}c_n(x;\beta|q)$$

in Ex. 7.19, show that

$$2x(1-\beta q^n)c_n(x;\beta|q) = (1-\beta^2 q^n)c_{n+1}(x;\beta|q) + (1-q^n)c_{n-1}(x;\beta|q),$$

for $n \geq 0$, with $c_{-1}(x;\beta|q) = 0$, $c_0(x;\beta|q) = 1$. Now set $\beta = s^{\lambda k}$ and $q = s\omega_k$, where $\omega_k = \exp(2\pi i/k)$ is a k-th root of unity, divide the above recurrence relation by $1 - s\omega_k^n$ and take the limit as $s \to 1$ to show that the limiting polynomials, $c_n^\lambda(x;k)$, called the *sieved ultraspherical polynomials of the first kind*, satisfy the recurrence relation

$$2xc_n^\lambda(x;k) = c_{n+1}^\lambda(x;k) + c_{n-1}^\lambda(x;k), \quad n \neq mk,$$
$$2x(m+\lambda)c_{mk}^\lambda(x;k) = (m+2\lambda)c_{mk+1}^\lambda(x;k) + mc_{mk-1}^\lambda(x;k)$$

where $c_0^\lambda(x;k) = 1$ and $c_1^\lambda(x;k) = x$.
(Al-Salam, Allaway and Askey [1984b])

7.24 Rewrite the orthogonality relation (7.4.15) in terms of the sieved orthogonal polynomial $c_n(x;\beta|q)$ defined in Ex. 7.23 and set $\beta = s^{\lambda k}$ and $q = s\omega_k$. By carefully taking the limits of the q-shifted factorials prove that

$$\int_{-1}^{1} c_m^\lambda(x;k)c_n^\lambda(x;k)w(x)\,dx = \frac{\delta_{m,n}}{h_n}.$$

where

$$w(x) = 2^{2\lambda(k-1)}(1-x^2)^{-\frac{1}{2}} \prod_{j=0}^{k-1} |x^2 - \cos^2(\pi j/k)|^\lambda$$

and

$$h_n = \frac{\Gamma(\lambda+1)}{\Gamma(\frac{1}{2})\Gamma(\lambda+\frac{1}{2})} \frac{(\lambda+1)_{\lfloor n/k \rfloor}(2\lambda)_{\lceil n/k \rceil}}{(1)_{\lfloor n/k \rfloor}(\lambda)_{\lceil n/k \rceil}},$$

where the *roof* and *floor* functions are defined by

$$\lceil a \rceil = \text{ smallest integer greater than or equal to } a,$$
$$\lfloor a \rfloor = \text{ largest integer less than or equal to } a.$$

(Al-Salam, Allaway and Askey [1984b])

7.25 The *sieved ultraspherical polynomials of the second kind* are defined by

$$B_n^\lambda(x;k) = \lim_{s\to 1} C_n(x; s^{\lambda k+1}\omega_k|s\omega_k), \ \omega_k = \exp(2\pi i/k).$$

Show that $B_n^\lambda(x;k)$ satisfies the recurrence relation

$$2xB_n^\lambda(x;k) = B_{n+1}^\lambda(x;k) + B_{n-1}^\lambda(x;k), \quad n+1 \neq mk,$$
$$2x(m+\lambda)B_{mk-1}^\lambda(x;k) = mB_{mk}^\lambda(x;k) + (m+2\lambda)B_{mk-2}^\lambda(x;k),$$

where $B_0^\lambda(x;k) = 1$; $B_1^\lambda(x;k) = 2x$ if $k \geq 2$; $B_1^\lambda(x;1) = 2(\lambda+1)x$. Show also that $B_n^\lambda(x;k)$ satisfies the orthogonality relation

$$\int_{-1}^{1} B_m^\lambda(x;k)B_n^\lambda(x;k)w(x)\,dx = \frac{\delta_{m,n}}{h_n},$$

where

$$w(x) = 2^{2\lambda(k-1)}(1-x^2)^{\frac{1}{2}} \prod_{j=0}^{k-1} |x^2 - \cos^2(\pi j/k)|^\lambda$$

and

$$h_n = \frac{2\Gamma(\lambda+1)}{\Gamma(\frac{1}{2})\Gamma(\lambda+\frac{1}{2})} \frac{(1)_{\lfloor n/k \rfloor}(\lambda+1)_{\lfloor \frac{n+1}{k} \rfloor}}{(\lambda+1)_{\lfloor n/k \rfloor}(2\lambda+1)_{\lfloor \frac{n+1}{k} \rfloor}}.$$

(Al-Salam, Allaway and Askey [1984b])

7.26 Using (2.5.1) show that

$$p_n(x;a,b,c,d|q) = \frac{(ab,ac,bc,q;q)_n}{(abcdq^{-1};q)_n}$$
$$\cdot \sum_{k=0}^{n} \frac{\left(abcq^{-1}e^{i\theta};q\right)_k \left(1-abce^{i\theta}q^{2k-1}\right)\left(ae^{i\theta},be^{i\theta},ce^{i\theta};q\right)_k}{(q;q)_k(1-abcq^{-1}e^{i\theta})(bc,ac,ab;q)_k}$$
$$\cdot \frac{\left(abcdq^{-1};q\right)_{n+k}\left(de^{-i\theta};q\right)_{n-k}}{\left(abce^{i\theta};q\right)_{n+k}(q;q)_{n-k}} e^{i(n-2k)\theta}.$$

Deduce that the polynomials

$$p_n(x) = \lim_{q\to 0} p_n(x;a,b,c,d|q)$$

are given by

$$\begin{aligned} p_0(x) &= 1 = U_0(x),\\ p_1(x) &= (1-s_4)U_1(x) + (s_3-s_1)U_0(x),\\ p_2(x) &= U_2(x) - s_1U_1(x) + (s_2-s_4)U_0(x),\\ p_n(x) &= \sum_{k=0}^{4}(-1)^k s_k U_{n-k}(x), \quad n\ge 3, \end{aligned}$$

where

$$s_0 = 1,\ s_1 = a+b+c+d,\ s_2 = ab+ac+ad+bc+bd+cd,$$
$$s_3 = abc+abd+acd+bcd,\ s_4 = abcd,$$

and $U_n(\cos\theta) = \sin(n+1)\theta/\sin\theta$, $U_{-1}(x) = 0$. When $\max(|a|,|b|,|c|,|d|) < 1$ show that these polynomials satisfy the orthogonality relation

$$\frac{2}{\pi}\int_{-1}^{1} \frac{p_m(x)p_n(x)(1-x^2)^{\frac12}\,dx}{(1-2ax+a^2)(1-2bx+b^2)(1-2cx+c^2)(1-2dx+d^2)}$$
$$= \begin{cases} 0, & m\ne n,\\ \dfrac{1-abcd}{(1-ab)(1-ac)(1-ad)(1-bc)(1-bd)(1-cd)}, & m=n=0,\\ 1-abcd, & m=n=1,\\ 1, & m=n\ge 2. \end{cases}$$

(Askey and Wilson [1985])

7.27 Prove that

(*i*) $$p_n(\cos\theta;q,-q,q^{\frac12},-q^{\frac12}|q) = \left(q^{n+2};q\right)_n \frac{\sin(n+1)\theta}{\sin\theta},$$

(*ii*) $$p_n(\cos\theta;1,-1,q^{\frac12},-q^{\frac12}|q) = 2(q^n;q)_n\cos n\theta, \quad n\ge 1,$$

(*iii*) $$p_n(\cos\theta;q,-1,q^{\frac12},-q^{\frac12}|q) = \left(q^{n+1};q\right)_n \frac{\sin(n+\frac12)\theta}{\sin\frac{\theta}{2}},$$

(*iv*) $$p_n(\cos\theta;1,-q,q^{\frac12},-q^{\frac12}|q) = \left(q^{n+1};q\right)_n \frac{\cos(n+\frac12)\theta}{\cos\frac{\theta}{2}}.$$

7.28 Use the orthogonality relations (7.5.28) and (7.5.29) to prove the quadratic transformation formula (7.5.26).

7.29 Verify the orthogonality relations (7.5.28) and (7.5.29).

7.30 Verify formula (7.5.36).

7.31 Suppose that a, b, c, d are complex parameters with $\max(|b|, |c|, |d|, |q|) < 1 < |a|$ such that $|aq^{N+1}| < 1 < |aq^N|$, where N is a nonnegative integer. Use (6.6.12) to prove that

$$\int_{-1}^{1} p_m(x)p_n(x)w(x; a, b, c, d|q)\, dx + 2\pi \sum_{k=0}^{N} p_m(x_k)p_n(x_k)w_k$$
$$= \frac{\delta_{m,n}}{h_n(a, b, c, d|q)},$$

where $p_n(x) = p_n(x; a, b, c, d|q)$, $x_k = \frac{1}{2}\left(aq^k + a^{-1}q^{-k}\right)$ and w_k is given by (7.5.22).

7.32 Prove that

$$(i) \qquad P_n^{(\alpha,\beta)}(-x; q) = (-1)^n P_n^{(\beta,\alpha)}(x; q),$$
$$(ii) \qquad P_n^{(\alpha,\beta)}(-x|q) = (-1)^n P_n^{(\beta,\alpha)}(x|q).$$

7.33 Using (7.4.1), (7.4.7) and (2.11.2) prove that

$$(i) \qquad \sum_{n=0}^{\infty} \frac{(q; q)_n}{(\beta^2; q)_n} C_n(x; \beta|q)C_n(y; \beta|q)t^n$$
$$= \frac{(t^2, \beta; q)_\infty}{(\beta t^2, \beta^2; q)_\infty} \left| \frac{\left(\beta te^{i\theta+i\phi}, \beta te^{i\theta-i\phi}; q\right)_\infty}{\left(te^{i\theta+i\phi}, te^{i\theta-i\phi}; q\right)_\infty} \right|^2$$
$$\cdot {}_8W_7\left(\beta t^2 q^{-1}; \beta, te^{i\theta+i\phi}, te^{-i\theta-i\phi}, te^{i\theta-i\phi}, te^{i\phi-i\theta}; q, \beta\right),$$

$$(ii) \qquad \sum_{n=0}^{\infty} \frac{(q; q)_n}{(\beta^2; q)_n} \frac{1-\beta q^n}{1-\beta} C_n(x; \beta|q)C_n(y; \beta|q)t^n$$
$$= \frac{\left(t^2, \beta; q\right)_\infty}{\left(q\beta t^2, \beta^2; q\right)_\infty} \left| \frac{\left(\beta te^{i\theta+i\phi}, q\beta te^{i\theta-i\phi}; q\right)_\infty}{\left(te^{i\theta+i\phi}, te^{i\theta-i\phi}; q\right)_\infty} \right|^2$$
$$\cdot {}_8W_7\left(\beta t^2; \beta, qte^{i\theta+i\phi}, qte^{-i\theta-i\phi}, te^{i\theta-i\phi}, te^{i\phi-i\theta}; q, \beta\right),$$

where $-1 < q, \beta, t < 1$ and $x = \cos\theta,\ y = \cos\phi$.
(Gasper and Rahman [1983a], Rahman and Verma [1986a])

7.34 Show that

$$p_n(\cos\theta; a, b, c, d|q) = A^{-1}(\theta)(ab, ac, bc; q)_n$$
$$\cdot \int_{qe^{i\theta}/d}^{qe^{-i\theta}/d} \frac{\left(due^{i\theta}, due^{-i\theta}, abcdu/q; q\right)_\infty}{(dau/q, dbu/q, dcu/q; q)_\infty} \frac{(q/u; q)_n}{(abcdu/q; q)_n} \left(\frac{du}{q}\right)^n d_q u,$$

where

$$A(\theta) = \frac{-iq(1-q)}{2d}(q, ab, ac, bc; q)_\infty h(\cos\theta; d)w(\cos\theta; a, b, c, d|q).$$

Hence show that

$$\sum_{n=0}^{\infty} \frac{(a^2c^2;q)_n p_n(\cos\theta; a, aq^{\frac{1}{2}}, -c, -cq^{\frac{1}{2}}|q)}{(q;q)_n(a^2q^{\frac{1}{2}}, -ac, -acq^{\frac{1}{2}};q)_n} t^n$$
$$= \frac{(at, ac^2t, -acte^{i\theta}, -acte^{-i\theta};q)_\infty}{(-ct, -a^2ct, te^{i\theta}, te^{-i\theta};q)_\infty}$$
$$\cdot {}_8W_7\left(-a^2ctq^{-1}; -ac, -a/c, -ctq^{-\frac{1}{2}}, ae^{i\theta}, ae^{-i\theta}; q, -ctq^{\frac{1}{2}}\right).$$

(Gasper and Rahman [1986])

7.35 Show that

$$\int_{-1}^{1} w(y; a, b, \mu e^{i\theta}, \mu e^{-i\theta}|q) p_n(y; a, b, c, d|q)\, dy$$
$$= \frac{2\pi(ab\mu^2;q)_\infty}{(q, ab, \mu^2, a\mu e^{i\theta}, a\mu e^{-i\theta}, b\mu e^{i\theta}, b\mu e^{-i\theta};q)_\infty} \frac{(ab;q)_n}{(ab\mu^2;q)_n} \mu^n$$
$$\cdot p_n(\cos\theta; a\mu, b\mu, c\mu^{-1}, d\mu^{-1}|q),$$

where $\max(|a|, |b|, |\mu|) < 1$.

7.36 Show that if for $|q| < 1$ we define

$$(a;q)_\nu = \frac{(a;q)_\infty}{(aq^\nu;q)_\infty},$$

where ν is a complex number and the principal value of q^ν is taken, then (7.7.6) extends to

$$D_q\left(ae^{i\theta}, ae^{-i\theta};q\right)_\nu = -\frac{2a(1-q^\nu)}{1-q}\left(aq^{\frac{1}{2}}e^{i\theta}, aq^{\frac{1}{2}}e^{-i\theta};q\right)_{\nu-1}.$$

7.37 Let $n = 1, 2, \ldots, r$, $x = \cos\theta$, and

$$U_n(x) = A_{n,r}\left(q^{\nu+1}e^{i\theta}, q^{\nu+1}e^{-i\theta};q\right)_n$$
$$\cdot {}_6\phi_5\left[\begin{matrix} q^{n-r}, q^{n+r+2\nu+2\lambda+1}, q^{n+\nu+\frac{1}{2}}, q^{n+2\nu}, q^{n+\nu+1}e^{i\theta}, q^{n+\nu+1}e^{-i\theta} \\ q^{2n+2\nu+1}, q^{n+\nu+\lambda+1}, q^{n+2\nu+1}, -q^{n+\nu+\lambda+1}, -q^{n+\nu+\frac{1}{2}} \end{matrix}; q, q\right]$$

with

$$A_{n,r} = \frac{(q;q)_n(q^{2\nu+2\lambda+1};q)_{n+r}q^{\frac{3}{2}n^2+n(2\nu+\frac{1}{2}-r)}}{(q;q)_{r-n}(q^{\nu+1}, q^{2\nu+1}, q^{\nu+\lambda+1}, -q^{\nu+1}, -q^{\nu+\frac{1}{2}}, -q^{\nu+\frac{1}{2}}, -q^{\nu+\lambda+1};q)_n}.$$

Show that $U_n(x)$ satisfies the q-differential equation

$$D_q[(q^{n+\nu+1}e^{i\theta}, q^{n+\nu+1}e^{-i\theta};q)_{-n}U_n(x)]$$
$$= -\frac{1-q^{n+2\nu}}{1-q^{n+1}}\left(q^{n+\nu+\frac{3}{2}}e^{i\theta}, q^{n+\nu+\frac{3}{2}}e^{-i\theta};q\right)_{-2n-2\nu-1}$$
$$\cdot D_q[(q^{-n-\nu}e^{i\theta}, q^{-n-\nu}e^{-i\theta};q)_{n+2\nu+1}U_{n+1}(x)].$$

(Gasper [1989b])

7.38 Show that the *discrete q-Hermite polynomials*

$$H_n(x;q) = \sum_{k=0}^{[n/2]} \frac{(q;q)_n}{(q^2;q^2)_k(q;q)_{n-2k}}(-1)^k q^{k(k-1)} x^{n-2k}$$

satisfy the recurrence relation

$$H_{n+1}(x;q) = xH_n(x;q) - q^{n-1}(1-q^n)H_{n-1}(x;q), \quad n \geq 1,$$

and the orthogonality relation

$$\int_{-1}^{1} H_m(x;q)H_n(x;q)\, d\psi(x) = q^{\binom{n}{2}}(q;q)_n\delta_{m,n},$$

where $\psi(x)$ is a step function with jumps

$$\frac{|x|}{2}\frac{(x^2q^2,q;q^2)_\infty}{(q^2;q^2)_\infty}$$

at the points $x = \pm q^j$, $j = 0, 1, 2, \ldots$.
(Al-Salam and Carlitz [1965], Al-Salam and Ismail [1988])

7.39 Let $a < 0$ and $0 < q < 1$. Show that

$$\int_{-\infty}^{\infty} U_m^{(a)}(x;q)U_n^{(a)}(x;q)\, d\alpha^{(a)}(x)$$
$$= (1-a)(-a)^n(q;q)_n q^{\binom{n}{2}}\delta_{m,n},$$

where

$$U_n^{(a)}(x;q) = (-a)^n q^{\binom{n}{2}}\, {}_2\phi_1(q^{-n}, x^{-1}; 0; q, qx/a)$$

and $\alpha^{(a)}(x)$ is a step function with jumps

$$\frac{q^k}{(aq;q)_\infty(q,q/a;q)_k}$$

at the points $x = q^k$, $k = 0, 1, \ldots$, and jumps

$$\frac{-aq^k}{(q/a;q)_\infty(q,aq;q)_k}$$

at the points $x = aq^k$, $k = 0, 1, \ldots$. Verify that when $a = -1$ this orthogonality relation reduces to the orthogonality relation for the discrete q-Hermite polynomials in Ex.7.38.
(Al-Salam and Carlitz [1957, 1965], Chihara [1978, (10.7)] and use the $z = q$ case of (3.3.5))

7.40 Show that if

$$h_n(x;q) = \sum_{k=0}^{n} \frac{(q;q)_n}{(q;q)_k(q;q)_{n-k}} x^k$$

then

$$h_n^2(x;q) - h_{n+1}(x;q)h_{n-1}(x;q)$$
$$= (1-q)(q;q)_{n-1}\sum_{k=0}^{n} h_{n-k}^2(x;q)\frac{q^{n-k}x^k}{(q;q)_{n-k}}, \quad n \geq 1.$$

Deduce that the polynomials $h_n(x;q)$ satisfy the Turán-type inequality

$$h_n^2(x;q) - h_{n+1}(x;q)h_{n-1}(x;q) \geq 0$$

for $x \geq 0$ when $0 < q < 1$ and $n = 1, 2, \ldots$.
(Carlitz [1957b])

7.41 Derive the addition formula

$$\begin{aligned}
&p_n(q^z;1,1;q)\, p_y(q^z;q^x,0;q) \\
&= p_n(q^{x+y};1,1;q)\, p_n(q^y;1,1;q)\, p_y(q^z;q^x,0;q) \\
&\quad + \sum_{k=1}^{n} \frac{(q;q)_{x+y+k}(q;q)_{n+k}\, q^{k(k+y-n)}}{(q;q)_{x+y}(q;q)_{n-k}(q;q)_k^2} \\
&\quad \cdot p_{n-k}(q^{x+y};q^k,q^k;q)\, p_{n-k}(q^y;q^k,q^k;q) \\
&\quad \cdot p_{y+k}(q^z;q^x,0;q) + \sum_{k=1}^{n} \frac{(q;q)_y(q;q)_{n+k}\, q^{k(x+y-n+1)}}{(q;q)_{y-k}(q;q)_{n-k}(q;q)_k^2} \\
&\quad \cdot p_{n-k}(q^{x+y-k};q^k,q^k;q)\, p_{n-k}(q^{y-k};q^k,q^k;q)\, p_{y-k}(q^z;q^x,0;q)
\end{aligned}$$

where $x, y, z, n = 0, 1, \ldots,$ and $p_n(t;a,b;q)$ is the little q-Jacobi polynomial defined in Ex. 1.32.
(Koornwinder [1989b])

7.42 Derive the product formula

$$p_n(q^x;1,1;q)\, p_n(q^y;1,1;q) = (1-q)\sum_{z=0}^{\infty} p_n(q^z;1,1;q)\, K(q^x,q^y,q^z;q)q^z$$

where $x, y, z, n = 0, 1, \ldots,$ $p_n(t;a,b;q)$ is the little q-Jacobi polynomial, and

$$\begin{aligned}
K(q^x,q^y,q^z;q) &= \frac{(q^{x+1},q^{y+1},q^{z+1};q)_\infty}{(1-q)(q,q;q)_\infty}\, q^{xy+xz+yz} \\
&\quad \cdot \left\{{}_3\phi_2(q^{-x},q^{-y},q^{-z};0,0;q,q)\right\}^2.
\end{aligned}$$

(Koornwinder [1989b])

7.43 The *q-Laguerre polynomials* are defined by

$$L_n^\alpha(x;q) = \frac{(q^{\alpha+1};q)_n}{(q;q)_n}\, {}_1\phi_1(q^{-n};q^{\alpha+1};q,-xq^{n+\alpha+1}).$$

Show that if $\alpha > -1$ then these polynomials satisfy the orthogonality relation

$$\begin{aligned}
(i) \qquad &\int_0^\infty L_m^\alpha(x;q)L_n^\alpha(x;q)\frac{x^\alpha\, dx}{(-(1-q)x;q)_\infty} \\
&= \frac{\Gamma(\alpha+1)\Gamma(-\alpha)(q^{\alpha+1};q)_n}{\Gamma_q(-\alpha)(q;q)_n q^n}\,\delta_{m,n}
\end{aligned}$$

and the discrete orthogonality relation

$$(ii) \quad \sum_{k=-\infty}^{\infty} L_m^\alpha(cq^k;q)L_n^\alpha(cq^k;q)\frac{q^{k(\alpha+1)}}{(-c(1-q)q^k;q)_\infty} = A\frac{(q^{\alpha+1};q)_n}{(q;q)_n q^n}\,\delta_{m,n},$$

where

$$A = \frac{(q, -c(1-q)q^{\alpha+1}, -1/cq^{\alpha}(1-q); q)_\infty}{(q^{\alpha+1}, -c(1-q), -q/c(1-q); q)_\infty}.$$

(Moak [1981])

7.44 Let

$$v(y; a_1, a_2, a_3, a_4, a_5|q)$$
$$= \frac{h(y; 1, -1, q^{\frac{1}{2}}, -q^{\frac{1}{2}}, a_1a_2a_3a_4a_5)}{h(y; a_1, a_2, a_3, a_4, a_5)} (1-y^2)^{-\frac{1}{2}}, \quad y = \cos\phi.$$

and

$$g(a_1, a_2, a_3, a_4, a_5|q)$$
$$= \int_{-1}^{1} v(y; a_1, a_2, a_3, a_4, a_5|q)\, dy.$$

Show that

$$\int_{-1}^{1} v(y; a, b, c, \mu e^{i\theta}, \mu e^{-i\theta}|q) \frac{(abc\mu e^{i\theta}, abc\mu e^{-i\theta}; q)_n}{(abc\mu^2 e^{i\phi}, abc\mu^2 e^{-i\phi}; q)_n}$$
$$\cdot\, p_n(y; a, b, c, d|q)\, dy$$
$$= g(a, b, c, \mu e^{i\theta}, \mu e^{-i\theta}|q) \frac{(ab, ac, bc; q)_n}{(ab\mu^2, ac\mu^2, bc\mu^2; q)_n} \mu^n$$
$$\cdot\, p_n(x; a\mu, b\mu, c\mu, d\mu^{-1}|q), \quad x = \cos\theta,$$

where $p_n(x; a, b, c, d|q)$ are the Askey-Wilson polynomials defined in (7.5.2) and $\max(|a|, |b|, |c|, |\mu|, |q|) < 1$.
(Rahman [1988a])

Notes 7

§7.1 See also Atakishiyev and Suslov [1987a,b], Nikiforov and Uvarov [1988], and Szegő [1968, 1982].

§7.2 Andrews and Bressoud [1984] used the concept of a crossing number to provide a combinatorial interpretation of the q-Hahn polynomials. Koelink and Koornwinder [1989] showed that the q-Hahn and dual q-Hahn polynomials admit a quantum group theoretic interpretation, analogous to an interpretation of (dual) Hahn polynomials in terms of Clebsch-Gordon coefficients for $SU(2)$. For how Clebsch-Gordon coefficients arise in quantum mechanics, see Biedenharn and Louck [1981a,b]. L. Chihara [1987] considered the locations of zeros of q-Racah polynomials and employed her results to prove non-existence of perfect codes and tight designs in the classical association schemes. For the relationship between orthogonal polynomials and association schemes, see Bannai and Ito [1984], L. Chihara and Stanton [1986], Delsarte [1976b], and Leonard [1982].

§7.3 Al-Salam and Ismail [1977] constructed a family of reproducing kernels (bilinear formulas) for the little q-Jacobi polynomials. In Al-Salam and Ismail [1983] they considered a related family of orthogonal polynomials associated with the Rogers-Ramanujan continued fraction. A biorthogonal extension of the little q-Jacobi polynomials is studied in Al-Salam and Verma [1983a]. When $b = 0$ the little q-Jacobi polynomials reduce (after changing variables and renormalizing) to the *Wall polynomials*

$$W_n(x;b,q) = (-1)^n (b;q)_n q^{\binom{n+1}{2}} \sum_{j=0}^{n} \begin{bmatrix} n \\ j \end{bmatrix}_q \frac{q^{\binom{j}{2}}(-q^{-n}x)^j}{(b;q)_j}$$

and to the *generalized Stieltjes-Wigert polynomials*

$$S_n(x;p,q) = (-1)^n q^{-n(2n+1)/2} (p;q)_n \sum_{j=0}^{n} \begin{bmatrix} n \\ j \end{bmatrix}_q \frac{q^{j^2}(-q^{\frac{1}{2}}x)^j}{(p;q)_j},$$

which are q-analogues of the Laguerre polynomials that are different from those considered in Ex. 7.43. Since the Hamburger and Stieltjes moment problems corresponding to these polynomials are both indeterminate, there are infinitely many nonequivalent measures on $[0,\infty)$ for which these polynomials are orthogonal. See Chihara [1978, Chapter VI], [1968b, 1971, 1979, 1982, 1985], Al-Salam and Verma [1982b], L. Chihara and T.S. Chihara [1987], and Shohat and Tamarkin [1950].

§7.4 An integral of the product of two continuous q-ultraspherical polynomials and a q-ultraspherical function of the second kind is evaluated in Askey, Koornwinder and Rahman [1986]. Al-Salam, Allaway and Askey [1984a] gave a characterization of the continuous q-ultraspherical polynomials as orthogonal polynomial solutions of certain integral equations. Askey [1989b] showed that the polynomials $C_n(ix;\beta|q)$, $0 \le n \le N$, are orthogonal on the real line with respect to a positive measure when $0 < q < 1$ and $\beta > q^{-N}$.

§7.5 Asymptotic formulas and generating functions for the Askey-Wilson polynomials and their special cases are derived in Ismail and Wilson [1982] and Ismail [1986c]. Kalnins and Miller [1989a] employed symmetry techniques to give an elementary proof of the orthogonality relation for the Askey-Wilson polynomials.

§7.6 For additional results on connection coefficients (and the corresponding projection formulas), see Andrews [1979a], Gasper [1974, 1975a].

Ex. 7.8 Stanton [1981b] showed that the q-Krawtchouk polynomials $K_n(x;a,N;q)$ are spherical functions for three different Chevalley groups over finite fields and derived three addition theorems for these polynomials by decomposing the irreducible representations with respect to maximal parabolic subgroups. In Koornwinder [1989c] it is shown that the orthogonality relation for the q-Krawtchouk polynomials $K_n(x;a,N|q)$ expresses the fact that the matrix representations of the quantum group $S_\mu U(2)$ are unitary.

Ex. 7.11 The affine q-Krawtchouk polynomials are the eigenvalues of the association schemes of bilinear, alternating, symmetric and hermitian forms over a finite field (see Carlitz and Hodges [1955], Delsarte [1978], Delsarte and Goethals [1975], and Stanton [1981a,b, 1984]). L. Chihara and Stanton [1987] showed that the zeros of the affine q-Krawtchouk polynomials are never zero at integral values of x, and they gave some interlacing theorems for the zeros of q-Krawtchouk polynomials.

Ex. 7.22 Askey [1989b] proved that the polynomials $H_n(ix|q)$ are orthogonal on the real lime with respect to a positive measure when $q > 1$.

Exercises 7.23–7.25 Other sieved orthogonal polynomials are considered in Al-Salam and Chihara [1987], Askey [1984b], Askey and Shukla [1989], Charris and Ismail [1986, 1987], and Ismail [1985a, 1986a,b].

Exercises 7.38–7.40 For additional material on q-analogues of Hermite polynomials, see Allaway [1980], Al-Salam and Chihara [1976], Al-Salam and Ismail [1988], Carlitz [1963b, 1972], Chihara [1968a, 1982, 1985], Dehesa [1979], Désarménien [1982], Ismail [1985b], Ismail, Stanton and Viennot [1987], Lubinsky and Saff [1987], and Szegő [1926].

Ex. 7.41 Rahman [1989a] gave a simple proof for this addition formula. For derivations of the addition formula for Jacobi polynomials, see Koornwinder [1974a,b] and Laine [1982].

Ex. 7.43 See also Cigler [1981] and Pastro [1985]. In view of the two different orthogonality relations for the q-Laguerre polynomials, it follows that there are infinitely many measures for which these polynomials are orthogonal. The *Stieltjes-Wigert polynomials* (see Chihara [1978, pp. 172-174], Szegő [1975, p. 33] and the above Notes for §7.3)

$$s_n(x) = (-1)^n q^{(2n+1)/4} (q;q)_n^{-\frac{1}{2}} \sum_{j=0}^{n} \begin{bmatrix} n \\ j \end{bmatrix}_q q^{j^2} (-q^{\frac{1}{2}} x)^j,$$

which are orthogonal with respect to the log normal weight function

$$w(x) = k\pi^{-\frac{1}{2}} \exp(-k^2 \log^2 x), \qquad 0 < x < \infty,$$

where $q = \exp[-(2k^2)^{-1}]$ and $k > 0$, are a limit case of the q-Laguerre polynomials. Askey [1986] gave the orthogonality relation for these polynomials (with a slightly different definition) that follows as a limit case of the first orthogonality relation in this exercise. Al-Salam and Verma [1983b,c] studied a pair of biorthogonal sets of polynomials, called the q-Konhauser polynomials, which were suggested by the q-Laguerre polynomials.

8

FURTHER APPLICATIONS

8.1 Introduction

In this chapter we derive some formulas that are related to products of the q-orthogonal polynomials introduced in the previous chapter and use these formulas to obtain q-analogues of various product formulas, Poisson kernels and linearization formulas for ultraspherical and Jacobi polynomials. The method in Gasper and Rahman [1984] originates with the observation that since

$$\left(q^{-x}, aq^{x}; q\right)_j = \prod_{k=0}^{j-1} \left(1 - q^k(q^{-x} + aq^x) + aq^{2k}\right)$$

is a polynomial of degree j in powers of $q^{-x} + aq^x$, there must exist an expansion of the form

$$\begin{aligned} &\left(q^{-x}, aq^{x}; q\right)_j \left(q^{-x}, aq^{x}; q\right)_k \\ &= \sum_{m=0}^{j+k} A_m(j,k,a;q) \left(q^{-x}, aq^{x}; q\right)_m . \end{aligned} \tag{8.1.1}$$

Since, for $k \geq j$,

$${}_3\phi_2\left(q^{-j}, q^{k-x}, aq^{k+x}; aq^k, q^{1+k-j}; q, q\right) = \frac{\left(q^{-x}, aq^x; q\right)_j}{\left(q^{-k}, aq^k; q\right)_j}$$

by the q-Saalschütz formula (1.7.2), it is easy to verify that

$$\begin{aligned} &\left(q^{-x}, aq^{x}; q\right)_j \left(q^{-x}, aq^{x}; q\right)_k = (q;q)_j (q;q)_k (a;q)_{j+k} \\ &\cdot \sum_{m=\max(j,k)}^{j+k} \frac{\left(q^{-x}, aq^x; q\right)_m q^{\binom{j}{2}+\binom{k}{2}+\binom{m+1}{2}-m(j+k)} (-1)^{j+k+m}}{(a;q)_m (q;q)_{m-j} (q;q)_{m-k} (q;q)_{j+k-m}} . \end{aligned} \tag{8.1.2}$$

This linearizes the product on the left side and forms the basis for the product formulas derived in the following section.

Suppose $\{B_j\}_{j=0}^{\infty}$ and $\{C_j\}_{j=0}^{\infty}$ are arbitrary complex sequences and b, c are complex numbers such that $(b;q)_k, (c;q)_k$ do not vanish for $k = 1, 2, \ldots$. Then, setting

$$F_n = \sum_{j=0}^{n} \frac{\left(q^{-n}, aq^n; q\right)_j}{(q,b;q)_j} B_j \sum_{k=0}^{n} \frac{\left(q^{-n}, aq^n; q\right)_k}{(q,c;q)_k} C_k, \tag{8.1.3}$$

we find by using (8.1.2) that

$$F_n = \sum_{m=0}^{n} \left(q^{-n}, aq^n; q\right)_m \sum_{k=0}^{m} \frac{C_k q^{k^2-mk}}{(q,c;q)_k (q,b;q)_{m-k}}$$

$$\cdot \sum_{j=0}^{k} \frac{(q^{-k}, aq^m; q)_j}{(q, bq^{m-k}; q)_j} B_{m-k+j} \tag{8.1.4}$$

for $n = 0, 1, 2, \ldots$. This formula does not extend to noninteger values of n because, in general, the triple sum on the right side does not converge.

8.2 A product formula for balanced ${}_4\phi_3$ polynomials

Since we are mainly interested in q-orthogonal polynomials which are expressible as balanced ${}_4\phi_3$ series or their limit cases, we shall now specialize (8.1.4) to such cases. Set

$$B_j = \frac{(b_1, b_2; q)_j}{(b_3, qab_1b_2/bb_3; q)_j} q^j, \ C_j = \frac{(c_1, c_2; q)_j}{(c_3, qac_1c_2/cc_3; q)_j} q^j, \tag{8.2.1}$$

where it is assumed that the parameters are such that no zero factors appear in the denominators. Then formula (8.1.4) gives

$$f_n = \sum_{m=0}^{n} \sum_{k=0}^{m} \frac{(q^{-n}, aq^n; q)_m (c_1, c_2; q)_{m-k}}{(q, b; q)_k (q, c, c_3, qac_1c_2/cc_3; q)_{m-k}} q^{k^2-mk+m}$$
$$\cdot \frac{(b_1, b_2; q)_k}{(b_3, qab_1b_2/bb_3; q)_k} {}_4\phi_3 \left[\begin{matrix} q^{k-m}, aq^m, b_1q^k, b_2q^k \\ bq^k, b_3q^k, ab_1b_2q^{k+1}/bb_3 \end{matrix} ; q, q \right], \tag{8.2.2}$$

where f_n is the right side of (8.1.3) with B_j and C_j as defined in (8.2.1). The cruical step in the next round of calculations is to convert the ${}_4\phi_3$ series in (8.2.2) into a very-well-poised ${}_8\phi_7$ series by Watson's formula (2.5.1), i.e.,

$${}_4\phi_3 \left[\begin{matrix} q^{k-m}, aq^m, b_1q^k, b_2q^k \\ bq^k, b_3q^k, ab_1b_2q^{k+1}/bb_3 \end{matrix} ; q, q \right]$$
$$= \frac{(bb_3q^{k-m}/ab_1, bb_3q^{k-m}/ab_2; q)_{m-k}}{(bb_3q^{2k-m}/a, bb_3q^{-m}/ab_1b_2; q)_{m-k}}$$
$$\cdot {}_8W_7 \left(\frac{bb_3q^{2k-m-1}}{a}; \frac{b_3q^{k-m}}{a}, \frac{bq^{k-m}}{a}, b_1q^k, b_2q^k, q^{k-m}; q, \frac{bb_3q^{m-k}}{b_1b_2} \right). \tag{8.2.3}$$

Substituting this into (8.2.2) gives

$$f_n = \sum_{m=0}^{n} \frac{(q^{-n}, aq^n, qab_1/bb_3, qab_2/bb_3; q)_m}{(q, c, qa/bb_3, qab_1b_2/bb_3; q)_m} q^m$$
$$\cdot \sum_{k=0}^{m} \sum_{j=0}^{m-k} \frac{(bb_3q^{-m-1}/a; q)_{2k+j} \left(1 - bb_3q^{2j+2k-m-1}/a\right) (c_1, c_2; q)_{m-k}}{(q; q)_j \left(1 - bb_3q^{-m-1}/a\right) (c_3, qac_1c_2/cc_3; q)_{m-k}}$$
$$\cdot \frac{\left(q^{1-m}/c; q\right)_k (b_1, b_2, bq^{-m}/a, b_3q^{-m}/a, q^{-m}; q)_{j+k}}{(q, bq^{-m}/a, b_3q^{-m}/a; q)_k (b, b_3, bb_3/a, bb_3q^{-m}/ab_1, bb_3q^{-m}/ab_2; q)_{j+k}}$$
$$\cdot (-1)^k q^{mk - \binom{k}{2}} \left(\frac{bb_3c}{qab_1b_2} \right)^k \left(\frac{bb_3q^{m-k}}{b_1b_2} \right)^j . \tag{8.2.4}$$

Then, replacing j by $j-k$ in the sum on the right side of (8.2.4), we obtain

$$
\begin{aligned}
&{}_4\phi_3\left[\begin{matrix} q^{-n},\ aq^n,\ b_1,\ b_2 \\ b,\ b_3,\ qab_1b_2/bb_3 \end{matrix};q,q\right] {}_4\phi_3\left[\begin{matrix} q^{-n},\ aq^n,\ c_1,\ c_2 \\ c,\ c_3,\ qac_1c_2/cc_3 \end{matrix};q,q\right] \\
&= \sum_{m=0}^{n} \frac{(q^{-n},aq^n,c_1,c_2,qab_1/bb_3,qab_2/bb_3;q)_m}{(q,c,c_3,qa/bb_3,qac_1c_2/cc_3,qab_1b_2/bb_3;q)_m} q^m \\
&\cdot \sum_{j=0}^{m} \frac{(bb_3q^{-m-1}/a;q)_j(1-bb_3q^{2j-m-1}/a)(b_1,b_2,bq^{-m}/a,b_3q^{-m}/a,q^{-m};q)_j}{(q;q)_j(1-bb_3q^{-m-1}/a)(bb_3q^{-m}/ab_1,bb_3q^{-m}/ab_2,b_3,b,bb_3/a;q)_j} \\
&\cdot \left(\frac{bb_3q^m}{b_1b_2}\right)^j {}_5\phi_4\left[\begin{matrix} q^{-j},q^{1-m}/c,q^{1-m}/c_3,bb_3q^{j-m-1}/a,cc_3q^{-m}/ac_1c_2 \\ q^{1-m}/c_1,q^{1-m}/c_2,bq^{-m}/a,b_3q^{-m}/a \end{matrix};q,q\right].
\end{aligned} \tag{8.2.5}
$$

Note that the ${}_5\phi_4$ series in (8.2.5) is balanced and, in the special case $c=aq/b$ and $c_3=aq/b_3$, becomes a ${}_3\phi_2$ which is summable by (1.7.2). Thus, we obtain the formula

$$
\begin{aligned}
&{}_4\phi_3\left[\begin{matrix} q^{-n},\ aq^n,\ b_1,b_2 \\ b,\ b_3,\ qab_1b_2/bb_3 \end{matrix};q,q\right] {}_4\phi_3\left[\begin{matrix} q^{-n},\ aq^n,\ c_1,\ c_2 \\ aq/b,\ aq/b_3,\ bb_3c_1c_2/aq \end{matrix};q,q\right] \\
&= \sum_{m=0}^{n} \frac{(q^{-n},aq^n,c_1,c_2,qab_1/bb_3,qab_2/bb_3;q)_m}{(q,aq/b,aq/b_3,aq/bb_3,qab_1b_2/bb_3,bb_3c_1c_2/aq;q)_m} q^m \\
&\cdot {}_{10}\phi_9\left[\begin{matrix} \lambda,\ q\lambda^{\frac12},\ -q\lambda^{\frac12},\ b_1,\ b_2,\ bb_3c_1/aq, \\ \lambda^{\frac12},\ -\lambda^{\frac12},\ bb_3q^{-m}/ab_1,\ bb_3q^{-m}/ab_2,\ q^{1-m}/c_1, \end{matrix}\right. \\
&\qquad \left.\begin{matrix} bb_3c_2/aq,\ bq^{-m}/a,\ b_3q^{-m}/a,\ q^{-m} \\ q^{1-m}/c_2,\ b_3,\ b,\ bb_3/a, \end{matrix};q,\frac{aq^2}{b_1b_2c_1c_2}\right],
\end{aligned} \tag{8.2.6}
$$

where $\lambda=bb_3q^{-m-1}/a$. This formula is a q-analogue of Bailey's [1933] product formula

$$
\begin{aligned}
&{}_2F_1(-n,a+n;b;x)\,{}_2F_1(-n,a+n;1+a-b;y) \\
&= F_4(-n,a+n;b,1+a-b;x(1-y),y(1-x)),
\end{aligned} \tag{8.2.7}
$$

where

$$
F_4(a,b;c,d;x,y) = \sum_{m=0}^{\infty}\sum_{n=0}^{\infty} \frac{(a)_{m+n}(b)_{m+n}}{m!n!(c)_m(d)_n} x^m y^n. \tag{8.2.8}
$$

However, even though (8.2.6) is valid only when the series on both sides terminate, (8.2.7) holds whether or not n is a nonnegative integer, subject to the absolute convergence of the two ${}_2F_1$ series on the left and the F_4 series on the right.

Application of Sears' transformation formula (2.10.4) enables us to transform one or both of the ${}_4\phi_3$ series on the left side of (8.2.6) and derive a number of equivalent forms. Two particularly interesting ones are

$$
{}_4\phi_3\left[\begin{matrix} q^{-n},\ aq^n,\ b_1,\ b_2 \\ b,\ b_3,\ qab_1b_2/bb_3 \end{matrix};q,q\right] {}_4\phi_3\left[\begin{matrix} q^{-n},\ aq^n,\ bb_3c_1/aq,\ bb_3c_2/aq \\ b,\ b_3,\ bb_3c_1c_2/aq \end{matrix};q,q\right]
$$

$$
= \frac{(aq/b, aq/b_3;q)_n}{(b,b_3;q)_n}\left(\frac{bb_3}{aq}\right)^n
$$

$$
\cdot \sum_{m=0}^{n} \frac{(q^{-n}, aq^n, c_1, c_2, qab_1/bb_3, qab_2/bb_3;q)_m}{(q, aq/b, aq/b_3, aq/bb_3, qab_1b_2/bb_3, bb_3c_1c_2/aq;q)_m} q^m
$$

$$
\cdot {}_{10}\phi_9\left[\begin{matrix} \lambda, & q\lambda^{\frac{1}{2}}, & -q\lambda^{\frac{1}{2}}, & b_1, & b_2, & bb_3c_1/aq, & bb_3c_2/aq, \\ & \lambda^{\frac{1}{2}}, & -\lambda^{\frac{1}{2}}, & bb_3q^{-m}/ab_1, & bb_3q^{-m}/ab_2, & q^{1-m}/c_1, & q^{1-m}/c_2, \end{matrix}\right.
$$

$$
\left.\begin{matrix} bq^{-m}/a, & b_3q^{-m}/a, & q^{-m} \\ b_3, & b, & bb_3/a \end{matrix}; q, \frac{aq^2}{b_1b_2c_1c_2}\right], \tag{8.2.9}
$$

and

$$
{}_4\phi_3\left[\begin{matrix} q^{-n}, & aq^n, & qab_1/bb_3, & qab_2/bb_3 \\ & aq/b, & aq/b_3, & qab_1b_2/bb_3 \end{matrix}; q, q\right] {}_4\phi_3\left[\begin{matrix} q^{-n}, & aq^n, & c_1, & c_2 \\ aq/b, & aq/b_3, & bb_3c_1c_2/aq \end{matrix}; q, q\right]
$$

$$
= \frac{(b,b_3;q)_n}{(aq/b, aq/b_3;q)_n}\left(\frac{aq}{bb_3}\right)^n
$$

$$
\cdot \sum_{m=0}^{n} \frac{(q^{-n}, aq^n, c_1, c_2, qab_1/bb_3, qab_2/bb_3;q)_m}{(q, aq/b, aq/b_3, aq/bb_3, qab_1b_2/bb_3, bb_3c_1c_2/aq;q)_m} q^m
$$

$$
\cdot {}_{10}\phi_9\left[\begin{matrix} \lambda, & q\lambda^{\frac{1}{2}}, & -q\lambda^{\frac{1}{2}}, & b_1, & b_2, & bb_3c_1/aq, & bb_3c_2/aq, \\ & \lambda^{\frac{1}{2}}, & -\lambda^{\frac{1}{2}}, & bb_3q^{-m}/ab_1, & bb_3q^{-m}/ab_2, & q^{1-m}/c_1, & q^{1-m}/c_2, \end{matrix}\right.
$$

$$
\left.\begin{matrix} bq^{-m}/a, & b_3q^{-m}/a, & q^{-m} \\ b_3, & b, & bb_3/a \end{matrix}; q, \frac{aq^2}{b_1b_2c_1c_2}\right], \tag{8.2.10}
$$

where $\lambda = bb_3q^{-m-1}/a$.

Either of the formulas (8.2.9) and (8.2.10) may be regarded as a q-analogue of Watson's [1922] product formula for the Jacobi polynomials

$$
{}_2F_1(-n, a+n; b; x)\, {}_2F_1(-n, a+n; b; y)
= (-1)^n \frac{(1+a-b)_n}{(b)_n} F_4\left(-n, a+n; b, 1+a-b; xy, (1-x)(1-y)\right), \tag{8.2.11}
$$

where $n = 0, 1, \ldots$.

The special case in which the $_{10}\phi_9$ series in (8.2.6), (8.2.9) or (8.2.10) become balanced is also of interest in some applications. Thus, if we set $c_2 = aq/b_1b_2c_1$, then by using Bailey's transformation formula (2.10.8) we may express (8.2.9) in the form

$$
{}_4\phi_3\left[\begin{matrix} q^{-n}, & aq^n, & b_1, & b_2 \\ b, & b_3, & qab_1b_2/bb_3 \end{matrix}; q, q\right] {}_4\phi_3\left[\begin{matrix} q^{-n}, & aq^n, & bb_3c_1/aq, & bb_3/b_1b_2c_1 \\ & b, & b_3, & bb_3/b_1b_2 \end{matrix}; q, q\right]
$$

$$
= \frac{(aq/b, aq/b_3;q)_n}{(b,b_3;q)_n}\left(\frac{bb_3}{aq}\right)^n \sum_{m=0}^{n} \frac{(q^{-n}, aq^n, b_1c_1, b_2c_1, aq/b_1b_2c_1;q)_m}{(q, aq/b, aq/b_3, b_1b_2c_1, bb_3/b_1b_2;q)_m} q^m
$$

$$
\cdot {}_{10}\phi_9\left[\begin{matrix} \mu, & q\mu^{\frac{1}{2}}, & -q\mu^{\frac{1}{2}}, & b_1, & b_2 & b_1b_2c_1/b, & b_1b_2c_1/b_3, \\ & \mu^{\frac{1}{2}}, & -\mu^{\frac{1}{2}}, & b_2c_1, & b_1c_1, & b, & b_3, \end{matrix}\right.
$$

$$\left.\begin{matrix} bb_3c_1/aq, & aq^m, & q^{-m} \\ qab_1b_2/bb_3, & b_1b_2c_1q^{-m}/a, & b_1b_2c_1q^{-m} \end{matrix} ;q,q\right], \tag{8.2.12}$$

where $\mu = b_1b_2c_1q^{-1}$. This provides a q-analogue of Bateman's [1932, p. 392] product formula

$$\begin{aligned} &{}_2F_1(-n,a+n;b;x)\ {}_2F_1(-n,a+n;b;y) \\ &= (-1)^n \frac{(1+a-b)_n}{(b)_n} \sum_{k=0}^{n} \frac{(-n)_k(a+n)_k}{k!(1+a-b)_k}(1-x-y)^k \\ &\quad \cdot {}_2F_1\left(-k,a+k;b;-xy/(1-x-y)\right). \end{aligned} \tag{8.2.13}$$

8.3 Product formulas for q-Racah and Askey-Wilson polynomials

Let us replace the parameters $a, b, b_1, b_2, b_3, c_1, c_2$ in (8.2.9) by $abq, aq, q^{-x}, cq^{x-N}, bcq, c^{-1}q^{-y}, q^{y-N}$, respectively, to obtain the following product formula for the q-Racah polynomials introduced in §7.2:

$$\begin{aligned} &W_n(x;a,b,c,N;q)\ W_n(y;a,b,c,N;q) \\ &= \frac{(bq,qac^{-1};q)_n}{(aq,bcq;q)_n} c^n \sum_{m=0}^{n} \frac{(q^{-n},abq^{n+1},q^{x-N},q^{y-N},c^{-1}q^{-x},c^{-1}q^{-y};q)_m}{(q,bq,qac^{-1},c^{-1},q^{-N},q^{-N};q)_m} q^m \\ &\quad \cdot {}_{10}\phi_9\left[\begin{matrix} cq^{-m}, q(cq^{-m})^{\frac{1}{2}}, -q(cq^{-m})^{\frac{1}{2}}, ca^{-1}q^{-m}, b^{-1}q^{-m}, q^{-m}, q^{-x}, \\ (cq^{-m})^{\frac{1}{2}}, -(cq^{-m})^{\frac{1}{2}}, aq, bcq, cq, cq^{x+1-m}, \end{matrix}\right. \\ &\qquad \left.\begin{matrix} q^{-y}, cq^{x-N}, cq^{y-N} \\ cq^{y+1-m}, q^{N-x+1-m}, q^{N-y+1-m} \end{matrix} ;q, abq^{2N+3}\right], \end{aligned} \tag{8.3.1}$$

where

$$\begin{aligned} &W_n(x;a,b,c,N;q) \\ &= {}_4\phi_3\left[\begin{matrix} q^{-n}, abq^{n+1}, q^{-x}, cq^{x-N} \\ aq, q^{-N}, bcq \end{matrix} ;q,q\right] \end{aligned} \tag{8.3.2}$$

is the q-Racah polynomial defined in (7.2.17). This is a Watson-type formula. Two additional Watson-type formulas are given in Ex. 8.1.

Letting $c \to 0$ in (8.3.1) gives a product formula for the q-Hahn polynomials defined in (7.2.21):

$$\begin{aligned} &Q_n(x;a,b,N;q)\ Q_n(y;a,b,N;q) \\ &= (-aq)^n q^{\binom{n}{2}} \frac{(bq;q)_n}{(aq;q)_n} \sum_{m=0}^{n} \frac{(q^{-n},abq^{n+1},q^{x-N},q^{y-N};q)_m}{(q,bq,q^{-N},q^{-N};q)_m} \left(aq^{x+y}\right)^{-m} \\ &\quad \cdot {}_4\phi_3\left[\begin{matrix} q^{-x}, q^{-y}, b^{-1}q^{-m}, q^{-m} \\ aq, q^{N-x+1-m}, q^{N-y+1-m} \end{matrix} ;q, abq^{2N+3}\right]. \end{aligned} \tag{8.3.3}$$

To obtain a Watson-type product formula for the Askey-Wilson polynomials defined in (7.5.2) we replace $a, b, b_1, b_2, b_3, c_1, c_2$ in (8.2.9) by

$abcdq^{-1}, ab, ae^{i\theta}, ae^{-i\theta}, ac, de^{i\phi}, de^{-i\phi}$, respectively, where $x = \cos\theta$, $y = \cos\phi$. This gives

$$\begin{aligned} &p_n(x;a,b,c,d|q)\; p_n(y;a,b,c,d|q) \\ &= (ab, ac, ad, ad, bd, cd; q)_n (ad)^{-n} \\ &\cdot \sum_{m=0}^{n} \frac{(q^{-n}, abcdq^{n-1}, de^{i\theta}, de^{-i\theta}, de^{i\phi}, de^{-i\phi}; q)_m}{(q, ad, ad, bd, cd, da^{-1}; q)_m} q^m \\ &\cdot {}_{10}\phi_9 \left[\begin{matrix} aq^{-m}/d, q(aq^{-m}/d)^{\frac12}, -q(aq^{-m}/d)^{\frac12}, q^{1-m}/bd, q^{1-m}/cd, q^{-m}, \\ (aq^{-m}/d)^{\frac12}, -(aq^{-m}/d)^{\frac12}, ab, ac, aq/d, \end{matrix} \right. \\ &\qquad \left. \begin{matrix} ae^{i\theta}, ae^{-i\theta}, ae^{i\phi}, ae^{-i\phi} \\ q^{1-m}e^{-i\theta}/d, q^{1-m}e^{i\theta}/d, q^{1-m}e^{-i\phi}/d, q^{1-m}e^{i\phi}/d \end{matrix} ; q, \frac{bcq}{ad} \right]. \end{aligned} \tag{8.3.4}$$

When $b = aq^{\frac12}$ and $d = cq^{\frac12}$, the ${}_{10}\phi_9$ series in (8.3.4) becomes balanced and hence can be transformed to another balanced ${}_{10}\phi_9$ via (2.9.1). This leads to a Bateman-type product formula

$$\begin{aligned} &p_n(x;a,aq^{\frac12},c,cq^{\frac12}|q)\; p_n(y;a,aq^{\frac12},c,cq^{\frac12}|q) \\ &= \left(a^2q^{\frac12}, ac, acq^{\frac12}, acq^{\frac12}, acq, c^2q^{\frac12}; q\right)_n \left(acq^{\frac12}\right)^{-n} \\ &\cdot \sum_{m=0}^{n} \frac{\left(q^{-n}, a^2c^2q^n, acq^{\frac12}e^{i\theta+i\phi}, acq^{\frac12}e^{i\phi-i\theta}, cq^{\frac12}e^{-i\phi}; q\right)_m}{\left(q, c^2q^{\frac12}, acq^{\frac12}, acq, a^2cq^{\frac12}e^{i\phi}; q\right)_m} q^m \\ &\cdot {}_{10}\phi_9 \left[\begin{matrix} \nu, & q\nu^{\frac12}, & -q\nu^{\frac12}, & ae^{i\phi}, & aq^{\frac12}e^{i\phi}, & ce^{i\phi}, & ae^{i\theta}, \\ & \nu^{\frac12}, & -\nu^{\frac12}, & acq^{\frac12}, & ac, & a^2q^{\frac12}, & acq^{\frac12}e^{i\phi-i\theta}, \end{matrix} \right. \\ &\qquad \left. \begin{matrix} ae^{-i\theta}, & a^2c^2q^m, & q^{-m} \\ acq^{\frac12}e^{i\phi+i\theta}, & q^{\frac12-m}e^{i\phi}/c, & a^2ce^{i\phi}q^{m+\frac12} \end{matrix} ; q, q \right], \end{aligned} \tag{8.3.5}$$

where $\nu = a^2ce^{i\phi}q^{-\frac12}$. In fact, if we replace a and c by $q^{(2\alpha+1)/4}$ and $-q^{(2\beta+1)/4}$, respectively, then this gives a Bateman-type product formula for the continuous q-Jacobi polynomials (7.5.24) which, on letting $q \to 1$, gives Bateman's [1932] product formula for the Jacobi polynomials:

$$\begin{aligned} \frac{P_n^{(\alpha,\beta)}(x)P_n^{(\alpha,\beta)}(y)}{P_n^{(\alpha,\beta)}(1)P_n^{(\alpha,\beta)}(1)} &= (-1)^n \frac{(\beta+1)_n}{(\alpha+1)_n} \sum_{k=0}^{n} \frac{(-n)_k(n+\alpha+\beta+1)_k}{k!(\beta+1)_k} \left(\frac{x+y}{2}\right)^k \\ &\cdot P_k^{(\alpha,\beta)}\left(\frac{1+xy}{x+y}\right) / P_k^{(\alpha,\beta)}(1), \end{aligned} \tag{8.3.6}$$

which is equivalent to (8.2.13).

For terminating series there is really no difference between the Watson formula (8.2.11) and the Bailey formula (8.2.7) since one can be transformed into the other in a trivial way. However, for the continuous q-ultraspherical polynomials given in (7.4.14), there is an interesting Bailey-type product formula that can be obtained from (8.2.6) by replacing $a, b, b_1, b_2, b_3, c_1, c_2$ by

$a^4, a^2q^{\frac{1}{2}}, ae^{i\theta}, ae^{-i\theta}, -a^2q^{\frac{1}{2}}, ae^{i\phi}$ and $ae^{-i\phi}$, respectively:

$$
{}_4\phi_3\left[\begin{matrix} q^{-n}, a^4q^n, ae^{i\theta}, ae^{-i\theta} \\ a^2q^{\frac{1}{2}}, -a^2q^{\frac{1}{2}}, -a^2 \end{matrix}; q, q\right] {}_4\phi_3\left[\begin{matrix} q^{-n}, a^4q^n, ae^{i\phi}, ae^{-i\phi} \\ a^2q^{\frac{1}{2}}, -a^2q^{\frac{1}{2}}, -a^2 \end{matrix}; q, q\right]
$$

$$
= \sum_{m=0}^{n} \frac{(q^{-n}, a^4q^n, ae^{i\phi}, ae^{-i\phi}, -ae^{i\theta}, -ae^{-i\theta}; q)_m}{\left(q, a^2q^{\frac{1}{2}}, -a^2q^{\frac{1}{2}}, -1, -a^2, -a^2; q\right)_m} q^m
$$

$$
\cdot {}_{10}\phi_9\left[\begin{matrix} -q^{-m}, & q(-q^{-m})^{\frac{1}{2}}, & -q(-q^{-m})^{\frac{1}{2}}, & q^{\frac{1}{2}-m}/a^2, & -q^{\frac{1}{2}-m}/a^2, & q^{-m}, \\ & (-q^{-m})^{\frac{1}{2}}, & -(-q^{-m})^{\frac{1}{2}}, & -a^2q^{\frac{1}{2}}, & a^2q^{\frac{1}{2}}, & -q, \end{matrix}\right.
$$

$$
\left.\begin{matrix} ae^{i\theta}, & ae^{-i\theta}, & -ae^{i\phi}, & -ae^{-i\phi} \\ -q^{1-m}e^{-i\theta}/a, & -q^{1-m}e^{i\theta}/a, & q^{1-m}e^{-i\phi}/a, & q^{1-m}e^{i\phi}/a \end{matrix}; q, q^2\right]. \tag{8.3.7}
$$

For further information about product formulas see Rahman [1982] and Gasper and Rahman [1984].

8.4 A product formula in integral form for the continuous q-ultraspherical polynomials

As an application of the Bateman-type product formula (8.3.5) for the Askey-Wilson polynomials we shall now derive a product formula for the continuous q-ultraspherical polynomials in the integral form

$$
\begin{aligned}
&C_n(x;\beta|q)C_n(y;\beta|q) \\
&= \frac{(\beta^2;q)_n}{(q;q)_n}\beta^{-n/2}\int_{-1}^{1} K(x,y,z;\beta|q)C_n(z;\beta|q)\,dz,
\end{aligned} \tag{8.4.1}
$$

where

$$
\begin{aligned}
K(x,y,z;\beta|q) &= \frac{(q,\beta,\beta;q)_\infty \left|\left(\beta e^{2i\theta}, \beta e^{2i\phi}; q\right)_\infty\right|^2}{2\pi(\beta^2;q)_\infty} \\
&\cdot w\left(z; \beta^{\frac{1}{2}}e^{i\theta+i\phi}, \beta^{\frac{1}{2}}e^{-i\theta-i\phi}, \beta^{\frac{1}{2}}e^{i\theta-i\phi}, \beta^{\frac{1}{2}}e^{i\phi-i\theta}\right)
\end{aligned} \tag{8.4.2}
$$

with $w(z;a,b,c,d)$ defined as in (6.3.1) and $x = \cos\theta, y = \cos\phi$.

First, we set $c = -a$ in (8.3.5) and rewrite it in the form

$$
r_n(x;a,aq^{\frac{1}{2}},-a,-aq^{\frac{1}{2}}|q)r_n(y;a,aq^{\frac{1}{2}},-a,-aq^{\frac{1}{2}}|q) = \frac{1+a^2q^n}{1+a^2}\left(-q^{-\frac{1}{2}}\right)^n
$$

$$
\cdot \sum_{m=0}^{n} \frac{\left(q^{-n}, a^4q^n, -a^2q^{\frac{1}{2}}e^{i\theta+i\phi}, -a^2q^{\frac{1}{2}}e^{i\phi-i\theta}, -aq^{\frac{1}{2}}e^{-i\phi}; q\right)_m}{\left(q, a^2q^{\frac{1}{2}}, -a^2q^{\frac{1}{2}}, -a^2q, -a^3q^{\frac{1}{2}}e^{i\phi}; q\right)_m} q^m
$$

$$
\cdot {}_{10}W_9\left(-a^3q^{-\frac{1}{2}}e^{i\phi}; ae^{i\phi}, -ae^{i\phi}, aq^{\frac{1}{2}}e^{i\phi}, ae^{i\theta}, ae^{-i\theta}, a^4q^m, q^{-m}; q, q\right), \tag{8.4.3}
$$

where

$$
\begin{aligned}
&r_n(x;a,b,c,d|q) \\
&= {}_4\phi_3\left[\begin{matrix} q^{-n}, abcdq^{n-1}, ae^{i\theta}, ae^{-i\theta} \\ ab, ac, ad \end{matrix}; q, q\right].
\end{aligned} \tag{8.4.4}
$$

The key step now is to use the $d = -(aq)^{\frac{1}{2}}$ case of Bailey's transformation formula (2.8.3) to transform the balanced ${}_{10}\phi_9$ series in (8.4.3) to a balanced ${}_4\phi_3$ series:

$$
\begin{aligned}
&{}_{10}W_9\left(-a^3q^{-\frac{1}{2}}e^{i\phi}; ae^{i\phi}, -ae^{i\phi}, aq^{\frac{1}{2}}e^{i\phi}, ae^{i\theta}, ae^{-i\theta}, a^4q^m, q^{-m}; q, q\right) \\
&= \frac{\left(a^2e^{-2i\phi}, -a^3q^{\frac{1}{2}}e^{i\phi}; q\right)_m}{\left(a^4, -aq^{\frac{1}{2}}e^{-i\phi}; q\right)_m} \\
&\quad \cdot {}_4\phi_3\left[\begin{matrix} a^2e^{2i\phi}, -q^{\frac{1}{2}}e^{i\theta+i\phi}, -q^{\frac{1}{2}}e^{i\phi-i\theta}, q^{-m} \\ -a^2q^{\frac{1}{2}}e^{i\phi-i\theta}, -a^2q^{\frac{1}{2}}e^{i\theta+i\phi}, q^{1-m}e^{2i\phi}/a^2 \end{matrix}; q, q\right].
\end{aligned} \tag{8.4.5}
$$

So (8.4.3) reduces to

$$
\begin{aligned}
&r_n(x; a, aq^{\frac{1}{2}}, -a, -aq^{\frac{1}{2}}|q)\, r_n(y; a, aq^{\frac{1}{2}}, -a, -aq^{\frac{1}{2}}|q) = \frac{1+a^2q^n}{1+a^2}\left(-q^{-\frac{1}{2}}\right)^n \\
&\quad \cdot \sum_{m=0}^{n} \frac{\left(q^{-n}, a^4q^n, a^2e^{-2i\phi}, -a^2q^{\frac{1}{2}}e^{i\theta+i\phi}, -a^2q^{\frac{1}{2}}e^{i\phi-i\theta}; q\right)_m}{\left(q, a^4, a^2q^{\frac{1}{2}}, -a^2q^{\frac{1}{2}}, -a^2q; q\right)_m} q^m \\
&\quad \cdot {}_4\phi_3\left[\begin{matrix} a^2e^{2i\phi}, -q^{\frac{1}{2}}e^{i\theta+i\phi}, -q^{\frac{1}{2}}e^{i\phi-i\theta}, q^{-m} \\ -a^2q^{\frac{1}{2}}e^{i\phi-i\theta}, -a^2q^{\frac{1}{2}}e^{i\theta+i\phi}, q^{1-m}e^{2i\phi}/a^2 \end{matrix}; q, q\right].
\end{aligned} \tag{8.4.6}
$$

Transforming this ${}_4\phi_3$ series by Sears' transformation formula (2.10.4), we obtain a further reduction

$$
\begin{aligned}
&r_n(x; a, aq^{\frac{1}{2}}, -a, -aq^{\frac{1}{2}}|q)\, r_n(y; a, aq^{\frac{1}{2}}, -a, -aq^{\frac{1}{2}}|q) \\
&= \frac{1+a^2q^n}{1+a^2}\left(-q^{-\frac{1}{2}}\right)^n \sum_{m=0}^{n} \frac{\left(q^{-n}, a^4q^n, -q^{\frac{1}{2}}e^{i\theta-i\phi}, -a^2q^{\frac{1}{2}}e^{i\phi-i\theta}; q\right)_m}{\left(q, a^2q^{\frac{1}{2}}, -a^2q^{\frac{1}{2}}, -a^2q; q\right)_m} q^m \\
&\quad \cdot {}_4\phi_3\left[\begin{matrix} q^{-m}, a^2, a^2e^{2i\phi}, a^2e^{-2i\theta} \\ a^4, -a^2q^{\frac{1}{2}}e^{i\phi-i\theta}, -q^{\frac{1}{2}-m}e^{i\phi-i\theta} \end{matrix}; q, q\right].
\end{aligned} \tag{8.4.7}
$$

Now observe that, by (6.1.1),

$$
\begin{aligned}
&\int_{-1}^{1} w\left(z; ae^{i\phi-i\theta}, ae^{i\theta-i\phi}, ae^{i\theta+i\phi}, ae^{-i\theta-i\phi}\right)\left(ae^{i\phi-i\theta+i\psi}, ae^{i\phi-i\theta-i\psi}; q\right)_j dz \\
&= \int_{-1}^{1} w\left(z; aq^je^{i\phi-i\theta}, ae^{i\theta-i\phi}, ae^{i\theta+i\phi}, ae^{-i\theta-i\phi}\right) dz \\
&= \frac{2\pi(a^4;q)_\infty}{(q, a^2, a^2; q)_\infty |(a^2e^{2i\phi}, a^2e^{-2i\theta}; q)_\infty|^2} \frac{\left(a^2, a^2e^{2i\phi}, a^2e^{-2i\theta}; q\right)_j}{(a^4;q)_j},
\end{aligned} \tag{8.4.8}
$$

where $|a| < 1$ and $z = \cos\psi$. Hence

$$
{}_4\phi_3\left[\begin{matrix} q^{-m}, a^2, a^2e^{2i\phi}, a^2e^{-2i\theta} \\ a^4, -a^2q^{\frac{1}{2}}e^{i\phi-i\theta}, -q^{\frac{1}{2}-m}e^{i\phi-i\theta} \end{matrix}; q, q\right]
$$

$$
\begin{aligned}
&= \frac{(q, a^2, a^2; q)_\infty \,|\, (a^2 e^{2i\phi}, a^2 e^{-2i\theta}; q)_\infty |^2}{2\pi (a^4; q)_\infty} \\
&\quad \cdot \int_{-1}^{1} w\left(z; a e^{i\phi - i\theta}, a e^{i\theta - i\phi}, a e^{i\theta + i\phi}, a e^{-i\theta - i\phi}\right) \\
&\quad \cdot {}_3\phi_2 \left[\begin{matrix} q^{-m}, a e^{i\phi - i\theta + i\psi}, a e^{i\phi - i\theta - i\psi} \\ -a^2 q^{\frac{1}{2}} e^{i\phi - i\theta}, -q^{\frac{1}{2} - m} e^{i\phi - i\theta} \end{matrix} ; q, q \right] dz \\
&= \frac{(q, a^2, a^2; q)_\infty \,|\, (a^2 e^{2i\phi}, a^2 e^{-2i\theta}; q)_\infty |^2}{2\pi (a^4; q)_\infty} \\
&\quad \cdot \int_{-1}^{1} w\left(z; a e^{i\phi - i\theta}, a e^{i\theta - i\phi}, a e^{i\theta + i\phi}, a e^{-i\theta - i\phi}\right) \\
&\quad \cdot \frac{\left(-a q^{\frac{1}{2}} e^{i\psi}, -a q^{\frac{1}{2}} e^{-i\psi}; q\right)_m}{\left(-a^2 q^{\frac{1}{2}} e^{i\phi - i\theta}, -q^{\frac{1}{2}} e^{i\theta - i\phi}; q\right)_m} dz. \qquad (8.4.9)
\end{aligned}
$$

Substituting (8.4.9) into (8.4.7) and using (2.10.4) we finally obtain

$$
\begin{aligned}
&r_n(x; a, a q^{\frac{1}{2}}, -a, -a q^{\frac{1}{2}} | q)\, r_n(y; a, a q^{\frac{1}{2}}, -a, -a q^{\frac{1}{2}} | q) \\
&= \int_{-1}^{1} K(x, y, z; a^2 | q)\, r_n(z; a, a q^{\frac{1}{2}}, -a, -a q^{\frac{1}{2}} | q)\, dz. \qquad (8.4.10)
\end{aligned}
$$

This yields (8.4.1) if we replace a by $\beta^{\frac{1}{2}}$ and use (7.4.14). By setting $\beta = q^\lambda$ in (8.4.1) and taking the limit $q \to 1$, Rahman and Verma [1986b] showed that (8.4.1) tends to Gegenbauer's [1874] product formula

$$
\frac{C_n^\lambda(x) C_n^\lambda(y)}{C_n^\lambda(1) C_n^\lambda(1)} = \int_{-1}^{1} K(x, y, z) \frac{C_n^\lambda(z)}{C_n^\lambda(1)} dz, \qquad (8.4.11)
$$

where

$$
K(x, y, z) = \frac{\Gamma(\lambda + \frac{1}{2})}{\Gamma(\lambda)\Gamma(\frac{1}{2})} \frac{\left(1 - x^2 - y^2 - z^2 + 2xyz\right)^{\lambda - 1}}{[(1 - x^2)(1 - y^2)]^{\lambda - \frac{1}{2}}} \text{ or } 0, \qquad (8.4.12)
$$

according as $1 - x^2 - y^2 - z^2 + 2xyz$ is positive or negative.

Rahman and Verma [1986b] were also able to derive an addition formula for the continuous q-ultraspherical polynomials corresponding to the product formula (8.4.1). This is left as an exercise (Ex. 8.11).

For an extension of (8.4.10) to the continuous q-Jacobi polynomials $r_n(x; a, a q^{\frac{1}{2}}, -b, -b q^{\frac{1}{2}} | q)$, see Rahman [1986d].

8.5 Rogers' linearization formula for the continuous q-ultraspherical polynomials

Rogers [1895] used an induction argument to prove the linearization formula

$$C_m(x;\beta|q)C_n(x;\beta|q)$$
$$= \sum_{k=0}^{\min(m,n)} \frac{(q;q)_{m+n-2k}(\beta;q)_{m-k}(\beta;q)_{n-k}(\beta;q)_k(\beta^2;q)_{m+n-k}}{(\beta^2;q)_{m+n-2k}(q;q)_{m-k}(q;q)_{n-k}(q;q)_k(\beta q;q)_{m+n-k}}$$
$$\cdot \frac{(1-\beta q^{m+n-2k})}{(1-\beta)} C_{m+n-2k}(x;\beta|q). \tag{8.5.1}$$

Different proofs of (8.5.1) have been given by Bressoud [1981d], Rahman [1981] and Gasper [1985]. We shall give Gasper's proof since it appers to be the simplest.

We use (7.4.2) for $C_n(x;\beta|q)$ and, via Heine's transformation formula (1.4.3),

$$C_m(x;\beta|q) = \frac{(\beta e^{-2i\theta};q)_\infty}{(q\beta^{-1}e^{-2i\theta};q)_\infty} \frac{(\beta;q)_m}{(q;q)_m} e^{im\theta}$$
$$\cdot {}_2\phi_1\left(q\beta^{-1}, \beta^{-2}q^{1-m}; \beta^{-1}q^{1-m}; q, \beta e^{-2i\theta}\right), \tag{8.5.2}$$

where $x=\cos\theta$. Then, temporarily assuming that $|q|<|\beta|<1$, we have

$$C_m(x;\beta|q)C_n(x;\beta|q) = A_{m,n}\sum_{r=0}^{n} \frac{(q^{-n},\beta;q)_r}{(q,\beta^{-1}q^{1-n};q)_r}\left(q\beta^{-1}e^{-2i\theta}\right)^r$$
$$\cdot \sum_{s=0}^{\infty} \frac{\left(q\beta^{-1},\beta^{-2}q^{1-m};q\right)_s}{(q,\beta^{-1}q^{1-m};q)_s}\left(\beta e^{-2i\theta}\right)^s$$
$$= A_{m,n}\sum_{k=0}^{\infty} \frac{\left(q\beta^{-1},\beta^{-2}q^{1-m};q\right)_k}{(q,\beta^{-1}q^{1-m};q)_k}\left(\beta e^{-2i\theta}\right)^k$$
$$\cdot {}_4\phi_3\left[\begin{matrix} q^{-k}, q^{-n}, \beta, \beta q^{m-k} \\ \beta^2 q^{m-k}, \beta q^{-k}, \beta^{-1}q^{1-n} \end{matrix}; q, q\right], \tag{8.5.3}$$

where

$$A_{m,n} = \frac{(\beta e^{-2i\theta};q)_\infty}{(q\beta^{-1}e^{-2i\theta};q)_\infty} \frac{(\beta;q)_m(\beta;q)_n}{(q;q)_m(q;q)_n} e^{i(m+n)\theta}. \tag{8.5.4}$$

The crucial point here is that the ${}_4\phi_3$ series in (8.5.3) is balanced and so, by (2.5.1),

$${}_4\phi_3\left[\begin{matrix} q^{-k}, q^{-n}, \beta, \beta q^{m-k} \\ \beta^2 q^{m-k}, \beta q^{-k}, \beta^{-1}q^{1-n} \end{matrix}; q, q\right] = \frac{\left(\beta^{-2}q^{1-m-n},\beta^{-1}q^{1-m};q\right)_k}{(\beta^{-1}q^{1-m-n},\beta^{-2}q^{1-m};q)_k}$$
$$\cdot {}_8W_7\left(\beta^{-1}q^{-m-n}; \beta, \beta^{-2}q^{1+k-m-n}, q^{-m}, q^{-n}, q^{-k}; q, q\beta^{-1}\right). \tag{8.5.5}$$

Substituting this into (8.5.3) and interchanging the order of summation, we obtain

$$C_m(x;\beta|q)C_n(x;\beta|q) = A_{m,n}$$
$$\cdot \sum_{k=0}^{\min(m,n)} \frac{(\beta^{-1}q^{-m-n};q)_k\left(1-\beta^{-1}q^{2k-m-n}\right)(\beta,q^{-m},q^{-n};q)_k}{(q;q)_k\left(1-\beta^{-1}q^{-m-n}\right)(\beta^{-2}q^{1-m-n},\beta^{-1}q^{1-n},\beta^{-1}q^{1-m};q)_k}$$
$$\cdot \frac{\left(\beta^{-2}q^{1-m-n};q\right)_{2k}}{\left(\beta^{-1}q^{1-m-n};q\right)_{2k}}\left(q\beta^{-1}e^{-2i\theta}\right)^k$$
$$\cdot\, {}_2\phi_1\left(q\beta^{-1},\beta^{-2}q^{1+2k-m-n};\beta^{-1}q^{1+2k-m-n};q,\beta e^{-2i\theta}\right), \tag{8.5.6}$$

which gives (8.5.1) by using (8.5.2) and observing that both sides of (8.5.1) are polynomials in x. Notice that the linearization coefficients in (8.5.1) are nonnegative when $-1<\beta<1$ and $-1<q<1$.

For an extension of the linearization formula to the continuous q-Jacobi polynomials, see Rahman [1981].

8.6 The Poisson kernel for $C_n(x;\beta|q)$

For a system of orthogonal polynomials $\{p_n(x)\}$ which satisfies an orthogonality relation of the form (7.1.6), the bilinear generating function

$$K_t(x,y) = \sum_{n=0}^{\infty} h_n p_n(x)p_n(y)t^n \tag{8.6.1}$$

is called a Poisson kernel for these polynomials provided that $h_n = cv_n$ for some constant $c>0$. The Poisson kernel for the continuous q-ultraspherical polynomials is defined by

$$K_t(x,y;\beta|q) = \sum_{n=0}^{\infty} \frac{(q;q)_n(1-\beta q^n)}{(\beta^2;q)_n(1-\beta)} C_n(x;\beta|q)C_n(y;\beta|q)t^n, \tag{8.6.2}$$

where $|t|<1$.

Gasper and Rahman [1983a] used (8.5.1) to show that

$$K_t(x,y;\beta|q) = \frac{(\beta,t^2;q)_\infty}{(\beta^2,q\beta t^2;q)_\infty}\left|\frac{\left(\beta te^{i\theta+i\phi},q\beta te^{i\theta-i\phi};q\right)_\infty}{(te^{i\theta+i\phi},te^{i\theta-i\phi};q)_\infty}\right|^2$$
$$\cdot\, {}_8\phi_7\left[\begin{matrix}\beta t^2, & qt\beta^{\frac12}, & -qt\beta^{\frac12}, & qte^{i\theta+i\phi}, & qte^{-i\theta-i\phi}, \\ & t\beta^{\frac12}, & -t\beta^{\frac12}, & \beta te^{-i\theta-i\phi}, & \beta te^{i\theta+i\phi},\end{matrix}\right.$$
$$\left.\begin{matrix}te^{i\theta-i\phi}, & te^{i\phi-i\theta}, & \beta \\ q\beta te^{i\phi-i\theta}, & q\beta te^{i\theta-i\phi}, & qt^2\end{matrix};q,\beta\right], \tag{8.6.3}$$

where $x=\cos\theta, y=\cos\phi$ and $\max(|q|,|t|,|\beta|)<1$. They also computed a closely related kernel

$$L_t(x,y;\beta|q) = \sum_{n=0}^{\infty} \frac{(q;q)_n}{(\beta^2;q)_n} C_n(x;\beta|q)C_n(y;\beta|q)t^n$$

$$= \frac{(\beta, t^2; q)_\infty}{(\beta^2, \beta t^2; q)_\infty} \left| \frac{(\beta t e^{i\theta+i\phi}, \beta t e^{i\theta-i\phi}; q)_\infty}{(t e^{i\theta+i\phi}, t e^{i\theta-i\phi}; q)_\infty} \right|^2$$

$$\cdot {}_8\phi_7 \left[\begin{matrix} \beta t^2 q^{-1}, & t(q\beta)^{\frac{1}{2}}, & -t(q\beta)^{\frac{1}{2}}, & t e^{i\theta+i\phi}, & t e^{-i\theta-i\phi}, \\ & t(\beta q^{-1})^{\frac{1}{2}}, & -t(\beta q^{-1})^{\frac{1}{2}}, & \beta t e^{-i\theta-i\phi}, & \beta t e^{i\theta+i\phi}, \end{matrix} \right.$$

$$\left. \begin{matrix} t e^{i\theta-i\phi}, & t e^{i\phi-i\theta}, & \beta \\ \beta t e^{i\phi-i\theta}, & \beta t e^{i\theta-i\phi}, & t^2 \end{matrix} ; q, \beta \right]. \tag{8.6.4}$$

Alternative derivations of (8.6.3) and (8.6.4) were given by Rahman and Verma [1986a]. In view of the product formula (8.4.1), however, one can now give a simpler proof. Let us assume, for the moment, that $|t\beta^{-\frac{1}{2}}| < 1$ and $|\beta| < 1$. Then, by (7.4.1) and (8.4.1), we find that

$$L_t(x, y; \beta|q) = \int_{-1}^{1} K(x, y, z; \beta|q) \frac{\left(t\beta^{\frac{1}{2}} e^{i\psi}, t\beta^{\frac{1}{2}} e^{-i\psi}; q\right)_\infty}{\left(t\beta^{-\frac{1}{2}} e^{i\psi}, t\beta^{-\frac{1}{2}} e^{-i\psi}; q\right)_\infty} dz$$

$$= \frac{(q, \beta, \beta; q)_\infty |\left(\beta e^{2i\theta}, \beta e^{2i\phi}; q\right)_\infty|^2}{2\pi(\beta^2; q)_\infty}$$

$$\cdot \int_0^\pi \frac{h\left(\cos\psi; 1, -1, q^{\frac{1}{2}}, -q^{\frac{1}{2}}, t\beta^{\frac{1}{2}}\right)}{h\left(\cos\psi; \beta^{\frac{1}{2}} e^{i\theta+i\phi}, \beta^{\frac{1}{2}} e^{-i\theta-i\phi}, \beta^{\frac{1}{2}} e^{i\theta-i\phi}, \beta^{\frac{1}{2}} e^{i\phi-i\theta}, t\beta^{-\frac{1}{2}}\right)} d\psi. \tag{8.6.5}$$

By (6.3.9),

$$\int_0^\pi \frac{h\left(\cos\psi; 1, -1, q^{\frac{1}{2}}, -q^{\frac{1}{2}}, t\beta^{\frac{1}{2}}\right)}{h\left(\cos\psi; \beta^{\frac{1}{2}} e^{i\theta+i\phi}, \beta^{\frac{1}{2}} e^{-i\theta-i\phi}, \beta^{\frac{1}{2}} e^{i\theta-i\phi}, \beta^{\frac{1}{2}} e^{i\phi-i\theta}, t\beta^{-\frac{1}{2}}\right)} d\psi$$

$$= \frac{2\pi\left(\beta, t^2; q\right)_\infty \left|\left(\beta t e^{i\theta+i\phi}, \beta t e^{i\theta-i\phi}; q\right)_\infty\right|^2}{(q, \beta, \beta, \beta t^2; q)_\infty \left|(\beta e^{2i\theta}, \beta e^{2i\phi}, t e^{i\theta+i\phi}, t e^{i\theta-i\phi}; q)_\infty\right|^2}$$

$$\cdot {}_8W_7\left(\beta t^2 q^{-1}; t e^{i\theta+i\phi}, t e^{-i\theta-i\phi}, t e^{i\theta-i\phi}, t e^{i\phi-i\theta}, \beta; q, \beta\right). \tag{8.6.6}$$

Formula (8.6.4) follows immediately from (8.6.5) and (8.6.6).

It is slightly more complicated to compute (8.6.1). Consider the generating function

$$G_t(z) = \sum_{n=0}^{\infty} \frac{1-\beta q^n}{1-\beta} C_n(z;\beta|q) t^n$$

$$= \frac{\left(\beta t e^{i\psi}, \beta t e^{-i\psi}; q\right)_\infty}{(1-\beta)(t e^{i\psi}, t e^{-i\psi}; q)_\infty} - \frac{\beta}{1-\beta} \frac{\left(\beta q t e^{i\psi}, \beta q t e^{-i\psi}; q\right)_\infty}{(q t e^{i\psi}, q t e^{-i\psi}; q)_\infty}$$

$$= (1-\beta t^2) \frac{\left(\beta q t e^{i\psi}, \beta q t e^{-i\psi}; q\right)_\infty}{\left(t e^{i\psi}, t e^{-i\psi}; q\right)_\infty}. \tag{8.6.7}$$

Then

$$
\begin{aligned}
&K_t(x,y;\beta|q) \\
&\quad = \int_{-1}^{1} K(x,y,z;\beta|q)\, G_{t\beta^{-\frac{1}{2}}}(z)\, dz \\
&\quad = \frac{(1-t^2)(q,\beta,\beta;q)_\infty \left|(\beta e^{2i\theta},\beta e^{2i\phi};q)_\infty\right|^2}{2\pi(\beta^2;q)_\infty} \\
&\quad \cdot \int_0^\pi \frac{h\left(\cos\psi;1,-1,q^{\frac{1}{2}},-q^{\frac{1}{2}},qt\beta^{\frac{1}{2}}\right)}{h\left(\cos\psi;\beta^{\frac{1}{2}}e^{i\theta+i\phi},\beta^{\frac{1}{2}}e^{-i\theta-i\phi},\beta^{\frac{1}{2}}e^{i\theta-i\phi},\beta^{\frac{1}{2}}e^{i\phi-i\theta},t\beta^{-\frac{1}{2}}\right)}\, d\psi.
\end{aligned} \tag{8.6.8}
$$

This gives (8.6.3) via (7.4.1) and an application of (2.10.1).

By analytic continuation, formulas (8.6.3) and (8.6.4) hold when $\max(|q|,|t|,|\beta|) < 1$.

Even though it is clear from (8.6.3) and (8.6.4) that these kernels are positive when $-1 < q,t < 1$ and $-1 \le x,y \le 1$ if $0 \le \beta < 1$, it is not clear what happens when $-1 < \beta < 0$ since both ${}_8\phi_7$ series in (8.6.3) and (8.6.4) become alternating series. It is shown in Gasper and Rahman [1983a] that the Poissson kernel $K_t(x,y;\beta|q)$ is also positive for $-1 < t < 1$ when $-1 < q < \beta < 0$ and when $2^{3/2} - 3 \le \beta < 0$, $-1 < q \le 0$.

For the nonnegativity of the Poisson kernel for the continuous q-Jacobi polynomials, see Gasper and Rahman [1986].

8.7 Poisson kernels for the q-Racah polynomials

For the q-Racah polynomials

$$W_n(x;q) \equiv W_n(x;a,b,c,N;q)$$

we shall give conditions under which the Poisson kernel

$$\sum_{n=0}^{N} h_n(q)W_n(x;q)W_n(y;q)t^n, \quad 0 \le t < 1, \tag{8.7.1}$$

and the so-called discrete Poisson kernel

$$\sum_{n=0}^{z} \frac{(-z;q)_n}{(-N;q)_n} h_n(q)W_n(x;q)W_n(y;q), \tag{8.7.2}$$

$z = 0,1,\ldots,N$, are nonnegative for $x,y = 0,1,\ldots,N$.

Let us first consider a more general bilinear sum

$$
\begin{aligned}
P_z(x,y) &\equiv P_z(x,y;a,b,c,\alpha,\gamma,K,M,N;q) \\
&= \sum_{n=0}^{z} \frac{(q^{-z};q)_n}{(q^{-K};q)_n} h_n(a,b,c,N;q)W_n(x;a,b,c,N;q) \\
&\quad \cdot W_n(y;\alpha,ab\alpha^{-1},\gamma,M;q),
\end{aligned} \tag{8.7.3}
$$

where $z = 0, 1, \ldots, \min(K, N)$ and $N \leq M$. If $\alpha = a, \gamma = c$ and $M = N$, then (8.7.3) reduces to (8.7.2) when $K = N$, and it has the Poisson kernel (8.7.1) as a limit case.

From the product formula (8.2.5) it follows that

$$W_n(x; a, b, c, N; q)W_n(y; \alpha, ab\alpha^{-1}, \gamma, M; q)$$
$$= \frac{(bq, aq/c; q)_n}{(aq, bc; q)_n} c^n \sum_{r=0}^{n} \sum_{s=0}^{n-r} \frac{(q^{-n}, abq^{n+1}; q)_{r+s} (q^{-x}, cq^{x-N}, q^{-y}, \gamma q^{y-N}; q)_r}{(q^{-N}; q)_{r+s} (q, \alpha q, q^{-M}, ab\gamma q\alpha^{-1}, cq^{1-s}; q)_r}$$
$$\cdot \frac{(q^{x-N}, c^{-1}q^{1-x}; q)_s}{(q, bq, ac^{-1}q; q)_s} \frac{1 - cq^{r-s}}{1 - cq^{-s}} q^{r+s-rs} A_{r,s}, \tag{8.7.4}$$

where

$$A_{r,s} = {}_5\phi_4 \left[\begin{matrix} q^{-s}, q^{r-y}, \gamma q^{r+y-M}, aq^{r+1}, bcq^{r+1} \\ \alpha q^{r+1}, q^{r-M}, ab\gamma\alpha^{-1}q^{r+1}, cq^{1+r-s} \end{matrix} ; q, q \right]. \tag{8.7.5}$$

Using (8.7.5) in (8.7.4) and changing the order of summation, we find that

$$P_z(x, y) = \frac{(bq, ac^{-1}q; q)_N}{(abq^2, c^{-1}; q)_N} \sum_{r=0}^{z} \sum_{s=0}^{z-r} \frac{(q^{-z}; q)_{r+s} (abq^2; q)_{2r+2s}}{(q^{-K}, abq^{N+2}; q)_{r+s}}$$
$$\cdot \frac{(q^{-x}, cq^{x-N}, q^{-y}, \gamma q^{y-M}; q)_r (q^{x-N}, c^{-1}q^{-x}; q)_s}{(q, \alpha q, q^{-M}, ab\gamma q\alpha^{-1}, cq^{1-s}; q)_r (q, bq, ac^{-1}q; q)_s} \frac{1 - cq^{r-s}}{1 - cq^{-s}}$$
$$\cdot (-1)^{r+s} q^{(r+s)(2N-r-s+1)/2-rs} A_{r,s} B_{r,s}, \tag{8.7.6}$$

where

$$B_{r,s} = {}_5\phi_4 \left[\begin{matrix} \lambda, q\lambda^{\frac{1}{2}}, -q\lambda^{\frac{1}{2}}, q^{r+s-N}, q^{r+s-z} \\ \lambda^{\frac{1}{2}}, -\lambda^{\frac{1}{2}}, abq^{N+r+s+2}, q^{r+s-K} \end{matrix} ; q, q^{N-r-s} \right], \tag{8.7.7}$$

with $\lambda = abq^{2r+2s+1}$. We shall now show that when $K = N$,

$$B_{r,s} = \frac{(q^{N-z}; q)_{z-r-s}}{(abq^{N+r+s+2}; q)_{z-r-s}} {}_2\phi_1 \left(q^{r+s-z}, q^{1+r+s-z}; q^{1+N-z}; q, abq^{N+z+2}\right). \tag{8.7.8}$$

To prove (8.7.8) it suffices to show that

$${}_4\phi_3 \left[\begin{matrix} a, & qa^{\frac{1}{2}}, & -qa^{\frac{1}{2}}, & b \\ & a^{\frac{1}{2}}, & -a^{\frac{1}{2}}, & w \end{matrix} ; q, \frac{w}{aq} \right]$$
$$= \frac{(wb/aq, w/b; q)_\infty}{(w/aq, w; q)_\infty} {}_2\phi_1(b, bq; wb/a; q, w/b) \tag{8.7.9}$$

whether or not b is a negative integer power of q, provided the series on both sides converge. Since $1 - aq^{2k} = 1 - q^k + q^k(1 - aq^k)$, the left side of (8.7.9) equals

$${}_2\phi_1(aq, b; w; q, w/a) + \frac{w(1 - b)}{aq(1 - w)} {}_2\phi_1(aq, bq; wq; q, w/aq)$$
$$= \frac{(wb/a, w/b; q)_\infty}{(w/a, w; q)_\infty} {}_2\phi_1(b, bq; wb/a; q, w/b)$$

$$+\frac{w(1-b)(wb/a,w/b;q)_\infty}{aq(w/aq,w;q)_\infty}\,{}_2\phi_1(b,bq;wb/a;q,w/b)$$

$$=\frac{(wb/aq,w/b;q)_\infty}{(w/aq,w;q)_\infty}\,{}_2\phi_1(b,bq;wb/a;q,w/b) \tag{8.7.10}$$

by (1.4.5). Also, for $\alpha = a$ the ${}_5\phi_4$ series in (8.7.5) reduces to a ${}_4\phi_3$ series which, by (2.10.4), equals

$$\frac{\left(q^{M+1-y-s},b^{-1}\gamma^{-1}q^{-y-s};q\right)_s}{\left(q^{r-M},b\gamma q^{r+1};q\right)_s}\left(b\gamma q^{r+s+y-M}\right)^s$$

$$\cdot\,{}_4\phi_3\left[\begin{matrix}q^{-s},q^{r-y},b^{-1}q^{-s},c\gamma^{-1}q^{M+1-y-s}\\ cq^{1+r-s},q^{M+1-y-s},b^{-1}\gamma^{-1}q^{-y-s}\end{matrix};q,q\right]. \tag{8.7.11}$$

From (8.7.6), (8.7.8) and (8.7.11) it follows that

$$P_z(x,y;a,b,c,a,\gamma,N,M,N;q)\geq 0 \tag{8.7.12}$$

for $x=0,1,\ldots,N,\ y=0,1,\ldots,M,\ z=0,1,\ldots,N$ when $0<q<1$, $0<aq<1$, $0\leq bq<1$, $0<c<aq^N$ and $cq\leq\gamma<q^{M-1}\leq q^{N-1}$. Hence the discrete Poisson kernel (8.7.2) is nonnegative for $x,y,z=0,1,\ldots,N$ when $0<q<1$, $0<aq<1$, $0\leq bq<1$ and $0<c<aq^N$.

If in (8.7.3) we write the sum with N as the upper limit of summation, replace $(q^{-z};q)_n$ by $(tq^{-K};q)_n$ and let $K\to\infty$, it follows from (8.7.6) that

$$L_t(x,y;a,b,c,\alpha,\gamma,M,N;q)$$

$$=\sum_{n=0}^{N}t^n h_n(a,b,c,N;q)W_n(x;a,b,c,N;q)W_n(y;\alpha,ab\alpha^{-1},\gamma,M;q)$$

$$=\frac{(bq,aq/c;q)_N}{(abq^2,c^{-1};q)_N}\sum_{r=0}^{x}\sum_{s=0}^{N-x}\frac{(abq^2;q)_{2r+2s}\left(q^{-x},cq^{x-N},q^{-y},\gamma q^{y-M};q\right)_r}{\left(abq^{N+2};q\right)_{r+s}\left(q,\alpha q,q^{-M},ab\gamma q\alpha^{-1},cq^{1-s};q\right)_r}$$

$$\cdot\frac{\left(q^{x-N},c^{-1}q^{-x};q\right)_s(1-cq^{r-s})}{(bq,ac^{-1}q;q)_s(1-cq^{-s})}A_{r,s}C_{r,s}(-t)^{r+s}q^{(r+s)(2N-r-s+1)/2-rs}, \tag{8.7.13}$$

for $x=0,1,\ldots,N,y=0,1,\ldots,M$ with $A_{r,s}$ defined in (8.7.5) and

$$C_{r,s}=\,{}_4\phi_3\left[\begin{matrix}\lambda,q\lambda^{\frac12},-q\lambda^{\frac12},q^{r+s-N}\\ \lambda^{\frac12},-\lambda^{\frac12},abq^{N+r+s+2}\end{matrix};q,tq^{N-r-s}\right], \tag{8.7.14}$$

where $\lambda=abq^{2r+2s+1}$. However, by Ex. 2.2,

$${}_4\phi_3\left[\begin{matrix}\lambda, & q\lambda^{\frac12}, & -q\lambda^{\frac12}, & b^{-1}\\ & \lambda^{\frac12}, & -\lambda^{\frac12}, & \lambda bq\end{matrix};q,tb\right]$$

$$=\frac{(t,\lambda q;q)_\infty}{(tb,\lambda bq;q)_\infty}\,{}_2\phi_1(b,tb;tq;q,\lambda q) \tag{8.7.15}$$

for $\max(|tb|,|\lambda q|)<1$. Use of this in (8.7.14) yields

$$C_{r,s}=\left(t,abq^{2r+2s+2};q\right)_{N-r-s}\,{}_2\phi_1\left(q^{N-r-s},tq^{N-r-s};tq;q,abq^{2r+2s+2}\right), \tag{8.7.16}$$

from which it is obvious that $C_{r,s} \geq 0$ for $0 \leq t < 1$, $r+s \leq N$ when $0 \leq abq^2 < 1$. Combining this with our previous observation that $A_{r,s}$ equals the expression in (8.7.11) when $\alpha = a$, it follows from (8.7.13) that

$$L_t(x, y; a, b, c, a, \gamma, M, N; q) > 0 \tag{8.7.17}$$

for $x = 0, 1, \ldots, N$, $y = 0, 1, \ldots, M$, $0 \leq t < 1$ when $0 < q < 1$, $0 < aq < 1$, $0 \leq bq < 1$, $0 < c < aq^N$ and $cq \leq \gamma < q^{M-1} \leq q^{N-1}$. In particular, the Poisson kernel (8.7.1) is positive for $x, y = 0, 1, \ldots, N$, $0 \leq t < 1$ when $0 < q < 1$, $0 < aq < 1$, $0 \leq bq < 1$ and $0 < c < aq^N$.

For further details on the nonnegative bilinear sums of discrete orthogonal polynomials, see Gasper and Rahman [1984] and Rahman [1982].

8.8 q-Analogues of Clausen's formula

Clausen's [1828] formula

$$\left\{ {}_2F_1\left[\begin{matrix} a, b \\ a+b+\frac{1}{2} \end{matrix}; z \right] \right\}^2 = {}_3F_2\left[\begin{matrix} 2a, 2b, a+b \\ 2a+2b, a+b+\frac{1}{2} \end{matrix}; z \right], \tag{8.8.1}$$

where $|z| < 1$, provides a rare example of the square of a hypergeometric series that is expressible as a hypergeometric series. Ramanujan's [1927, pp. 23-39] rapidly convergent series representations of $1/\pi$, which have been used to compute π to millions of decimal digits, are based on special cases of (8.8.1); see the Chudnovskys' [1988] survey paper. Clausen's formula was used in Askey and Gasper [1976] to prove that

$${}_3F_2\left[\begin{matrix} -n, n+\alpha+2, \frac{1}{2}(\alpha+1) \\ \alpha+1, \frac{1}{2}(\alpha+3) \end{matrix}; \frac{1-x}{2} \right] \geq 0 \tag{8.8.2}$$

when $\alpha > -2$, $-1 \leq x \leq 1$, $n = 0, 1, \ldots$, which was then used to prove the positivity of certain important kernels involving sums of Jacobi polynomials; see Askey [1975] and the extensions in Gasper [1975a, 1977]. The special cases $\alpha = 2, 4, 6, \ldots$ of (8.8.2) turned out to be the inequalities de Branges [1985] needed to complete the last step in his celebrated proof of the Bieberbach conjecture. In this section we consider q-analogues of (8.8.1).

Jackson [1940, 1941] derived the product formula given in Ex. 3.11 and additional proofs of it have been given by Singh [1959], Nassrallah [1982], and Jain and Srivastava [1986]. But, unfortunately, the left side of it is not a square and so Jackson's formula cannot be used to write certain basic hypergeometric series as sums of squares as was done with Clausen's formula in Askey and Gasper [1976] to prove (8.8.2).

In order to obtain a q-analogue of Clausen's formula which expressed the square of a basic hypergeometric series as a basic hypergeometric series, the authors derived the formula

$$\left\{ {}_4\phi_3\left[\begin{matrix} a, b, abz, ab/z \\ abq^{\frac{1}{2}}, -abq^{\frac{1}{2}}, -ab \end{matrix}; q, q \right] \right\}^2 = {}_5\phi_4\left[\begin{matrix} a^2, b^2, ab, abz, ab/z \\ a^2b^2, abq^{\frac{1}{2}}, -abq^{\frac{1}{2}}, -ab \end{matrix}; q, q \right], \tag{8.8.3}$$

which holds when the series terminate. See Gasper [1989b], where it was pointed out that there are several ways of proving (8.8.3), such as using the Rogers' linearization formula (8.5.1), the product formula in §8.2, or the Rahman and Verma integral (8.4.10).

In this section we derive a nonterminating q-analogue of Clausen's formula which reduces to (8.8.3) when it terminates. The key to the discovery of this formula is the observation that the proof of Rogers' linearization formula given in §8.5 is independent of the fact that the parameter n in the ${}_2\phi_1$ series in (7.4.2) is a nonnegative integer. In view of (7.4.2) let

$$f(z) = {}_2\phi_1(\alpha, \beta; \alpha q/\beta; q, zq/\beta), \tag{8.8.4}$$

which reduces to the ${}_2\phi_1$ series in (7.4.2) when $\alpha = q^{-n}$ and $z = e^{-2i\theta}$. Temporarily assume that $|q| < |\beta| < 1$ and $|z| \le 1$. From Heine's transformation (1.4.3),

$$f(z) = \frac{(\beta z; q)_\infty}{(zq/\beta; q)_\infty} \, {}_2\phi_1(\alpha q/\beta^2, q/\beta; \alpha q/\beta; q, \beta z). \tag{8.8.5}$$

Hence, if we multiply the two ${}_2\phi_1$ series in (8.8.4) and (8.8.5) and collect the coefficients of z^j, we get

$$f^2(z) = \frac{(\beta z; q)_\infty}{(zq/\beta; q)_\infty} \sum_{k=0}^{\infty} A_k \frac{(\alpha q/\beta^2, q/\beta; q)_k}{(q, \alpha q/\beta; q)_k} (\beta z)^k, \tag{8.8.6}$$

where

$$A_k = {}_4\phi_3 \left[\begin{matrix} q^{-k}, \beta, \beta q^{-k}/\alpha, \alpha \\ \beta q^{-k}, \beta^2 q^{-k}/\alpha, \alpha q/\beta \end{matrix} ; q, q \right] \tag{8.8.7}$$

is a terminating balanced series. As in (8.5.5) we now apply (2.5.1) to the ${}_4\phi_3$ series in (8.8.7) to obtain that

$$A_k = \frac{(\alpha q/\beta, \alpha^2 q/\beta^2; q)_\infty}{(\alpha^2 q/\beta, \alpha q/\beta^2; q)_\infty} \, {}_8W_7(\alpha^2/\beta; \alpha, \alpha, \beta, \alpha^2 q^{k+1}/\beta^2, q^{-k}; q, q/\beta). \tag{8.8.8}$$

Using (8.8.8) in (8.8.6) and changing the order of summation we get the formula

$$\begin{aligned} f^2(z) = {} & \frac{(\beta z; q)_\infty}{(zq/\beta; q)_\infty} \sum_{k=0}^{\infty} \frac{(1 - \alpha^2 q^{2k}/\beta)(\alpha^2/\beta, \alpha, \alpha, \beta; q)_k}{(1 - \alpha^2/\beta)(q, \alpha q/\beta, \alpha q/\beta, \alpha^2 q/\beta^2; q)_k} \\ & \cdot \frac{(\alpha^2 q/\beta^2; q)_{2k}}{(\alpha^2 q/\beta; q)_{2k}} \left(\frac{zq}{\beta} \right)^k {}_2\phi_1 \left[\begin{matrix} \alpha^2 q^{2k+1}/\beta^2, q/\beta \\ \alpha^2 q^{2k+1}/\beta \end{matrix} ; q, \beta z \right]. \end{aligned} \tag{8.8.9}$$

Observe that since the ${}_2\phi_1$ series in (8.8.9) is well-poised we may transform it by applying the quadratic transformation formula (3.4.7) to express it as an ${}_8\phi_7$ series and then apply (2.10.10) to get the transformation formula

$$\begin{aligned} {}_2\phi_1 \left[\begin{matrix} \alpha^2 q^{2k+1}/\beta^2, q/\beta \\ \alpha^2 q^{2k+1}/\beta \end{matrix} ; q, \beta z \right] = {} & \frac{(\alpha z q/\beta; q)_\infty (-\alpha/\beta z)^k q^{\binom{k+1}{2}}}{(\beta z/\alpha; q)_\infty (\alpha q/\beta z, \alpha z q/\beta; q)_k} \\ & \cdot {}_4\phi_3 \left[\begin{matrix} \alpha q^k, \alpha q^{k+\frac{1}{2}}/\beta, -\alpha q^{k+\frac{1}{2}}/\beta, -\alpha q^{k+1}/\beta \\ \alpha^2 q^{2k+1}/\beta, \alpha z q^{k+1}/\beta, \alpha q^{k+1}/\beta z \end{matrix} ; q, q \right] \end{aligned}$$

$$+\frac{(zq,\alpha,\alpha zq,\alpha^2q/\beta^2;q)_\infty}{(\beta z,\alpha q/\beta,\alpha/\beta z,\alpha^2q/\beta;q)_\infty}\frac{(\alpha q/\beta,\alpha/\beta z;q)_k(\alpha^2q/\beta;q)_{2k}}{(\alpha,\alpha zq;q)_k(\alpha^2q/\beta^2;q)_{2k}}$$

$$\cdot\,{}_4\phi_3\left[\begin{matrix}\beta z,zq^{\frac12},-zq^{\frac12},-zq\\ qz^2,\alpha zq^{k+1},\beta zq^{1-k}/\alpha\end{matrix};q,q\right]. \tag{8.8.10}$$

We can now substitute (8.8.10) into (8.8.9) and change the orders of summation to find that

$$f^2(z)=\frac{(\beta z,\alpha zq/\beta;q)_\infty}{(zq/\beta,\beta z/\alpha;q)_\infty}\sum_{m=0}^\infty\frac{(\alpha;q)_m(\alpha^2q/\beta^2;q)_{2m}}{(q,\alpha zq/\beta,\alpha q/\beta z,\alpha^2q/\beta,\alpha q/\beta;q)_m}$$

$$\cdot\,q^m\,{}_6W_5(\alpha^2/\beta;\alpha,\beta,q^{-m};q,\alpha q^{m+1}/\beta^2)$$

$$+\frac{(\alpha,\alpha^2q/\beta,zq,\alpha zq;q)_\infty}{(\alpha q/\beta,\alpha^2q/\beta,zq/\beta,\alpha/\beta z;q)_\infty}\sum_{m=0}^\infty\frac{(\beta z,zq^{\frac12},-zq^{\frac12},-zq;q)_m}{(q,qz^2,\alpha zq,\beta zq/\alpha;q)_m}q^m$$

$$\cdot\,{}_6W_5(\alpha^2/\beta;\alpha,\beta,\alpha q^{-m}/\beta z;q,zq^{m+1}/\beta). \tag{8.8.11}$$

Summing the above ${}_6W_5$ series by means of (2.7.1), we obtain the formula

$$\{{}_2\phi_1(\alpha,\beta;\alpha q/\beta;q,zq/\beta)\}^2=\frac{(\beta z,\alpha zq/\beta;q)_\infty}{(zq/\beta,\beta z/\alpha;q)_\infty}$$

$$\cdot\,{}_5\phi_4\left[\begin{matrix}\alpha,\alpha q/\beta^2,\alpha q^{\frac12}/\beta,-\alpha q^{\frac12}/\beta,-\alpha q/\beta\\ \alpha q/\beta,\alpha^2q/\beta^2,\alpha q/\beta z,\alpha zq/\beta\end{matrix};q,q\right]$$

$$+\frac{(\alpha,\alpha q/\beta^2,zq,zq,\alpha zq/\beta;q)_\infty}{(\alpha q/\beta,\alpha q/\beta,zq/\beta,zq/\beta,\alpha/\beta z;q)_\infty}$$

$$\cdot\,{}_5\phi_4\left[\begin{matrix}\beta z,zq/\beta,zq^{\frac12},-zq^{\frac12},-zq\\ zq,z^2q,\beta zq/\alpha,\alpha zq/\beta\end{matrix};q,q\right], \tag{8.8.12}$$

which gives the square of a well-poised ${}_2\phi_1$ series as the sum of the two balanced ${}_5\phi_4$ series. By analytic continuation, (8.8.12) holds when $|q|<1$ and $|zq/\beta|<1$.

To derive (8.8.3) from (8.8.12), observe that if $\alpha=q^{-n}$, $n=0,1,\ldots$, then $(\alpha;q)_\infty=0$ and (8.8.12) gives

$$f^2(z)$$

$$=\frac{(\beta z,zq^{1-n}/\beta;q)_\infty}{(\beta zq^n,zq/\beta;q)_\infty}\,{}_5\phi_4\left[\begin{matrix}q^{-n},q^{\frac12-n}/\beta,q^{1-n}/\beta^2,-q^{\frac12-n}/\beta,-q^{1-n}/\beta\\ q^{1-n}/\beta,q^{1-2n}/\beta^2,zq^{1-n}/\beta,q^{1-n}/\beta z\end{matrix};q,q\right]$$

$$=\frac{(\beta^2,\beta^2;q)_n}{(\beta,\beta;q)_n}\left(\frac{z}{\beta}\right)^n\,{}_5\phi_4\left[\begin{matrix}q^{-n},\beta^2q^n,\beta,\beta z,\beta/z\\ \beta^2,\beta q^{\frac12},-\beta q^{\frac12},-\beta\end{matrix};q,q\right] \tag{8.8.13}$$

by reversing the order of summation. Since

$$f(z)=\frac{(\beta^2;q)_n}{(\beta;q)_n}\left(\frac{z}{\beta}\right)^n\,{}_4\phi_3\left[\begin{matrix}q^{-n},\beta^2q^n,(\beta z)^{\frac12},(\beta/z)^{\frac12}\\ \beta q^{\frac12},-\beta q^{\frac12},-\beta\end{matrix};q,q\right] \tag{8.8.14}$$

by (7.4.14), it follows from (8.8.13) that

$$\left\{{}_4\phi_3\left[\begin{matrix}q^{-n},\beta^2q^n,(\beta z)^{\frac12},(\beta/z)^{\frac12}\\ \beta q^{\frac12},-\beta q^{\frac12},-\beta\end{matrix};q,q\right]\right\}^2$$

$$= {}_5\phi_4\left[\begin{matrix} q^{-n}, \beta^2 q^n, \beta, \beta z, \beta/z \\ \beta^2, \beta q^{\frac{1}{2}}, -\beta q^{\frac{1}{2}}, -\beta \end{matrix} ; q, q\right] \tag{8.8.15}$$

for $n = 0, 1, \ldots$, which is formula (8.8.3) written in an equivalent form.

Now note that

$$\begin{aligned} &{}_2\phi_1(\alpha, \beta; \alpha q/\beta; q, zq/\beta) \\ &= \frac{(z(\alpha q)^{\frac{1}{2}}, -z(\alpha q)^{\frac{1}{2}}, zq\alpha^{\frac{1}{2}}/\beta, -zq\alpha^{\frac{1}{2}}/\beta; q)_\infty}{(zq^{\frac{1}{2}}, -zq^{\frac{1}{2}}, zq/\beta, -\alpha zq/\beta; q)_\infty} \\ &\quad \cdot {}_8W_7(-\alpha z/\beta; \alpha^{\frac{1}{2}}, -\alpha^{\frac{1}{2}}, (\alpha q)^{\frac{1}{2}}/\beta, -(\alpha q)^{\frac{1}{2}}/\beta, -z; q, -zq), \end{aligned} \tag{8.8.16}$$

by (3.4.7) and (2.10.1), and set $a = \alpha^{\frac{1}{2}}, b = (\alpha q)^{\frac{1}{2}}/\beta$ to obtain from (8.8.12) the following q-analogue of Clausen's formula:

$$\begin{aligned} &\left\{\frac{(a^2z^2q, b^2z^2q; q^2)_\infty}{(z^2q, a^2b^2z^2q; q^2)_\infty}\, {}_8W_7(-abzq^{-\frac{1}{2}}; a, -a, b, -b, -z; q, -zq)\right\}^2 \\ &= \frac{(azq^{\frac{1}{2}}/b, bzq^{\frac{1}{2}}/a; q)_\infty}{(zq^{\frac{1}{2}}/ab, abzq^{\frac{1}{2}}; q)_\infty}\, {}_5\phi_4\left[\begin{matrix} a^2, b^2, ab, -ab, -abq^{\frac{1}{2}} \\ a^2b^2, abq^{\frac{1}{2}}, abzq^{\frac{1}{2}}, abq^{\frac{1}{2}}/z \end{matrix}; q, q\right] \\ &\quad + \frac{(zq, zq, a^2, b^2; q)_\infty}{(abq^{\frac{1}{2}}, abq^{\frac{1}{2}}, ab/zq^{\frac{1}{2}}, abzq^{\frac{1}{2}}; q)_\infty} \\ &\quad \cdot {}_5\phi_4\left[\begin{matrix} azq^{\frac{1}{2}}/b, bzq^{\frac{1}{2}}/a, zq^{\frac{1}{2}}, -zq^{\frac{1}{2}}, -zq \\ zq, z^2q, abzq^{\frac{1}{2}}, zq^{\frac{3}{2}}/ab \end{matrix}; q, q\right], \end{aligned} \tag{8.8.17}$$

where $|q| < 1$ and $|zq| < 1$.

To see that (8.8.17) is a nonterminating q-analogue of Clausen's formula, it suffices to replace a by q^a, b by q^b and let $q \to 1^-$; then the left side and the first term on the right side of (8.8.17) tend to the left and right sides of (8.8.1) with z replaced by $-4z(1-z)^{-2}$ and so, by (8.8.1), the second term on the right side of (8.8.17) must tend to zero.

It is shown in Gasper and Rahman [1989b] that the nonterminating extension (3.4.1) of the Sears-Carlitz quadratic transformation can be used in place of (8.8.10) to derive the product formula

$$\begin{aligned} &{}_2\phi_1(a, b; c; q, z)\ {}_2\phi_1(a, aq/c; aq/b; q, z) = \frac{(az, abz/c; q)_\infty}{(z, bz/c; q)_\infty} \\ &\cdot {}_6\phi_5\left[\begin{matrix} a,\ c/b,\ (ac/b)^{\frac{1}{2}},\ -(ac/b)^{\frac{1}{2}},\ (acq/b)^{\frac{1}{2}},\ -(acq/b)^{\frac{1}{2}} \\ aq/b,\ c,\ ac/b,\ az,\ cq/bz \end{matrix}; q, q\right] \\ &+ \frac{(a, c/b, az, bz, azq/c; q)_\infty}{(c, aq/b, z, z, c/bz; q)_\infty} \\ &\cdot {}_6\phi_5\left[\begin{matrix} z,\ abz/c,\ z(ab/c)^{\frac{1}{2}},\ -z(ab/c)^{\frac{1}{2}},\ z(abq/c)^{\frac{1}{2}},\ -z(abq/c)^{\frac{1}{2}} \\ az,\ bz,\ azq/c,\ bzq/c,\ abz^2/c \end{matrix}; q, q\right], \end{aligned} \tag{8.8.18}$$

where $|z| < 1$ and $|q| < 1$. This formula reduces to (8.8.12) when $a = \alpha,\ b = \beta,\ c = \alpha q/\beta$ and z is replaced by zq/β.

By applying various transformation formulas to the ${}_2\phi_1$ series in (8.8.12) and (8.8.18), these formulas can be written in many equivalent forms. For instance, by replacing b in (8.8.18) by c/b and applying (1.5.4) we obtain

$$\begin{aligned}
&{}_2\phi_2(a,b;c,az;q,cz/b)\ {}_2\phi_2(a,b;abq/c,az;q,azq/c)\\
&= \frac{(z,az/b;q)_\infty}{(az,z/b;q)_\infty}\ {}_6\phi_5\left[\begin{matrix} a,b,(ab)^{\frac12},-(ab)^{\frac12},(abq)^{\frac12},-(abq)^{\frac12} \\ abq/c,c,ab,az,bq/z\end{matrix};q,q\right]\\
&\quad+\frac{(a,b,cz/b,azq/c;q)_\infty}{(c,abq/c,az,b/z;q)_\infty}\\
&\quad\cdot {}_6\phi_5\left[\begin{matrix} z,abz,z(a/b)^{\frac12},-z(a/b)^{\frac12},z(aq/b)^{\frac12},-z(aq/b)^{\frac12} \\ az,cz/b,aqz/c,zq/b,az^2/b\end{matrix};q,q\right],
\end{aligned}\tag{8.8.19}$$

where $\max(|q|,|azq/c|,|cz/b|)<1$. If we replace a,b,z in (8.8.19) by q^a,b^b, $z/(z-1)$, respectively, and let $q\to 1^-$, we obtain the Ramanujan [1957, Vol. 2] and Bailey [1933, 1935a] product formula

$$\begin{aligned}
&{}_2F_1(a,b;c;z)\ {}_2F_1(a,b;a+b-c+1;z)\\
&=\ {}_4F_3(a,b,(a+b)/2,(a+b+1)/2;c,a+b,a+b-c+1;4z(1-z)),
\end{aligned}\tag{8.8.20}$$

where $|z|<1$ and $|4z(1-z)|<1$. This is an extension of Clausen's formula in the sense that by replacing $a,b,c,4z(1-z)$ in (8.8.20) by $2a,2b,a+b+\frac12,z$, respectively, and using the quadratic transformation (Erdélyi [1953, 2.11 (2)])

$${}_2F_1(2a,2b;a+b+\frac{1}{2};z)=\ {}_2F_1(a,b;a+b+\frac{1}{2};4z(1-z)),\tag{8.8.21}$$

we get (8.8.1). See Askey [1989d].

8.9 Nonnegative basic hypergeometric series

Our main aim in this section is to show how the terminating q-Clausen formula (8.8.3) can be used to derive q-analogues of the Askey-Gasper inequalities (8.8.2) and of the nonnegative hypergeometric series in Gasper [1975a, Equations (8.19), (8.20), (8.22)].

As in Gasper [1989b], let us set

$$\gamma=q^{2b},a_1=q^b,a_2=q^be^{i\theta},a_3=q^be^{-i\theta},b_1=q^{2b},b_2=-q^b,$$

$$c_1=q^{-n},c_2=q^{n+a},d_1=q^{\frac12(a+1)}=-d_2,e_1=q^{b+\frac12}=-e_3,x=q,w=1,$$

in the $r=3,s=t=u=k=2$ case of (3.7.9) to obtain the expansion

$$\begin{aligned}
&{}_5\phi_4\left[\begin{matrix} q^{-n},q^{n+a},q^b,q^be^{i\theta},q^be^{-i\theta} \\ q^{2b},q^{\frac12(a+1)},-q^{\frac12(a+1)},-q^b\end{matrix};q,q\right]\\
&=\sum_{j=0}^{n}\frac{(q^{-n},q^{n+a},q^{b+\frac12},-q^{b+\frac12};q)_j}{(q,q^{\frac12(a+1)},-q^{\frac12(a+1)},q^{j+2b};q)_j}(-1)^jq^{j+\binom{j}{2}}
\end{aligned}$$

$$\cdot {}_4\phi_3\left[\begin{matrix} q^{j-n}, q^{j+n+a}, q^{j+b+\frac{1}{2}}, -q^{j+b+\frac{1}{2}} \\ q^{2j+2b+1}, q^{j+\frac{1}{2}(a+1)}, -q^{j+\frac{1}{2}(a+1)} \end{matrix} ; q, q\right]$$

$$\cdot {}_5\phi_4\left[\begin{matrix} q^{-j}, q^{j+2b}, q^b, q^b e^{i\theta}, q^b e^{-i\theta} \\ q^{2b}, q^{b+\frac{1}{2}}, -q^{b+\frac{1}{2}}, -q^b \end{matrix} ; q, q\right], \tag{8.9.1}$$

where, as through this section, $n = 0, 1, \ldots$. By Ex. 2.8 the ${}_4\phi_3$ series in (8.9.1) equals zero when $n - j$ is odd and equals

$$\frac{(q, q^{a-2b}; q^2)_k}{(q^{2n-4k+a+1}, q^{2n-4k+2b+2}; q^2)_k} q^{2k(n-2k+b+1/2)}$$

when $n - j = 2k$ and $k = 0, 1, \ldots$. Hence, using (8.8.3) to write the ${}_5\phi_4$ as the square of a ${}_4\phi_3$ series, we have

$${}_5\phi_4\left[\begin{matrix} q^{-n}, q^{n+a}, q^b, q^b e^{i\theta}, q^b e^{-i\theta} \\ q^{2b}, q^{\frac{1}{2}(a+1)}, -q^{\frac{1}{2}(a+1)}, -q^b \end{matrix} ; q, q\right]$$

$$= \sum_{k=0}^{[n/2]} \frac{(-1)^n (q^{-n}, q^{n+a}, q^{b+\frac{1}{2}}, -q^{b+\frac{1}{2}}; q)_{n-2k}}{(q, q^{\frac{1}{2}(a+1)}, -q^{\frac{1}{2}(a+1)}, q^{n-2k+2b}; q)_{n-2k}}$$

$$\cdot \frac{(q, q^{a-2b}; q^2)_k}{(q^{2n-4k+a+1}, q^{2n-4k+2b+2}; q^2)_k} q^{2k(n-2k+b+\frac{1}{2})+\frac{1}{2}(n-2k)(n-2k+1)}$$

$$\cdot \left\{ {}_4\phi_3\left[\begin{matrix} q^{2k-n}, q^{n-2k+2b}, q^{\frac{1}{2}b} e^{\frac{1}{2}i\theta}, q^{\frac{1}{2}b} e^{-\frac{1}{2}i\theta} \\ q^{b+\frac{1}{2}}, -q^{b+\frac{1}{2}}, -q^b \end{matrix} ; q, q\right]\right\}^2. \tag{8.9.2}$$

Since $(-1)^n (q^{-n}; q)_{n-2k} \geq 0$, it is clear from (8.9.2) that

$${}_5\phi_4\left[\begin{matrix} q^{-n}, q^{n+a}, q^b, q^b e^{i\theta}, q^b e^{-i\theta} \\ q^{2b}, q^{\frac{1}{2}(a+1)}, -q^{\frac{1}{2}(a+1)}, -q^b \end{matrix} ; q, q\right] \geq 0 \tag{8.9.3}$$

when $a \geq 2b > -1$ and $0 < q < 1$. By setting $x = \cos\theta$ and letting $q \to 1^-$, it follows from (8.9.3) that

$${}_3F_2\left[\begin{matrix} -n, n+a, b \\ 2b, \frac{1}{2}(a+1) \end{matrix} ; \frac{1-x}{2}\right] \geq 0, \quad -1 \leq x \leq 1, \tag{8.9.4}$$

when $a \geq 2b > -1$ which shows that (8.9.3) is a q-analogue of (8.9.4). When $a = \alpha + 2$ and $b = \frac{1}{2}(\alpha + 1)$, (8.9.4) reduces to (8.8.2). Special cases of (8.9.4) were used by de Branges [1986] in his work on coefficient estimates for Riemann mapping functions.

Another q-analogue of (8.9.4) can be derived by using (8.9.1), (8.8.3) and Ex. 2.8 to obtain

$${}_6\phi_5\left[\begin{matrix} q^{-n}, q^{n+a}, q^b, -q^b, q^{\frac{1}{2}a} e^{i\theta}, q^{\frac{1}{2}a} e^{-i\theta} \\ q^{2b}, q^{\frac{1}{2}(a+1)}, -q^{\frac{1}{2}(a+1)}, -q^{\frac{1}{2}a}, -q^{\frac{1}{2}a} \end{matrix} ; q, q\right]$$

$$= \sum_{j=0}^{n} \frac{(q^{-n}, q^{n+a}, q^{\frac{1}{2}a}, q^{\frac{1}{2}a} e^{i\theta}, q^{\frac{1}{2}a} e^{-i\theta}; q)_j}{(q, q^{\frac{1}{2}(a+1)}, -q^{\frac{1}{2}(a+1)}, -q^{\frac{1}{2}a}, q^{j+a-1}; q)_j} (-1)^j q^{j+\binom{j}{2}}$$

$$\cdot {}_5\phi_4\left[\begin{matrix} q^{j-n}, q^{n+j+a}, q^{j+\frac{1}{2}a}, q^{j+\frac{1}{2}a} e^{i\theta}, q^{j+\frac{1}{2}a} e^{-i\theta} \\ q^{2j+a}, q^{j+\frac{1}{2}(a+1)}, -q^{j+\frac{1}{2}(a+1)}, -q^{j+\frac{1}{2}a} \end{matrix} ; q, q\right]$$

$$\cdot {}_4\phi_3\left[\begin{matrix} q^{-j}, q^{j+a-1}, q^b, -q^b \\ q^{2b}, q^{\frac{1}{2}a}, -q^{\frac{1}{2}a} \end{matrix}; q, q\right]$$

$$= \sum_{k=0}^{[n/2]} \frac{(q^{-n}, q^{n+a}, q^{\frac{1}{2}a}, q^{\frac{1}{2}a}e^{i\theta}, q^{\frac{1}{2}a}e^{-i\theta}; q)_{2k}}{(q, q^{\frac{1}{2}(a+1)}, -q^{\frac{1}{2}(a+1)}, -q^{\frac{1}{2}a}, q^{2k+a-1}; q)_{2k}}$$

$$\cdot \frac{(q, q^{a-2b}; q^2)_k}{(q^a, q^{2b+1}; q^2)_k} q^{2k^2+k+2kb}$$

$$\cdot \left\{ {}_4\phi_3\left[\begin{matrix} q^{2k-n}, q^{n+2k+a}, q^{k+\frac{1}{4}a}e^{\frac{1}{2}i\theta}, q^{k+\frac{1}{4}a}e^{-\frac{1}{2}i\theta} \\ q^{2k+\frac{1}{2}(a+1)}, -q^{2k+\frac{1}{2}(a+1)}, -q^{2k+\frac{1}{2}a} \end{matrix}; q, q\right]\right\}^2, \tag{8.9.5}$$

which shows that

$${}_6\phi_5\left[\begin{matrix} q^{-n}, q^{n+a}, q^b, -q^b, q^{\frac{1}{2}a}e^{i\theta}, q^{\frac{1}{2}a}e^{-i\theta} \\ q^{2b}, q^{\frac{1}{2}(a+1)}, -q^{\frac{1}{2}(a+1)}, -q^{\frac{1}{2}a}, -q^{\frac{1}{2}a} \end{matrix}; q, q\right] \geq 0 \tag{8.9.6}$$

when $a \geq 2b > -1$ and $0 < q < 1$.

The expansions (8.12) and (8.17) in Gasper [1975a] are special cases of the $q \to 1^-$ limit cases of (8.9.2) and (8.9.5), respectively, when (7.4.14) and Gasper [1975a, (8.10)] are used. A q-analogue of the expansion (Gasper [1975a, 1989b])

$${}_3F_2\left[\begin{matrix} -n, n+\alpha+2, \frac{1}{2}(\alpha+1) \\ \alpha+1, \frac{1}{2}(\alpha+3) \end{matrix}; (1-x^2)(1-y^2)\right]$$

$$= \sum_{j=0}^{n} \frac{n!(n+\alpha+2)_j \left(\frac{\alpha+2}{2}\right)_j}{j!(n-j)! \left(\frac{\alpha+3}{2}\right)_j (j+\alpha+1)_j} (1-y^2)^j$$

$$\cdot \left\{ \frac{j!(n-j)!}{(\alpha+1)_j (2j+\alpha+2)_{n-j}} C_j^{\frac{1}{2}(\alpha+1)}(x) C_{n-j}^{j+\frac{1}{2}(\alpha+2)}(y) \right\}^2 \tag{8.9.7}$$

is easily derived by employing (3.7.9), (8.8.3) and (7.4.14) to obtain

$${}_7\phi_6\left[\begin{matrix} q^{-n}, q^{n+\alpha+2}, q^{\frac{1}{2}(\alpha+1)}, q^{\frac{1}{2}(\alpha+1)}e^{2i\theta}, q^{\frac{1}{2}(\alpha+1)}e^{-2i\theta}, \\ q^{\alpha+1}, q^{\frac{1}{2}(\alpha+3)}, -q^{\frac{1}{2}(\alpha+3)}, -q^{\frac{1}{2}(\alpha+2)}, \end{matrix}\right.$$

$$\left.\begin{matrix} q^{\frac{1}{2}(\alpha+2)}e^{2i\tau}, q^{\frac{1}{2}(\alpha+2)}e^{-2i\tau} \\ -q^{\frac{1}{2}(\alpha+2)}, -q^{\frac{1}{2}(\alpha+1)} \end{matrix}; q, q\right]$$

$$= \sum_{j=0}^{n} \frac{(q^{-n}, q^{n+\alpha+2}, q^{\frac{1}{2}(\alpha+2)}, q^{\frac{1}{2}(\alpha+2)}e^{2i\tau}, q^{\frac{1}{2}(\alpha+2)}e^{-2i\tau}; q)_j}{(q, q^{\frac{1}{2}(\alpha+3)}, -q^{\frac{1}{2}(\alpha+3)}, -q^{\frac{1}{2}(\alpha+2)}, q^{j+\alpha+1}; q)_j} (-1)^j q^{j+\binom{j}{2}}$$

$$\cdot \left\{ \frac{(q;q)_j (q;q)_{n-j}}{(q^{\alpha+1};q)_j (q^{2j+\alpha+2};q)_{n-j}} q^{\frac{1}{2}(j+\alpha+\frac{3}{2})} \right.$$

$$\left. \cdot C_j(\cos\theta; q^{\frac{1}{2}(\alpha+1)}|q) C_{n-j}(\cos\tau; q^{j+\frac{1}{2}(\alpha+2)}|q) \right\}^2, \tag{8.9.8}$$

which is obviously nonnegative for real θ and τ when $\alpha > -2$.

If we proceed as in (8.9.5), but use the q-Saalschütz summation formula (1.7.2) instead of Ex. 2.8, we find that

$$
{}_6\phi_5\left[\begin{matrix} q^{-n}, q^{n+a}, q^{\frac{1}{2}a}, q^b, q^{\frac{1}{2}a}e^{i\theta}, q^{\frac{1}{2}a}e^{-i\theta} \\ q^{\frac{1}{2}(a+1)}, q^c, q^{a+b-c}, -q^{\frac{1}{2}(a+1)}, -q^{\frac{1}{2}a} \end{matrix}; q, q\right]
$$

$$
= \sum_{j=0}^{n} \frac{(q^{-n}, q^{n+a}, q^{\frac{1}{2}a}, q^{a-c}, q^{c-b}, q^{\frac{1}{2}a}e^{i\theta}, q^{\frac{1}{2}a}e^{-i\theta}; q)_j}{(q, q^{\frac{1}{2}(a+1)}, -q^{\frac{1}{2}(a+1)}, -q^{\frac{1}{2}a}, q^{j+a-1}, q^c, q^{a+b-c}; q)_j}
$$

$$
\cdot (-1)^j q^{j(b+1)+\binom{j}{2}} \left\{ {}_4\phi_3\left[\begin{matrix} q^{j-n}, q^{j+n+a}, q^{\frac{1}{2}j+\frac{1}{4}a}e^{\frac{1}{2}i\theta}, q^{\frac{1}{2}j+\frac{1}{4}a}e^{-\frac{1}{2}i\theta} \\ q^{j+\frac{1}{2}(a+1)}, -q^{j+\frac{1}{2}(a+1)}, -q^{j+\frac{1}{2}a} \end{matrix}; q, q\right]\right\}^2 \tag{8.9.9}
$$

and hence

$$
{}_6\phi_5\left[\begin{matrix} q^{-n}, q^{n+a}, q^{\frac{1}{2}a}, q^b, q^{\frac{1}{2}a}e^{i\theta}, q^{\frac{1}{2}a}e^{-i\theta} \\ q^{\frac{1}{2}(a+1)}, q^c, q^{a+b-c}, -q^{\frac{1}{2}(a+1)}, -q^{\frac{1}{2}a} \end{matrix}; q; q\right] \geq 0 \tag{8.9.10}
$$

for real θ when $a \geq c \geq b$, $a+b > c > 0$ and $0 < q < 1$. By letting $q \to 1^-$ in (8.9.10) it follows that

$$
{}_4F_3\left[\begin{matrix} -n, n+a, \frac{1}{2}a, b \\ \frac{1}{2}(a+1), c, a+b-c \end{matrix}; \frac{1-x}{2}\right] \geq 0 \tag{8.9.11}
$$

for $-1 \leq x \leq 1$ when $a \geq c \geq b$ and $a+b > c > 0$, which gives the nonnegativity of the ${}_3F_2$ series in Gasper [1975a, (8.22)] when $a = \alpha + \frac{3}{2}, b = \frac{1}{2}\alpha + \frac{5}{4}$ and $c = \alpha + 1$. For nonnegative sums of Askey-Wilson polynomials, see Gasper [1989d].

8.10 Applications in the theory of partitions of positive integers

In the aplications given so far we have dealt almost exclusively with orthogonal polynomials which are representable as basic hypergeometric series. These are important results and most of them have appeared in print during the last twenty years. They constitute the main bulk of applications as far as this book is concerned. However, our task would remain incomplete if we failed to mention some of the earliest examples where basic hypergeometric series played crucial roles. The simplest among these examples is Euler's [1748] enumeration $p(n)$ of the partitions of a positive integer n, where a *partition* of a positive integer n is a finite monotone decreasing sequence of positive integers (called the parts of the partition) whose sum is n. To illustrate how $p(n)$ arises in a q-series let us consider a formal series expansion of the infinite product $(q;q)_\infty^{-1}$ in powers of q:

$$
\begin{aligned}
(q;q)_\infty^{-1} &= \prod_{k=0}^{\infty} (1-q^{k+1})^{-1} \\
&= \sum_{k_1 \geq 0} \sum_{k_2 \geq 0} \cdots q^{k_1 \cdot 1 + k_2 \cdot 2 + \cdots} \\
&= \sum_{n=0}^{\infty} p(n) q^n.
\end{aligned} \tag{8.10.1}
$$

where

$$n = k_1 \cdot 1 + k_2 \cdot 2 + \cdots + k_n \cdot n, \tag{8.10.2}$$

$p(0) = 1$ and, for a positive integer n, $p(n)$ is the number of partitions of n into parts $\leq n$. In the partition (8.10.2) of n there are k_m m's and hence $0 \leq k_m \leq n/m, 1 \leq m \leq n$. For small values of n, $p(n)$ can be calculated quite easily, but the number increases very rapidly. For example, $p(3) = 3$, $p(4) = 5$, $p(5) = 7$, but $p(243) = 133978259344888$. Hardy and Ramanujan [1918] found the following asymptotic formula for large n:

$$p(n) \sim \frac{1}{4n\sqrt{3}} \exp\left[\pi\left(\frac{2n}{3}\right)^{\frac{1}{2}}\right]. \tag{8.10.3}$$

Also of interest are the enumerations of partitions of a positive integer n into parts restricted in certain ways such as:

(i) $p_N(n)$, the number of partitions of n into parts $\leq N$, which is given by the generating function

$$(q;q)_N^{-1} = \sum_{n=0}^{\infty} p_N(n) q^n, \tag{8.10.4}$$

where $n = k_1 \cdot 1 + k_2 \cdot 2 + \cdots + k_N \cdot N$ has k_m m's ;

(ii) $p_e(n)$, the number of partitions of an even integer n into even parts, generated by

$$\begin{aligned}(q^2;q^2)_\infty^{-1} &= \prod_{k=0}^{\infty} (1-q^{2k+2})^{-1} \\ &= \sum_{n=0}^{\infty} p_e(n) q^n; \end{aligned} \tag{8.10.5}$$

(iii) $p_{\text{dist}}(n)$, the number of partitions of n into distinct positive integers, generated by

$$\begin{aligned}(-q;q)_\infty &= \sum_{k_1=0}^{1} \sum_{k_2=0}^{1} \cdots q^{k_1 \cdot 1 + k_2 \cdot 2 + \cdots} \\ &= \sum_{n=0}^{\infty} p_{\text{dist}}(n) q^n, \end{aligned} \tag{8.10.6}$$

where $n = k_1 \cdot 1 + k_2 \cdot 2 + \cdots + k_n \cdot n$, $0 \leq k_i \leq 1$, $1 \leq i \leq n$, and

(iv) $p_O(n)$, the number of partitions of n into odd parts, generated by

$$\begin{aligned}(q;q^2)_\infty^{-1} &= \prod_{k=0}^{\infty} (1-q^{2k+1})^{-1} \\ &= \sum_{n=0}^{\infty} p_O(n) q^n. \end{aligned} \tag{8.10.7}$$

Euler's partition identity

$$p_{\text{dist}}(n) = p_O(n) \tag{8.10.8}$$

follows from (8.10.6), (8.10.7) and the fact that

$$(-q;q)_\infty = \left(q^{\frac{1}{2}}, -q^{\frac{1}{2}}; q\right)_\infty^{-1} = (q;q^2)_\infty^{-1}. \tag{8.10.9}$$

Other combinatorial identities of this type can be discovered from q-series identities similar to, but perhaps somewhat more complicated than, (8.10.9). For example, let us consider Euler's [1748] identity involving the *pentagonal numbers* $n(3n \pm 1)/2$:

$$\begin{aligned}(q;q)_\infty &= \sum_{n=-\infty}^{\infty} (-1)^n q^{n(3n+1)/2} \\ &= 1 + \sum_{n=1}^{\infty} (-1)^n q^{n(3n+1)/2} + \sum_{n=1}^{\infty} (-1)^n q^{n(3n-1)/2},\end{aligned} \tag{8.10.10}$$

which is given in Ex. 2.18. A formal power series expansion gives

$$\begin{aligned}(q;q)_\infty &= \prod_{k=0}^{\infty} (1 - q^{k+1}) \\ &= \sum_{k_1=0}^{1} \sum_{k_2=0}^{1} \cdots (-1)^{k_1+k_2+\cdots} q^{k_1\cdot 1 + k_2 \cdot 2 + \cdots},\end{aligned} \tag{8.10.11}$$

which differs from the multiple series in (8.10.6) only in the factor $(-1)^{k_1+k_2+\cdots}$. This factor is ± 1 according as the partition has an even or odd number of parts. Denoting these numbers by $p_{\text{even}}(n)$ and $p_{\text{odd}}(n)$, respectively, we find that

$$(q;q)_\infty = 1 + \sum_{n=1}^{\infty} \left[p_{\text{even}}(n) - p_{\text{odd}}(n)\right] q^n. \tag{8.10.12}$$

From (8.10.10) and (8.10.12) it follows that

$$p_{\text{even}}(n) - p_{\text{odd}}(n) = \begin{cases} (-1)^k & \text{for } n = k(3k \pm 1)/2, \\ 0 & \text{otherwise.} \end{cases} \tag{8.10.13}$$

Thus Euler's identity (8.10.10) expresses the important property that a positive integer n which is not a pentagonal number of the form $k(3k \pm 1)/2$ can be partitioned as often into an even number of parts as into an odd number of parts. However, if $n = k(3k \pm 1)/2$, $k = 1, 2, \ldots$, then $p_{\text{even}}(n)$ exceeds $p_{\text{odd}}(n)$ by $(-1)^k$. See Hardy and Wright [1979], Rademacher [1973] and Andrews [1976, 1983] for related results.

Two of the most celebrated identities in combinatorial analysis are the so-called Rogers-Ramanujan identities (2.7.3) and (2.7.4) which, for the purposes of the present discussion, we rewrite in the following form:

$$1 + \sum_{n=1}^{\infty} \frac{q^{n^2}}{(q;q)_n} = \frac{1}{\prod_{k=0}^{\infty} (1 - q^{5k+1})(1 - q^{5k+4})}, \tag{8.10.14}$$

$$1+\sum_{n=1}^{\infty}\frac{q^{n(n+1)}}{(q;q)_n}=\frac{1}{\prod\limits_{k=0}^{\infty}(1-q^{5k+2})(1-q^{5k+3})}. \tag{8.10.15}$$

It is clear that the infinite product on the right side of (8.10.14) enumerates the partitions of a positive integer n into parts of the form $5k+1$ and $5k+4$, while that on the right side of (8.10.15) enumerates partitions of n into parts of the form $5k+2$ and $5k+3$, $k=0,1,\ldots$.

Following Hardy and Wright [1979] we shall now give combinatorial interpretations of the left sides of the above identities. Since

$$k^2 = 1+3+5+\cdots+(2k-1),$$

we can exhibit this square in a graph of k rows of dots, each row having 2 more dots than the lower one. We then take any partition of $n-k^2$ into at most k parts with the parts in descending order, marked with $\times$'s and placed at the ends of the rows of dots to obtain a partion of n into parts with minimal difference 2. For example, when $k=5$ and $n=32=5^2+7$ we add 4 $\times$'s to the top row, 1 each to the 2nd, 3rd and 4th rows, counted from above. This gives the partition $32=13+8+6+4+1$ displayed in the graph below.

```
•  •  •  •  •  •  •  •  •  ×  ×  ×  ×

•  •  •  •  •  •  •  ×

•  •  •  •  •  ×

•  •  •  ×

•
```

The identity (8.10.14) states that the number of partitions of n in which the differences between parts are at least 2 is equal to the number of partitions of n into parts congruent to 1 or 4 (mod 5).

Observing that

$$k(k+1) = 2+4+6+\cdots+2k,$$

a similar interpretation can be given to the left side of (8.10.15). Since the first number in the above sum is 2, one deduces that the partitions of n into parts not less than 2 and with minimal difference 2 are equinumerous with the partitions of n into parts congruent to 2 or 3 (mod 5).

For more applications of basic hypergeometric series to partition theory, see Andrews [1976–1988], Fine [1948, 1988], Andrews and Askey [1977], and Andrews, Dyson and Hickerson [1988]. Additional results on Rogers-Ramanujan type identities are given in Slater [1951, 1952a], Jain and Verma [1980, 1982] and Andrews [1975a, 1984a,b,c,d].

8.11 Representations of positive integers as sums of squares

One of the most interesting problems in number theory is the representations of positive integers as sums of squares of integers. Fermat proved that all primes of the form $4n+1$ can be uniquely expressed as the sum of 2 squares.

Lagrange showed in 1770 that all positive integers can be represented by sums of 4 squares and that this number is minimal. Earlier in the same year Waring posed the general problem of representing a positive integer as a sum of a fixed number of nonnegative k-th powers of integers (positive, negative, or zero) with order taken into account and stated without proof that every integer is the sum of 4 squares, of 9 cubes, of 19 biquadrates, 'and so on'. More than 100 years later Hilbert [1909] proved that all positive integers are representable by s kth powers where $s = s(k)$ depends only on k. For an historical account of the Waring problem, see Dickson [1920], Grosswald [1985], and Hua [1982].

To illustrate the usefulness of basic hypergeometric series in the study of such representations we shall restrict ourselves to the simplest cases: sums of 2 and sums of 4 squares, where it is understood, for example, that $n = x_1^2 + x_2^2 = y_1^2 + y_2^2$ are two different representations of n as a sum of 2 squares if $x_1 \neq y_1$ or $x_2 \neq y_2$.

Let $r_{2k}(n)$ be the number of different representations of n as a sum of $2k$ squares, $k = 1, 2, \ldots$. We will show by basic hypergeometric series techniques that, for $n \geq 1$,

$$r_2(n) = 4(d_1(n) - d_3(n)), \tag{8.11.1}$$

$$r_4(n) = 8 \sum_{d|n, 4\nmid d} d, \tag{8.11.2}$$

where $d_i(n)$, $i = 1, 3$, is the number of (positive) divisors of n congruent to i (mod 4) and the summation in (8.11.2) indicates the sum over all divisors of n not divisible by 4. The numbers 4 and 8 in (8.11.1) and (8.11.2), respectively, reflect the fact that $r_2(1) = 4$ since $1 = 0^2 + (\pm 1)^2 = (\pm 1)^2 + 0^2$, and $r_4(1) = 8$ since $1 = 0^2 + 0^2 + 0^2 + (\pm 1)^2 = 0^2 + 0^2 + (\pm 1)^2 + 0^2 = 0^2 + (\pm 1)^2 + 0^2 + 0^2 = (\pm 1)^2 + 0^2 + 0^2 + 0^2$. Both of these results were proved by Jacobi by means of the theory of elliptic functions, but the proofs below are based, as in Andrews [1974a], on the formulas stated in Ex. 5.1, 5.2, and 5.3. Combining Ex. 5.1 and 5.2 we have

$$\left[\sum_{n=-\infty}^{\infty} (-1)^n q^{n^2}\right]^2 = 1 + 4 \sum_{n=1}^{\infty} \frac{(-1)^n q^{n(n+1)/2}}{1 + q^n}. \tag{8.11.3}$$

However, the bilateral sum on the left side is clearly a generating function of $(-1)^n r_2(n)$ and so it suffices to prove that

$$\sum_{n=1}^{\infty} [d_1(n) - d_3(n)](-q)^n = \sum_{n=1}^{\infty} \frac{(-1)^n q^{n(n+1)/2}}{1 + q^n}. \tag{8.11.4}$$

By splitting into odd and even parts and then by formal series manipulations, we find that

$$\begin{aligned}\sum_{n=1}^{\infty} \frac{(-1)^n q^{n(n+1)/2}}{1 + q^n} &= \sum_{r=1}^{\infty} \frac{q^{r(2r+1)}}{1 + q^{2r}} - \sum_{m=0}^{\infty} \frac{q^{(m+1)(2m+1)}}{1 + q^{2m+1}} \\ &= \sum_{r=1}^{\infty} \sum_{m=r}^{\infty} (-1)^{m+r} q^{(2m+1)r} + \sum_{m=0}^{\infty} \sum_{r=m+1}^{\infty} (-1)^{m+r} q^{(2m+1)r}\end{aligned}$$

$$= \left(\sum_{r=1}^{\infty}\sum_{m=r}^{\infty} + \sum_{r=1}^{\infty}\sum_{m=0}^{r-1}\right)(-1)^{m+r}q^{(2m+1)r}$$

$$= \sum_{m=0}^{\infty}\sum_{r=1}^{\infty}(-1)^{m+r}q^{(2m+1)r}$$

$$= \sum_{m=0}^{\infty}\sum_{r=1}^{\infty}(-1)^{r}\left(q^{(4m+1)r} - q^{(4m+3)r}\right)$$

$$= \sum_{n=1}^{\infty}(-q)^{n}(d_1(n) - d_3(n)), \tag{8.11.5}$$

which completes the proof of (8.11.4).

To prove (8.11.2) we first replace q by $-q$ in Ex. 5.1 and 5.3 and find that

$$\sum_{n=0}^{\infty} r_4(n)q^n = \left[\sum_{n=-\infty}^{\infty} q^{n^2}\right]^4$$

$$= 1 + 8\sum_{n=1}^{\infty}\frac{q^n}{(1+(-q)^n)^2}$$

$$= 1 + 8\sum_{n=1}^{\infty}\frac{nq^n}{1+(-q)^n}, \tag{8.11.6}$$

where the last line is obtained from the previous one by expanding $(1+(-q)^n)^{-2}$ and interchanging the order of summation. Now,

$$\sum_{n=1}^{\infty}\frac{nq^n}{1+(-q)^n}$$

$$= \sum_{\text{odd } n\geq 1}\frac{nq^n}{1-q^n} + \sum_{\text{even } n\geq 2}\frac{nq^n}{1+q^n}$$

$$= \sum_{n=1}^{\infty}\frac{nq^n}{1-q^n} + \sum_{\text{even } n\geq 2} nq^n\left(\frac{1}{1+q^n} - \frac{1}{1-q^n}\right)$$

$$= \sum_{n=1}^{\infty}\frac{nq^n}{1-q^n} - \sum_{\text{even } n\geq 2}\frac{2nq^{2n}}{1-q^{2n}}$$

$$= \sum_{n=1}^{\infty}\frac{nq^n}{1-q^n} - \sum_{n=1}^{\infty}\frac{4nq^{4n}}{1-q^{4n}}$$

$$= \sum_{4\nmid n}\frac{nq^n}{1-q^n}. \tag{8.11.7}$$

Thus,

$$\sum_{n=0}^{\infty} r_4(n)q^n = 1 + 8\sum_{4\nmid n}\frac{nq^n}{1-q^n}. \tag{8.11.8}$$

Since $r_4(0) = 1$, this immediately leads to (8.11.2). For a direct proof of (8.11.8) based only on Jacobi's triple product identity (1.6.1), see Hirschhorn [1987].

Ex. (5.1) and (5.4) can be employed in a similar manner to show that

$$r_8(n) = 16(-1)^n \sum_{d|n} (-1)^d d^3, \quad n \geq 1. \tag{8.11.9}$$

See Andrews [1974a]. Some remarks on other applications are given in Notes 8.

Exercises 8

8.1 Prove for the little q-Jacobi polynomials $p_n(x; a, b; q)$ defined in (7.3.1) that

(*i*)

$$\begin{aligned} &p_n(x;a,b;q)p_n(y;a,b;q) \\ &= (-aq)^n q^{\binom{n}{2}} \frac{(bq;q)_n}{(aq;q)_n} \sum_{m=0}^{n} \frac{(q^{-n}, abq^{n+1}, x^{-1}, y^{-1}; q)_m}{(q, bq; q)_m} \left(\frac{xyq^{1-m}}{a} \right)^m \\ &\quad \cdot \sum_{k=0}^{m} \frac{(q^{-m}, b^{-1}q^{-m}; q)_k}{(q, aq, xq^{1-m}, yq^{1-m}; q)_k} (abxy)^k q^{k^2+2k}, \end{aligned}$$

(*ii*)

$$\begin{aligned} &p_n(x;a,b;q)p_n(y;a,b;q) \\ &= (-bq)^{-n} q^{-\binom{n}{2}} \frac{(bq;q)_n}{(aq;q)_n} \sum_{m=0}^{n} \frac{(q^{-n}, abq^{n+1}; q)_m}{(q, bq; q)_m} (-byq^2)^m q^{\binom{m}{2}} \\ &\quad \cdot \sum_{k=0}^{m} \frac{(q^{-m}, abq^{m+1}; q)_k}{(q, aq; q)_k} (xq)^k \, {}_2\phi_1(q^{k-m}, bxq; 0; q, b^{-1}y^{-1}). \end{aligned}$$

8.2 Derive the following product formula for the big q-Jacobi polynomials defined in (7.3.10):

$$\begin{aligned} &P_n(x;a,b,c;q)P_n(y;a,b,c;q) \\ &= (-aq)^n q^{\binom{n}{2}} \frac{(bq;q)_n}{(aq;q)_n} \sum_{m=0}^{n} \frac{(q^{-n}, abq^{n+1}, cqx^{-1}, cqy^{-1}; q)_m}{(q, bq, cq, cq; q)_m} \left(\frac{xy}{a} \right)^m \\ &\quad \cdot {}_4\phi_3 \left[\begin{matrix} q^{-m}, b^{-1}q^{-m}, x, y \\ aq, xc^{-1}q^{-m}, yc^{-1}q^{-m} \end{matrix} ; q, abqc^{-2} \right]. \end{aligned}$$

8.3 Prove that

$$\begin{aligned} &\sum_{n=0}^{\infty} \frac{(abq, aq; q)_n (1 - abq^{2n+1})}{(q, bq; q)_n (1 - abq)} \left(\frac{t}{aq} \right)^n p_n(q^x; a, b; q) p_n(q^y; a, b; q) \\ &= (t, abq^2; q)_\infty \sum_{r=0}^{\infty} \sum_{s=0}^{\min(x,y)} \frac{(q^{-x}, q^{-y}; q)_s \, a^{-s}}{(q, bq; q)_s (q, aq; q)_r} q^{(x+y)(r+s)} \\ &\quad \cdot {}_2\phi_1 \left(0, 0; qt; q, abq^{2r+2s+2}\right) q^{-2rs-s^2} t^{r+s}, \end{aligned}$$

and that this gives the positivity of the Poisson kernel for the little q-Jacobi polynomials for $x, y = 0, 1, \ldots,\ 0 \le t < 1$ when $0 < q < 1, 0 < aq < 1$ and $0 < bq < 1$.

8.4 Prove for the q-Hahn polynomials that

(i)

$$Q_n(x;a,b,N;q)Q_n(y;a,b,N;q) = \frac{(bq, abq^{N+2};q)_n}{(aq,q^{-N};q)_n}\left(bq^{N+1}\right)^{-n}$$

$$\cdot \sum_{m=0}^{n} \frac{\left(q^{-n}, abq^{n+1}, bq^{N-x+1}, bq^{N-y+1};q\right)_m}{(q, bq, abq^{N+2}, bq^{N+1};q)_m} q^m$$

$$\cdot \sum_{k=0}^{m} \frac{(b^{-1}q^{-N-m-1};q)_k\left(1-b^{-1}q^{2k-N-m-1}\right)\left(a^{-1}b^{-1}q^{-N-m-1}, b^{-1}q^{-m};q\right)_k}{(q;q)_k(1-b^{-1}q^{-N-m-1})(aq,q^{-N};q)_k}$$

$$\cdot \frac{(q^{-m}, q^{-x}, q^{-y};q)_k}{\left(b^{-1}q^{-N}, b^{-1}q^{x-N-m}, b^{-1}q^{y-N-m};q\right)_k}(ab)^k q^{(x+y+2m-k)k},$$

(ii)

$$Q_n(x;a,b,N;q)Q_n(y;a,b,N;q) = \frac{\left(bq, abq^{N+2};q\right)_n}{\left(aq,q^{-N};q\right)_n}\left(bq^{N+1}\right)^{-n}$$

$$\cdot \sum_{m=0}^{n} \frac{\left(q^{-n}, abq^{n+1}, bq^{n+1-x-y};q\right)_m}{(q, bq, abq^{N+2};q)_m} q^{(x+1)m}$$

$$\cdot \sum_{k=0}^{m} \frac{\left(q^{-m}, abq^{m+1}, q^{-x}, q^{-y};q\right)_k}{\left(q, aq, q^{-N}, bq^{N+1-x-y};q\right)_k}\left(bq^{N-x+2}\right)^k$$

$$\cdot\ {}_3\phi_2\left[\begin{matrix} q^{k-m}, q^{k-x}, bq^{N+1-x} \\ 0,\ bq^{N+1-x-y+k} \end{matrix};q,q\right].$$

8.5 Prove for the q-Racah polynomials that

(i)

$$W_n(x;a,b,c,N;q)W_n(y;a,b,c,N;q) = \frac{\left(bq, abq^{N+2};q\right)_n}{(aq,q^{-N};q)_n}\left(bq^{N+1}\right)^{-n}$$

$$\cdot \sum_{m=0}^{n} \frac{\left(q^{-n}, abq^{n+1}, bq^{N-x+1}, bq^{N-y+1};q\right)_m}{(q, bq, abq^{N+2}, bcq;q)_m}$$

$$\cdot \frac{\left(bcq^{x+1}, bcq^{y+1};q\right)_m}{\left(bcq, bq^{N+1};q\right)_m} q^m$$

$$\cdot\ {}_{10}W_9\left(b^{-1}q^{-N-m-1}; a^{-1}b^{-1}q^{-N-m-1}, b^{-1}q^{-m},\right.$$
$$\left. q^{-m}, q^{-x}, cq^{x-N}, q^{-y}, cq^{y-N}; q, ab^{-1}c^{-2}q\right),$$

(*ii*)

$$W_n(x;a,b,c,N;q)W_n(y;a,b,c,N;q) = \frac{(aqc^{-1}, abq^{N+2};q)_n}{(bcq,q^{-N};q)_n}\left(ac^{-1}q^{N+1}\right)^{-n}$$

$$\cdot \sum_{m=0}^{n} \frac{(q^{-n}, abq^{n+1}, ac^{-1}q^{N-x+1}, ac^{-1}q^{N-y+1};q)_m}{(q, abq^{N+2}, aqc^{-1}, ac^{-1}q^{N+1};q)_m}$$

$$\cdot \frac{(aq^{x+1}, aq^{y+1};q)_m}{(aq,aq;q)_m} q^m$$

$$\cdot {}_{10}W_9\left(ca^{-1}q^{-N-m-1}; ca^{-1}q^{-m}, a^{-1}b^{-1}q^{-N-m-1},\right.$$
$$\left. q^{-m}, q^{-x}, cq^{x-N}, q^{-y}, cq^{y-N}; q, qba^{-1}\right).$$

8.6 Show that

(*i*)

$$W_n(x;a,b,c,N;q)W_n(y;a,b,c,M;q)$$

$$= \frac{(bq, aqc^{-1};q)_n}{(aq,bcq;q)_n} c^n \sum_{m=0}^{n} \frac{(q^{-n}, abq^{n+1}, q^{x-N}, q^{y-M};q)_m}{(q, bq, aqc^{-1}, c^{-1};q)_m}$$

$$\cdot \frac{(c^{-1}q^{-x}, c^{-1}q^{-y};q)_m}{(q^{-N}, q^{-M};q)_m} q^m$$

$$\cdot {}_{10}W_9\left(cq^{-m}; ca^{-1}q^{-m}, b^{-1}q^{-m}, q^{-m}, q^{-x}, cq^{x-N}, q^{-y}, cq^{y-M}; q, abq^{M+N+3}\right),$$

where $n = 0, 1, \ldots, \min(M, N)$, $x = 0, 1, \ldots, N$, and $y = 0, 1, \ldots, M$;

(*ii*)

$$W_n(x;a,a,c,N;q)W_n(y;a,a,c,N;q)$$

$$= \frac{(aqc^{-1}, a^2q^{N+2};q)_n}{(acq, q^{-N};q)_n}\left(ac^{-1}q^{N+1}\right)^{-n} \sum_{m=0}^{n} \frac{(q^{-n}, a^2q^{n+1}, aq^{x+1};q)_m}{(q, a^2q^{N+2}, aq;q)_m}$$

$$\cdot \frac{(aq^{1+y-x}, ac^{-1}q^{N-x-y+1};q)_m}{(aqc^{-1}, aq^{1-x};q)_m} q^m$$

$$\cdot {}_{10}W_9\left(aq^{-x}; a^2q^{m+1}, q^{-m}, aq^{N-x+1}, c^{-1}q^{-x}, q^{-x}, q^{-y}, cq^{y-N}; q, q\right).$$

8.7 For the q-Hahn polynomials prove that

(*i*)

$$\sum_{n=0}^{z} \frac{(abq, aq, q^{-z};q)_n\,(1-abq^{2n+1})}{(q, bq, abq^{N+2};q)_n(1-abq)}(-aq)^{-n} q^{Nn-\binom{n}{2}}$$

$$\cdot\, Q_n(x;a,b,N;q)Q_n(y;a,b,M;q)$$

$$= \sum_{r=0}^{z}\sum_{s=0}^{z-r} \frac{(q^{-z};q)_{r+s}\,(abq^2;q)_{2r+2s}(q^{-x}, q^{-y};q)_r(q^{x-N}, q^{y-M};q)_s}{(q^{-M}, q^{-N}, abq^{N+2};q)_{r+s}\,(q, aq;q)_r(q, bq;q)_s}$$

$$\cdot \frac{(q^{N-z};q)_{z-r-s}}{(abq^{N+r+s+2};q)_{z-r-s}}(-1)^{r+s}a^{-s}q^{(r+s)(2N-r-s+1)/2-(x+y+1)s}$$

$$\cdot\, {}_2\phi_1\left(q^{r+s-z}, q^{1+r+s-z}; q^{1+N-z}; q, abq^{N+z+2}\right) \geq 0.$$

for $x = 0, 1, \ldots, N,\ y = 0, 1, \ldots, M,\ z = 0, 1, \ldots, N,\ 0 \leq t < 1$ when $0 < q < 1,\ 0 < aq < 1,\ 0 \leq bq < 1$ and $N \leq M$;

(ii)

$$\sum_{n=0}^{N} \frac{(abq, aq, q^{-N}; q)_n (1 - abq^{2n+1})}{(q, bq, abq^{N+2}; q)_n (1 - abq)} \left(-\frac{t}{aq}\right)^n q^{Nn-\binom{n}{2}}$$

$$\cdot\, Q_n(x; a, b, N; q) Q_n(y; a, b, M; q)$$

$$= \sum_{r=0}^{x} \sum_{s=0}^{N-x} \frac{(abq^2; q)_{2r+2s} (q^{-x}, q^{-y}; q)_r (q^{x-N}, q^{y-M}; q)_s}{(q^{-M}, abq^{N+2}; q)_{r+s} (q, aq; q)_r (q, bq; q)_s}$$

$$\cdot\, (t, abq^{2r+2s+2}; q)_{N-r-s}\, {}_2\phi_1\left(q^{N-r-s}, tq^{N-r-s}; qt; q, abq^{2r+2s+2}\right)$$

$$\cdot\, a^{-s} q^{(r+s)(2N-r-s+1)/2-(x+y+1)s} (-t)^{r+s} > 0,$$

for $x = 0, 1, \ldots, N,\ y = 0, 1, \ldots, M,\ 0 \leq t < 1$ when $0 < q < 1,\ 0 < aq < 1,\ 0 \leq bq < 1$ and $N \leq M$.

(Gasper and Rahman [1984], Jain and Verma [1981])

8.8 Prove that

$$\sum_{n=0}^{z} \frac{(aq, abq, bcq; q)_n (-c)^{-n}}{(q, bq, aqc^{-1}; q)_n (abq; q)_{2n}} \lambda_n(z) q^{\binom{n}{2}}$$

$$\cdot\, W_n(x; a, b, c, N; q) W_n(y; \alpha, ab\alpha^{-1}, \gamma, M; q)$$

$$= \sum_{r=0}^{z} \sum_{s=0}^{z-r} \frac{(q^{-z}; q)_{r+s} (q^{-x}, cq^{x-N}, q^{-y}, \gamma q^{y-M}; q)_r}{(q^{-N}; q)_{r+s} (q, \alpha q, q^{-M}, ab\gamma q\alpha^{-1}, cq^{1-s}; q)_r}$$

$$\cdot\, \frac{(q^{x-N}, c^{-1}q^{-x}; q)_s (1 - cq^{r-s})}{(q, bq, aqc^{-1}; q)_s (1 - cq^{-s})} q^{-rs} A_{r,s}\, \mu_{r+s},$$

where $z = 0, 1, \ldots, N,\quad N \leq M$,

$$\lambda_n(z) = \sum_{k=0}^{z-n} \frac{(q^{-z}; q)_{n+k}\, \mu_{n+k}}{(q, abq^{2n+2}; q)_k},$$

and $\{\mu_r\}$ is an arbitrary complex sequence, $A_{r,s}$ being the same as in (8.7.5).

8.9 Deduce from Ex. 8.8 that

$$\sum_{n=0}^{z} \frac{(abq, aq, d, e, bcq, q^{-z}; q)_n (1 - abq^{2n+1})}{(q, bq, abq^2/d, abq^2/e, aq/c, abq^{z+2}; q)_n (1 - abq)} \left(\frac{abq^{z+2}}{cde}\right)^n$$

$$\cdot\, W_n(x; a, b, c, N; q) W_n(y; \alpha, ab\alpha^{-1}, \gamma, M; q)$$

$$= \frac{(abq^2, abq^2/de; q)_z}{(abq^2/d, abq^2/e; q)_z} \sum_{r=0}^{z} \sum_{s=0}^{z-r} \frac{(q^{-z}, d, e; q)_{r+s} (q^{-x}, cq^{x-N}; q)_r}{(q^{-N}, deq^{-z-1}/ab; q)_{r+s} (q, q^{-M}; q)_r}$$

$$\cdot\, \frac{(q^{-y}, \gamma q^{y-M}; q)_r (q^{x-N}, c^{-1}q^{-x}; q)_s (1 - cq^{r-s})}{(\alpha q, ab\gamma q\alpha^{-1}, cq^{1-s}; q)_r (q, bq, aq/c; q)_s (1 - cq^{-s})} q^{r+s-rs} A_{r,s}.$$

8.10 Use (7.4.14) to show that (8.8.3) is equivalent to the formula

$$\{C_n(\cos\theta;\beta|q)\}^2 = \frac{(\beta^2,\beta^2;q)_n}{(q,q;q)_n}\beta^{-n}\, {}_5\phi_4\left[\begin{matrix} q^{-n},\beta^2q^n,\beta,\beta e^{2i\theta},\beta e^{-2i\theta} \\ \beta^2,\beta q^{\frac{1}{2}},-\beta q^{\frac{1}{2}},-\beta \end{matrix};q,q\right],$$

where $n=0,1,\ldots$.

8.11 In view of the product formula (8.4.10) and Gegenbauer's [1874, 1893] addition formula for ultraspherical polynomials (see also Erdélyi [1953, 10.9 (34)] and Szegő [1975, p. 98]) it is natural to look for an expansion of the form

$$p_n(z;a,aq^{\frac{1}{2}},-a,-aq^{\frac{1}{2}}|q) = \sum_{k=0}^{n} A_{k,n}(\theta,\phi)\, p_k(z;ae^{i\theta+i\phi},ae^{-i\theta-i\phi},ae^{i\theta-i\phi},ae^{i\phi-i\theta}|q),$$

where $p_n(z;a,b,c,d|q)$ is the Askey-Wilson polynomial. By multiplying both sides by

$$w(z;ae^{i\theta+i\phi},ae^{-i\theta-i\phi},ae^{i\theta-i\phi},ae^{i\phi-i\theta}) \cdot p_m(z;ae^{i\theta+i\phi},ae^{-i\theta-i\phi},ae^{i\theta-i\phi},ae^{i\phi-i\theta}|q)$$

and then integrating over z from -1 to 1, show that

$$A_{m,n}(\theta,\phi) = \frac{(q;q)_n\left(a^4q^n,a^4q^{-1},a^2q^{\frac{1}{2}},-a^2q^{\frac{1}{2}},-a^2;q\right)_m a^{n-m}}{(q;q)_m(q;q)_{n-m}\,(a^4q^{-1};q)_{2m}\left(a^2q^{\frac{1}{2}},-a^2q^{\frac{1}{2}},-a^2;q\right)_n} \cdot p_{n-m}\left(x;aq^{m/2},aq^{(m+1)/2},-aq^{m/2},-aq^{(m+1)/2}|q\right) \cdot p_{n-m}\left(y;aq^{m/2},aq^{(m+1)/2},-aq^{m/2},-aq^{(m+1)/2}|q\right).$$

(Rahman and Verma [1986b])

8.12 Prove the inverse of the linearization formula (8.5.1), namely,

$$C_{m+n}(x;\beta|q) = \sum_{k=0}^{\min(m,n)} b(k,m,n)C_{m-k}(x;\beta|q)C_{n-k}(x;\beta|q),$$

where

$$b(k,m,n) = \frac{(q;q)_m(q;q)_n(\beta;q)_{m+n}}{(\beta;q)_m(\beta;q)_n(q;q)_{m+n}} \cdot \frac{(\beta^{-2}q^{-m-n},\beta^{-1};q)_k\,(1-\beta^{-2}q^{2k-m-n})}{(q,\beta^{-1}q^{1-m-n};q)_k\,(1-\beta^{-2}q^{-m-n})}\left(\beta^2q^{-1}\right)^k.$$

8.13 Give alternate derivations of (8.6.3) and (8.6.4) by using the q-integral representation (7.4.7) of $C_n(x;\beta|q)$ and the q-integral formula (2.10.19) for an ${}_8\phi_7$ series.

8.14 By equating the coefficients of $e^{i(m+n-2k)\theta}$ on both sides of the linearization formula (8.5.1) show that

(*i*)

$$
{}_4\phi_3\left[\begin{matrix} q^{-k}, q^{-m}, \beta, \beta q^{n-k} \\ \beta^{-1}q^{1-k}, \beta^{-1}q^{1-m}, q^{1+n-k} \end{matrix}; q, \frac{q^2}{\beta^2}\right] = \frac{(q^{-m-n}, \beta^{-1}q^{1-n}; q)_k}{(q^{-n}, \beta^{-1}q^{1-m-n}; q)_k}
$$

$$
\cdot \Phi\left[\begin{matrix} \beta, & q^{-k}, & q^{-m}, & q^{-n}, & \beta^{-1}q^{-m-n}, \\ & \beta^{-1}q^{1-k}, & \beta^{-1}q^{1-m}, & \beta^{-1}q^{1-n}, & \beta^{-2}q^{1-m-n}, \end{matrix}\right.
$$

$$
\left.\begin{matrix} q^{k-m-n}: & \beta^{-2}q^{1-m-n}, & \beta^{-2}q^{2-m-n}, & \beta^{-1}q^{2-m-n} \\ \beta^{-1}q^{1+k-m-n}: & q^{1-m-n}, & q^{-m-n}, & \beta^{-1}q^{-m-n} \end{matrix}; q, q^2; \frac{q}{\beta}\right],
$$

for $k = 0, 1, \ldots, n$, and

(*ii*)

$$
{}_4\phi_3\left[\begin{matrix} q^{k-m-n}, q^{-n}, \beta, \beta q^{k-n} \\ \beta^{-1}q^{1-n}, \beta^{-1}q^{1+k-m-n}, q^{1+k-n} \end{matrix}; q, \frac{q^2}{\beta^2}\right]
$$

$$
= \frac{(q^{m+1}, \beta^{-1}q^{1-k}; q)_n}{(q^{-k}, \beta q^m; q)_n} (\beta q^{-1})^n
$$

$$
\cdot \Phi\left[\begin{matrix} \beta, & q^{-k}, & q^{-m}, & q^{-n}, & \beta^{-1}q^{-m-n}, \\ & \beta^{-1}q^{1-k}, & \beta^{-1}q^{1-m}, & \beta^{-1}q^{1-n}, & \beta^{-2}q^{1-m-n}, \end{matrix}\right.
$$

$$
\left.\begin{matrix} q^{k-m-n}: & \beta^{-2}q^{1-m-n}, & \beta^{-2}q^{2-m-n}, & \beta^{-1}q^{2-m-n} \\ \beta^{-1}q^{1+k-m-n}: & q^{1-m-n}, & q^{-m-n}, & \beta^{-1}q^{-m-n} \end{matrix}; q, q^2; \frac{q}{\beta}\right]
$$

for $k = n, n+1, \ldots, n+m$, where Φ is the bibasic series defined in §3.9. (Gasper [1985])

8.15 Show that, by analytic continuation, it follows from Ex. 8.14 that

$$
{}_4\phi_3\left[\begin{matrix} a, & b, & c, & d \\ & bq/a, & cq/a, & dq/a \end{matrix}; q, \frac{q^2}{a^2}\right]
$$

$$
= \frac{(a/d, bq/d, cq/d, abc/d; q)_\infty}{(q/d, ab/d, ac/d, bcq/d; q)_\infty}
$$

$$
\cdot {}_{12}W_{11}(bc/d; (bcq/ad)^{\frac{1}{2}}, -(bcq/ad)^{\frac{1}{2}}, q(bc/ad)^{\frac{1}{2}}, -q(bc/ad)^{\frac{1}{2}},
$$

$$
ab/d, ac/d, a, b, c; q, q/a),
$$

where at least one of a, b, c is of the form $q^{-n}, n = 0, 1, \ldots$. (Gasper [1985])

8.16 From Ex. 8.15 deduce that

$$
{}_{10}W_9\left(a; q^{\frac{1}{2}}, -q^{\frac{1}{2}}, -q, a^2/b, a^2/c, b, c; q, q/a\right)
$$

$$
= \frac{(aq, a^2/b, a^2/c, aq/bc; q)_\infty}{(a^2, aq/b, aq/c, a^2/bc; q)_\infty} {}_4\phi_3\left[\begin{matrix} a, & b, & c, & bc/a \\ & bq/a, & cq/a, & bcq/a^2 \end{matrix}; q, \frac{q^2}{a^2}\right]
$$

where one of a, b, c is of the form $q^{-n}, n = 0, 1, \ldots$.

8.17 Prove that

$$\left\{ {}_4\phi_3 \left[\begin{matrix} a^2, b^2, abz, ab/z \\ a^2b^2q, -ab, -abq \end{matrix} ; q^2, q^2 \right] \right\}^2$$

$$= {}_5\phi_4 \left[\begin{matrix} a^2, b^2, ab, abz, ab/z \\ a^2b^2, abq^{\frac{1}{2}}, -abq^{\frac{1}{2}}, -ab \end{matrix} ; q, q \right]$$

and

$$\left\{ {}_4\phi_3 \left[\begin{matrix} a^2, b^2, abz, ab/z \\ abq^{\frac{1}{2}}, -abq^{\frac{1}{2}}, -a^2b^2 \end{matrix} ; q, q \right] \right\}^2$$

$$= {}_5\phi_4 \left[\begin{matrix} a^4, b^4, a^2b^2, a^2b^2z^2, z^2b^2/z^2 \\ a^4b^4, a^2b^2q, -a^2b^2q, -a^2b^2 \end{matrix} ; q^2, q^2 \right]$$

when the series terminate.
(Gasper [1989b])

8.18 With the notation of Ex. 7.37 prove that

$$D_q \left[(q^{1-\nu}e^{i\theta}, q^{1-\nu}e^{-i\theta}; q)_{2\nu} U_n(\cos\theta) \right] \leq 0$$

when $\nu > -\frac{1}{2}, \lambda \geq 0, \theta$ is real and $n = 1, \ldots, r$.
(Gasper [1989b])

Notes 8

§8.5 For additional nonnegativity results for the coefficients in the linearization of the product of orthogonal polynomials and their applications to convolution structures, Banach algebras, multiplier theory, heat and diffusion equations, maximal principles, stochastic processes, etc., see Askey and Gasper [1977], Gasper [1970, 1971, 1972, 1975a, 1983], Gasper and Rahman [1983b], Gasper and Trebels [1977, 1979], Ismail and Mulla [1987], and Rahman [1986d].

§8.6 and 8.7 The nonnegativity of other Poisson kernels and their applications to probability theory and other fields are considered in Beckmann [1973] and Gasper [1973, 1975a,b, 1976, 1977].

§8.8 and 8.9 A historical summary of related inequalities is given in the survey paper Askey and Gasper [1986]. For additional material related to de Branges' proof of the Bieberbach conjecture, see de Branges [1968–1986], Duren [1983], Gasper [1986], Koornwinder [1984, 1986], and Milin [1977].

§8.10 Additional applications of q-series are given in A.K. Agarwal, Kalnins and Miller [1987], Alder [1969], Andrews and Askey [1989], Andrews, Dyson and Hickerson [1988], Andrews and Forrester [1986], Andrews and Onofri [1984], Askey [1984a,b, 1988a, 1989a,e], Askey, Koornwinder and Schempp [1984], Berndt [1988, 1989], Berndt and Joshi [1983], Bressoud and Goulden [1985, 1987], Cohen [1988], Comtet [1974], Geronimo [1989], Gustafson and Milne [1986], Kadell [1987a], Kirillov and Reshetikhin [1988], Milne [1980–1989c], Rota and Goldman [1969], and Rota and Mullin [1970].

§8.11 Mordell[1917] considered the representation of numbers as the sum of $2r$ squares. Another proof of (8.11.1) is given in Hirschhorn [1985].

Appendix I

IDENTITIES INVOLVING q-SHIFTED FACTORIALS, q-GAMMA FUNCTIONS AND q-BINOMIAL COEFFICIENTS

q-Shifted factorials:

$$(a;q)_n = \begin{cases} 1, & n=0, \\ (1-a)(1-aq)\cdots(1-aq^{n-1}), & n=1,2,\ldots, \\ \left[(1-aq^{-1})(1-aq^{-2})\cdots(1-aq^{n})\right]^{-1}, & n=-1,-2,\ldots. \end{cases} \tag{I.1}$$

$$(a;q)_{-n} = \frac{1}{(aq^{-n};q)_n} = \frac{(-q/a)^n}{(q/a;q)_n} q^{\binom{n}{2}} \tag{I.2}$$

and

$$(a;q^{-1})_n = (a^{-1};q)_n(-a)^n q^{-\binom{n}{2}}, \tag{I.3}$$

where $\binom{n}{2} = n(n-1)/2$.

$$(a;q)_\infty = \prod_{k=0}^{\infty}(1-aq^k), \tag{I.4}$$

$$(a;q)_n = \frac{(a;q)_\infty}{(aq^n;q)_\infty}, \tag{I.5}$$

and, for any complex number α,

$$(a;q)_\alpha = \frac{(a;q)_\infty}{(aq^\alpha;q)_\infty}, \tag{I.6}$$

where the principal value of q^α is taken and it is assumed that $|q| < 1$.

$$(a;q)_n = \left(q^{1-n}/a;q\right)_n (-a)^n q^{\binom{n}{2}}. \tag{I.7}$$

$$\left(aq^{-n};q\right)_n = (q/a;q)_n \left(-\frac{a}{q}\right)^n q^{-\binom{n}{2}}. \tag{I.8}$$

$$\frac{(aq^{-n};q)_n}{(bq^{-n};q)_n} = \frac{(q/a;q)_n}{(q/b;q)_n}\left(\frac{a}{b}\right)^n. \tag{I.9}$$

$$(a;q)_{n-k} = \frac{(a;q)_n}{\left(q^{1-n}/a;q\right)_k}\left(-\frac{q}{a}\right)^k q^{\binom{k}{2}-nk}. \tag{I.10}$$

$$\frac{(a;q)_{n-k}}{(b;q)_{n-k}} = \frac{(a;q)_n}{(b;q)_n}\frac{\left(q^{1-n}/b;q\right)_k}{\left(q^{1-n}/a;q\right)_k}\left(\frac{b}{a}\right)^k. \tag{I.11}$$

$$\left(q^{-n};q\right)_k = \frac{(q;q)_n}{(q;q)_{n-k}}(-1)^k q^{\binom{k}{2}-nk}. \tag{I.12}$$

$$\left(aq^{-n};q\right)_k = \frac{(a;q)_k(q/a;q)_n}{\left(q^{1-k}/a;q\right)_n} q^{-nk}. \tag{I.13}$$

$$\left(aq^{-n};q\right)_{n-k} = \frac{(q/a;q)_n}{(q/a;q)_k}\left(-\frac{a}{q}\right)^{n-k} q^{\binom{k}{2}-\binom{n}{2}}. \tag{I.14}$$

$$\left(aq^{-2n};q\right)_n = \frac{(q/a;q)_{2n}}{(q/a;q)_n}\left(-\frac{a}{q^2}\right)^n q^{-3\binom{n}{2}}. \tag{I.15}$$

$$\left(aq^{-kn};q\right)_n = \frac{(q/a;q)_{kn}}{(q/a;q)_{(k-1)n}}(-a)^n q^{\binom{n}{2}-kn^2}. \tag{I.16}$$

$$(a;q)_{n+k} = (a;q)_n(aq^n;q)_k. \tag{I.17}$$

$$(aq^n;q)_k = \frac{(a;q)_k(aq^k;q)_n}{(a;q)_n}. \tag{I.18}$$

$$(aq^{kn};q)_n = \frac{(a;q)_{(k+1)n}}{(a;q)_{kn}}. \tag{I.19}$$

$$(aq^k;q)_{n-k} = \frac{(a;q)_n}{(a;q)_k}. \tag{I.20}$$

$$(aq^{2k};q)_{n-k} = \frac{(a;q)_n(aq^n;q)_k}{(a;q)_{2k}}. \tag{I.21}$$

$$\left(aq^{jk};q\right)_{n-k} = \frac{(a;q)_n(aq^n;q)_{(j-1)k}}{(a;q)_{jk}}. \tag{I.22}$$

$$(a_1,a_2,\ldots,a_k;q)_n = (a_1;q)_n(a_2;q)_n\cdots(a_k;q)_n. \tag{I.23}$$

$$(a_1,a_2,\ldots,a_k;q)_\infty = (a_1;q)_\infty(a_2;q)_\infty\cdots(a_k;q)_\infty. \tag{I.24}$$

$$(a;q)_{2n} = (a,aq;q^2)_n, \tag{I.25}$$

$$(a;q)_{3n} = (a,aq,aq^2;q^3)_n, \tag{I.26}$$

and, in general,

$$(a;q)_{kn} = (a,aq,\ldots,aq^{k-1};q^k)_n. \tag{I.27}$$

$$(a^2;q^2)_n = (a,-a;q)_n, \tag{I.28}$$

$$(a^3;q^3)_n = (a,a\omega,a\omega^2;q)_n, \quad \omega = e^{2\pi i/3}, \tag{I.29}$$

and, in general,

$$(a^k;q^k)_n = (a,a\omega_k,\ldots,a\omega_k^{k-1};q)_n, \quad \omega_k = e^{2\pi i/k}. \tag{I.30}$$

$$\frac{(qa^{\frac{1}{2}},-qa^{\frac{1}{2}},q)_n}{(a^{\frac{1}{2}},-a^{\frac{1}{2}};q)_n} = \frac{(aq^2;q^2)_n}{(a;q^2)_n} = \frac{1-aq^{2n}}{1-a}, \tag{I.31}$$

$$\frac{\left(qa^{\frac{1}{3}},q\omega a^{\frac{1}{3}},q\omega^2 a^{\frac{1}{3}};q\right)_n}{\left(a^{\frac{1}{3}},\omega a^{\frac{1}{3}},\omega^2 a^{\frac{1}{3}};q\right)_n} = \frac{(aq^3;q^3)}{(a;q^3)_n} = \frac{1-aq^{3n}}{1-a}, \tag{I.32}$$

and, in general,

$$\frac{\left(qa^{\frac{1}{k}},q\omega_k a^{\frac{1}{k}},\ldots q\omega_k^{k-1}a^{\frac{1}{k}};q\right)_n}{\left(a^{\frac{1}{k}},\omega_k a^{\frac{1}{k}},\ldots,\omega_k^{k-1}a^{\frac{1}{k}};q\right)_n} = \frac{(aq^k;q^k)_n}{(a;q^k)_n} = \frac{1-aq^{kn}}{1-a}, \tag{I.33}$$

where $\omega = e^{2\pi i/3}$ and $\omega_k = e^{2\pi i/k}$.

$$\lim_{q\to 1^-}\frac{(zq^\alpha;q)_\infty}{(z;q)_\infty} = (1-z)^{-\alpha}, \quad |z|<1. \tag{I.34}$$

q-Gamma function:

$$\Gamma_q(x) = \begin{cases} \dfrac{(q;q)_\infty}{(q^x;q)_\infty}(1-q)^{1-x}, & 0 < q < 1, \\ \dfrac{(q^{-1};q^{-1})_\infty}{(q^{-x};q^{-1})_\infty}(q-1)^{1-x}q^{\binom{x}{2}}, & q > 1. \end{cases} \tag{I.35}$$

$$\lim_{q\to 1}\Gamma_q(x) = \Gamma(x). \tag{I.36}$$

$$\Gamma_q(2x)\Gamma_{q^2}\left(\frac{1}{2}\right) = \Gamma_{q^2}(x)\Gamma_{q^2}\left(x+\frac{1}{2}\right)(1+q)^{2x-1}. \tag{I.37}$$

$$\begin{aligned} &\Gamma_q(nx)\Gamma_r\left(\frac{1}{n}\right)\Gamma_r\left(\frac{2}{n}\right)\cdots\Gamma_r\left(\frac{n-1}{n}\right) \\ &= \left(1+q+\ldots+q^{n-1}\right)^{nx-1}\Gamma_r(x)\Gamma_r\left(x+\frac{1}{n}\right)\cdots\Gamma_r\left(x+\frac{n-1}{n}\right), \end{aligned} \tag{I.38}$$

with $r = q^n$.

q-Binomial coefficient:

$$\begin{bmatrix} n \\ k \end{bmatrix}_q = \begin{bmatrix} n \\ n-k \end{bmatrix}_q = \frac{(q;q)_n}{(q;q)_k(q;q)_{n-k}} \tag{I.39}$$

and, for $|q| < 1$ and complex α and β,

$$\begin{bmatrix} \alpha \\ \beta \end{bmatrix}_q = \frac{\left(q^{\beta+1}, q^{\alpha-\beta+1}; q\right)_\infty}{\left(q, q^{\alpha+1}; q\right)_\infty}, \tag{I.40}$$

$$\begin{bmatrix} \alpha \\ \beta \end{bmatrix}_q = \frac{\Gamma_q(\alpha+1)}{\Gamma_q(\beta+1)\Gamma_q(\alpha-\beta+1)}, \tag{I.41}$$

$$\begin{bmatrix} \alpha \\ k \end{bmatrix}_q = \frac{(q^{-\alpha};q)_k}{(q;q)_k}\left(-q^\alpha\right)^k q^{-\binom{k}{2}}, \tag{I.42}$$

$$\begin{bmatrix} k+\alpha \\ k \end{bmatrix}_q = \frac{\left(q^{\alpha+1};q\right)_k}{(q;q)_k}, \tag{I.43}$$

$$\begin{bmatrix} -\alpha \\ k \end{bmatrix}_q = \begin{bmatrix} \alpha+k-1 \\ k \end{bmatrix}_q \left(-q^{-\alpha}\right)^k q^{-\binom{k}{2}}, \tag{I.44}$$

$$\begin{bmatrix} \alpha+1 \\ k \end{bmatrix}_q = \begin{bmatrix} \alpha \\ k \end{bmatrix}_q q^k + \begin{bmatrix} \alpha \\ k-1 \end{bmatrix}_q = \begin{bmatrix} \alpha \\ k \end{bmatrix}_q + \begin{bmatrix} \alpha \\ k-1 \end{bmatrix}_q q^{\alpha+1-k}. \tag{I.45}$$

Appendix II

SELECTED SUMMATION FORMULAS

Sums of basic hypergeometric series:

The two q-exponential functions,

$$e_q(z) = \sum_{n=0}^{\infty} \frac{z^n}{(q;q)_n} = \frac{1}{(z;q)_\infty}, \quad |z| < 1, \tag{II.1}$$

$$E_q(z) = \sum_{n=0}^{\infty} \frac{q^{\binom{n}{2}} z^n}{(q;q)_n} = (-z;q)_\infty. \tag{II.2}$$

The q-binomial theorem,

$$ {}_1\phi_0(a;—;q,z) = \frac{(az;q)_\infty}{(z;q)_\infty}, \quad |z| < 1, \tag{II.3}$$

or, when $a = q^{-n}$, where, as elsewhere in this appendix, n denotes a nonnegative integer,

$$ {}_1\phi_0(q^{-n};—;q,z) = (zq^{-n};q)_n. \tag{II.4}$$

The sum of a ${}_1\phi_1$ series,

$$ {}_1\phi_1(a;c;q,c/a) = \frac{(c/a;q)_\infty}{(c;q)_\infty}. \tag{II.5}$$

The q-Vandermonde (q-Chu-Vandermonde) sums,

$$ {}_2\phi_1(a,q^{-n};c;q,q) = \frac{(c/a;q)_n}{(c;q)_n} a^n \tag{II.6}$$

and, reversing the order of summation,

$$ {}_2\phi_1(a,q^{-n};c;q,cq^n/a) = \frac{(c/a;q)_n}{(c;q)_n}. \tag{II.7}$$

The q-Gauss sum,

$$ {}_2\phi_1(a,b;c;q,c/ab) = \frac{(c/a,c/b;q)_\infty}{(c,c/ab;q)_\infty}. \tag{II.8}$$

The q-Kummer (Bailey-Daum) sum,

$$ {}_2\phi_1(a,b;aq/b;q,-q/b) = \frac{(-q;q)_\infty (aq,aq^2/b^2;q^2)_\infty}{(-q/b,aq/b;q)_\infty}. \tag{II.9}$$

A q-analogue of Bailey's ${}_2F_1(-1)$ sum,

$$ {}_2\phi_2(a,q/a;-q,b;q,-b) = \frac{(ab,bq/a;q^2)_\infty}{(b;q)_\infty}. \tag{II.10}$$

A q-analogue of Gauss' ${}_2F_1(-1)$ sum,

$$ {}_2\phi_2\left(a^2, b^2; abq^{\frac{1}{2}}, -abq^{\frac{1}{2}}; q, -q\right) = \frac{(a^2q, b^2q; q^2)_\infty}{(q, a^2b^2q; q^2)_\infty}. \tag{II.11} $$

The q-Saalschütz (q-Pfaff-Saalschütz) sum,

$$ {}_3\phi_2\left[\begin{matrix} a, b, q^{-n} \\ c, abc^{-1}q^{1-n} \end{matrix}; q, q\right] = \frac{(c/a, c/b; q)_n}{(c, c/ab; q)_n}. \tag{II.12} $$

The q-Dixon sum,

$$ {}_4\phi_3\left[\begin{matrix} a, -qa^{\frac{1}{2}}, b, c \\ -a^{\frac{1}{2}}, aq/b, aq/c \end{matrix}; q, \frac{qa^{\frac{1}{2}}}{bc}\right] = \frac{\left(aq, qb^{-1}a^{\frac{1}{2}}, qc^{-1}a^{\frac{1}{2}}, aq/bc; q\right)_\infty}{\left(aq/b, aq/c, qa^{\frac{1}{2}}, qa^{\frac{1}{2}}/bc; q\right)_\infty}, \tag{II.13} $$

or, when $c = q^{-n}$,

$$ {}_4\phi_3\left[\begin{matrix} a, -qa^{\frac{1}{2}}, b, q^{-n} \\ -a^{\frac{1}{2}}, aq/b, aq^{1+n} \end{matrix}; q, \frac{q^{1+n}a^{\frac{1}{2}}}{b}\right] = \frac{\left(aq, qa^{\frac{1}{2}}/b; q\right)_n}{\left(qa^{\frac{1}{2}}, aq/b; q\right)_n}. \tag{II.14} $$

Jackson's terminating q-analogue of Dixon's sum,

$$ {}_3\phi_2\left[\begin{matrix} q^{-2n}, & b, & c \\ & q^{1-2n}/b, & q^{1-2n}/c \end{matrix}; q, \frac{q^{2-n}}{bc}\right] = \frac{(b, c; q)_n (q, bc; q)_{2n}}{(q, bc; q)_n (b, c; q)_{2n}}. \tag{II.15} $$

A q-analogue of Watson's ${}_3F_2$ sum,

$$ {}_8\phi_7\left[\begin{matrix} \lambda, q\lambda^{\frac{1}{2}}, -q\lambda^{\frac{1}{2}}, a, b, c, -c, \lambda q/c^2 \\ \lambda^{\frac{1}{2}}, -\lambda^{\frac{1}{2}}, \lambda q/a, \lambda q/b, \lambda q/c, -\lambda q/c, c^2 \end{matrix}; q, -\frac{\lambda q}{ab}\right] $$

$$ = \frac{(\lambda q, c^2/\lambda; q)_\infty (aq, bq, c^2q/a, c^2q/b; q^2)_\infty}{(\lambda q/a, \lambda q/b; q)_\infty (q, abq, c^2q, c^2q/ab; q^2)_\infty}, \tag{II.16} $$

where $\lambda = -c(ab/q)^{\frac{1}{2}}$; and Andrews' terminating q-analogue,

$$ {}_4\phi_3\left[\begin{matrix} q^{-n}, aq^n, c, -c \\ (aq)^{\frac{1}{2}}, -(aq)^{\frac{1}{2}}, c^2 \end{matrix}; q, q\right] $$

$$ = \begin{cases} 0, & \text{if } n \text{ is odd}, \\ \dfrac{c^n(q, aq/c^2; q^2)_{n/2}}{(aq, c^2q; q^2)_{n/2}}, & \text{if } n \text{ is even}. \end{cases} \tag{II.17} $$

A q-analogue of Whipple's ${}_3F_2$ sum,

$$ {}_8\phi_7\left[\begin{matrix} -c, & q(-c)^{\frac{1}{2}}, & -q(-c)^{\frac{1}{2}}, & a, & q/a, & c, & -d, & -q/d \\ & (-c)^{\frac{1}{2}}, & -(-c)^{\frac{1}{2}}, & -cq/a, & -ac, & -q, & cq/d, & cd \end{matrix}; q, c\right] $$

$$ = \frac{(-c, -cq; q)_\infty \left(acd, acq/d, cdq/a, cq^2/ad; q^2\right)_\infty}{(cd, cq/d, -ac, -cq/a; q)_\infty}, \tag{II.18} $$

and a terminating q-analogue,

$$ {}_4\phi_3\left[\begin{matrix} q^{-n}, q^{n+1}, c, -c \\ e, c^2q/e, -q \end{matrix}; q, q\right] $$

$$ = \frac{\left(eq^{-n}, eq^{n+1}, c^2q^{1-n}/e, c^2q^{n+2}/e; q^2\right)_\infty}{(e, c^2q/e; q)_\infty}. \tag{II.19} $$

The sum of a very-well-poised ${}_6\phi_5$ series,

$$
{}_6\phi_5\left[\begin{matrix} a, & qa^{\frac{1}{2}}, & -qa^{\frac{1}{2}}, & b, & c, & d \\ & a^{\frac{1}{2}}, & -a^{\frac{1}{2}}, & aq/b, & aq/c, & aq/d \end{matrix}; q, \frac{aq}{bcd}\right]
= \frac{(aq, aq/bc, aq/bd, aq/cd; q)_\infty}{(aq/b, aq/c, aq/d, aq/bcd; q)_\infty} \tag{II.20}
$$

or, when $d = q^{-n}$,

$$
{}_6\phi_5\left[\begin{matrix} a, qa^{\frac{1}{2}}, -qa^{\frac{1}{2}}, b, c, q^{-n} \\ a^{\frac{1}{2}}, -a^{\frac{1}{2}}, aq/b, aq/c, aq^{n+1} \end{matrix}; q, \frac{aq^{n+1}}{bc}\right] = \frac{(aq, aq/bc; q)_n}{(aq/b, aq/c; q)_n}. \tag{II.21}
$$

Jackson's q-analogue of Dougall's ${}_7F_6$ sum,

$$
{}_8\phi_7\left[\begin{matrix} a, & qa^{\frac{1}{2}}, & -qa^{\frac{1}{2}}, & b, & c, & d, & e, & q^{-n} \\ & a^{\frac{1}{2}}, & -a^{\frac{1}{2}}, & aq/b, & aq/c, & aq/d, & aq/e, & aq^{n+1} \end{matrix}; q, q\right]
= \frac{(aq, aq/bc, aq/bd, aq/cd; q)_n}{(aq/b, aq/c, aq/d, aq/bcd; q)_n}, \tag{II.22}
$$

where $a^2q = bcdeq^{-n}$.

A nonterminating form of the q-Vandermonde sum,

$$
{}_2\phi_1(a, b; c; q, q) + \frac{(q/c, a, b; q)_\infty}{(c/q, aq/c, bq/c; q)_\infty}
\cdot {}_2\phi_1(aq/c, bq/c; q^2/c; q, q) = \frac{(q/c, abq/c; q)_\infty}{(aq/c, bq/c; q)_\infty}. \tag{II.23}
$$

A nonterminating form of the q-Saalschütz sum,

$$
{}_3\phi_2\left[\begin{matrix} a, b, c \\ e, f \end{matrix}; q, q\right] + \frac{(q/e, a, b, c, qf/e; q)_\infty}{(e/q, aq/e, bq/e, cq/e, f; q)_\infty}
\cdot {}_3\phi_2\left[\begin{matrix} aq/e, & bq/e, & cq/e \\ & q^2/e, & qf/e \end{matrix}; q, q\right] = \frac{(q/e, f/a, f/b, f/c; q)_\infty}{(aq/e, bq/e, cq/e, f; q)_\infty}, \tag{II.24}
$$

where $ef = abcq$.

Bailey's nonterminating extension of Jackson's ${}_8\phi_7$ sum,

$$
{}_8\phi_7\left[\begin{matrix} a, & qa^{\frac{1}{2}}, & -qa^{\frac{1}{2}}, & b, & c, & d, & e, & f \\ & a^{\frac{1}{2}}, & -a^{\frac{1}{2}}, & aq/b, & aq/c, & aq/d, & aq/e, & aq/f \end{matrix}; q, q\right]
- \frac{b}{a}\,\frac{(aq, c, d, e, f, bq/a, bq/c, bq/d, bq/e, bq/f; q)_\infty}{(aq/b, aq/c, aq/d, aq/e, aq/f, bc/a, bd/a, be/a, bf/a, b^2q/a; q)_\infty}
\cdot {}_8\phi_7\left[\begin{matrix} b^2/a, qba^{-\frac{1}{2}}, -qba^{-\frac{1}{2}}, b, bc/a, bd/a, be/a, bf/a \\ ba^{-\frac{1}{2}}, -ba^{-\frac{1}{2}}, bq/a, bq/c, bq/d, bq/e, bq/f \end{matrix}; q, q\right]
= \frac{(aq, b/a, aq/cd, aq/ce, aq/cf, aq/de, aq/df, aq/ef; q)_\infty}{(aq/c, aq/d, aq/e, aq/f, bc/a, bd/a, be/a, bf/a; q)_\infty}, \tag{II.25}
$$

where $qa^2 = bcdef$.

q-Analogues of the Karlsson-Minton sums,

$$ {}_{r+2}\phi_{r+1}\left[\begin{matrix} a, b, b_1q^{m_1}, \ldots, b_rq^{m_r} \\ bq, b_1, \ldots, b_r \end{matrix}; q, a^{-1}q^{1-(m_1+\cdots+m_r)}\right] $$
$$ = \frac{(q, bq/a; q)_\infty (b_1/b; q)_{m_1} \cdots (b_r/b; q)_{m_r}}{(bq, q/a; q)_\infty (b_1; q)_{m_1} \cdots (b_r; q)_{m_r}} b^{m_1+\cdots+m_r} \tag{II.26} $$

and

$$ {}_{r+1}\phi_r\left[\begin{matrix} a, b_1q^{m_1}, \ldots, b_rq^{m_r} \\ b_1, \ldots, b_r \end{matrix}; q, a^{-1}q^{-(m_1+\cdots+m_r)}\right] = 0, \tag{II.27} $$

where $m_1, \ldots, m_r$ are arbitrary nonnegative integers.

Sums of bilateral basic series:

Jacobi's triple product,

$$ \sum_{k=-\infty}^{\infty} q^{k^2} z^k = \left(q^2, -qz, -q/z; q^2\right)_\infty. \tag{II.28} $$

Ramanujan's sum,

$$ {}_1\psi_1(a; b; q, z) = \frac{(q, b/a, az, q/az; q)_\infty}{(b, q/a, z, b/az; q)_\infty}. \tag{II.29} $$

The sum of a well-poised ${}_2\psi_2$ series,

$$ {}_2\psi_2(b, c; aq/b, aq/c; q, -aq/bc) $$
$$ = \frac{(aq/bc; q)_\infty (aq^2/b^2, aq^2/c^2, q^2, aq, q/a; q^2)_\infty}{(aq/b, aq/c, q/b, q/c, -aq/bc; q)_\infty}. \tag{II.30} $$

Bailey's sum of a well-poised ${}_3\psi_3$,

$$ {}_3\psi_3\left[\begin{matrix} b, c, d \\ q/b, q/c, q/d \end{matrix}; q, \frac{q}{bcd}\right] $$
$$ = \frac{(q, q/bc, q/bd, q/cd; q)_\infty}{(q/b, q/c, q/d, q/bcd; q)_\infty}. \tag{II.31} $$

A basic bilaterial analogue of Dixon's sum,

$$ {}_4\psi_4\left[\begin{matrix} -qa^{\frac{1}{2}}, & b, & c, & d \\ -a^{\frac{1}{2}}, & aq/b, & aq/c, & aq/d \end{matrix}; q, \frac{qa^{\frac{3}{2}}}{bcd}\right] $$
$$ = \frac{(aq, aq/bc, aq/bd, aq/cd, qa^{\frac{1}{2}}/b, qa^{\frac{1}{2}}/c, qa^{\frac{1}{2}}/d, q, q/a; q)_\infty}{(aq/b, aq/c, aq/d, q/b, q/c, q/d, qa^{\frac{1}{2}}, qa^{-\frac{1}{2}}, qa^{\frac{3}{2}}/bcd; q)_\infty}. \tag{II.32} $$

The sum of a very-well-poised ${}_6\psi_6$ series,

$$ {}_6\psi_6\left[\begin{matrix} qa^{\frac{1}{2}}, & -qa^{\frac{1}{2}}, & b, & c, & d, & e \\ a^{\frac{1}{2}}, & -a^{\frac{1}{2}}, & aq/b, & aq/c, & aq/d, & aq/e \end{matrix}; q, \frac{qa^2}{bcde}\right] $$
$$ = \frac{(aq, aq/bc, aq/bd, aq/be, aq/cd, aq/ce, aq/de, q, q/a; q)_\infty}{(aq/b, aq/c, aq/d, aq/e, q/b, q/c, q/d, q/e, qa^2/bcde; q)_\infty}. \tag{II.33} $$

Bibasic sums:

Gosper's indefinite bibasic sum,

$$\sum_{k=0}^{n} \frac{1-ap^kq^k}{1-a} \frac{(a;p)_k(c;q)_k}{(q;q)_k(ap/c;p)_k} c^{-k} = \frac{(ap;p)_n(cq;q)_n}{(q;q)_n(ap/c;p)_n} c^{-n}. \tag{II.34}$$

An extension of (II.34),

$$\sum_{k=0}^{n} \frac{(1-ap^kq^k)(1-bp^kq^{-k})}{(1-a)(1-b)} \frac{(a,b;p)_k(c,a/bc;q)_k}{(q,aq/b;q)_k(ap/c,bcp;p)_k} q^k$$
$$= \frac{(ap,bp;p)_n(cq,aq/bc;q)_n}{(q,aq/b;q)_n(ap/c,bcp;p)_n} \tag{II.35}$$

and, more generally,

$$\sum_{k=-m}^{n} \frac{(1-adp^kq^k)(1-bp^k/dq^k)}{(1-ad)(1-b/d)} \frac{(a,b;p)_k(c,ad^2/bc;q)_k}{(dq,adq/b;q)_k(adp/c,bcp/d;p)_k} q^k$$
$$= \frac{(1-a)(1-b)(1-c)(1-ad^2/bc)}{d(1-ad)(1-b/d)(1-c/d)(1-ad/bc)} \left\{ \frac{(ap,bp;p)_n(cq,ad^2q/bc;q)_n}{(dq,adq/b;q)_n(adp/c,bcp/d;p)_n} \right.$$
$$\left. - \frac{(c/ad,d/bc;p)_{m+1}(1/d,b/ad;q)_{m+1}}{(1/c,bc/ad^2;q)_{m+1}(1/a,1/b;p)_{m+1}} \right\}, \tag{II.36}$$

where m is an integer or $+\infty$.

An extension of the formula for the n-th q-difference of $(ap^k;q)_{n-1}$,

$$\left(1-\frac{a}{q}\right)\left(1-\frac{b}{q}\right) \sum_{k=0}^{n} \frac{(ap^k,bp^{-k};q)_{n-1}(1-ap^{2k}/b)}{(p;p)_n(p;p)_{n-k}(ap^k/b;p)_{n+1}} (-1)^k p^{\binom{k}{2}} = \delta_{n,0}. \tag{II.37}$$

Appendix III

SELECTED TRANSFORMATION FORMULAS

Heine's transformations of $_2\phi_1$ series:

$$ {}_2\phi_1(a,b;c;q,z) = \frac{(b,az;q)_\infty}{(c,z;q)_\infty}\, {}_2\phi_1(c/b,z;az;q,b) \tag{III.1} $$

$$ = \frac{(c/b,bz;q)_\infty}{(c,z;q)_\infty}\, {}_2\phi_1(abz/c,b;bz;q,c/b) \tag{III.2} $$

$$ = \frac{(abz/c;q)_\infty}{(z;q)_\infty}\, {}_2\phi_1(c/a,c/b;c;q,abz/c). \tag{III.3} $$

Jackson's transformations of $_2\phi_1$, $_2\phi_2$ and $_3\phi_2$ series:

$$ {}_2\phi_1(a,b;c;q,z) = \frac{(az;q)_\infty}{(z;q)_\infty}\, {}_2\phi_2(a,c/b;c,az;q,bz) \tag{III.4} $$

$$ = \frac{(abz/c;q)_\infty}{(bz/c;q)_\infty}\, {}_3\phi_2\left[\begin{matrix} a, c/b, 0 \\ c, cq/bz \end{matrix}; q, q\right]. \tag{III.5} $$

Transformations of terminating $_2\phi_1$ series:

$$ {}_2\phi_1(q^{-n},b;c;q,z) = \frac{(c/b;q)_n}{(c;q)_n}\left(\frac{bz}{q}\right)^n \cdot {}_3\phi_2(q^{-n},q/z,c^{-1}q^{1-n};bc^{-1}q^{1-n},0;q,q) \tag{III.6} $$

$$ = \frac{(c/b;q)_n}{(c;q)_n}\, {}_3\phi_2\left[\begin{matrix} q^{-n}, b, bzq^{-n}/c \\ bq^{1-n}/c, 0 \end{matrix}; q, q\right] \tag{III.7} $$

$$ = \frac{(c/b;q)_n}{(c;q)_n} b^n\, {}_3\phi_1\left[\begin{matrix} q^{-n}, b, q/c \\ bq^{1-n}/c \end{matrix}; q, \frac{z}{c}\right], \tag{III.8} $$

where, as elsewhere in this appendix, n denotes a non-negative integer.

Transformations of $_3\phi_2$ series:

$$ {}_3\phi_2\left[\begin{matrix} a,\ b,\ c \\ d,\ e \end{matrix}; q, \frac{de}{abc}\right] $$

$$ = \frac{(e/a,de/bc;q)_\infty}{(e,de/abc;q)_\infty}\, {}_3\phi_2\left[\begin{matrix} a,\ d/b,\ d/c \\ d,\ de/bc \end{matrix}; q, \frac{e}{a}\right] \tag{III.9} $$

$$ = \frac{(b,de/ab,de/bc;q)_\infty}{(d,e,de/abc;q)_\infty}\, {}_3\phi_2\left[\begin{matrix} d/b, e/b, de/abc \\ de/ab, de/bc \end{matrix}; q, b\right], \tag{III.10} $$

$$ {}_3\phi_2\left[\begin{matrix} q^{-n},\ b,\ c \\ d,\ e \end{matrix}; q, q\right] $$

$$ = \frac{(de/bc;q)_n}{(e;q)_n}\left(\frac{bc}{d}\right)^n {}_3\phi_2\left[\begin{matrix} q^{-n},\ d/b,\ d/c \\ d,\ de/bc \end{matrix}; q, q\right] \tag{III.11} $$

$$= \frac{(e/c;q)_n}{(e;q)_n} c^n \, {}_3\phi_2 \left[\begin{matrix} q^{-n}, c, d/b \\ d, cq^{1-n}/e \end{matrix} ; q, \frac{bq}{e} \right], \tag{III.12}$$

$${}_3\phi_2 \left[\begin{matrix} q^{-n}, b, c \\ d, e \end{matrix} ; q, \frac{deq^n}{bc} \right] = \frac{(e/c;q)_n}{(e;q)_n} \, {}_3\phi_2 \left[\begin{matrix} q^{-n}, \ c, \ d/b \\ d, \ cq^{1-n}/e \end{matrix} ; q, q \right]. \tag{III.13}$$

The Sears-Carlitz transformation of a terminating well-poised ${}_3\phi_2$ series,

$${}_3\phi_2 \left[\begin{matrix} a, & b, & c \\ & aq/b, & aq/c \end{matrix} ; q, \frac{aqz}{bc} \right]$$

$$= \frac{(az;q)_\infty}{(z;q)_\infty} \, {}_5\phi_4 \left[\begin{matrix} a^{\frac{1}{2}}, -a^{\frac{1}{2}}, (aq)^{\frac{1}{2}}, -(aq)^{\frac{1}{2}}, aq/bc \\ aq/b, aq/c, az, q/z \end{matrix} ; q, q \right] \tag{III.14}$$

provided that $a = q^{-n}$. See (III.35) for a nonterminating case.

Sears' transformations of terminating balanced ${}_4\phi_3$ series:

$${}_4\phi_3 \left[\begin{matrix} q^{-n}, a, b, c \\ d, e, f \end{matrix} ; q, q \right]$$

$$= \frac{(e/a, f/a;q)_n}{(e, f;q)_n} a^n \, {}_4\phi_3 \left[\begin{matrix} q^{-n}, \ a, \ d/b, \ d/c \\ d, \ aq^{1-n}/e, \ aq^{1-n}/f \end{matrix} ; q, q \right] \tag{III.15}$$

$$= \frac{(a, ef/ab, ef/ac;q)_n}{(e, f, ef/abc;q)_n} \, {}_4\phi_3 \left[\begin{matrix} q^{-n}, e/a, f/a, ef/abc \\ ef/ab, ef/ac, q^{1-n}/a \end{matrix} ; q, q \right], \tag{III.16}$$

where $def = abcq^{1-n}$.

Watson's transformation formulas:

$${}_8\phi_7 \left[\begin{matrix} a, & qa^{\frac{1}{2}}, & -qa^{\frac{1}{2}}, & b, & c, & d, & e, & f \\ & a^{\frac{1}{2}}, & -a^{\frac{1}{2}}, & aq/b, & aq/c, & aq/d, & aq/e, & aq/f \end{matrix} ; q, \frac{a^2q^2}{bcdef} \right]$$

$$= \frac{(aq, aq/de, aq/df, aq/ef;q)_\infty}{(aq/d, aq/e, aq/f, aq/def;q)_\infty} \, {}_4\phi_3 \left[\begin{matrix} aq/bc, d, e, f \\ aq/b, aq/c, def/a \end{matrix} ; q, q \right] \tag{III.17}$$

when the ${}_8\phi_7$ series converges and the ${}_4\phi_3$ series terminates, and, when $f = q^{-n}$,

$${}_8\phi_7 \left[\begin{matrix} a, & qa^{\frac{1}{2}}, & -qa^{\frac{1}{2}}, & b, & c, & d, & e, & q^{-n} \\ & a^{\frac{1}{2}}, & -a^{\frac{1}{2}}, & aq/b, & aq/c, & aq/d, & aq/e, & aq^{n+1} \end{matrix} ; q, \frac{a^2q^{n+2}}{bcde} \right]$$

$$= \frac{(aq, aq/de;q)_n}{(aq/d, aq/e;q)_n} \, {}_4\phi_3 \left[\begin{matrix} aq/bc, & d, & e, & q^{-n} \\ & aq/b, & aq/c, & deq^{-n}/a \end{matrix} ; q, q \right] \tag{III.18}$$

or, equivalently,

$${}_4\phi_3 \left[\begin{matrix} q^{-n}, a, b, c \\ d, e, f \end{matrix} ; q, q \right] = \frac{(d/b, d/c;q)_n}{(d, d/bc;q)_n}$$

$$\cdot \, {}_8\phi_7 \left[\begin{matrix} \sigma, & q\sigma^{\frac{1}{2}}, & -q\sigma^{\frac{1}{2}}, & f/a, & e/a, & b, & c, & q^{-n} \\ & \sigma^{\frac{1}{2}}, & -\sigma^{\frac{1}{2}}, & e, & f, & ef/ab, & ef/ac, & efq^n/a \end{matrix} ; q, \frac{efq^n}{bc} \right], \tag{III.19}$$

where $def = abcq^{1-n}$ and $\sigma = ef/aq$.

Another transformation of a terminating balanced ${}_4\phi_3$ series to a very-well-poised ${}_8\phi_7$ series,

$$
{}_4\phi_3\left[\begin{matrix} q^{-n}, a, b, c \\ d, e, f \end{matrix}; q, q\right] = \frac{(abq/f, acq/f, bcq/f, q/f; q)_\infty}{(aq/f, bq/f, cq/f, abcq/f; q)_\infty}
$$

$$
\cdot\, {}_8\phi_7\left[\begin{matrix} \mu, & q\mu^{\frac{1}{2}}, & -q\mu^{\frac{1}{2}}, & a, & b, & c, & dq^n, & eq^n \\ & \mu^{\frac{1}{2}}, & -\mu^{\frac{1}{2}}, & \mu q/a, & \mu q/b, & \mu q/c, & e, & d \end{matrix}; q, \frac{de}{abc}\right], \tag{III.20}
$$

where $def = abcq^{1-n}$ and $\mu = abc/f$.

Singh's quadratic transformation:

$$
{}_4\phi_3\left[\begin{matrix} a^2, b^2, c, d \\ abq^{\frac{1}{2}}, -abq^{\frac{1}{2}}, -cd \end{matrix}; q, q\right]
$$

$$
= {}_4\phi_3\left[\begin{matrix} a^2, b^2, c^2, d^2 \\ a^2b^2q, -cd, -cdq \end{matrix}; q^2, q^2\right], \tag{III.21}
$$

provided the series terminate.

A q-analogue of Clausen's formula:

$$
\left\{ {}_4\phi_3\left[\begin{matrix} a, b, abz, ab/z \\ abq^{\frac{1}{2}}, -abq^{\frac{1}{2}}, -ab \end{matrix}; q, q\right]\right\}^2
$$

$$
= {}_5\phi_4\left[\begin{matrix} a^2, b^2, ab, abz, ab/z \\ abq^{\frac{1}{2}}, -abq^{\frac{1}{2}}, -ab, a^2b^2 \end{matrix}; q, q\right] \tag{III.22}
$$

provided both series terminate. A non-terminating q-analogue of Clausen's formula is given in (8.8.17).

Transformations of very-well-poised ${}_8\phi_7$ series:

$$
{}_8\phi_7\left[\begin{matrix} a, & qa^{\frac{1}{2}}, & -qa^{\frac{1}{2}}, & b, & c, & d, & e, & f \\ & a^{\frac{1}{2}}, & a^{\frac{1}{2}}, & aq/b, & aq/c, & aq/d, & aq/e, & aq/f \end{matrix}; q, \frac{a^2q^2}{bcdef}\right]
$$

$$
= \frac{(aq, aq/ef, \lambda q/e, \lambda q/f; q)_\infty}{(aq/e, aq/f, \lambda q, \lambda q/ef; q)_\infty}
$$

$$
\cdot\, {}_8\phi_7\left[\begin{matrix} \lambda, & q\lambda^{\frac{1}{2}}, & -q\lambda^{\frac{1}{2}}, & \lambda b/a, & \lambda c/a, & \lambda d/a, & e, & f \\ & \lambda^{\frac{1}{2}}, & -\lambda^{\frac{1}{2}}, & aq/b, & aq/c, & aq/d, & \lambda q/e, & \lambda q/f \end{matrix}; q, \frac{aq}{ef}\right] \tag{III.23}
$$

$$
= \frac{(aq, b, bc\mu/a, bd\mu/a, be\mu/a, bf\mu/a; q)_\infty}{(aq/c, aq/d, aq/e, aq/f, \mu q, b\mu/a; q)_\infty}
$$

$$
\cdot\, {}_8\phi_7\left[\begin{matrix} \mu, & q\mu^{\frac{1}{2}}, & -q\mu^{\frac{1}{2}}, & aq/bc, & aq/bd, & aq/be, & aq/bf, & b\mu/a \\ & \mu^{\frac{1}{2}}, & -\mu^{\frac{1}{2}}, & bc\mu/a, & bd\mu/a, & be\mu/a, & bf\mu/a, & aq/b \end{matrix}; q, b\right], \tag{III.24}
$$

where $\lambda = qa^2/bcd$ and $\mu = q^2a^3/b^2cdef$.

Transformations of a nearly-poised $_5\phi_4$ series:

$$
{}_5\phi_4\left[\begin{matrix} a, & b, & c, & d, & q^{-n} \\ & aq/b, & aq/c, & aq/d, & a^2q^{-n}/\lambda^2 \end{matrix};q,q\right]
$$
$$
=\frac{(\lambda q/a,\lambda^2q/a;q)_n}{(\lambda q,\lambda^2q/a^2;q)_n}\,{}_{12}\phi_{11}\left[\begin{matrix} \lambda, & q\lambda^{\frac12}, & -q\lambda^{\frac12}, & b\lambda/a, & c\lambda/a, & d\lambda/a, \\ & \lambda^{\frac12}, & -\lambda^{\frac12}, & aq/b, & aq/c, & aq/d, \end{matrix}\right.
$$
$$
\left.\begin{matrix} a^{\frac12}, & -a^{\frac12}, & (aq)^{\frac12}, & -(aq)^{\frac12}, & \lambda^2q^{n+1}/a, & q^{-n} \\ \lambda q/a^{\frac12}, & -\lambda q/a^{\frac12}, & \lambda(q/a)^{\frac12}, & -\lambda(q/a)^{\frac12}, & aq^{-n}/\lambda, & \lambda q^{n+1} \end{matrix};q,q\right], \tag{III.25}
$$

$$
{}_5\phi_4\left[\begin{matrix} q^{-n}, & b, & c, & d, & e \\ & q^{1-n}/b, & q^{1-n}/c, & q^{1-n}/d, & eq^{-2n}/\mu^2 \end{matrix};q,q\right]
$$
$$
=\frac{(\mu^2q^{n+1},\mu q/e;q)_n}{(\mu^2q^{n+1}/e,\mu q;q)_n}
$$
$$
\cdot\,{}_{12}\phi_{11}\left[\begin{matrix} \mu, & q\mu^{\frac12}, & -q\mu^{\frac12}, & \mu bq^n, & \mu cq^n, & \mu dq^n, \\ & \mu^{\frac12}, & -\mu^{\frac12}, & q^{1-n}/b, & q^{1-n}/c, & q^{1-n}/d, \end{matrix}\right.
$$
$$
\left.\begin{matrix} q^{-n/2}, & -q^{-n/2}, & q^{(1-n)/2}, & -q^{(1-n)/2}, & e, & \mu^2q^{n+1}/e \\ \mu q^{(n+2)/2}, & -\mu q^{(n+2)/2}, & \mu q^{(n+1)/2}, & -\mu q^{(n+1)/2}, & \mu q/e, & eq^{-n}/\mu \end{matrix};q,q\right], \tag{III.26}
$$

where $\lambda=qa^2/bcd$ and $\mu=q^{1-2n}/bcd$.

Transformation of a nearly-poised $_7\phi_6$ series:

$$
{}_7\phi_6\left[\begin{matrix} a, & qa^{\frac12}, & -qa^{\frac12}, & b, & c, & d, & q^{-n} \\ & a^{\frac12}, & -a^{\frac12}, & aq/b, & aq/c, & aq/d, & a^2q^{2-n}/\lambda^2 \end{matrix};q,q\right]
$$
$$
=\frac{(\lambda/aq,\lambda^2/aq;q)_n}{(\lambda q,\lambda^2/a^2q;q)_n}\frac{1-\lambda^2q^{2n-1}/a}{(1-\lambda^2/aq)}\,{}_{12}\phi_{11}\left[\begin{matrix} \lambda, & q\lambda^{\frac12}, & -q\lambda^{\frac12}, & b\lambda/a, & c\lambda/a, \\ & \lambda^{\frac12}, & -\lambda^{\frac12}, & aq/b, & aq/c, \end{matrix}\right.
$$
$$
\left.\begin{matrix} d\lambda/a, & (aq)^{\frac12}, & -(aq)^{\frac12}, & qa^{\frac12}, & -qa^{\frac12}, & \lambda^2q^{n-1}/a, & q^{-n} \\ aq/d, & \lambda(q/a)^{\frac12}, & -\lambda(q/a)^{\frac12}, & \lambda/a^{\frac12}, & -\lambda/a^{\frac12}, & aq^{2-n}/\lambda, & \lambda q^{n+1} \end{matrix};q,q\right], \tag{III.27}
$$

where $\lambda=qa^2/bcd$.

Bailey's $_{10}\phi_9$ transformation formula:

$$
{}_{10}\phi_9\left[\begin{matrix} a,qa^{\frac12},-qa^{\frac12},b,c,d,e,f,\lambda aq^{n+1}/ef,q^{-n} \\ a^{\frac12},-a^{\frac12},aq/b,aq/c,aq/d,aq/e,aq/f,efq^{-n}/\lambda,aq^{n+1} \end{matrix};q,q\right]
$$
$$
=\frac{(aq,aq/ef,\lambda q/e,\lambda q/f;q)_n}{(aq/e,aq/f,\lambda q/ef,\lambda q;q)_n}\,{}_{10}\phi_9\left[\begin{matrix} \lambda, & q\lambda^{\frac12}, & -q\lambda^{\frac12}, & \lambda b/a, & \lambda c/a, \\ & \lambda^{\frac12}, & -\lambda^{\frac12}, & aq/b, & aq/c, \end{matrix}\right.
$$
$$
\left.\begin{matrix} \lambda d/a, & e, & f, & \lambda aq^{n+1}/ef, & q^{-n}, \\ aq/d, & \lambda q/e, & \lambda q/f, & efq^{-n}/a, & \lambda q^{n+1} \end{matrix};q,q\right], \tag{III.28}
$$

where $\lambda = qa^2/bcd$.

Transformations of ${}_{r+2}\phi_{r+1}$ series:

$${}_{r+2}\phi_{r+1}\left[\begin{matrix} a, b, b_1q^{m_1}, \ldots, b_rq^{m_r} \\ bq^{1+m}, b_1, \ldots, b_r \end{matrix}; q, a^{-1}q^{m+1-(m_1+\cdots+m_r)}\right]$$

$$= \frac{(q, bq/a; q)_\infty}{(bq, q/a; q)_\infty} \frac{(bq; q)_m (b_1/b; q)_{m_1} \cdots (b_r/b; q)_{m_r}}{(q; q)_m (b_1; q)_{m_1} \cdots (b_r; q)_{m_r}} b^{m_1+\cdots+m_r-m}$$

$$\cdot {}_{r+2}\phi_{r+1}\left[\begin{matrix} q^{-m}, b, bq/b_1, \ldots, bq/b_r \\ bq/a, bq^{1-m_1}/b_1, \ldots, bq^{1-m_r}/b_r \end{matrix}; q, q\right] \tag{III.29}$$

and

$${}_{r+2}\phi_{r+1}\left[\begin{matrix} a, b, b_1q^{m_1}, \ldots, b_rq^{m_r} \\ bcq, b_1, \ldots, b_r \end{matrix}; q, a^{-1}q^{1-(m_1+\cdots+m_r)}\right]$$

$$= \frac{(bq/a, cq; q)_\infty}{(bcq, q/a; q)_\infty} \frac{(b_1/b; q)_{m_1} \cdots (b_r/b; q)_{m_r}}{(b_1; q)_{m_1} \cdots (b_r; q)_{m_r}} b^{m_1+\cdots+m_r}$$

$$\cdot {}_{r+2}\phi_{r+1}\left[\begin{matrix} c^{-1}, b, bq/b_1, \ldots, bq/b_r \\ bq/a, bq^{1-m_1}/b_1, \ldots, bq^{1-m_r}/b_r \end{matrix}; q, cq\right], \tag{III.30}$$

where $m, m_1, \ldots, m_r$ are arbitrary nonnegative integers.

Three-term transformation formulas:

$${}_2\phi_1(a, b; c; q, z) = \frac{(abz/c, q/c; q)_\infty}{(az/c, q/a; q)_\infty} {}_2\phi_1(c/a, cq/abz; cq/az; q, bq/c)$$

$$- \frac{(b, q/c, c/a, az/q, q^2/az; q)_\infty}{(c/q, bq/c, q/a, az/c, cq/az; q)_\infty} {}_2\phi_1(aq/c, bq/c; q^2/c; q, z). \tag{III.31}$$

$${}_2\phi_1(a, b; c; q, z) = \frac{(b, c/a, az, q/az; q)_\infty}{(c, b/a, z, q/z; q)_\infty} {}_2\phi_1(a, aq/c; aq/b; q, cq/abz)$$

$$+ \frac{(a, c/b, bz, q/bz; q)_\infty}{(c, a/b, z, q/z; q)_\infty} {}_2\phi_1(b, bq/c; bq/a; q, cq/abz). \tag{III.32}$$

$${}_3\phi_2\left[\begin{matrix} a, b, c \\ d, e \end{matrix}; q, \frac{de}{abc}\right]$$

$$= \frac{(e/b, e/c, cq/a, q/d; q)_\infty}{(e, cq/d, q/a, e/bc; q)_\infty} {}_3\phi_2\left[\begin{matrix} c, d/a, cq/e \\ cq/a, bcq/e \end{matrix}; q, \frac{bq}{d}\right]$$

$$- \frac{(q/d, eq/d, b, c, d/a, de/bcq, bcq^2/de; q)_\infty}{(d/q, e, bq/d, cq/d, q/a, e/bc, bcq/e; q)_\infty} {}_3\phi_2\left[\begin{matrix} aq/d, bq/d, cq/d \\ q^2/d, eq/d \end{matrix}; q, \frac{de}{abc}\right]. \tag{III.33}$$

$${}_3\phi_2\left[\begin{matrix} a, b, c \\ d, e \end{matrix}; q, \frac{de}{abc}\right] = \frac{(e/b, e/c; q)_\infty}{(e, e/bc; q)_\infty} {}_3\phi_2\left[\begin{matrix} d/a, b, c \\ d, bcq/e \end{matrix}; q, q\right]$$

$$+ \frac{(d/a, b, c, de/bc; q)_\infty}{(d, e, bc/e, de/abc; q)_\infty} {}_3\phi_2\left[\begin{matrix} e/b, e/c, de/abc \\ de/bc, eq/bc \end{matrix}; q, q\right]. \tag{III.34}$$

$$
{}_3\phi_2\left[\begin{matrix} a, & b, & c \\ & aq/b, & aq/c \end{matrix};q,\frac{aqx}{bc}\right]
$$

$$
= \frac{(ax;q)_\infty}{(x;q)_\infty}\,{}_5\phi_4\left[\begin{matrix} a^{\frac{1}{2}}, & -a^{\frac{1}{2}}, & (aq)^{\frac{1}{2}}, & -(aq)^{\frac{1}{2}}, & aq/bc \\ & aq/b, & aq/c, & ax, & q/x \end{matrix};q,q\right]
$$

$$
+\frac{(a,aq/bc,aqx/b,aqx/c;q)_\infty}{(aq/b,aq/c,aqx/bc,x^{-1};q)_\infty}
$$

$$
\cdot\,{}_5\phi_4\left[\begin{matrix} xa^{\frac{1}{2}},-xa^{\frac{1}{2}},x(aq)^{\frac{1}{2}},-x(aq)^{\frac{1}{2}},aqx/bc \\ aqx/b,aqx/c,xq,ax^2 \end{matrix};q,q\right]. \tag{III.35}
$$

$$
{}_8\phi_7\left[\begin{matrix} a, & qa^{\frac{1}{2}}, & -qa^{\frac{1}{2}}, & b, & c, & d, & e, & f \\ & a^{\frac{1}{2}}, & -a^{\frac{1}{2}}, & aq/b, & aq/c, & aq/d, & aq/e, & aq/f \end{matrix};q,\frac{a^2q^2}{bcdef}\right]
$$

$$
= \frac{(aq,aq/de,aq/df,aq/ef;q)_\infty}{(aq/d,aq/e,aq/f,aq/def;q)_\infty}\,{}_4\phi_3\left[\begin{matrix} aq/bc,d,e,f \\ aq/b,aq/c,def/a \end{matrix};q,q\right]
$$

$$
+\frac{(aq,aq/bc,d,e,f,a^2q^2/bdef,a^2q^2/cdef;q)_\infty}{(aq/b,aq/c,aq/d,aq/e,aq/f,a^2q^2/bcdef,def/aq;q)_\infty}
$$

$$
\cdot\,{}_4\phi_3\left[\begin{matrix} aq/de,aq/df,aq/ef,a^2q^2/bcdef \\ a^2q^2/bdef,a^2q^2/cdef,aq^2/def \end{matrix};q,q\right]. \tag{III.36}
$$

$$
{}_8\phi_7\left[\begin{matrix} a, & qa^{\frac{1}{2}}, & -qa^{\frac{1}{2}}, & b, & c, & d, & e, & f \\ & a^{\frac{1}{2}}, & -a^{\frac{1}{2}}, & aq/b, & aq/c, & aq/d, & aq/e, & aq/f \end{matrix};q,\frac{a^2q^2}{bcdef}\right]
$$

$$
= \frac{(aq,aq/de,aq/df,aq/ef,eq/c,fq/c,b/a,bef/a;q)_\infty}{(aq/d,aq/e,aq/f,aq/def,q/c,efq/c,be/a,bf/a;q)_\infty}
$$

$$
\cdot\,{}_8\phi_7\left[\begin{matrix} ef/c,q\,(ef/c)^{\frac{1}{2}},-q\,(ef/c)^{\frac{1}{2}},aq/bc,aq/cd,ef/a,e,f \\ (ef/c)^{\frac{1}{2}},-(ef/c)^{\frac{1}{2}},bef/a,def/a,aq/c,fq/c,eq/c \end{matrix};q,\frac{bd}{a}\right]
$$

$$
+\frac{b}{a}\frac{(aq,bq/a,bq/c,bq/d,bq/e,bq/f,d,e,f,aq/bc;q)_\infty}{(aq/b,aq/c,aq/d,aq/e,aq/f,bd/a,be/a;q)_\infty}
$$

$$
\cdot\frac{(bdef/a^2,a^2q/bdef;q)_\infty}{(bf/a,def/a,aq/def,q/c,b^2q/a;q)_\infty}
$$

$$
\cdot\,{}_8\phi_7\left[\begin{matrix} b^2/a,bqa^{-\frac{1}{2}},-bqa^{-\frac{1}{2}},b,bc/a,bd/a,be/a,bf/a \\ ba^{-\frac{1}{2}},-ba^{-\frac{1}{2}},bq/a,bq/c,bq/d,bq/e,bq/f \end{matrix};q,\frac{a^2q^2}{bcdef}\right]. \tag{III.37}
$$

Transformation of an ${}_8\psi_8$:

$$
\frac{(aq/b,aq/c,aq/d,aq/e,q/ab,q/ac,q/ad,q/ae;q)_\infty}{(fa,ga,f/a,g/a,qa^2,q/a^2;q)_\infty}
$$

$$
\cdot\,{}_8\psi_8\left[\begin{matrix} qa, & -qa & ba, & ca, & da, & ea, & fa, & ga \\ a, & -a, & aq/b, & aq/c, & aq/d, & aq/e, & aq/f, & aq/g \end{matrix};q,\frac{q^2}{bcdefg}\right]
$$

$$= \frac{(q, q/bf, q/cf, q/df, q/ef, qf/b, qf/c, qf/d, qf/e; q)_\infty}{(fa, q/fa, aq/f, f/a, g/f, fg, qf^2; q)_\infty}$$

$$\cdot {}_8\phi_7 \left[\begin{matrix} f^2, & qf, & -qf, & fb, & fc, & fd, & fe, & fg \\ & f, & -f, & fq/b, & fq/c, & fq/d, & fq/e, & fq/g \end{matrix} ; q, \frac{q^2}{bcdefg} \right]$$

$$+ \text{ idem } (f; g). \tag{III.38}$$

Bailey's four-term ${}_{10}\phi_9$ transformation:

$${}_{10}\phi_9 \left[\begin{matrix} a, qa^{\frac{1}{2}}, -qa^{\frac{1}{2}}, b, c, d, e, f, g, h \\ a^{\frac{1}{2}}, -a^{\frac{1}{2}}, aq/b, aq/c, aq/d, aq/e, aq/f, aq/g, aq/h \end{matrix} ; q, q \right]$$

$$+ \frac{(aq, b/a, c, d, e, f, g, h, bq/c; q)_\infty}{(b^2q/a, a/b, aq/c, aq/d, aq/e, aq/f, aq/g, aq/h, bc/a; q)_\infty}$$

$$\cdot \frac{(bq/d, bq/e, bq/f, bq/g, bq/h; q)_\infty}{(bd/a, be/a, bf/a, bg/a, bh/a; q)_\infty}$$

$$\cdot {}_{10}\phi_9 \left[\begin{matrix} b^2/a, & qba^{-\frac{1}{2}}, & -qba^{-\frac{1}{2}}, & b, & bc/a, & bd/a, & be/a, & bf/a, & bg/a, & bh/a \\ & ba^{-\frac{1}{2}}, & -ba^{-\frac{1}{2}}, & bq/a, & bq/c, & bq/d, & bq/e, & bq/f, & bq/g, & bq/h \end{matrix} ; q, q \right]$$

$$= \frac{(aq, b/a, \lambda q/f, \lambda q/g, \lambda q/h, bf/\lambda, bg/\lambda, bh/\lambda; q)_\infty}{(\lambda q, b/\lambda, aq/f, aq/g, aq/h, bf/a, bg/a, bh/a; q)_\infty}$$

$$\cdot {}_{10}\phi_9 \left[\begin{matrix} \lambda, q\lambda^{\frac{1}{2}}, -q\lambda^{\frac{1}{2}}, b, \lambda c/a, \lambda d/a, \lambda e/a, f, g, h \\ \lambda^{\frac{1}{2}}, -\lambda^{\frac{1}{2}}, \lambda q/b, aq/c, aq/d, aq/e, \lambda q/f, \lambda q/g, \lambda q/h \end{matrix} ; q, q \right]$$

$$+ \frac{(aq, b/a, f, g, h, bq/f, bq/g, bq/h, \lambda c/a, \lambda d/a, \lambda e/a, abq/\lambda c; q)_\infty}{(b^2q/\lambda, \lambda/b, aq/c, aq/d, aq/e, aq/f, aq/g, aq/h, bc/a, bd/a, be/a, bf/a; q)_\infty}$$

$$\cdot \frac{(abq/\lambda d, abq/\lambda e; q)_\infty}{(bg/a, bh/a; q)_\infty} {}_{10}\phi_9 \left[\begin{matrix} b^2/\lambda, qb\lambda^{-\frac{1}{2}}, -qb\lambda^{-\frac{1}{2}}, b, bc/a, bd/a, \\ b\lambda^{-\frac{1}{2}}, -b\lambda^{-\frac{1}{2}}, bq/\lambda, abq/c\lambda, abq/d\lambda, \end{matrix} \right.$$

$$\left. \begin{matrix} be/a, bf/\lambda, bg/\lambda, bh/\lambda \\ abq/e\lambda, bq/f, bq/g, bq/h \end{matrix} ; q, q \right], \tag{III.39}$$

where $a^3q^2 = bcdefgh$ and $\lambda = qa^2/cde$.

Transformation of a ${}_{10}\psi_{10}$:

$$\frac{(aq/b, aq/c, aq/d, aq/e, aq/f, q/ab, q/ac, q/ad, q/ae, q/af; q)_\infty}{(ag, ah, ak, g/a, h/a, k/a, qa^2, q/a^2; q)_\infty}$$

$$\cdot {}_{10}\psi_{10} \left[\begin{matrix} qa, -qa, ba, ca, da, ea, fa, ga, ha, ka \\ a, -a, aq/b, aq/c, aq/d, aq/e, aq/f, aq/g, aq/h, aq/k \end{matrix} ; q, \frac{q^2}{bcdefghk} \right]$$

$$= \frac{(q, q/bg, q/cg, q/dg, q/eg, q/fg, qg/b, qg/c, qg/d, qg/e, qg/f; q)_\infty}{(gh, gk, h/g, ag, q/ag, g/a, aq/g, qg^2; q)_\infty}$$

$$\cdot {}_{10}\phi_9 \left[\begin{matrix} g^2, qg, -qg, gb, gc, gd, ge, gf, gh, gk \\ g, -g, qg/b, qg/c, qg/d, qg/e, qg/f, qg/h, qg/k \end{matrix} ; q, \frac{q^2}{bcdefghk} \right]$$

$$+ \text{ idem } (g; h, k). \tag{III.40}$$

References

Adams, C. R. (1931). Linear q-difference equations, *Bull. Amer. Math. Soc.* **37**, 361-400.

Adiga, C., Berndt, B. C., Bhargava, S. and Watson, G. N. (1985). Chapter 16 of Ramanujan's second notebook: Theta-functions and q-series, *Memoirs Amer. Math. Soc.* **315**.

Agarwal, A. K., Andrews, G. and Bressoud, D. (1987). The Bailey Lattice, *J. Indian Math. Soc.*, **51**, 57-73.

Agarwal, A. K., Kalnins, E. G. and Miller, W. (1987). Canonical equations and symmetry techniques for q-series, *SIAM J. Math. Anal.* **18**, 1519-1538.

Agarwal, N. (1959). Certain basic hypergeometric identities of the Cayley-Orr type, *J. London Math. Soc.* **34**, 37-46.

Agarwal, R. P. (1953a). Some basic hypergeometric identities, *Ann. Soc. Sci. Bruxelles* **67**, 1-21.

Agarwal, R. P. (1953b). On integral analogues of certain transformations of well-poised basic hypergeometric series, *Quart. J. Math.* (Oxford) (2) **4**, 161-167.

Agarwal, R. P. (1953c). Some transformations of well-poised basic hypergeometric series of the type ${}_8\phi_7$, *Proc. Amer. Math. Soc.* **4**, 678-685.

Agarwal, R. P. (1953d). Associated basic hypergeometric series, *Proc. Glasgow Math. Assoc.* **1**, 182-184.

Agarwal, R. P. (1953e). On the partial sums of series of hypergeometric type, *Proc. Camb. Phil. Soc.* **49**, 441-445.

Agarwal, R. P. (1963). *Generalized Hypergeometric Series*, Asia Publishing House, Bombay, London and New York.

Agarwal, R. P. (1969a). Certain basic hypergeometric identities associated with mock theta functions, *Quart. J. Math.* (Oxford) (2) **20**, 121-128.

Agarwal, R. P. (1969b). Certain fractional q-integrals and q-derivatives, *Proc. Camb. Phil. Soc.* **66**, 365-370.

Agarwal, R. P. and Verma, A. (1967a). Generalized basic hypergeometric series with unconnected bases, *Proc. Camb. Phil. Soc.* **63**, 727-734.

Agarwal, R. P. and Verma, A. (1967b). Generalized basic hypergeometric series with unconnected bases (II), *Quart. J. Math.* (Oxford) (2) **18**, 181-192; Corrigenda, ibid. **21** (1970), 384.

Aigner, M. (1979). *Combinatorial Theory*, Springer, Berlin and New York.

Alder, H. L. (1969). Partition identities from Euler to the present, *Amer. Math. Monthly*, **76**, 733-746.

Allaway, Wm. R. (1980). Some properties of the q-Hermite polynomials, *Canad. J. Math.* **32**, 686-694.

Al-Salam, W. A. (1966). Some fractional q-integrals and q-derivatives, *Proc. Edin. Math. Soc.* **15**, 135-140.

Al-Salam, W. A., Allaway, Wm. R. and Askey, R. (1984a). A characterization of the continuous q-ultraspherical polynomials, *Canad. Math. Bull.* **27** (3), 329-336.

Al-Salam, W. A., Allaway, Wm. R. and Askey, R. (1984b). Sieved ultraspherical polynomials, *Trans. Amer. Math. Soc.* **284**, 39-55.

Al-Salam, W. A. and Carlitz, L. (1957). A q-analogue of a formula of Toscano, *Boll. Unione Math. Ital.* **12**, 414-417.

Al-Salam, W. A. and Carlitz, L. (1965). Some orthogonal q-polynomials, *Math. Nachr.* **30**, 47-61.

Al-Salam, W. A. and Chihara, T. S. (1976). Convolution of orthogonal polynomials, *SIAM J. Math. Anal.* **7**, 16-28.

Al-Salam, W. A. and Chihara, T. S. (1987). q-Pollaczek polynomials and a conjecture of Andrews and Askey, *SIAM J. Math. Anal.* **18**, 228-242.

Al-Salam, W. A. and Ismail, M. E. H. (1977). Reproducing kernels for q-Jacobi polynomials, *Proc. Amer. Math. Soc.* **67**, 105-110.

Al-Salam, W. A. and Ismail, M. E. H. (1983). Orthogonal polynomials associated with the Rogers-Ramanujan continued fraction, *Pacific J. Math.* **104**, 269-283.

Al-Salam, W. A. and Ismail, M. E. H. (1988). q-Beta integrals and the q-Hermite polynomials, *Pacific J. Math.*, **135**, 209-221.

Al-Salam, W. A. and Verma, A. (1975a). A fractional Leibniz q-formula, *Pacific J. Math.* **60**, 1-9.

Al-Salam, W. A. and Verma, A. (1975b). Remarks on fractional q-integrals, *Bull. Soc. Roy. Sci. Liège* **44**, 600-607.

Al-Salam, W. A. and Verma, A. (1982a). Some remarks on q-beta integral, *Proc. Amer. Math. Soc.* **85**, 360-362.

Al-Salam, W. A. and Verma, A. (1982b). On an orthogonal polynomial set, *Indag. Math.* **44**, 335-340.

Al-Salam, W. A. and Verma, A. (1983a). A pair of biorthogonal sets of polynomials, *Rocky Mtn. J. Math.* **13**, 273-279.

Al-Salam, W. A. and Verma, A. (1983b). q-Konhauser polynomials, *Pacific J. Math.* **108**, 1-7.

Al-Salam, W. A. and Verma, A. (1983c). q-Analogues of some biorthogonal functions, *Canad. Math. Bull.* **26** (2), 225-227.

Al-Salam, W. A. and Verma, A. (1984). On quadratic transformations of basic series, *SIAM J. Math. Anal.* **15**, 414-420.

Andrews, G. E. (1965). A simple proof of Jacobi's triple product identity, *Proc. Amer. Math. Soc.* **16**, 333-334.

Andrews, G. E. (1966a). On basic hypergeometric series, mock theta functions, and partitions (I), *Quart. J. Math.* (Oxford) (2) **17**, 64-80.

Andrews, G. E. (1966b). On basic hypergeometric series, mock theta functions, and partitions (II), *Quart. J. Math.* (Oxford) (2) **17**, 132-143.

Andrews, G. E. (1968). On q-difference equations for certain well-poised basic hypergeometric series, *Quart. J. Math.* (Oxford) (2) **19**, 433-447.

Andrews, G. E. (1969). On Ramanujan's summation of ${}_1\psi_1(a,b,z)$, *Proc. Amer. Math. Soc.* **22**, 552-553.

Andrews, G. E. (1970a). On a transformation of bilateral series with applications, *Proc. Amer. Math. Soc.* **25**, 554-558.

Andrews, G. E. (1970b). A polynomial identity which implies the Rogers-Ramanujan identities, *Scripta Math.* **28**, 297-305.

Andrews, G. E. (1971). *Number Theory*, W. B. Saunders, Philadelphia; reprinted by Hindustan Publishing Co., New Delhi, 1984.

Andrews, G. E. (1971a). On the foundations of combinatorial theory V, Eulerian differential operators, *Studies in Appl. Math.* **50**, 345-375.

Andrews, G. E. (1972). Summations and transformations for basic Appell series, *J. London Math. Soc.* (2) **4**, 618-622.

Andrews, G. E. (1973). On the q-analogue of Kummer's theorem and applications, *Duke Math. J.* **40**, 525-528.

Andrews, G. E. (1974a). Applications of basic hypergeometric functions, *SIAM Rev.* **16**, 441-484.

Andrews, G. E. (1974b). An analytic generalization of the Rogers-Ramanujan identities for odd moduli, *Proc. Nat. Acad. Sci. USA* **71**, 4082-4085.

Andrews, G. E. (1974c). A general theory of identities of the Rogers-Ramanujan type, *Bull. Amer. Math. Soc.* **80**, 1033-1052.

Andrews, G. E. (1975a). Problems and prospects for basic hypergeometric functions, *Theory and Applications of Special Functions* (R. Askey, ed.), Academic Press, New York, pp. 191-224.

Andrews, G. E. (1975b). Identities in combinatorics. II: A q-analog of the Lagrange inversion theorem, *Proc. Amer. Math. Soc.* **53**, 240-245.

Andrews, G. E. (1976). *The Theory of Partitions*, Encyclopedia of Mathematics and Its Applications, Vol. 2, Addison-Wesley, Reading, Mass.; reissued by Cambridge University Press, Cambridge, 1985.

Andrews, G. E. (1976a). On q-analogues of the Watson and Whipple summations, *SIAM J. Math. Anal.* **7**, 332-336.

Andrews, G. E. (1976b). On identities implying the Rogers-Ramanujan identities, *Houston J. Math.* **2**, 289-298.

Andrews, G. E. (1979a). Connection coefficient problems and partitions, *Proc. Sympos. Pure Math.* **34**, Amer. Math. Soc., Providence, R. I., pp. 1-24.

Andrews, G. E. (1979b). Partitions: yesterday and today, *New Zealand Math. Soc.*, Wellington.

Andrews, G. E. (1979c). An introduction to Ramanujan's "lost" notebook, *Amer. Math. Monthly* **86**, 89-108.

Andrews, G. E. (1981a). The hard-hexagon model and the Rogers-Ramanujan identities, *Proc. Nat. Acad. Sci. USA* **78**, 5290-5292.

Andrews, G. E. (1981b). Ramanujan's "lost" notebook I. Partial θ-functions, *Advances in Math.* **41**, 137-172.

Andrews, G. E. (1981c). Ramanujan's "lost" notebook II. θ-function expansions, *Advances in Math.* **41**, 173-185.

Andrews, G. E. (1983). Euler's pentagonal number theorem, *Math. Magazine* **56**, 279-284.

Andrews, G. E. (1984a). Generalized Frobenius partitions, *Memoirs Amer. Math. Soc.* **301**.

Andrews, G. E. (1984b). Multiple series Rogers-Ramanujan type identities, *Pacific J. Math.* **114**, 267-283.

Andrews, G. E. (1984c). Ramanujan's "lost" notebook IV. Stacks and alternating parity in partitions, *Advances in Math.* **53**, 55-74.

Andrews, G. E. (1984d). Ramanujan and SCRATCHPAD, *Proc. of the 1984 MACSYMA Users' Conference*, General Electric, Schenectady, N.Y., pp. 384-408.

Andrews, G. E. (1986). *q-Series: Their Development and Application in Analysis, Number Theory, Combinatorics, Physics, and Computer Algebra*, CBMS Regional Conference Lecture Series **66**, Amer. Math. Soc., Providence, R. I.

Andrews, G. E. (1987a). The Rogers-Ramanujan identities without Jacobi's triple product, *Rocky Mtn. J. Math.* **17**, 659-672.

Andrews, G. E. (1987b). Physics, Ramanujan, and computer algebra, *Computer Algebra as a Tool for Research in Mathematics and Physics*, Proceedings of the International Conference, New York University, 1984, M. Dekker.

Andrews, G. E. (1988). Ramanujan's fifth order mock theta functions as constant terms, *Ramanujan Revisited* (G. E. Andrews *et al.*, eds.), Academic Press, New York, pp. 47-56.

Andrews, G. E. and Askey, R. (1977). Enumeration of partitions: the role of Eulerian series and q-orthogonal polynomials, *Higher Combinatorics* (M. Aigner, ed.), Reidel, Boston, Mass., pp. 3-26.

Andrews, G. E. and Askey, R. (1978). A simple proof of Ramanujan's summation of the ${}_1\psi_1$, *Aequationes Math.* **18**, 333-337.

Andrews, G. E. and Askey, R. (1981). Another q-extension of the beta function, *Proc. Amer. Math. Soc.* **81**, 97-100.

Andrews, G. E. and Askey, R. (1985). Classical orthogonal polynomials, *Polynômes orthogonaux et applications*, Lecture Notes in Math. **1171**, Springer, Berlin and New York, pp. 36-62.

Andrews, G. E. and Askey, R. (1989). *The Classical and Discrete Orthogonal Polynomials and their q-Analogues*, in preparation.

Andrews, G. E., Askey, R., Berndt, B. C., Ramanathan, K. G. and Rankin R. A., eds. (1988). *Ramanujan Revisited*, Academic Press, New York.

Andrews, G. E. and Baxter, R. J. (1986). Lattice gas generalization of the hard hexagon model. II. The local densities as elliptic functions, *J. Stat. Phys.* **44**, 713-728.

Andrews, G. E. and Baxter, R. J. (1987). Lattice gas generalization of the hard hexagon model. III. q-Trinomial coefficients, *J. Stat. Phys.* **47**, 297-330.

Andrews, G. E., Baxter, R. J., Bressoud, D. M., Burge, W. H., Forrester, P. J. and Viennot, G. (1987). Partitions with prescribed hook differences, *Europ. J. Combinatorics* **8**, 341-350.

Andrews, G. E., Baxter, R. J., Forrester, P. J. (1984). Eight-vertex SOS model and generalized Rogers-Ramanujan-type identities, *J. Stat. Phys.* **35**, 193-266.

Andrews, G. E. and Bressoud, D. M. (1984). Identities in combinatorics III: Further aspects of ordered set sorting, *Discrete Math.* **49**, 223-236.

Andrews, G. E., Dyson, F. J. and Hickerson, D. (1988). Partitions and indefinite quadratic forms, *Invent. Math.* **91**, 391-407.

Andrews, G.E. and Foata, D. (1980). Congruences for the q-secant numbers, *Europ. J. Combinatorics* **1**, 283-287.

Andrews, G. E. and Forrester, P. J. (1986). Height probabilities in solid-on-solid models: I, *J. Phys. A* **19**, L923-L926.

Andrews, G. E., Goulden, I. P. and Jackson, D. M. (1986). Shank's convergence acceleration transform, Padé approximants and partitions, *J. Comb. Thy. A* **43**, 70-84.

Andrews, G. E. and Onofri, E. (1984). Lattice gauge theory, orthogonal polynomials and q-hypergeometric functions, *Special Functions: Group Theoretical Aspects and Applications* (R. Askey *et al.*, eds.), Reidel, Boston, Mass., pp. 163-188.

Aomoto, K. (1987). Jacobi polynomials associated with Selberg integrals, *SIAM J. Math. Anal.* **18**, 545-549.

Appell, P. and Kampé de Fériet, J. (1926). *Fonctions Hypergéométriques et Hypersphériques*, Gauthier-Villars, Paris.

Artin, E. (1964). *The Gamma function*, translated by M. Butler, Holt, Rinehart and Winston, New York.

Askey, R. (1970). Orthogonal polynomials and positivity, *Studies in Applied Mathematics* **6**, *Special Functions and Wave Propagation* (D. Ludwig and F. W. J. Olver, eds.), SIAM, Philadelphia, pp. 64-85.

Askey, R., ed. (1975). *Orthogonal Polynomials and Special Functions*, Regional Conference Series in Applies Mathematics **21**, SIAM, Philadelphia.

Askey, R. (1978). The q-gamma and q-beta functions, *Applicable Analysis* **8**, 125-141.

Askey, R. (1980a). Ramanujan's extensions of the gamma and beta functions, *Amer. Math. Monthly* **87**, 346-359.

Askey, R. (1980b). Some basic hypergeometric extensions of integrals of Selberg and Andrews, *SIAM J. Math. Anal.* **11**, 938-951.

Askey, R. (1981). A q-extension of Cauchy's form of the beta integral, *Quart. J. Math.* Oxford (2) **32**, 255-266.

Askey, R. (1982a). Two integrals of Ramanujan, *Proc. Amer. Math. Soc.* **85**, 192-194.

Askey, R. (1982b). A q-beta integral associated with BC_1, *SIAM J. Math. Anal.* **13**, 1008-1010.

Askey, R. (1984a). Orthogonal polynomials and some definite integrals, *Proc. Int. Congress of Mathematicians*, Aug. 16-24, 1983, Warsaw, pp. 935-943.

Askey, R. (1984b). Orthogonal polynomials old and new, and some combinatorial connections, *Enumeration and Design* (D. M. Jackson and S. A. Vanstone, eds.), Academic Press, New York, pp. 67-84.

Askey, R. (1984c). The very well poised ${}_6\psi_6$. II, *Proc. Amer. Math. Soc.* **90**, 575-579.

Askey, R. (1985). Some problems about special functions and computations, *Rend. del Sem. Mat. di Torino*, pp. 1-22.

Askey, R. (1986). Limits of some q-Laguerre polynomials, *J. Approx. Thy.* **46**, 213-216.

Askey, R. (1987). Ramanujan's ${}_1\psi_1$ and formal Laurent series, *Indian J. Math.* **29**, 101-105.

Askey, R. (1988a). Beta integrals in Ramanujan's papers, his unpublished work and further examples, *Ramanujan Revisited* (G. E. Andrews *et al.*, eds.), Academic Press, New York, pp. 561-590.

Askey, R. (1988b). Beta integrals and q-extensions, *Papers of the Ramanujan Centennial International Conference, Ramanujan Mathematical Society*, 1987, pp. 85-102.

Askey, R. (1989a). Ramanujan and hypergeometric and basic hypergeometric series, *Proc. Ramanujan Symposium on Classical Analysis*, Pune, India, Dec. 26-28, 1987, to appear.

Askey, R. (1989b). Continuous q-Hermite polynomials when $q > 1$, *Workshop on q-Series and Partitions* (D. Stanton, ed.), IMA Volumes in Mathematics and its Applications, Springer, Berlin and New York, to appear.

Askey, R. (1989c). Orthogonal polynomials and theta functions, *Theta Functions Bowdoin. 1987* (R. Gunning and L. Ehrenpreis, eds.), Amer. Math. Soc., Providence, R. I., to appear.

Askey, R. (1989d). Variants of Clausen's formula for the square of a special ${}_2F_1$, to appear.

Askey, R. (1989e). Beta integrals and the associated orthogonal polynomials, to appear.

Askey, R. (1989f). Computer algebra and definite integrals, *Computer Algebra* (D. Chudnovsky and R. Jenks, eds.), M. Dekker, New York, to appear.

Askey, R. (1989g). Integration and computers, *Computers and Mathematics Conference*, Stanford University, July 29-Aug. 2, 1986, to appear.

Askey, R. and Gasper, G. (1971). Jacobi polynomial expansions of Jacobi polynomials with non-negative coefficients, *Proc. Camb. Phil. Soc.* **70**, 243-255.

Askey, R. and Gasper, G. (1976). Positive Jacobi polynomial sums. II, *Amer. J. Math.* **98**, 709-737.

Askey, R. and Gasper, G. (1977). Convolution structures for Laguerre polynomials, *J. d'Analyse Math.* **31**, 48-68.

Askey, R. and Gasper, G. (1986). Inequalities for polynomials, *The Bieberbach Conjecture: Proc. of the Symposium on the Occasion of the Proof* (A. Baernstein, D. Drasin, P. Duren, and A. Marden, eds.), Math. Surveys and Monographs **21**, Amer. Math. Soc., Providence, R. I., pp. 7-32.

Askey, R. and Ismail, M. E. H. (1979). The very well poised ${}_6\psi_6$, *Proc. Amer. Math. Soc.* **77**, 218-222.

Askey, R. and Ismail, M. E. H. (1980). The Rogers q-ultraspherical polynomials, *Approximation Theory* III (E. W. Cheny, ed.), Academic Press, New York, pp. 175-182.

Askey, R. and Ismail, M. E. H. (1983). A generalization of ultraspherical polynomials, *Studies in Pure Mathematics* (P. Erdős, ed.), Birkhäuser, Boston, Mass., pp. 55-78.

Askey, R. and Ismail, M. E. H. (1984). Recurrence relations, continued fractions and orthogonal polynomials, *Memoirs Amer. Math. Soc.* **300**.

Askey, R., Koornwinder, T. H. and Rahman, M. (1986). An integral of products of ultraspherical functions and a q-extension, *J. London Math. Soc.* (2) **33**, 133-148.

Askey, R., Koornwinder, T. H. and Schempp, W., eds. (1984). *Special Functions: Group Theoretical Aspects and Applications*, Reidel, Boston, Mass.

Askey, R. and Roy, R. (1986). More q-beta integrals, *Rocky Mtn. J. Math.* **16**, 365-372.

Askey, R. and Shukla, D. P. (1989). Sieved Jacobi polynomials, in preparation.

Askey, R. and Wilson, J.A. (1979). A set of orthogonal polynomials that generalized the Racah coefficients or 6-j symbols, *SIAM J. Math. Anal.* **10**, 1008-1016.

Askey, R. and Wilson, J.A. (1982). A set of hypergeometric orthogonal polynomials, *SIAM J. Math. Anal.* **13**, 651-655.

Askey, R. and Wilson, J.A. (1985). Some basic hypergeometric polynomials that generalize Jacobi polynomials, *Memoirs Amer. Math. Soc.* **319**.

Atakishiyev, N. M. and Suslov, S. K. (1987a). About one class of special functions, *Comunicaciones Técnicas, Serie naranja: investigaciones* **497**.

Atakishiyev, N. M. and Suslov, S. K. (1987b). Continuous orthogonality property for some classical polynomials of a discrete variable, *Comunicaciones Técnicas, Serie naranja: investigaciones* **498**.

Atkin, A. O. L. and Swinnerton-Dyer, P. (1954). Some properties of partitions, *Proc. London Math. Soc.* (3) **4**, 84-106.

Atkinson, F. V. (1964). *Discrete and Continuous Boundary Problems*, Academic Press, New York.

Bailey, W. N. (1929). An identity involving Heine's basic hypergeometric series, *J. London Math. Soc.* **4**, 254-257.

Bailey, W. N. (1933). A reducible case of the fourth type of Appell's hypergeometric functions of two variables, *Quart. J. Math.* (Oxford) **4**, 305-308.

Bailey, W. N. (1935). *Generalized Hypergeometric Series*, Cambridge University Press, Cambridge, reprinted by Stechert-Hafner, New York, 1964.

Bailey, W. N. (1935a). Some theorems concerning products of hypergeometric series, *Proc. London Math. Soc.* (2) **38**, 377-384.

Bailey, W. N. (1936). Series of hypergeometric type which are infinite in both directions, *Quart. J. Math.* (Oxford) **7**, 105-115.

Bailey, W. N. (1941). A note on certain q-identities, *Quart. J. Math.* (Oxford) **12**, 173-175.

Bailey, W. N. (1947a). Some identities in combinatory analysis, *Proc. London Math. Soc.* (2) **49**, 421-435.

Bailey, W. N. (1947b). Well-poised basic hypergeometric series, *Quart. J. Math.* (Oxford) **18**, 157-166.

Bailey, W. N. (1947c). A transformation of nearly-poised basic hypergeometric series, *J. London Math. Soc.* **22**, 237-240.

Bailey, W. N. (1949). Identities of the Rogers-Ramanujan type, *Proc. London Math. Soc.* (2) **50**, 1-10.

Bailey, W. N. (1950a). On the basic bilateral hypergeometric series ${}_2\psi_2$, *Quart. J. Math.* (Oxford) (2) **1**, 194-198.

Bailey, W. N. (1950b). On the analogue of Dixon's theorem for bilateral basic hypergeometric series, *Quart. J. Math.* (Oxford) (2) **1**, 318-320.

Bailey, W. N. (1951). On the simplification of some identities of the Rogers-Ramanujan type, *Proc. London Math. Soc.* (3) **1**, 217-221.

Note: For a complete list of W. N. Bailey's publications, see his obituary notice Slater [1963].

Baker, M. and Coon, D. (1970). Dual resonance theory with nonlinear trajectories, *Phys. Rev. D* **2**, 2349-2358.

Bannai, E. and Ito, T. (1984). *Algebraic Combinatorics I*, Benjamin/Cummings Pub. Co., London.

Barnes, E. W. (1908). A new development of the theory of the hypergeometric functions, *Proc. London Math. Soc.* (2) **6**, 141-177.

Barnes, E. W. (1910). A transformation of generalized hypergeometric series, *Quart. J. Math.* **41**, 136-140.

Bateman, H. (1932). *Partial Differential Equations of Mathematical Physics*, 1959 edition, Cambridge University Press, Cambridge.

Baxter, R. J. (1980). Hard hexagons: exact solution, *J. Phys.* **A 13**, 161-170.

Baxter, R. J. (1981). Rogers-Ramanujan identities in the hard hexagon model, *J. Stat. Phys.* **26**, 427-452.

Baxter, R. J. (1982). *Exactly Solved Models in Statistical Mechanics*, Academic Press, New York.

Baxter, R. J. (1988). Ramanujan's identities in statistical mechanics, *Ramanujan Revisited* (G. E. Andrews *et al.*, eds.), Academic Press, New York, pp. 69-84.

Baxter, R. J. and Andrews, G. E. (1986). Lattice gas generalization of the hard hexagon model I. Star-triangle relation and local densities, *J. Stat. Phys.* **44**, 249-271.

Baxter, R. J. and Pearce, P. A. (1983). Hard squares with diagonal attractions, *J. Phys. A* **16**, 2239-2255.

Baxter, R. J. and Pearce, P. A. (1984). Deviations form critical density in the generalized hard hexagon model, *J. Phys. A* **17**, 2095-2108.

Beckmann, P. (1973). *Orthogonal Polynomials for Engineers and Physicists*, Golem Press, Boulder, Colo.

Bellman, R. (1961). *A Brief Introduction to Theta Functions*, Holt, Rinehart and Winston, New York.

Bender, E. A. (1971). A generalized q-binomial Vandermonde convolution, *Discrete Math.* **1**, 115-119.

Berman, G. and Fryer, K. D. (1972). *Introduction to Combinatorics*, Academic Press, New York.

Berndt, B. C. (1985). *Ramanujan's Notebooks*, Part I, Springer, Berlin and New York.

Berndt, B. C. (1988). Ramanujan's modular equations, *Ramanujan Revisited* (G.E. Andrews *et al.*, eds.), Academic Press, New York, pp. 313-333.

Berndt, B. C. (1989). *Ramanujan's Notebooks*, Part 2, Springer, Berlin and New York, to appear.

Berndt, B. C. and Joshi, P. T. (1983). *Chapter 9 of Ramanujan's second notebook: infinite series identities, transformations, and evaluations*, Contemporary Mathematics **23**, Amer. Math. Soc., Providence, R. I.

Biedenharn, L. C. and Louck, J. D. (1981a). *Angular Momentum in Quantum Physics: Theory and Application*, Encyclopedia of Mathematics and Its Applications, Vol. 8, Addison-Wesley, Reading, Mass.

Biedenharn, L. C. and Louck, J. D. (1981b). *The Racah-Wigner Algebra in Quantum Theory*, Encyclopedia of Mathematics and Its Applications, Vol. 9, Addison-Wesley, Reading, Mass.

Bohr, H. and Mollerup, J. (1922). *Laerebog i matematisk Analyse*, Vol. III, Copenhagen, pp. 149-164; see Artin [1964, pp. 14-15].

Borwein, J. M. and Borwein, P. B. (1988). More Ramanujan-type series, *Ramanujan Revisited* (G. E. Andrews *et al.*, eds.), Academic Press, New York, pp. 359-366.

de Branges, L. (1968). *Hilbert Spaces of Entire Functions*, Prentice-Hall, Englewood Cliffs, N.J.

de Branges, L. (1978). Quantum Cesáro operators, *Topics in Functional Analysis*, Advances in Mathematics Supplementary Studies, Vol. 3, (I. Gohberg and M. Kac, eds.), Academic Press, New York.

de Branges, L. (1985). A proof of the Bieberbach conjecture, *Acta Math.* **154**, 137-152.

de Branges, L. (1986). Powers of Riemann mapping functions, *The Bieberbach Conjecture: Proc. of the Symposium on the Occasion of the Proof* (A. Baernstein, *et al.*, eds.), Math. Surveys and Monographs **21**, Amer. Math. Soc., Providence, R. I., pp. 51-67.

Bressoud, D. M. (1978). Applications of Andrews' basic Lauricella transformation, *Proc. Amer. Math. Soc.* **72**, 89-94.

Bressoud, D. M. (1980a). Analytic and combinatorial generalizations of the Rogers-Ramanujan identities, *Memoirs Amer. Math. Soc.* **227**.

Bressoud, D. M. (1980b). A simple proof of Mehler's formula for q-Hermite polynomials, *Indiana Univ. Math. J.* **29**, 577-580.

Bressoud, D. M. (1981a). On partitions, orthogonal polynomials and the expansion of certain infinite products, *Proc. London Math. Soc.* (3) **42**, 478-500.

Bressoud, D. M. (1981b). Some identities for terminating q-series, *Math. Proc. Camb. Phil. Soc.* **81**, 211-223.

Bressoud, D. M. (1981c). On the value of Gaussian sums, *J. Numb. Thy.* **13**, 88-94.

Bressoud, D. M. (1981d). Linearization and related formulas for q-ultraspherical polynomials, *SIAM J. Math. Anal.* **12**, 161-168.

Bressoud, D. M. (1983a). An easy proof of the Rogers-Ramanujan identities, *J. Numb. Thy.* **16**, 235-241.

Bressoud, D. M. (1983b). A matrix inverse, *Proc. Amer. Math. Soc.* **88**, 446-448.

Bressoud, D. M. (1986). Hecke modular forms and q-Hermite polynomials, *Ill. J. Math.* **30**, 185-196.

Bressoud, D. M. (1987). Almost poised basic hypergeometric series, *Proc. Indian Acad. Sci. (Math. Sci.)* **96**, 61-66.

Bressoud, D. M. (1988). The Bailey Lattice: an introduction, *Ramanujan Revisited* (G. E. Andrews *et al.*, eds.), Academic Press, New York, pp. 57-67.

Bressoud, D. M. (1989). Unimodality of Gaussian polynomials, to appear.

Bressoud, D. M. and Goulden, I. P. (1985). Constant term identities extending the q-Dyson theorem, *Trans. Amer. Math. Soc.* **291**, 203-228.

Bressoud, D. M. and Goulden, I. P. (1987). The generalized plasma in one dimension: Evaluation of a partition function, *Comm. Math. Phys.* **110**, 287-291.

Bromwich, T. J. I'A. (1959). *An Introduction to the Theory of Infinite Series*, 2nd edition, Macmillan, New York.

Burchnall, J. L. and Chaundy, T. W. (1940). Expansions of Appell's double hypergeometric functions, *Quart. J. Math.* (Oxford) **11**, 249-270.

Burchnall, J. L. and Chaundy, T. W. (1941). Expansions of Appell's double hypergeometric functions (II), *Quart. J. Math.* (Oxford) **12**, 112-128.

Burchnall, J. L. and Chaundy, T. W. (1949). The hypergeometric identities of Cayley, Orr and Bailey, *Proc. London Math. Soc.* (2) **50**, 56-74.

Carlitz, L. (1955). Some polynomials related to theta functions, *Annali di Matematica Pura ed Applicata* (4) **41**, 359-373.

Carlitz, L. (1957a). Some polynomials related to theta functions, *Duke Math. J.* **24**, 521-527.

Carlitz, L. (1957b). A q-analogue of a formula of Toscano, *Boll. Unione Mat. Ital.* **12**, 414-417.

Carlitz, L. (1958). Note on orthogonal polynomials related to theta functions, *Publicationes Mathematicae* **5**, 222-228.

Carlitz, L. (1960). Some orthogonal polynomials related to elliptic functions, *Duke Math. J.* **27**, 443-460.

Carlitz, L. (1963a). A basic analog of the multinomial theorem, *Scripta Mathematica* **26**, 317-321.

Carlitz, L. (1963b). A q-identity, *Monatsh. für Math.* **67**, 305-310.

Carlitz, L. (1969a). Some formulas of F. H. Jackson, *Monatsh. für Math.* **73**, 193-198.

Carlitz, L. (1969b). A q-identity, *Boll. Unione Math. Ital.* (4) **1**, 100-101.

Carlitz, L. (1972). Generating functions for certain q-orthogonal polynomials, Seminario Matematico de Barcelona, *Collectanea Mathematica* **23**, 91-104.

Carlitz, L. (1973). Some inverse relations, *Duke Math. J.* **40**, 893-901.

Carlitz, L. (1974). A q-identity, *Fibonacci Quarterly* **12**, 369-372.

Carlitz, L. and Hodges, J. (1955). Representations by Hermitian forms in a finite field, *Duke Math. J.* **22**, 393-405.

Carlitz, L. and Subbarao, M. V. (1972). A simple proof of the quintuple product identity, *Proc. Amer. Math. Soc.* **32**, 42-44.

Carlson, B. C. (1977). *Special Functions of Applied Mathematics*, Academic Press, New York.

Carmichael, R. D. (1912). The general theory of linear q-difference equations, *Amer. J. Math.* **34**, 147-168.

Cauchy, A.-L. (1825). Sur les intégrales définies prises entre des limites imaginaires, *Bulletin de Ferussoc*, T. III, 214-221, *Oeuvres de Cauchy*, 2e série, T. II, Gauthier-Villars, Paris, 1958, 57-65.

Cauchy, A.-L. (1843). Mémoire sur les fonctions dont plusieurs valeurs sont liées entre elles par une équation linéaire, et sur diverses transformations de produits composés d'un nombre indéfini de facteurs, *C. R. Acad. Sci. Paris,* T. XVII, p. 523, *Oeuvres de Cauchy*, 1re série, T. VIII, Gauthier-Villars, Paris, 1893, pp. 42-50.

Cayley, A. (1858). On a theorem relating to hypergeometric series, *Phil. Mag.* (4) **16**, 356-357; reprinted in *Collected Papers* **3**, pp. 268-269.

Charris, J. and Ismail, M. E. H. (1986). On sieved orthogonal polynomials II: random walk polynomials, *Canad. J. Math.* **38**, 397-415.

Charris, J. and Ismail, M. E. H. (1987). On sieved orthogonal polynomials V: sieved Pollaczek polynomials, *SIAM J. Math. Anal.* **18**, 1177-1218.

Chaundy, T. W. (1962). Frank Hilton Jackson, *J. London Math. Soc.* **37**, 126-128.

Cheema, M. S. (1964). Vector partitions and combinatorial identities, *Math. Comp.* **18**, 414-420.

Chihara, L. (1987). On the zeros of the Askey-Wilson polynomials, with applications to coding theory, *SIAM J. Math. Anal.* **18**, 191-207.

Chihara, L. and Chihara, T. S. (1987). A class of nonsymmetric orthogonal polynomials, *J. Math. Anal. Appl.* **126**, 275-291.

Chihara, L. and Stanton, D. (1986). Association schemes and quadratic transformations for orthogonal polynomials, *Graphs and Combinatorics* **2**, 101-112.

Chihara, L. and Stanton, D. (1987). Zeros of generalized Krawtchouk polynomials, *IMA Preprint Series* **361**, University of Minnesota, Minn.

Chihara, T. S. (1968a). On indeterminate Hamburger moment problems, *Pacific J. Math.* **27**, 475-484.

Chihara, T. S. (1968b). Orthogonal polynomials with Brenke type generating functions, *Duke Math. J.* **35**, 505-518.

Chihara, T. S. (1971). Orthogonality relations for a class of Brenke polynomials, *Duke Math. J.* **38**, 599-603.

Chihara, T. S. (1978). *An Introduction to Orthogonal Polynomials*, Gordon and Breach, New York.

Chihara, T. S. (1979). On generalized Stieltjes-Wigert and related orthogonal polynomials, *J. Comput. Appl. Math.* **5**, 291-297.

Chihara, T. S. (1982). Indeterminate symmetric moment problems, *J. Math. Anal. Appl.* **85**, 331-346.

Chihara, T. S. (1985). Orthogonal polynomials and measures with end point masses, *Rocky Mtn. J. Math.* **15**, 705-719.

Chu Shih-Chieh (1303). *Ssu Yuan Yü Chien* (*Precious Mirror of the Four Elements*) (in Chinese); see Askey [1975, p. 59], Needham [1959, p. 138], and Takács [1973].

Chudnovsky, D. V. and Chudnovsky, G. V. (1988). Approximations and complex multiplication according to Ramanujan, *Ramanujan Revisited* (G. E. Andrews *et al.*, eds.), Academic Press, New York, pp. 375-472.

Cigler, J. (1979). Operatormethoden für q-Identitäten, *Monatsh. für Math.* **88**, 87-105.

Cigler, J. (1980). Operatormethoden für q-Identitäten III: Umbrale Inversion und die Lagrangesche Formel, *Arch. Math.* **35**, 533-543.

Cigler, J. (1981). Operatormethoden für q-Identitäten II: q-Laguerre-Polynome, *Monatsh. Math.* **91**, 105-117.

Clausen T. (1828). Ueber die Fälle, wenn die Reihe von der Form ... ein Quadrat von der Form ... hat, *J. reine angew. Math.* **3**, 89-91.

Cohen, H. (1988). q-Identities for Maass waveforms, *Invent. Math.* **91**, 409-422.

Comtet, L. (1974). *Advanced Combinatorics*, Reidel, Boston, Mass.

Daum, J. A. (1942). The basic analog of Kummer's theorem, *Bull. Amer. Math. Soc.* **48**, 711-713.

Dehesa, J. S. (1979). On a general system of orthogonal q-polynomials, *J. Computational Appl. Math.* **5**, 37-45.

Delsarte, P. (1976a). Properties and applications of the recurrence $F(i+1, k+1, n+1) = q^{k+1} F(i,k+1, n) - q^k F(i,k,n)$, *SIAM J. Appl. Math.* **31**, 262-270.

Delsarte, P. (1976b). Association schemes and t-designs in regular semilattices, *J. Comb. Thy.* **A 20**, 230-243.

Delsarte, P. (1978). Bilinear forms over a finite field, with applications to coding theory, *J. Comb. Thy.* **A 25**, 226-241.

Delsarte, P. and Goethals, J. M. (1975). Alternating bilinear forms over $GF(q)$, *J. Comb. Thy.* **A 19**, 26-50.

Désarménien, J. (1982). Les q-analogues des polyônmes d'Hermite, *Séminaire lotharingien de combinatoire* (V. Strehl, ed.), pp. 39-56.

Désarménien, J. and Foata, D. (1985). Fonctions symétriques et séries hypergéométriques basiques multivariées, *Bull. Soc. Math. France* **113**, 3-22.

Désarménien, J. and Foata, D. (1986). Fonctions symétriques et hypergéométriques basiques multivariées, II, in *Combinatoire énumérative*, Lecture Notes in Math. **1234**, Springer, Berlin and New York.

Désarménien, J. and Foata, D. (1988). Statistiques d'ordre sur les permutations colorées, *Publ. I.R.M.A. Strasbourg* **372**/S-20, Actes 20e Séminaire Lotharingien, 5-22.

Dickson, L. E. (1920). *History of the Theory of Numbers*, Vol. 2. Carnegie Institute of Washington, Publ. 256; reprinted by Chelsea, New York.

Dixon, A. C. (1903). Summation of a certain series, *Proc. London Math. Soc.* (1) **35**, 285-289.

Dobbie, J. M. (1962). A simple proof of the Rogers-Ramanujan identities, *Quart. J. Math.* (Oxford) (2) **13**, 31-34.

Dougall, J. (1907). On Vandermonde's theorem and some more general expansions, *Proc. Edin. Math. Soc.* **25**, 114-132.

Dowling, T. A. (1973). A q-analog of the partition lattice, *A Survey of Combinatorial Theory* (J. N. Srivastava *et al.* eds.), North-Holland, Amsterdam, pp. 101-115.

Dunkl, C. F. (1977). An addition theorem for some q-Hahn polynomials, *Monatsh. für Math.* **85**, 5-37.

Dunkl, C. F. (1979a). Orthogonal functions on some permutation groups, *Proc. Symp. Pure Math.* **34**, 129-147.

Dunkl, C. F. (1979b). Discrete quadrature and bounds on t-designs, *Michigan Math. J.* **26**, 102.

Dunkl, C. F. (1980). Orthogonal polynomials in two variables of q-Hahn and q-Jacobi type, *SIAM J. Alg. Disc. Meth.* **1**, 137-151.

Dunkl, C. F. (1981). The absorption distribution and the q-binomial theorem, *Commun. Statist.-Theor. Meth.* **A10** (19), 1915-1920.

Duren, P. L. (1983). *Univalent Functions*, Springer, Berlin and New York.

Dyson, F. J. (1962). Statistical theory of the energy levels of complex systems. I, *J. Math. Phys.* **3**, 140-156.

Dyson, F. J. (1988). A walk through Ramanujan's garden, *Ramanujan Revisited* (G. E. Andrews *et al.*, eds.), Academic Press, New York, pp. 7-28.

Edwards, D. (1923). An expansion in factorials similar to Vandermonde's theorem, and applications, *Messenger of Math.* **52**, 129-136.

Erdélyi, A., ed. (1953). *Higher Transcendental Functions*, Vols. I & II, McGraw-Hill, New York.

Euler, L. (1748). *Introductio in Analysis Infinitorum*, M-M Bousquet, Lausanne.

Evans, R., Ismail, M. E. H. and Stanton, D. (1982). Coefficients in expansions of certain rational functions, *Canad. J. Math.* **34**, 1011-1024.

Ewell J. A. (1981). An easy proof of the triple-product identity, *Amer. Math. Monthly* **88**, 270-272.

Favard, J. (1935). Sur les polynômes de Tchebicheff, *C. R. Acad. Sci.* (Paris) **200**, 2052-2053.

Feinsilver, P. (1982). Commutators, anti-commutators and Eulerian calculus, *Rocky Mtn. J. Math.* **12**, 171-183.

Fields, J. L. and Ismail, M. E. H. (1975). Polynomial expansions, *Math. Comp.* **29**, 894-902.

Fields, J. L. and Wimp, J. (1961). Expansions of hypergeometric functions in hypergeometric functions, *Math. Comp.* **15**, 390-395.

Fine, N. J. (1948). Some new results on partitions, *Proc. Nat. Acad. Sci. USA* **34**, 616-618.

Fine, N. J. (1988). *Basic Hypergeometric Series and Applications*, Mathematical Surveys and Monographs, Vol. 27, Amer. Math. Soc., Providence, R. I.

Foata, D. (1981). Further divisibility properties of the q-tangent numbers, *Proc. Amer. Math. Soc.* **81**, 143- 148.

Freud, G. (1971). *Orthogonal Polynomials*, Pergamon Press, New York.

Fürlinger, J. and Hofbauer, J. (1985). q-Catalan numbers, *J. Comb. Thy.* A **40**, 248-264.

Gangolli, R. (1967). Positive definite kernels on homogeneous spaces and certain stochastic processes related to Lévy's Brownian motion of several parameters, *Ann. Inst. H. Poincaré*, Sect. B, Vol. III, 121-226.

Garsia, A. M. (1981). A q-analogue of the Lagrange inversion formula, *Houston J. Math.* **7**, 205-237.

Garsia, A.M. and Milne, S.C. (1981). A Rogers-Ramanujan bijection, *J. Comb. Thy.* A, **31**, 289-339.

Garsia, A. and Remmel, J. (1986). A novel form of q-Lagrange inversion, *Houston J. Math.* **12**, 503-523.

Garvan, F. G. (1988). Combinatorial interpretations of Ramanujan's partition congruences, *Ramanujan Revisited* (G. E. Andrews *et al.*, eds.), Academic Press, New York, pp. 29-45.

Garvan, F.G. (1989). A proof of the Macdonald-Morris root system for F_4, to appear.

Garvan, F.G. and Gonnet, G.H. (1989). A proof of the two parameter q-case of the Macdonald-Morris constant term root system conjecture for F_4 via Zeilberger's method, to appear.

Garvan, F. G. and Stanton, D. (1989). Sieved partition functions and q-binomial coefficients, to appear.

Gasper, G. (1970). Linearization of the product of Jacobi polynomials. II, *Canad. J. Math.* **22**, 582-593.

Gasper, G. (1971). Positivity and the convolution structure for Jacobi series, *Annals of Math.* **93**, 112-118.

Gasper, G. (1972). Banach algebras for Jacobi series and positivity of a kernel, *Annals of Math.* **95**, 261-280.

Gasper, G. (1973). Nonnegativity of a discrete Poisson kernel for the Hahn polynomials, *J. Math. Anal. Appl.* **42**, 438-451.

Gasper, G. (1974). Projection formulas for orthogonal polynomials of a discrete variable, *J. Math. Anal. Appl.* **45**, 176-198.

Gasper, G. (1975a). Positivity and special functions, *Theory and Applications of Special Functions* (R. Askey, ed.), Academic Press, New York, pp. 375-433.

Gasper, G. (1975b). Positive integrals of Bessel functions, *SIAM J. Math. Anal.* **6**, 868-881.

Gasper, G. (1976). Solution to problem 74-21* (Two-dimensional discrete probability distributions, by P. Beckmann), *SIAM Review* **18**, 126-129.

Gasper, G. (1977). Positive sums of the classical orthogonal polynomials, *SIAM J. Math. Anal.* **8**, 423-447.

Gasper, G. (1981a). Summation formulas for basic hypergeometric series, *SIAM J. Math. Anal.* **12**, 196-200.

Gasper, G. (1981b). Orthogonality of certain functions with respect to complex valued weights, *Canad. J. Math.* **33**, 1261-1270.

Gasper, G. (1983). A convolution structure and positivity of a generalized translation operator for the continuous q-Jacobi polynomials, *Conference on Harmonic Analysis in Honor of Antoni Zygmund*, Wadsworth International Group, Belmont, Calif., pp. 44-59.

Gasper, G. (1985). Rogers' linerization formula for the continuous q-ultra-spherical polynomials and quadratic transformation formulas, *SIAM J. Math. Anal.* **16**, 1061-1071.

Gasper, G. (1986). A short proof of an inequality used by de Branges in his proof of the Bieberbach, Robertson and Milin conjectures, *Complex Variables* **7**, 45-50.

Gasper, G. (1987). Solution to problem #6497 (q-Analogues of a gamma function identity, by R. Askey), *Amer. Math. Monthly* **94**, 199-201.

Gasper, G. (1989a). Summation, transformation, and expansion formulas for bibasic series, *Trans. Amer. Math. Soc.*,**312**, 257-277.

Gasper, G. (1989b). q-Extensions of Clausen's formula and of the inequalities used by de Branges in his proof of the Bieberbach, Robertson, and Millin conjectures, *SIAM J. Math. Anal.*, **20**, 1019-1034.

Gasper, G. (1989c). q-Extensions of Barnes', Cauchy's, and Euler's beta integrals, *Topics in Mathematical Analysis*, T.M. Rassias, ed., World Scientific Pub. Co., London, Singapore and Teaneck, N.J., 294-314.

Gasper, G. (1989d). Bibasic summation, transformation and expansion formulas, q-analogues of Clausen's formula, and nonnegative basic hypergeometric series, *Workshop on q-Series and Partitions* (D. Stanton, ed.), IMA Volumes in Mathematics and its Applications, Springer, Berlin and New York, to appear.

Gasper, G. (1989e). Using symbolic computer algebraic systems to derive formulas involving orthogonal polynomials and other special functions, *Proceedings of the NATO Advanced Study Institute on "Orthogonal Polynomials and their Applications"*, to appear.

Gasper, G. and Rahman, M. (1983a). Positivity of the Poisson kernel for the continuous q-ultraspherical polynomials, *SIAM J. Math. Anal.* **14**, 409-420.

Gasper, G. and Rahman, M. (1983b). Nonnegative kernels in product formulas for q-Racah polynomials, *J. Math. Anal. Appl.* **95**, 304-318.

Gasper, G. and Rahman, M. (1984). Product formulas of Watson, Bailey and Bateman types and positivity of the Poisson kernel for q-Racah polynomials, *SIAM J. Math. Anal.* **15**, 768-789.

Gasper, G. and Rahman, M. (1986). Positivity of the Poisson kernel for the continuous q-Jacobi polynomials and some quadratic transformation formulas for basic hypergeometric series, *SIAM J. Math. Anal.* **17**, 970-999.

Gasper, G. and Rahman, M. (1989a). An indefinite bibasic summation formula and some quadratic, cubic, and quartic summation and transformation formulas, *Canad. J. Math.* to appear.

Gasper, G. and Rahman, M. (1989b). A nonterminating q-Clausen formula and some related product formulas, *SIAM J. Math. Anal.*, to appear.

Gasper, G. and Trebels, W. (1977). Multiplier criteria of Marcinkiewicz type for Jacobi expansions, *Trans. Amer. Math. Soc.* **231**, 117-132.

Gasper, G. and Trebels, W. (1979). A characterization of localized Bessel potential spaces and applications to Jacobi and Hankel multipliers, *Studia Math.* **65**, 243-278.

Gauss, C. F. (1813). Disquisitiones generales circa seriem infinitam ..., *Comm. soc. reg. sci. Gött. rec.*, Vol. II; reprinted in *Werke* **3** (1876), pp. 123-162.

Gegenbauer, L. (1874).Über einige bestimmte Integrale, *Sitz. Math. Natur. Kl. Akad. Wiss. Wien* (IIa) **70**, 433-443.

Gegenbauer, L. (1893). Das Additionstheorem der Functionen $C_n^\nu(x)$, *Sitz. Math. Natur. Kl. Akad. Wiss. Wien* (IIa) **102**, 942-950.

Geronimo, J. S. (1989). Scattering theory, orthogonal polynomials and q-series, *SIAM J. Math. Anal.*, to appear.

Gessel, I. (1980). A noncommutative generalization and q-analog of the Lagrange inversion formula, *Trans. Amer. Math. Soc.* **257**, 455-482.

Gessel, I. and Stanton, D. (1982). Strange evaluations of hypergeometric series, *SIAM J. Math. Anal.* **13**, 295-308.

Gessel, I. and Stanton, D. (1983). Applications of q-Lagrange inversion to basic hypergeometric series, *Trans. Amer. Math. Soc.* **277**, 173-201.

Gessel, I. and Stanton, D. (1986). Another family of q-Lagrange inversion formulas, *Rocky Mtn. J. Math.* **16**, 373-384.

Goldman, J. and Rota, G.-C. (1970). On the foundations of combinatorial theory IV: Finite vector spaces and Eulerian generating functions, *Studies in Appl. Math.* **49**, 239-258.

Gordon, B. (1961). Some identities in combinatorial analysis, *Quart. J. Math.* (Oxford) (2) **12**, 285-290.

Gosper, R. Wm. (1988a). Some identities, for your amusement, *Ramanujan Revisited* (G. E. Andrews *et al.*, eds.), Academic Press, New York, pp. 607-609.

Gosper, R. Wm. (1988b). March 27 and July 31 letters to R. Askey.

Greiner, P. C. (1980). Spherical harmonics on the Heisenberg group, *Canad. Math. Bull.* **23**, 383-396.

Grosswald, E. (1985). *Representations of Integers as Sums of Squares*, Springer, Berlin and New York.

Gustafson, R. A. (1987a). A Whipple's transformation for hypergeometric series in $U(n)$ and multivariable hypergeometric orthogonal polynomials, *SIAM J. Math. Anal.* **18**, 495-530.

Gustafson, R. A. (1987b). Multilateral summation theorems for ordinary and basic hypergeometric series in $U(n)$, *SIAM J. Math. Anal.*, **18**, 1576-1596.

Gustafson, R. A. (1989a). A summation theorem for hypergeometric series very-well-poised on G_2, to appear.

Gustafson, R. A. (1989b). The Macdonald identities for affine root systems of classical type and hypergeometric series very-well-poised on semisimple Lie algebras, *Proc. Ramanujan Symposium on Classical Analysis*, Pune, India, Dec. 26-28, 1987, to appear.

Gustafson, R. A. and Milne, S. C. (1986). A q-analogue of transposition symmetry for invariant G-functions, *J. Math. Anal. Appl.* **114**, 210-240.

Habsieger, L. (1989). Une q-intégrale de Selberg-Askey, *SIAM J. Math. Anal.*, to appear.

Hahn, W. (1949a). Über Orthogonalpolynome, die q-Differenzengleichungen genügen, *Math. Nachr.* **2**, 4-34.

Hahn, W. (1949b). Über Polynome, die gleichzeitig zwei verschiedenen Orthogonalsystemen angehören, *Math. Nachr.* **2**, 263-278.

Hahn, W. (1949c). Beiträge zur Theorie der Heineschen Reihen, die 24 Integrale der hypergeometrischen q-Differenzengleichung, das q-Analogon der Laplace-Transformation, *Math. Nachr.* **2**, 340-379.

Hahn, W. (1950). Über die höheren Heineschen Reihen und eine einheitliche Theorie der sogenannten speziellen Funktionen, *Math. Nachr.* **3**, 257-294.

Hahn, W. (1952). Über uneigentliche Lösungen linearer geometrischer Differenzengleichungen, *Math. Annalen* **125**, 67-81.

Hahn, W. (1953). Die mechanische Deutung einer geometrischen Differenzengleichung, *Zeitschr. angew. Math. Mech.* **33**, 270-272.

Hall, N.A. (1936). An algebraic identity, *J. London Math. Soc.* **11**, 276.

Handa, B. R. and Mohanty, S. G. (1980). On q-binomial coefficients and some statistical applications, *SIAM J. Math. Anal.* **11**, 1027-1035.

Hardy, G. H. (1937). The Indian mathematician Ramanujan, *Amer. Math. Monthly* **44**, 137-155; reprinted in *Collected Papers* **7**, pp. 612-630.

Hardy, G. H. (1940). *Ramanujan*, Cambridge University Press, Cambridge; reprinted by Chelsea, New York, 1978.

Hardy, G. H. and Ramanujan, S. (1918). Asymptotic formulae in combinatory analysis, *Proc. London Math. Soc.* (2) **17**, 75-115; in the *Collected Papers of G. H. Hardy I*, Oxford University Press, Oxford, pp. 306-339.

Hardy, G. H. and Wright, E. M. (1979). *An Introduction to the Theory of Numbers*, 5th edition, Oxford University Press, Oxford.

Heine, E. (1846). Über die Reihe ..., *J. reine angew. Math.* **32**, 210-212.

Heine, E. (1847). Untersuchungen über die Reihe ..., *J. reine angew. Math.* **34**, 285-328.

Heine, E. (1878). *Handbuch der Kugelfunctionen, Theorie und Anwendungen*, Vol. 1, Reimer, Berlin.

Henrici, P. (1974). *Applied and Computational Complex Analysis*, Vol. I, John Wiley & Sons, New York.

Hilbert, D. (1909). Beweis für die Darstellbarkeit der ganzen Zahlen durch eine feste Anzahl n-ter Potenzen (Waringsches Problem), *Math. Annalen* **67**, 281-300.

Hirschhorn, M. D. (1985). A simple proof of Jacobi's two-square theorem, *Amer. Math. Monthly* **92**, 579-580.

Hirschhorn, M. D. (1987). A simple proof of Jacobi's four-square theorem, *Proc. Amer. Math. Soc.* **101**, 436-438.

Hirschhorn, M. D. (1988). A generalization of the quintuple product identity, *J. Austral. Math. Soc.* **A 44**, 42-45.

Hofbauer, J. (1982). Lagrange Inversion, *Séminaire lotharingien de combinatoire* (V. Strehl, ed.), pp. 1-38.

Hofbauer, J. (1984). A q-analogue of the Lagrange expansion, *Arch. Math.* **42**, 536-544.

Hua, L. K. (1982). *Introduction to Number Theory*, Springer, Berlin and New York.

Ihrig, E. and Ismail, M. E. H. (1981). A q-umbral calculus, *J. Math. Anal. Appl.* **84**, 178-207.

Ismail, M. E. H. (1977). A simple proof of Ramanujan's ${}_1\psi_1$ sum, *Proc. Amer. Math. Soc.* **63**, 185-186.

Ismail, M. E. H. (1981). The basic Bessel functions and polynomials, *SIAM J. Math. Anal.* **12**, 454-468.

Ismail, M. E. H. (1982). The zeros of basic Bessel functions, the functions $J_{\nu+ax}(x)$, and associated orthogonal polynomials, *J. Math. Anal. Appl.* **86**, 1-19.

Ismail, M. E. H. (1985a). On sieved orthogonal polynomials. I: symmetric Pollaczek analogues, *SIAM J. Math. Anal.* **16**, 1093-1113.

Ismail, M. E. H. (1985b). A queueing model and a set of orthogonal polynomials, *J. Math. Anal. Appl.* **108**, 575-594.

Ismail, M. E. H. (1986a). On sieved orthogonal polynomials. III: Orthogonality on several intervals, *Trans. Amer. Math. Soc.* **294**, 89-111.

Ismail, M. E. H. (1986b). On sieved orthogonal polynomials. IV: Generating functions, *J. Approx. Thy.* **46**, 284-296.

Ismail, M. E. H. (1986c). Asymptotics of the Askey-Wilson and q-Jacobi polynomials, *SIAM J. Math. Anal.* **17**, 1475-1482.

Ismail, M. E. H., Lorch, L. and Muldoon, M. E. (1986). Completely monotonic functions associated with the gamma function and its q-analogues, *J. Math. Anal. Appl.* **116**, 1-9.

Ismail, M. E. H., Merkes, E. and Styer, D. (1989). A generalization of starlike functions, to appear.

Ismail, M. E. H. and Muldoon, M. E. (1988). On the variation with respect to a parameter of zeros of Bessel and q-Bessel functions, *J. Math. Anal. Appl.* **135**, 187-207.

Ismail, M. E. H., Mulla, F. S. (1987). On the generalized Chebyshev polynomials, *SIAM J. Math. Anal.* **18**, 243-258.

Ismail, M. E. H. and Stanton, D. (1988). On the Askey-Wilson and Rogers polynomials, *Canad. J. Math.* **40**, 1025-1045.

Ismail, M. E. H., Stanton, D. and Viennot, G. (1987). The combinatorics of q-Hermite polynomials and the Askey-Wilson integral, *European J. Combinatorics* **8**, 379-392.

Ismail, M. E. H. and Wilson, J. A. (1982). Asymptotic and generating relations for the q-Jacobi and ${}_4\phi_3$ polynomials, *J. Approx. Thy.* **36**, 43-54.

Jackson, F. H. (1904a). A basic-sine and cosine with symbolical solutions of certain differential equations, *Proc. Edin. Math. Soc.* **22**, 28-38.

Jackson, F. H. (1904b). Note of a theorem of Lommel, *Proc. Edin. Math. Soc.* **22**, 80-85.

Jackson, F. H. (1904c). Theorems relating to a generalization of the Bessel-function, *Trans. Roy. Soc. Edin.* **41**, 105-118.

Jackson, F. H. (1904d). On generalized functions of Legendre and Bessel, *Trans. Roy. Soc. Edin.* **41**, 1-28.

Jackson, F. H. (1904e). A generalization of the functions $\Gamma(n)$ and x^n, *Proc. Roy. Soc. London* **74**, 64-72.

Jackson, F. H. (1905a). The application of basic numbers to Bessel's and Legendre's functions, *Proc. London Math. Soc.* (2) **2**, 192-220.

Jackson, F. H. (1905b). The application of basic numbers to Bessel's and Legendre's functions (Second paper), *Proc. London Math. Soc.* (2) **3**, 1-23.

Jackson, F. H. (1905c). Some properties of a generalized hypergeometric function, *Amer. J. Math.* **27**, 1-6.

Jackson, F. H. (1905d). The basic gamma-function and the elliptic functions, *Proc. Roy. Soc. London A* **76**, 127-144.

Jackson, F. H. (1905e). Theorems relating to a generalization of Bessel's function. II., *Trans. Roy. Soc. Edin.* **41**, 399-408.

Jackson, F. H. (1908). On q-functions and a certain difference operator, *Trans. Roy. Soc. Edin.* **46**, 253-281.

Jackson, F. H. (1909a). Generalization of the differential operative symbol with an extended form of Boole's equation ..., *Messenger of Math.* **38**, 57-61.

Jackson, F. H. (1909b). A q-form of Taylor's theorem, *Messenger of Math.* **38**, 62-64.

Jackson, F. H. (1909c). The q-series corresponding to Taylor's series, *Messenger of Math.* **39**, 26-28.

Jackson, F. H. (1910a). Transformations of q-series, *Messenger of Math.* **39**, 145-153.

Jackson, F. H. (1910b). A q-generalization of Abel's series, *Rendiconti Palermo* **29**, 340-346.

Jackson, F. H. (1910c). On q-definite integrals, *Quart. J. Pure and Appl. Math.* **41**, 193-203.

Jackson, F. H. (1910d). q-Difference equations, *Amer. J. Math.* **32**, 305-314.

Jackson, F. H. (1910e). Borel's integral and q-series, *Proc. Roy. Soc. Edin.* **30**, 378-385.

Jackson, F. H. (1911). The products of q-hypergeometric functions, *Messenger of Math.* **40**, 92-100.

Jackson, F. H. (1917). The q-integral analogous to Borel's integral, *Messenger of Math.* **47**, 57-64.

Jackson, F. H (1921). Summation of q-hypergeometric series, *Messenger of Math.* **50**, 101-112.

Jackson, F. H (1927). A new transformation of Heinean series, *Quart. J. Pure and Appl. Math.* **50**, 377-384.

Jackson, F. H (1928). Examples of a generalization of Euler's transformation for power series, *Messenger of Math.* **57**, 169-187.

Jackson, F. H (1940). The q^θ equations whose solutions are products of solutions of q^θ equations of lower order, *Quart. J. Math.* (Oxford) **11**, 1-17.

Jackson, F. H. (1941). Certain q-identities, *Quart. J. Math.* (Oxford) **12**, 167-172.

Jackson, F. H. (1942). On basic double hypergeometric functions, *Quart. J. Math.* (Oxford) **13**, 69-82.

Jackson, F. H. (1944). Basic double hypergeometric functions (II), *Quart. J. Math.* (Oxford) **15**, 49-61.

Jackson, F. H. (1951). Basic integration, *Quart. J. Math.* (Oxford) (2) **2**, 1-16.

Note: For additional publications of F. H. Jackson, see his obituary notice Chaundy [1962].

Jackson, M. (1949). On some formulae in partition theory, and bilateral basic hypergeometric series, *J. London Math. Soc.* **24**, 233-237.

Jackson, M. (1950a). On well-poised bilateral hypergeometric series of the type ${}_8\psi_8$, *Quart. J. Math.* (Oxford) (2) **1**, 63-68.

Jackson, M. (1950b). On Lerch's transcendant and the basic bilateral hypergeometric series ${}_2\psi_2$, *J. London Math. Soc.* **25**, 189-196.

Jackson, M. (1954). Transformations of series of the type ${}_3\psi_3$, *Pac. J. Math.* **4**, 557-562.

Jacobi, C. G. J. (1829). Fundamenta Nova Theoriae Functionum Ellipticarum, Regiomonti. Sumptibus fratrum Bornträger; reprinted in *Gesammelte Werke* **1** (1881), 49-239, Reimer, Berlin; reprinted by Chelsea, New York, 1969.

Jacobi, C. G. J. (1846). Über einige der Binomialreihe analoge Reihen, *J. reine angew. Math.* **32**, 197-204; reprinted in *Gesammelte Werke* **6** (1881), 163-173, Reimer, Berlin.

Jain, V. K. (1980a). Some expansions involving basic hypergeometric functions of two variables, *Pac. J. Math.* **91**, 349-361.

Jain, V. K. (1980b). Some transformations of basic hypergeometric series and their applications, *Proc. Amer. Math. Soc.* **78**, 375-384.

Jain, V. K. (1980c). Summations of basic hypergeometric series and Rogers-Ramanujan identities, *Houston J. Math.* **6**, 511-522.

Jain, V. K. (1981). Some transformations of basic hypergeometric functions. Part II, *SIAM J. Math. Anal.* **12**, 957-961.

Jain, V. K. (1982). Certain transformations of basic hypergeometric series and their applications, *Pac. J. Math.* **101**, 333-349.

Jain, V. K. and Srivastava, H. M. (1986). q-Series identities and reducibility of basic double hypergeometric functions, *Canad. J. Math.* **38**, 215-231.

Jain, V. K. and Verma, A. (1980). Transformations between basic hypergeometric series on different bases and identities of Rogers-Ramanujan type, *J. Math. Anal. Appl.* **76**, 230-269.

Jain, V. K. and Verma, A. (1981). Some transformations of basic hypergeometric functions, Part I, *SIAM J. Math. Anal.* **12**, 943-956.

Jain, V. K. and Verma, A. (1982). Transformations of non-terminating basic hypergeometric series, their contour integrals and applications to Rogers-Ramanujan identities, *J. Math. Anal. Appl.* **87**, 9-44.

Jain, V. K. and Verma, A. (1985). Some summation formulae for nonterminating basic hypergeometric series, *SIAM J. Math. Anal.* **16**, 647-655.

Jain, V. K. and Verma, A. (1986). Basic analogues of transformations of nearly-poised basic hypergeometric series, *Number Theory*, Lecture Notes on Math. **1122**, Springer, Berlin and New York, pp. 206-217.

Jain, V. K. and Verma, A. (1987). On transformations of nearly-poised basic hypergeometric series and their applications, *Indian J. Pure Appl. Math.* **18** (1), 55-64.

Jimbo, M. (1985). A q-difference analogue of $U(g)$ and the Yang-Baxter equation, *Letters in Math. Phys.* **10**, 63-69.

Jimbo, M. (1986). A q-analogue of $U(gl(N+1))$, Hecke algebra, and the Yang-Baxter equation, *Letters in Math. Phys.* **11**, 247-252.

Joichi, J. T. and Stanton, D. (1987). Bijective proofs of basic hypergeometric series identities, *Pacific J. Math.* **127**, 103-120.

Joichi, J. T. and Stanton, D. (1989). An involution for Jacobi's identity, *Discrete Math.*, to appear.

Joshi, C. M. and Verma, A. (1979). Some remarks on summation of basic hypergeometric series, *Houston J. Math.* **5**, 277-294.

Kac, V. G. (1978). Infinite-dimensional algebras, Dedekind's η-function, classical Möbius function and the very strange formula, *Advances in Math.* **30**, 85-136.

Kac, V. G. (1985). *Infinite Dimensional Lie Algebras*, 2nd edition, Cambridge University Press, Cambridge.

Kadell, K. W. J. (1985a). Weighted inversion numbers, restricted growth functions, and standard Young tableaux, *J. Comb. Thy. A* **40**, 22-44.

Kadell, K. W. J. (1985b). A proof of Andrews' q-Dyson conjecture for $n=4$, *Trans. Amer. Math. Soc.* **290**, 127-144.

Kadell, K. W. J. (1987a). Path functions and generalized basic hypergeometric functions, *Memoirs Amer. Math. Soc.* **360**.

Kadell, K. W. J. (1987b). A probabilistic proof of Ramanujan's ${}_1\psi_1$ sum, *SIAM J. Math. Anal.* **18**, 1539-1548.

Kadell, K. W. J. (1988a). A proof of some q-analogues of Selberg's integral for $k=1$, *SIAM J. Math. Anal.* **19**, 944-968.

Kadell, K. W. J. (1988b). A proof of Askey's conjectured q-analogue of Selberg's integral and a conjecture of Morris, *SIAM J. Math. Anal.* **19**, 969-986.

Kadell, K. W. J. (1988c). The q-Selberg polynomials for $n=2$, *Trans. Amer. Math. Soc.*, **310**, 535-553.

Kadell, K. W. J. (1989a). Andrews' q-Dyson conjecture II: Symmetry, *Pacific J. Math.*, to appear.

Kadell, K. W. J. (1989b). The Selberg-Jack symmetric functions, *Advances in Math.*, to appear.

Kadell, K. W. J. (1989c). A proof of the q-Macdonald-Morris conjecture for BC_n, to appear.

Kairies, H.-H and Muldoon, M. E. (1982). Some characterizations of q-factorial functions, *Aequationes Math.* **25**, 67-76.

Kalnins, E. G. and Miller, W. (1989a). Symmetry techniques for q-series: Askey-Wilson polynomials, *Rocky Mtn. J. Math.* **19**, to appear.

Kalnins, E. G. and Miller, W. (1989b). q-Series and orthogonal polynomials associated with Barnes' first lemma, *IMA Preprint Series* **306**, April, 1987, to appear.

van Kampen, N. G. (1961). Exact calculation of the fluctuation spectrum for a nonlinear model system, *J. Math. Phys.* **2**, 592-601.

Karlsson, Per W. (1971). Hypergeometric functions with integral parameter differences, *J. Math. Phys.* **12**, 270-271.

Kendall, M. G. and Stuart, A. (1979). *The Advanced Theory of Statistics*, Vol. 2, *Inference and Relationship*, 4th edition, C. Griffin & Co., London.

Kirillov, A.N. and Reshetikhin, N. Yu. (1988). Representations of the algebra $U_q($ orthogonal polynomials and invariants of links, *LOMI preprint E-9-88, USSR Academy of Sciences*, Steklov Math. Inst., Leningrad.

Klein, F. (1933). *Vorlesungen über die Hypergeometrische Funktion*, Springer, Berlin.

Knuth, D. (1971). Subspaces, subsets, and partitions, *J. Comb. Thy.* **A 10**, 178-180.

Knuth, D. (1973). *The Art of Computer Programming*, Vols. 1-3, Addison-Wesley, Reading, Mass.

Koelink, H. T. and Koornwinder, T. H. (1989). The Clebsch-Gordan coefficients for the quantum group $S_\mu U(2)$ and q-Hahn polynomials, *Proc. Kon. Nederl. Akad. Wetensch. Series A*, to appear.

Koornwinder, T. H. (1974a). Jacobi polynomials, II. An analytic proof of the product formula, *SIAM J. Math. Anal.* **5**, 125-137.

Koornwinder, T. H. (1974b). Jacobi polynomials, III. An analytic proof of the addition formula, *SIAM J. Math. Anal.* **6**, 533-543.

Koornwinder, T. H. (1984). Squares of Gegenbauer polynomials and Milin type inequalities, *Report PM-R8412*, Centre for Math. and Computer Science, Amsterdam.

Koornwinder, T. H. (1986). A group theoretic interpretation of the last part of de Branges' proof of the Bieberbach conjecture, *Complex Variables* **6**, 309-321.

Koornwinder, T. H. (1989a). Jacobi functions as limit cases of q-ultraspherical polynomials, *J. Math. Anal. Appl.*, to appear.

Koornwinder, T. H. (1989b). The addition formula for little q-Legendre polynomials and the twisted $SU(2)$ quantum group, CWI Report AM-R8906.

Koornwinder, T. H. (1989c). Representations of the twisted $SU(2)$ quantum group and some q-hypergeometric orthogonal polynomials, *Proc. Kon. Nederl. Akad. Wetensch. Series A* **92**, 97-117.

Krattenthaler, C. (1984). A new q-Lagrange formula and some applications, *Proc. Amer. Math. Soc.* **90**, 338-344.

Krattenthaler, C. (1988). Operator methods and Lagrange inversion: a unified approach to Lagrange formulas, *Trans. Amer. Math. Soc.* **305**, 431-465.

Krattenthaler, C. (1989a). q-analogue of a two-variable inverse pair of series with applications to basic double hypergeometric series, *Canad. J. Math.*, to appear.

Krattenthaler, C. (1989b). A new matrix inverse, to appear.

Krattenthaler, C. (1989c). Some quadratic, cubic and quartic very-well-poised summation formulas for basic hypergeometric series, to appear.

Kummer, E.E. (1836), Über die hypergeometrische Reihe ..., *J. für Math.* **15**, 39-83 and 127-172.

Laine, T. P. (1982). Projection formulas and a new proof of the addition formula for the Jacobi polynomials, *SIAM J. Math. Anal.* **13**, 324-330.

Leonard, D. A. (1982). Orthogonal polynomials, duality and association schemes, *SIAM J. Math. Anal.* **13**, 656-663.

Lepowsky, J. (1982). Affine Lie algebras and combinatorial identities, *Lie Algebras and Related Topic*, Lecture Notes in Math. **933**, Springer, Berlin and New York, pp. 130-156.

Lepowsky, J. and Milne, S. (1978). Lie algebraic approaches to classical partition identities, *Advances in Math.* **29**, 15-59.

Lepowsky, J. and Wilson, R. L. (1982). A Lie theoretic interpretation and proof of the Rogers-Ramanujan identities, *Advances in Math.* **45**, 21-72.

Lewis, R. P. (1984). A combinatorial proof of the triple product identity, *Amer. Math. Monthly* **91**, 420-423.

Littlewood, J. E. (1907). On the asymptotic approximation to integral functions of zero order, *Proc. London Math. Soc.* (2) **5**, 361-410.

Lubinsky, D. S. and Saff, E. B. (1987). Convergence of Padé approximants of partial theta functions and the Rogers-Szegő polynomials, *Constr. Approx.* **3**, 331-361.

Luke, Y. L. (1969). *The Special Functions and Their Approximations*, Vols. I and II, Academic Press, New York.

Macdonald, I. G. (1972). Affine root systems and Dedekind's h-function, *Invent. Math.* **15**, 91-143 .

Macdonald, I. G. (1979). *Symmetric Functions and Hall Polynomials*, Oxford University Press, Oxford (also see Chapter VI of the 2nd edition, to appear).

Macdonald, I. G. (1982). Some conjectures for root systems, *SIAM J. Math. Anal.* **13**, 988-1007.

Macdonald, I. G. (1989). Orthogonal polynomials associated with root systems, to appear.

MacMahon, P. A. (1916). *Combinatory Analysis*, Cambridge University Press, Cambridge; reprinted by Chelsea, New York, 1960.

Masuda, T., Mimachi, K., Nakagami, Y., Noumi, M. and Ueno, K. (1989). Representations of quantum group $SU_q(2)$ and the little q-Jacobi polynomials, to appear.

Menon, P.K. (1965). On Ramanujan's continued fraction and related identities, *J. London Math. Soc.* **40**, 49-54.

Milin, I. M. (1977). *Univalent Functions and Orthogonal Systems*, Transl. Math. Monographs, Vol. 29, Amer. Math. Soc., Providence, R. I.

Miller, W. (1968). *Lie Theory and Special Functions*, Academic Press, New York.

Miller, W. (1970). Lie theory and q-difference equations, *SIAM J. Math. Anal.* **1**, 171-188.

Milne, S. C. (1980). A multiple series transformation of the very well poised ${}_{2k+4}\psi_{2k+4}$, *Pacific J. Math.* **91**, 419-430.

Milne, S. C. (1985a). An elementary proof of the Macdonald identities for $A_\ell^{(1)}$, *Advances in Math.* **57**, 34-70.

Milne, S. C. (1985b). A new symmetry related to $SU(n)$ classical basic hypergeometric series, *Advances in Math.* **57**, 71-90.

Milne, S. C. (1985c). A q-analogue of hypergeometric series well-poised in $SU(n)$ and invariant G-functions, *Advances in Math.* **58**, 1-60.

Milne, S. C. (1985d). A q-analogue of the ${}_5F_4$ (1) summation theorem for hypergeometric series well-poised in $SU(n)$, *Advances in Math.* **57**, 14-33.

Milne, S. C. (1985e). Basic hypergeometric series very-well-poised in $U(n)$, *J. Math. Anal. Appl.* **122**, 223-256.

Milne, S. C. (1986). A $U(n)$ generalization of Ramanujan's ${}_1\psi_1$ summation, *J. Math. Anal. Appl.* **118**, 263-277.

Milne, S. C. (1988a). Multiple q-series and $U(n)$ generalizations of Ramanujan's ${}_1\psi_1$ sum, *Ramanujan Revisited* (G. E. Andrews *et al.*, eds.), Academic Press, New York, pp. 473-524.

Milne, S. C. (1988b). A q-analog of the Gauss summation theorem for hypergeometric series in $U(n)$, *Advances in Math.* **72**, 59-131.

Milne, S. C. (1989a). A q-analog of the balanced ${}_3F_2$ summation theorem for hypergeometric series in $U(n)$, *Advances in Math.*, to appear.

Milne, S. C. (1989b). A q-analog of Whipple's transformation for hypergeometric series in $U(n)$, *Advances in Math.*, to appear.

Milne, S. C. (1989c). The multidimensional ${}_1\psi_1$ sum and Macdonald identities for $A_\ell^{(1)}$, *Proc. of the 1987 Summer Research Institute of Theta Functions*, Amer. Math. Soc., Providence, R. I.

Mimachi, K. (1989). Connection problem in holonomic q-difference system associated with a Jackson integral of Jordan-Pochhammer type, to appear.

Minton, B. M. (1970). Generalized hypergeometric function of unit argument, *J. Math. Phys.* **11**, 1375-1376.

Misra, K. C. (1988). Specialized characters for affine Lie algebras and the Rogers-Ramanujan identities, *Ramanujan Revisited* (G. E. Andrews *et al.*, eds.), Academic Press, New York, pp. 85-109.

Moak, D. S. (1980a). The q-gamma function for $q > 1$, *Aequationes Math.* **20**, 278-285.

Moak, D. S. (1980b). The q-gamma function for $x < 0$, *Aequationes Math.* **21**, 179-191.

Moak, D. S. (1981). The q-analogue of the Laguerre polynomials, *J. Math. Anal. Appl.* **81**, 20-47.

Mordell, L. J. (1917). On the representation of numbers as the sum of $2r$ squares, *Quart. J. Pure and Appl. Math.* **48**, 93-104.

Morris, W. (1982). Constant term identities for finite and affine root systems, Thesis, University of Wisconsin, Madison.

Nassrallah, B. (1982). Some quadratic transformations and projection formulas for basic hypergeometric series, Thesis, Carleton University, Ottawa.

Nassrallah, B. and Rahman, R. (1981). On the q-analogues of some transformations of nearly-poised hypergeometric series, *Trans. Amer. Math. Soc.* **268**, 211-229.

Nassrallah, B. and Rahman, R. (1985). Projection formulas, a reproducing kernel and a generating function for q-Wilson polynomials, *SIAM J. Math. Anal.* **16**, 186-197.

Nassrallah, B. and Rahman, R. (1986). A q-analogue of Appell's F_1 function and some quadratic transformation formulas for non-terminating basic hypergeometric series, *Rocky Mtn. J. of Math.* **16**, 63-82.

Needham, J. (1959). *Science and Civilization in China*, Vol. 3, *Mathematics and the Sciences of the Heavens and the Earth*, Cambridge University Press, Cambridge.

Nevai, P. G. (1979). Orthogonal polynomials, *Memoirs Amer. Math. Soc.* **213**.

Nikiforov, A. F. and Suslov, S. K. (1986). Classical orthogonal polynomials of a discrete variable on nonuniform lattices, *Letters in Mathematical Physics*, II, 27-34.

Nikiforov, A. F., Suslov, S. K. and Uvarov, V. B. (1985). *Classical Orthogonal Polynomials of a Discrete Variable* (in Russian), Nauka, Moscow.

Nikiforov, A. F. and Uvarov, V. B. (1988). *Special Functions of Mathematical Physics: A Unified Introduction with Applications*, Translated from the Russian by R. P. Boas, Birkhäuser, Boston, Mass.

O'Hara, K. M. (1989). Unimodality of the Gaussian coefficients: a constructive proof, *J. Comb. Thy. Ser. A*, to appear.

Orr, W. McF. (1899). Theorems relating to the product of two hypergeometric series, *Trans. Camb. Phil. Soc.* **17**, 1-15.

Pastro, P. I. (1985). Orthogonal polynomials and some q-beta integrals of Ramanujan, *J. Math. Anal. Appl.* **112**, 517-540.

Paule, P. (1985). On identities of the Rogers-Ramanujan type, *J. Math. Anal. Appl.* **107**, 255-284.

Paule, P. and Rother, W. (1985). Ein neuer Weg zur q-Lagrange inversion, *Bayreuther Math. Schriften* **18**, 1-37.

Perron, O. (1929). *Die Lehre von den Kettenbrüchen*, 2nd edition, Teubner, Leipzig.

Pfaff, J. F. (1797). Observationes analyticae ad L. Euler Institutiones Calculi Integralis, Vol. IV, Supplem. II et IV, Historia de 1793, *Nova acta acad. sci. Petropolitanae* **11** (1797), pp. 38-57.

Pólya, G. (1927). Über die algebraisch-funktionentheoretischen Untersuchungen von J. L. W. V. Jensen, *Kgl. Danske Videnskabernes Selskab. Math.-Fys. Medd.* **7**, No. 17, pp. 3-33; reprinted in *Collected Papers*, Vol. II, The MIT Press, Cambridge, Mass., pp. 278-308.

Pólya, G. (1970). Gaussian binomial coefficients and the enumeration of inversions, *Proceedings of the Second Chapel Hill Conference on Combinatorial Mathematics and its Applications*, Aug. 1970, Univ. of North Carolina, Chapel Hill, pp. 381-384.

Pólya, G. and Alexanderson, G. L. (1970). Gaussian binomial coefficients, *Elemente der Mathematik* **26**, 102-108.

Rademacher, H. (1973). *Topics in Analytic Number Theory*, Springer, Berlin and New York.

Rahman, M. (1981). The linearization of the product of continuous q-Jacobi polynomials, *Canad. J. Math.* **33**, 255-284.

Rahman, M. (1982). Reproducing kernels and bilinear sums for q-Racah and q-Wilson polynomials, *Trans. Amer. Math. Soc.* **273**, 483-508.

Rahman, M. (1984). A simple evaluation of Askey and Wilson's q-beta integral, *Proc. Amer. Math. Soc.* **92**, 413-417.

Rahman, M. (1985). A q-extension of Feldheim's bilinear sum for Jacobi polynomials and some applications, *Canad. J. Math.* **37**, 551-576.

Rahman, M. (1986a). Another conjectured q-Selberg integral, *SIAM J. Math. Anal.* **17**, 1267-1279.

Rahman, M. (1986b). An integral representation of a ${}_{10}\phi_9$ and continuous bi-orthogonal ${}_{10}\phi_9$ rational functions, *Canad. J. Math.* **38**, 605-618.

Rahman, M. (1986c). q-Wilson functions of the second kind, *SIAM J. Math. Anal.* **17**, 1280-1286.

Rahman, M. (1986d). A product formula for the continuous q-Jacobi polynomials, *J. Math. Anal. Appl.* **118**, 309-322.

Rahman, M. (1987). An integral representation and some transformation properties of q-Bessel functions, *J. Math. Anal. Appl.* **125**, 58-71.

Rahman, M. (1988a). A projection formula for the Askey-Wilson polynomials and an application, *Proc. Amer. Math. Soc.* **103**, 1099-1107.

Rahman, M. (1988b). Some extensions of Askey-Wilson's q-beta integral and the corresponding orthogonal systems, *Canad. Math. Bull.*, **33** (4), 111-120.

Rahman, M. (1988c). An addition theorem and some product formulas for q-Bessel functions, *Canad. J. Math.* **40**, 1203-1221.

Rahman, M. (1989a). A simple proof of Koornwinder's addition formula for the little q-Legendre polynomials, to appear.

Rahman, M. (1989b). Some infinite integrals of q-Bessel functions, *Proc. Ramanujan Centenary Symposium on Classical Analysis*, 1987, Pune, India, to appear.

Rahman, M. (1989c). A note on the biorthogonality of q-Bessel functions, *Canad. Math. Bull.*, to appear.

Rahman, M. (1989d). Some quadratic and cubic summation formulas for basic hypergeometric series, to appear.

Rahman, M. (1989e). Some cubic summation formulas for basic hypergeometric series, to appear.

Rahman, M. and Verma, A. (1986a). A q-integral representation of Rogers' q-ultraspherical polynomials and some applications, *Const. Approx.* **2**, 1-10.

Rahman, M. and Verma, A. (1986b). Product and addition formulas for the continuous q-ultraspherical polynomials, *SIAM J. Math. Anal.* **17**, 1461-1474.

Rahman, M. and Verma, A. (1987). Infinite sums of products of continuous q-ultraspherical functions, *Rocky Mtn. J. Math.* **17**, 371-384.

Rainville, E. D. (1960). *Special Functions*, Macmillan, New York.

Ramanujan, S. (1915). Some definite integrals, *Messenger of Math.* **44**, 10-18; reprinted in Ramanujan [1927], pp. 53-58.

Ramanujan, S. (1919). Proof of certain identities in combinatory analysis, *Proc. Camb. Phil. Soc.* **19**, 214-216; reprinted in Ramanujan [1927], pp. 214-215.

Ramanujan, S. (1927). *Collected Papers* (G. H. Hardy *et al.*, eds.), Cambridge University Press, Cambridge; reprinted by Chelsea, New York, 1962.

Ramanujan, S. (1957). *Notebooks* (2 volumes), Tata Institute of Fundamental Research, Bombay; reprinted by Narosa, New Delhi, 1984.

Ramanujan, S. (1988). *The lost notebook and other unpublished papers* (Introduction by G. E. Andrews), Narosa Publishing House, New Delhi.

Rogers, L. J. (1893a). On a three-fold symmetry in the elements of Heine's series, *Proc. London Math. Soc.* **24**, 171-179.

Rogers, L. J. (1893b). On the expansion of some infinite products, *Proc. London Math. Soc.* **24**, 337-352.

Rogers, L. J. (1894). Second memoir on the expansion of certain infinite products, *Proc. London Math. Soc.* **25**, 318-343.

Rogers, L. J. (1895). Third memoir on the expansion of certain infinite products, *Proc. London Math. Soc.* **26**, 15-32.

Rogers, L. J. (1917). On two theorems of combinatory analysis and some allied identities, *Proc. London Math. Soc.* (2) **16**, 315-336.

Rogers, L. J. and Ramanujan, S. (1919). Proof of certain identities in combinatory analysis (with a prefatory note by G. H. Hardy), *Proc. Camb. Phil. Soc.* **19**, 211-216.

Rota, G.-C. and Goldman, J. (1969). The number of subspaces of a vector space, *Recent Progress in Combinatorics* (W. T. Tutte, ed.), Academic Press, New York, pp. 75-83.

Rota, G.-C. and Mullin, R. (1970). On the foundations of combinatorial theory, III: Theory of binomial enumeration, *Graph Theory and Its Applications* (B. Harris, ed.), Academic Press, New York, pp. 167-213.

Saalschütz, L. (1890). Eine Summationsformel, *Zeitschr. Math. Phys.* **35**, 186-188.

Schur, I. J. (1917). Ein Beitrag zur additiven Zahlentheorie und zur Theorie der Kettenbrüche, *Sitz. Preuss. Akad. Wiss. Phys.-Math. Kl.*, pp. 302-321.

Schützenberger, M.-P. (1953). Une interprétation de certaines solutions de l'équation fonctionnelle: $F(x+y)=F(x)F(y)$, *C.R. Acad. Sci. Paris* **236**, 352-353.

Sears, D. B. (1951a). Transformations of basic hypergeometric functions of special type, *Proc. London Math. Soc.* (2) **52**, 467-483.

Sears, D. B. (1951b). On the transformation theory of hypergeometric functions and cognate trigonometric series, *Proc. London Math. Soc.* (2) **53**, 138-157.

Sears, D. B. (1951c). On the transformation theory of basic hypergeometric functions, *Proc. London Math. Soc.* (2) **53**, 158-180.

Sears, D. B. (1951d). Transformations of basic hypergeometric functions of any order, *Proc. London Math. Soc.* (2) **53**, 181-191.

Sears, D. B. (1952). Two identities of Bailey, *J. London Math. Soc.* **27**, 510-511.

Selberg, A. (1944). Bemerkninger om et multiplet integral, *Norske Mat. Tidsskr.* **26**, 71-78.

Shohat, J. and Tamarkin, T. (1950). *The Problem of Moments*, Mathematical Surveys **1**, Amer. Math. Soc., Providence, R. I.

Singh, V. N. (1959). The basic analogues of identities of the Cayley-Orr type, *J. London Math. Soc.* **34**, 15-22.

Slater, L. J. (1951). A new proof of Rogers' transformations of infinite series, *Proc. London Math. Soc.* (2) **53**, 460-475.

Slater, L. J. (1952a). Further identities of the Rogers-Ramanujan type, *Proc. London Math. Soc.* (2) **54**, 147-167.

Slater, L. J. (1952b). General transformations of bilateral series, *Quart. J. Math.* (Oxford) (2) **3**, 73-80.

Slater, L. J. (1952c). Integrals representing general hypergeometric transformations, *Quart. J. Math.* (Oxford) (2) **3**, 206-216.

Slater, L. J. (1952d). An integral of hypergeometric type, *Proc. Camb. Phil. Soc.* **48**, 578-582.

Slater, L. J. (1954a). A note on equivalent product theorems, *Math. Gazette* **38**, 127-128.

Slater, L. J. (1954b). Some new results on equivalent products, *Proc. Camb. Phil. Soc.* **50**, 394-403.

Slater, L. J. (1954c). The evaluation of the basic confluent hypergeometric functions, *Proc. Camb. Phil. Soc.* **50**, 404-413.

Slater, L. J. (1955). Some basic hypergeometric transforms, *J. London Math. Soc.* **30**, 351-360.

Slater, L. J. (1963). Wilfrid Norman Bailey (obituary), *J. London Math. Soc.* **37**, 504-512.

Slater, L. J. (1966). *Generalized Hypergeometric Functions*, Cambridge University Press, Cambridge.

Slater, L. J. and Lakin, A. (1956). Two proofs of the ${}_6\psi_6$ summation theorem, *Proc. Edin. Math. Soc.* **9**, 116-121.

Srivastava, H. M. (1984). Certain q-polynomial expansions for functions of several variables. II, *IMA J. Appl. Math.* **33**, 205-209.

Stanton, D. (1977). Some basic hypergeometric polynomials arising from finite classical groups, *Thesis, University of Wisconsin*, Madison.

Stanton, D. (1980a). Product formulas for q-Hahn polynomials, *SIAM J. Math. Anal.* **11**, 100-107.

Stanton, D. (1980b). Some q-Krawtchouk polynomials on Chevalley groups, *Amer. J. Math.* **102**, 625-662.

Stanton, D. (1980c). Some Erdös-Ko-Rado theorems for Chevalley groups, *SIAM J. Alg. Disc. Meth.* **1**, 160-163.

Stanton, D. (1981a). A partially ordered set and q-Krawtchouk polynomials, *J. Comb. Thy.* **A 30**, 276-284.

Stanton, D. (1981b). Three addition theorems for some q-Krawtchouk polynomials, *Geometriae Dedicata* **10**, 403-425.

Stanton, D. (1983). Generalized n-gons and Chebychev polynomials, *J. Comb. Thy.* **A 34**, 15-27.

Stanton, D. (1984). Orthogonal polynomials and Chevalley groups, *Special Functions: Group Theoretical Aspects and Applications* (R. Askey *et al.*, eds.), Reidel, Boston, Mass., pp. 87-128.

Stanton, D. (1986a). Harmonics on posets, *J. Comb. Thy.* **A 40**, 136-149.

Stanton, D. (1986b). Sign variations of the Macdonald identities, *SIAM J. Math. Anal.* **17**, 1454-1460.

Stanton, D. (1986c). t-designs in classical association schemes, *Graphs and Combinatorics* **2**, 283-286.

Stanton, D. (1988). Recent results for the q-Lagrange inversion formula, *Ramanujan Revisited* (G. E. Andrews *et al.*, eds.), Academic Press, New York, pp. 525-536.

Stanton, D. (1989). An elementary approach to the Macdonald identities, *Workshop on q-Series and Partitions* (D. Stanton, ed.), IMA Volumes in Mathematics and its Applications **18**, Springer, Berlin and New York, to appear.

Starcher, G. W. (1931). On identities arising from solutions of q-difference equations and some interpretations in number theory, *Amer. J. Math.* **53**, 801-816.

Stone, M. H. (1932). *Linear Transformations in Hilbert Spaces*, Amer. Math. Soc. Colloq. Publ. **15**, Providence, R.I.

Subbarao, M. V. and Vidyasagar, M. (1970). On Watson's quintuple product identity, *Proc. Amer. Math. Soc.* **26**, 23-27.

Sudler, C. (1966). Two enumerative proofs of an identity of Jacobi, *Proc. Edin. Math. Soc.* **15**, 67-71.

Suslov, S. K. (1982). Matrix elements of Lorentz boosts and the orthogonality of Hahn polynomials on a contour, *Sov. J. Nucl. Phys.* **36**, 621-622.

Suslov, S. K. (1987). Classical orthogonal polynomials of a discrete variable continuous orthogonality relation, *Letters in Math. Phys.* **14**, 77-88.

Sylvester, J.J. (1878). Proof of the hitherto undemonstrated fundamental theorem of invariants, *Philosophical Magazine* V, 178-188; reprinted in *Collected Mathematical Papers* **3**, pp. 117-126; reprinted by Chelsea, New York, 1973.

Sylvester, J. J. (1882). A constructive theory of partitions in three acts, an interact and an exodion, *Amer. J. Math.* **5**, 251-330 (and ibid. **6** (1884), 334-336); reprinted in *Collected Mathematical Papers* **4**, pp. 1-83; reprinted by Chelsea, New York, 1974.

Szegő, G. (1926). Ein Beitrag zur Theorie der Thetafunktionen, *Sitz. Preuss. Akad. Wiss. Phys.-Math. Kl*, 242-252; reprinted in Collected Papers **1**, pp. 793-805.

Szegő, G. (1968). An outline of the history of orthogonal polynomials, *Proc. Conf. on Orthogonal Expansions and their Continuous Analogues* (1967) (D. Haimo, ed.), Southern Illinois Univ. Press, Carbondale, 3-11; reprinted in *Collected Papers* **3**, pp. 857-865, also see the comments on pp. 866-869.

Szegő, G. (1975). *Orthogonal Polynomials*, 4th edition, Amer. Math. Soc. Colloq. Publ. **23**, Providence, R.I.

Szegő, G. (1982). *Collected Papers* (R. Askey, ed.), Vols. 1-3, Birkhäuser, Boston, Mass.

Takács, L. (1973). On an identity of Shih-Chieh Chu, *Acta. Sci. Math.* (Szeged) **34**, 383-391.

Thomae, J. (1869). Beiträge zur Theorie der durch die Heinesche Reihe ..., *J. reine angew. Math.* **70**, 258-281.

Thomae, J. (1870). Les séries Heinéennes supérieures, ou les séries de la forme ..., *Annali di Matematica Pura ed Applicata* **4**, 105-138.

Thomae, J. (1879). Ueber die Funktionen, welche durch Reihen von der Form dergestellt werden ... , *J. reine angew. Math.* **87**, 26-73.

Toeplitz, O. (1963). *The Calculus: A Genetic Approach*, University of Chicago Press, Chicago.

Trjitzinsky, W. J. (1933). Analytic theory of linear q-difference equations, *Acta Math.* **61**, 1-38.

Vandermonde, A. T. (1772). Mémoire sur des irrationnelles de différens ordres avec une application au cercle, *Mém. Acad. Roy. Sci. Paris*, 489-498.

Verma, A. (1966). Certain expansions of the basic hypergeometric functions, *Math. Comp.* **20**, 151-157.

Verma, A. (1972). Some transformations of series with arbitrary terms, *Instituto Lombardo* (Rend. Sc.) A **106**, 342-353.

Verma, A. (1980). A quadratic transformation of a basic hypergeometric series, *SIAM J. Math. Anal.* **11**, 425-427.

Vilenkin, N. Ja. (1968). *Special Functions and the Theory of Group Representations*, Amer. Math. Soc. Transl. of Math. Monographs **22**, Amer. Math. Soc., Providence, R. I.

Wallisser, R. (1985). Über ganze Funktionen, die in einer geometrischen Folge ganze Werte annehmen, *Monatsh. für Math.* **100**, 329-335.

Watson, G. N. (1910). The continuations of functions defined by generalized hypergeometric series, *Trans. Camb. Phil. Soc.* **21**, 281-299.

Watson, G. N. (1922). The product of two hypergeometric functions, *Proc. London Math. Soc.* (2) **20**, 189-195.

Watson, G. N. (1924). The theorems of Clausen and Cayley on products of hypergeometric functions, *Proc. London Math. Soc.* (2) **22**, 163-170.

Watson, G. N. (1929a). A new proof of the Rogers-Ramanujan identities, *J. London Math. Soc.* **4**, 4-9.

Watson, G. N. (1929b). Theorems stated by Ramanujan. VII: Theorems on continued fractions, *J. London Math. Soc.* **4**, 39-48.

Watson, G. N. (1931). Ramanujan's notebooks, *J. London Math. Soc.* **6**, 137-153.

Watson, G. N. (1936). The final problem: an account of the mock theta functions, *J. London Math. Soc.* **11**, 55-80.

Watson, G. N. (1937). The mock theta functions (2), *Proc. London Math. Soc.* (2) **42**, 274-304.

Watson, G. N. (1952). *A Treatise on the Theory of Bessel Functions*, 2nd edition, Cambridge University Press, Cambridge.

Whipple, F. J. W. (1926a). On well-poised series, generalized hypergeometric series having parameters in pairs, each pair with the same sum, *Proc. London Math. Soc.* (2) **24**, 247-263.

Whipple, F. J. W. (1926b). Well-poised series and other generalized hypergeometric series, *Proc. London Math. Soc.* (2) **25**, 525-544.

Whipple, F. J. W. (1927). Algebraic proofs of the theorems of Cayley and Orr concerning the products of certain hypergeometric series, *J. London Math. Soc.* **2**, 85-90.

Whipple, F. J. W. (1929). On a formula implied in Orr's theorems concerning the product of hypergeometric series, *J. London Math. Soc.* **4**, 48-50.

Whittaker, E. T. and Watson, G. N. (1965). *A Course of Modern Analysis*, 4th edition, Cambridge University Press, Cambridge.

Wilson, J. A. (1978). Hypergeometric series, recurrence relations and some new orthogonal polynomials, Thesis, Univ. of Wisconsin, Madison.

Wilson, J. A. (1980). Some hypergeometric orthogonal polynomials, *SIAM J. Math. Anal.* **11**, 690-701.

Wilson, J. A. (1985). Solution to problem 84-7 (A q-extension of Cauchy's beta integral, by R. Askey), *SIAM Review* **27**, 252-253.

Wintner, A. (1929). *Spektraltheorie der unendlichen Matrizen, Einführung in den analytischen Apparat der Quantenmechanik*, Hirzel, Leipzig.

Wright, E. M. (1965). An enumerative proof of an identity of Jacobi, *J. London Math. Soc.* **40**, 55-57.

Wright, E. M. (1968). An identity and applications, *Amer. Math. Monthly* **75**, 711-714.

Yang, K.-W. (1989), q-Algebras, to appear.

Zaslavsky, T. (1987). The Möbius function and the characteristic polynomial, *Combinatorial Geometries* (N. White, ed.), Cambridge University Press, Cambridge, pp. 114-138.

Zeilberger, D. (1987). A proof of the G_2 case of Macdonald's root system-Dyson conjecture, *SIAM J. Math. Anal.* **18**, 880-883.

Zeilberger, D. (1988). A unified approach to Macdonald's root-system conjectures, *SIAM J. Math. Anal.* **19**, 987-1013.

Zeilberger, D. (1989a). A Stembridge-Stanton style elementary proof of the Habsieger-Kadell q-Morris identity, *Discrete Math.*, to appear.

Zeilberger, D. (1989b). Kathy O'Hara's constructive proof of the unimodality of the Gaussian polynomials, *Amer. Math. Monthly*, to appear.

Zeilberger, D. (1989c). A holonomic systems approach to special functions identities, to appear.

Zeilberger, D. (1989d). Identities, *Workshop on q-Series and Partitions* (D. Stanton, ed.), IMA Volumes in Mathematics and Its Applications, Springer, Berlin and New York, to appear.

Zeilberger, D. and Bressoud, D. M. (1985). A proof of Andrews' q-Dyson conjecture, *Discrete Math.* **54**, 201-224.

Author Index

Symbol Index

Subject Index